THE BIOLOGY OF THE FIRST 1,000 DAYS

OXIDATIVE STRESS AND DISEASE

Series Editors

Lester Packer, PhD
Enrique Cadenas, MD, PhD

University of Southern California School of Pharmacy
Los Angeles, California

THE BIOLOGY OF THE FIRST 1,000 DAYS

EDITED BY
CRYSTAL D. KARAKOCHUK, KYLY C. WHITFIELD, TIM J. GREEN, AND KLAUS KRAEMER

CRC Press
Taylor & Francis Group
Boca Raton London New York

CRC Press is an imprint of the
Taylor & Francis Group, an **informa** business

CRC Press
Taylor & Francis Group
6000 Broken Sound Parkway NW, Suite 300
Boca Raton, FL 33487-2742

First issued in paperback 2020

© 2018 by Taylor & Francis Group, LLC
CRC Press is an imprint of Taylor & Francis Group, an Informa business

No claim to original U.S. Government works

ISBN-13: 978-1-4987-5679-2 (hbk)
ISBN-13: 978-0-367-65769-7 (pbk)

Library of Congress Cataloging-in-Publication Data

Names: Karakochuk, Crystal D., editor. | Whitfield, Kyly C., editor. | Green, Tim J., 1967- editor | Kraemer, Klaus, 1960- editor.
Title: The biology of the first 1,000 days / [edited by] Crystal D. Karakochuk, Kyly C. Whitfield, Tim J. Green, and Klaus Kraemer. Other titles: Biology of the first thousand days | Oxidative stress and disease ; 42.
Description: Boca Raton : Taylor & Francis, 2018. | Series: Oxidative stress and disease; 42 | Includes bibliographical references.
Identifiers: LCCN 2017013475 | ISBN 9781498756792 (hardback : alk. paper)
Subjects: | MESH: Pregnancy--physiology | Pregnancy Trimester, First--physiology | Maternal Nutritional Physiological Phenomena | Fetal Development | Nutritional Requirements | Pregnancy Complications--etiology | Pregnancy Complications--prevention & control
Classification: LCC RG525 | NLM WQ 200.1 | DDC 618.2--dc23
LC record available at https://lccn.loc.gov/2017013475

Visit the Taylor & Francis Web site at
http://www.taylorandfrancis.com

and the CRC Press Web site at
http://www.crcpress.com

Contents

SECTION I   Introduction

SECTION II   What Is Normal Growth?

SECTION III   Nutritional Requirements in the Life Stages

SECTION IV   Endocrinology in the Regulation of Growth

SECTION V   Adverse Pregnancy and Birth Outcomes (Pathophysiology and Consequences)

SECTION VI   Pathophysiology and Nutrition Requirements in Child Malnutrition

SECTION VII   Body Composition

SECTION VIII   The Gut Microbiome

SECTION IX   Effects of Early Life Exposures and Nutrition

SECTION X   Effective Interventions during the First 1,000 Days

SECTION XI   Before and Beyond the First 1,000 Days

SECTION XII Discovery Research

Series Preface

OXIDATIVE STRESS IN HEALTH AND DISEASE

Oxidative stress is an underlying factor in health and disease. In this series of books the importance of oxidative stress and disease associated with cell and organ systems of the body is highlighted by exploring the scientific evidence and the clinical applications of this knowledge. This series is intended for researchers in the biomedical sciences and clinicians and all persons with interest in the health sciences. The potential of such knowledge for healthy development, healthy aging, and disease prevention warrants further knowledge about how oxidants and antioxidants modulate cell and tissue function.

Crystal D. Karakochuk, Kyly C. Whitfield, Tim J. Green, and Klaus Kraemer are to be congratulated for producing this very excellent and timely book, *The Biology of the First 1,000 Days*, in the ever-growing field of importance of early nutrition to human health.

Lester Packer
Enrique Cadenas

Preface

The first 1,000 days, from conception to 2 years of age, is a critical window of growth and development. Exposures to dietary, environmental, hormonal, and other stressors during this period have been associated with an increased risk of adverse health outcomes. Researchers using cell culture, animal models, and humans have identified this time as a period of rapid physiological change and plasticity with significant potential for lasting effects. As such, interventions during the first 1,000 days will have the greatest impact on outcomes, particularly in low- and middle-income countries where the need is greatest.

To date, there is no single resource that compiles our knowledge of the biology of the first 1,000 days. Our knowledge and understanding of the biology behind the first 1,000 days is still limited. This greater understanding is helping us inform effective nutrition policy and programming. The strength of this book lies in its cross-disciplinary nature that encompasses the full range of human biology, providing a more holistic perspective during this critical time frame. Moreover, we have broadened the scope and included important periods before and after the 1,000 days.

We have designed this book as a comprehensive resource for those involved in global health and nutrition policy, strategy, programming, or research. This book will also be a resource for students learning about nutrition and health across the 1,000 days. The book includes an exceptional group of contributors who are experts in their given fields. As biology underlies the core of each discussion, it allows the readers to answer the what and why, and, we hope, the how for new discovery research and more effective interventions.

Each chapter in this volume provides insight into a specific life stage, disease state, nutrient, and stressor in the first 1,000 days. As such, each chapter can be read independently, providing a comprehensive overview of that subject. However, there is continuity between chapters allowing this collection of chapters to be read cover to cover. The first chapters set the stage, providing a succinct resource to understand the well established biological mechanisms that underlie growth regulation and nutrient recommendations throughout the first 1,000 days. The next chapters move on to the evidence behind nutrition-specific and nutrition-sensitive interventions to combat adverse outcomes and disease states in the first 1,000 days. This book also features emerging research areas, such as the gut microbiome, environmental enteric dysfunction, and the role of epigenetics in health and development. The final chapter pushes the boundaries of discovery research, exploring novel areas such as proteomics and metabolomics, and how insults such as environmental enteric dysfunction affect metabolism in the first 1,000 days.

We approached this book with the ambition to shed more light on the biology during 1,000 days, but there was also a need to put the biology into a broader context of nutrition and health. There are still many gaps in our understanding of the biology of the first 1,000 days. It is only by bridging this knowledge gap through research that we can inform effective interventions to improve outcomes during the first 1,000 days.

Crystal D. Karakochuk
Kyly C. Whitfield
Tim J. Green
Klaus Kraemer

Acknowledgments

We extend our sincere thanks to the contributing authors. We are grateful for your dedication to this project and it was an honor to work with each of you. We owe special thanks to Susie Lunt and Sandra Elias for their editorial assistance. Last, we thank Chuck Crumly and Jennifer Blaise for their support and leadership in the creation of this landmark publication.

Editors

Crystal D. Karakochuk, PhD, RD, is an assistant professor in the Department of Food, Nutrition, and Health at the University of British Columbia, and an investigator in Healthy Starts at BC Children's Hospital, Vancouver, British Columbia, Canada. She has worked internationally as a nutritionist for the United Nations World Food Programme (Rwanda, Malawi, Ethiopia, and Rome) and UNICEF (New York and Timor-Leste). Her research focuses on anemia, nutritional biomarkers (namely, iron and zinc), the effect of inflammation on nutritional biomarkers, and genetic hemoglobinopathies and blood disorders (e.g., sickle cell).

Kyly C. Whitfield, PhD, is an assistant professor in the Department of Applied Human Nutrition at Mount Saint Vincent University, Halifax, Nova Scotia, Canada. Her research focuses on identifying culturally appropriate public health interventions to combat micronutrient deficiencies in low-resource settings, particularly among lactating mothers and their infants. Her current work explores fortification to address thiamin deficiency among breastfed infants in Southeast Asia. She is also interested in exploring the long-term effects of infant feeding behaviors on disease risk later in life.

Tim J. Green, PhD, is a principal nutritionist in the Healthy Mothers, Babies, and Children Theme at the South Australian Health and Medical Research Institute, and an affiliate professor in the Discipline of Paediatrics at the University of Adelaide, Adelaide, Australia. His research focuses on micronutrients in prepregnancy, pregnancy, lactation, and early life with studies conducted in Canada, Oceania, Asia, and Africa. His group seeks to identify micronutrient deficiencies through nutrition surveys, better define micronutrient requirements and pregnancy outcomes in these groups through randomized control studies, and develop sustainable strategies to improve micronutrient status.

Klaus Kraemer, PhD, is the managing director of the Sight and Life Foundation, a nutrition think tank working toward a world free from malnutrition, headquartered in Basel, Switzerland; and an adjunct associate professor in the Department of International Health of Johns Hopkins Bloomberg School of Public Health, Baltimore, Maryland. With over 30 years of experience in research and advocacy in the field of nutrition and health, he has developed an expertise in nutrition and safety of micronutrients, and translating discovery research into effective and tailored nutrition solutions at scale. He is a member of the Steering Committee of the Micronutrient Forum, Executive Committee of the Home Fortification Technical Advisory Group, Executive Board of the Mongolian Health Initiative, and a founding member of the Society for Implementation Science in Nutrition, among others.

Contributors

Eman Allam
School of Dentistry
Indiana University
Indianapolis, Indiana

Tom Arnold
Institute of International and European
    Affairs
Dublin, Ireland

Fayrouz A. Sakr Ashour
Department of Nutrition and Food
    Science
University of Maryland
College Park, Maryland

Pamela L. Barrios
Department of Nutritional Sciences
New Jersey Institute for Food,
    Nutrition, and Health
Rutgers University
New Brunswick, New Jersey

Zulfiqar A. Bhutta
The Hospital for Sick Children
The Centre for Global Child Health
Toronto, Ontario, Canada

Robert E. Black
Department of International Health
Bloomberg School of Public Health
Johns Hopkins University
Baltimore, Maryland

André Briend
Department of Nutrition, Exercise
    and Sports
University of Copenhagen
Frederiksberg, Denmark

Anne Bush
Independent Consultant
Surrey, United Kingdom

Bianca Carducci
The Hospital for Sick Children
The Centre for Global Child Health
Toronto, Ontario, Canada

Adrienne Clermont
Global Disease and Epidemiology
    Program
Bloomberg School of Public Health
Johns Hopkins University
Baltimore, Maryland

William R. Crowley
Department of Pharmacology
    and Toxicology
University of Utah
Salt Lake City, Utah

Stephen P. Cummings
School of Science and Engineering
Teesside University
Middlesbrough, United Kingdom

Mercedes de Onis
Growth Assessment and Surveillance
    Unit
Department of Nutrition for Health
    and Development
World Health Organization
Geneva, Switzerland

Saskia de Pee
Nutrition Division
World Food Programme
Rome, Italy

Luz Maria De-Regil
Micronutrient Initiative
Ottawa, Ontario, Canada

Kathryn G. Dewey
Department of Nutrition
University of California, Davis
Davis, California

Jessica Fanzo
School of Advanced International
Studies
Johns Hopkins University
Baltimore, Maryland

Jessica Farebrother
Laboratory for Human Nutrition
Institute of Food, Nutrition and Health
ETH Zürich
Zürich, Switzerland

Maria Farren
University College Dublin Centre for
Human Reproduction
Coombe Women and Infants University
Hospital
Dublin, Ireland

Elisabet Forsum
Department of Clinical and
Experimental Medicine
Linköping University
Linköping, Sweden

Ahmed Ghoneima
School of Dentistry
Indiana University
Indianapolis, Indiana

Marta Gonzalez-Freire
National Institute on Aging
National Institutes of Health
Baltimore, Maryland

Tim J. Green
Healthy Mothers, Babies, and Children
South Australian Health and Medical
Research Institute
Adelaide, South Australia, Australia

Rebecca Heidkamp
Department of International Health
Bloomberg School of Public Health
Johns Hopkins University
Baltimore, Maryland

Daniel J. Hoffman
Department of Nutritional Sciences
New Jersey Institute for Food,
Nutrition, and Health
Rutgers University
New Brunswick, New Jersey

Philip T. James
MRC Unit The Gambia and MRC
International Nutrition Group
London School of Hygiene and Tropical
Medicine
London, United Kingdom

Marko Kerac
Department of Population Health
London School of Hygiene & Tropical
Medicine
London, United Kingdom

Amira M. Khan
The Hospital for Sick Children
The Centre for Global Child Health
Toronto, Ontario, Canada

Nancy F. Krebs
Section of Nutrition
Department of Pediatrics
University of Colorado
Aurora, Colorado

Herculina Salome Kruger
Centre of Excellence for Nutrition
North-West University
Potchefstroom, South Africa

Katherine Kula
School of Dentistry
Indiana University
Indianapolis, Indiana

Naomi S. Levitt
Chronic Disease Initiative for Africa
Division of Diabetic Medicine
    and Endocrinology
University of Cape Town
Cape Town, South Africa

Magali Leyvraz
Global Alliance for Improved Nutrition
Geneva, Switzerland

Julian C. Lui
Section on Growth and Development
Eunice Kennedy Shriver National
    Institute of Child Health and Human
    Development
National Institutes of Health
Bethesda, Maryland

Maria Makrides
Healthy Mothers, Babies, and Children
South Australian Health and Medical
    Research Institute
Adelaide, South Australia, Australia

Homero Martinez
Hospital Infantil de Mexico Federico
    Gómez
Mexico City, Mexico

Emily Mates
Emergency Nutrition Network
Oxford, United Kingdom

Christine M. McDonald
Children's Hospital Oakland Research
    Institute
Oakland, California

Marie McGrath
Emergency Nutrition Network
Oxford, United Kingdom

Anne M. Molloy
School of Medicine and School of
    Biochemistry and Immunology
Trinity College Dublin
Dublin, Ireland

Victoria Hall Moran
Maternal and Infant Nutrition
    and Nurture Unit
University of Central Lancashire
Preston, United Kingdom

Merryn Netting
Healthy Mothers, Babies, and Children
South Australian Health and Medical
    Research Institute
Adelaide, South Australia, Australia

Lynnette M. Neufeld
Global Alliance for Improved Nutrition
Geneva, Switzerland

Andrew M. Prentice
MRC Unit The Gambia and MRC
    International Nutrition Group
London School of Hygiene and Tropical
    Medicine
London, United Kingdom

Usha Ramakrishnan
Rollins School of Public Health
Emory University
Atlanta, Georgia

Fabian Rohner
GroundWork
Fläsch, Switzerland

Richard D. Semba
Wilmer Eye Institute
Johns Hopkins University
Baltimore, Maryland

Antonia W. Shand
Menzies Centre for Health Policy
University of Sydney
Sydney, New South Wales, Australia

Matt J. Silver
MRC Unit The Gambia and MRC
    International Nutrition Group
London School of Hygiene and Tropical
    Medicine
London, United Kingdom

Taylor Marie Snyder
Maternal and Infant Health Consulting
Salt Lake City, Utah

Christopher J. Stewart
Alkek Center for Metagenomics and
    Microbiome Research
Department of Molecular Virology and
    Microbiology
Baylor College of Medicine
Houston, Texas

Haley Swartz
Berman Institute of Bioethics
Johns Hopkins University
Baltimore, Maryland

Minghua Tang
University of Colorado, Denver
School of Medicine
Aurora, Colorado

Andrew L. Thorne-Lyman
Center for Human Nutrition
Department of International Health
Bloomberg School of Public Health
John Hopkins University
Baltimore, Maryland

Michael J. Turner
University College Dublin Centre for
    Human Reproduction
Coombe Women and Infants University
    Hospital
Dublin, Ireland

Amanda Wendt
Unit of Epidemiology and Biostatistics
Institute of Public Health
University of Heidelberg
Heidelberg, Germany

Sara Wuehler
Micronutrient Initiative
Ottawa, Ontario, Canada

Section I

Introduction

1 The Importance of the First 1,000 Days
An Epidemiological Perspective

Christine M. McDonald and
Andrew L. Thorne-Lyman

CONTENTS

EPIDEMIOLOGICAL BURDEN AND DISTRIBUTION OF THE PROBLEM OF MALNUTRITION

According to current estimates, 23% of the world's children under 5 years of age are stunted, a condition that is measured using short height-for-age (see Box 1.1) [1]. Although this represents a decline from 33% in 2000, the fact that 156 million children globally still suffer from chronic undernutrition underscores the continued need for renewed efforts and innovative approaches for growth promotion. Progress has been particularly slow in Africa, where one out of every three children is stunted (Figure 1.1) [1]. In fact, despite a decline in stunting prevalence, the absolute number of stunted children in Africa increased from 50.4 million in 2000 to 58.5 million in 2015 [1]. Although Asia has seen an average annual decline in the prevalence of child stunting of about 1.5%, from 38% in 2000 to 24% in 2015, countries in East Asia have accounted for most of this progress [1]. Reductions have been much slower among countries in South Asia where more than one out of every three children under age 5 is stunted. At a subregional level, more than 30% of children under 5 in Western Africa, Middle Africa, Eastern Africa, Southern Asia, and Oceania are stunted [1]. In addition to these geographical differences, there are also drastic sociodemographic disparities in the prevalence of chronic undernutrition. An analysis of 79 population-based surveys has illustrated that the prevalence of stunting is, on average, 2.5 times higher among children living in the poorest quintile of households than the richest quintile [1]. Similarly, the child stunting prevalence is 1.45 times higher in rural versus urban areas [2].

BOX 1.1   ANTHROPOMETRIC INDICATORS OF MALNUTRITION

- *Stunting* refers to impaired linear growth and can reflect chronic or recurrent undernutrition. The indicator can also be defined as having a short height for a given age compared with a reference population or standard. A cutoff of <–2 height-for-age z-scores (HAZ) is most commonly used to define stunting among children under 5 years of age.
- *Wasting* is a type of acute malnutrition resulting from recent weight loss or failure to gain weight. A cutoff of <–2 weight-for-height z-scores (WHZ) is most commonly used to define wasting among children under 5 years of age, and <–3 z-scores are used to define severe wasting.
- *Underweight* is used to describe a child whose weight is low in relation to his or her age. Underweight was historically used as a composite indicator of malnutrition but is now used less often since it fails to differentiate between chronic and acute malnutrition. A cutoff of <–2 weight-for-age z-scores (WAZ) is most commonly used to define underweight among children under 5 years of age.
- *Low birth weight* is typically defined as a birth weight <2,500 g.
- *Small for gestational age* (SGA) is usually defined by a birth weight below the 10th percentile for gestational age based on a reference population. Small for gestational age is an indicator that reflects an infant's growth rate *in utero*, as well as his or her gestational age at birth.

Globally, an estimated 50 million children are wasted, an indicator of thinness that is strongly associated with mortality [1]. Moderately and severely wasted children are, respectively, 3.0 and 9.4 times more likely to die than children with a weight-for-height z-score >–1 [3]. Although the global burden of child wasting is considerably smaller than that of stunting, efforts to enhance the coverage of interventions to prevent and treat acute malnutrition are urgently needed, particularly given the increased risk of mortality faced by wasted children. Approximately two-thirds of the world's wasted children live in Asia and one-quarter live in Africa. At 14%, the prevalence of child wasting in South Asia is near the 15% threshold used to define a critical public health emergency [1,4].

Of equal concern is the growing burden of overweight children. Worldwide, there are now 42 million overweight children under 5 years of age [1]. Since 2000, the number of overweight children under 5 has increased by more than 50% in Africa,

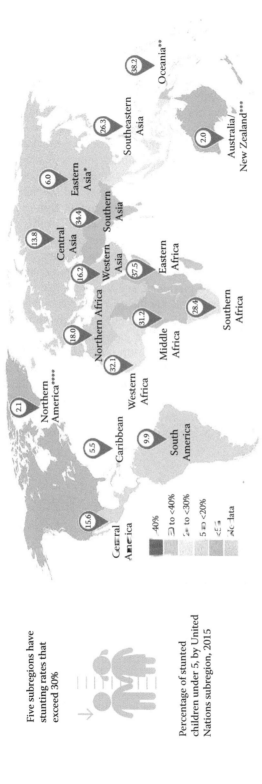

FIGURE 1.1 Percent of stunted children under age 5, by United Nations subregions, 2015. *excluding Japan, **excluding Australia and New Zealand, ***regional average based on Australian data, and ****regional average based on United States data. (Reproduced from UNICEF/WHO/World Bank Group, *Levels and trends in child malnutrition—Key findings of the 2016 edition*, New York: UNICEF, 2016. With permission.)

with gains being particularly pronounced in low- and middle-income countries [1]. At a subregional level, the prevalence of overweight among children under 5 now exceeds 10% in Central Asia, Northern Africa, and Southern Africa [1].

Intrauterine growth restriction is a key risk factor for stunting, wasting, and underweight in childhood. Recent estimates indicate that 15% to 20% of all births globally are low birth weight (LBW), which translates to more than 20 million LBW births per year [5]. Half of these births occur in only three countries: India, Pakistan, and Nigeria. India alone makes up 38% of the global total [5]. The prevalence of small-for-gestational-age (SGA) births is approximately twice the prevalence of LBW in all regions of the world [6]. In 2010, more than 32 million infants were born SGA, which represents more than one-quarter of all births in low- and middle-income countries [6]. The highest prevalence of SGA is found in South Asia and the Sahelian countries of Africa. As with LBW, the burden of SGA is particularly high in India; in 2010, 12.8 million infants, representing ~47% of all births, were SGA [6].

In addition to anthropometric indicators that reflect physical growth during the first 1,000 days, indicators of the micronutrient status of pregnant women and young children provide useful insight into the dietary quality of vulnerable subgroups, as they influence the risk of mortality, morbidity, and adverse developmental outcomes. Iron, vitamin A, iodine, zinc, and folate are the micronutrient deficiencies that have received the greatest attention in terms of their public health burden, and are estimated to account for approximately 7% of the global burden of disease every year [7]. Globally, an estimated 2 billion people are affected by deficiencies in at least one essential micronutrient. Figure 1.2 summarizes the global prevalence of vitamin A deficiency and iron deficiency anemia among pregnant women and children under 5.

The estimated global prevalence of iodine deficiency and zinc deficiency is 29% and 17%, respectively [2]. Unfortunately, limited data are available on the global prevalence of folate deficiency. Not surprisingly, there is a great deal of overlap in

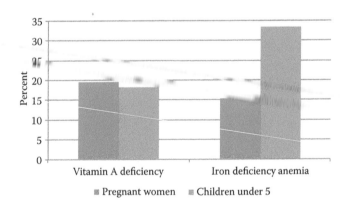

FIGURE 1.2 Global prevalence of vitamin A deficiency and iron deficiency anemia among pregnant women and children under 5. (Adapted from Black RE, Victora CG, Walker SP et al., *Lancet* 2013, 382(9890):427–51.)

the regional distribution of these micronutrient deficiencies and the anthropometric deficits previously described. Africa exhibits the highest levels of iron deficiency anemia among pregnant women and children under 5, vitamin A deficiency among children under 5, iodine deficiency and zinc deficiency, while Asia exhibits the highest prevalence of vitamin A deficiency among pregnant women [2]. Just as various forms of an anthropometric deficit can coexist, it is also common for women and children to suffer from multiple micronutrient deficits simultaneously, although the extent of overlap is often uncertain.

WINDOW OF OPPORTUNITY

One of the main epidemiological drivers for the focus on the first 1,000 days was the realization that height-for-age z-scores in countries throughout the world plummet during the first 24 months of life, as presented in blue in Figure 1.3 [3,8]. Such patterns suggest a rapid acceleration of growth faltering, followed by a leveling out of height-for-age z-scores and the prevalence of child stunting around 2 years of age. This pattern is apparent in all World Health Organization (WHO) regions. One striking difference, however, is that in South Asia, where 38% of under-5s are stunted, the average height-for-age z-score is −0.75 at birth. This dynamic reflects the high prevalence of intrauterine growth restriction and preterm delivery in South Asia, both of which contribute to LBW and small size at birth.

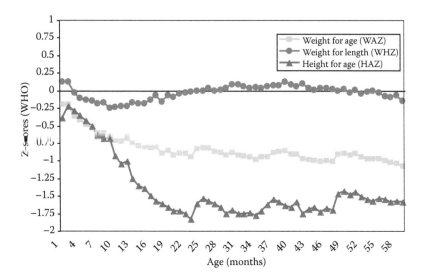

FIGURE 1.3 Mean anthropometric z-scores by age and WHO region from 54 studies. WHO Regions: Euro, Europe; EMRO, Eastern Mediterranean; AFRO, Africa; PAHO, Americas; SEARO, South East Asia. (Reproduced from Victora CG, de Onis M, Hallal PC et al., *Pediatrics* 2010, 125(3):e473–80. With permission.)

THE FIRST 300 DAYS

Starting around the time of conception, the first 300 days span a window that encapsulates the time of an average human female pregnancy (280 days), as well as the first several weeks of life. A woman's nutritional status as she enters pregnancy (the periconceptional period) is an important focus of growing scientific interest. Certain micronutrients have roles in regulating aspects of implantation, placentation, and cell differentiation, processes that can have both short- and long-term effects on pregnancy outcomes, fetal growth, and development [9]. During this window, a deficiency or excess of certain micronutrients may also lead to birth defects including teratogenicity. Perhaps the most prominent example of the effects of early micronutrient deficiencies is that folic acid supplementation can reduce the risk of neural tube defects, by an average of 41% [10]. While observational studies have suggested that the status of certain micronutrients (notably tocopherols) can influence the risk of miscarriage, and there is some suggestion of a protective effect of micronutrient supplementation on stillbirth risk, evidence remains inconsistent [11,12].

There are many nutritional influences on fetal growth. During pregnancy, a woman's energy and protein needs are 13% and 54% higher compared with a nonpregnant woman [13]. During the second and third trimesters, energy and protein intake appear to have greater importance, supported by meta-analyses of five pooled trials suggesting that balanced protein-energy supplementation trials in pregnancy reduces the risk of LBW by 32%, with greater effects on undernourished women; SGA by 34%; and stillbirth by 38% [14,15]. Studies show that a woman's weight gain during pregnancy and body mass entering pregnancy are also important factors influencing pregnancy outcomes [16].

According to a recent systematic review and meta-analysis, iron supplementation in pregnancy significantly decreases maternal anemia, iron deficiency, and LBW, but no significant effect was observed on risk of preterm birth, gestational length, SGA, or birth length [17]. Evidence of the benefits of multiple micronutrient supplementation on birth and child health outcomes is inconsistent, with a recent review of 20 randomized trials finding no significant benefits on mortality, growth at follow-up, head circumference, or cognitive function [18]. However, the largest of such trials to date, involving nearly 45,000 pregnancies from northwest Bangladesh, found slightly protective effects of antenatal and maternal postnatal micronutrient supplementation on stunting at birth, 1 month, and 3 months postpartum, but no effects on cognitive, language, or motor development [18,19]. Antenatal care guidelines from the World Health Organization issued in 2016 did not recommend micronutrient supplementation over iron and folic acid alone for improving maternal and perinatal health outcomes, but did identify the need for more research to identify potential benefits of individual nutrients and combinations of micronutrients [20].

Globally, nearly 15 million children are estimated to be born preterm and it is the leading cause of neonatal mortality, the second leading cause of death for children under 5, and an important cause of disability and cognitive impairment [21–23]. Numerous risk factors for preterm birth have been identified, including low or high maternal age, maternal undernutrition, micronutrient deficiencies, infections, diabetes, and hypertension [16,23].

Infants born small, or early, tend to stay small, although these effects appear to be stronger for the relationship between SGA and stunting than preterm. Infants born preterm have an approximately twofold greater risk of becoming stunted (OR 1.93, 95% CI: 1.71, 2.18); those born SGA have a 2.43 (95% CI: 2.22, 2.66) greater risk; and those born SGA and preterm a have 4.51-fold greater risk (95% CI: 3.42, 5.93) compared with appropriate-for-gestational-age term infants. It is clear that addressing nutritional deficiencies and other risk factors for SGA and preterm birth during and entering pregnancy should be an essential part of stunting prevention [24].

THE NEXT 700 DAYS OF LIFE

The first day of life is a period of transition and a critical window of risk for the newborn. During this time, successful establishment of breastfeeding is essential, and challenging. Only 45% of the 140 million live-born newborns born each year are breastfed in the first year of life, and little improvement has been observed in this indicator over the past 15 years (Figure 1.4). In many countries, cultural practices of early newborn ritual feeding with sugar water, tea, honey, or animal milk are prevalent, and there is emerging evidence that these interfere with the provision of colostrum and delay the timely initiation of breastfeeding [25,26]. In a three-country study examining the timing of the initiation of breastfeeding and the risk of neonatal mortality, the risk of death was 41% higher in neonates who initiated breastfeeding at 2 to 23 hours, and 79% higher in infants initiating breastfeeding at 24 to 96 hours, compared with those fed in the first hour of life [27]. The mechanisms through which

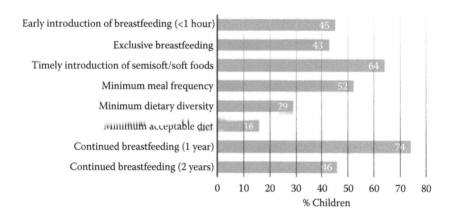

FIGURE 1.4 Infant and young child feeding practices on a global level. Age groups for denominators: Early introduction of breastfeeding, exclusive breastfeeding (0–5 months); introduced to solid, semisolid, or soft foods (6–8 months); minimum meal frequency, minimum diet diversity, and minimum acceptable diet (6–23 months); and continued breastfeeding at 1 year (12–15 months) and 2 years (20–23 months). Figures reflect UNICEF 2016 global averages, largely from low- and middle-income countries, 2010–2016. (Data from United Nations Children's Fund [UNICEF], *From the first hour of life: Making the case for improved infant and young child feeding everywhere*, New York: UNICEF, 2016.)

the early initiation of breastfeeding influences the risk of mortality require more exploration, but may include greater consumption of colostrum (the first milk, which is rich in certain micronutrients and immune substances), warmth from contact with the mother, and strengthened gastrointestinal barrier integrity [27].

WHO recommends that all infants are exclusively breastfed for the first 6 months of life, a practice defined as the exclusive consumption of breast milk, and medicines or vitamins/minerals as needed. Only 43% of infants were exclusively breastfed in 2015 [28] (Figure 1.4). The same three-country study described earlier found that compared with exclusive breastfeeding, partial breastfeeding and no breastfeeding at 1 month were associated with a 1.8 and 10.9 times greater risk of mortality during the first 6 months of life [27]. Nearly two-thirds of children are transitioned to semisolid or soft foods by 6 to 8 months, but far fewer children are transitioned to diets in which they are fed with the frequency or diversity recommended for healthy growth, or are breastfed well into the second year of life, as recommended by WHO and UNICEF (United Nations Children's Fund) (Figure 1.4).

RISK FACTORS FOR CHILD STUNTING

A recent review consolidated data from 137 developing countries to examine risk factors for stunting among children 2 years of age [29]. The findings, illustrated in Figure 1.5, underscore the importance of the first 300 days in the etiology of stunting, and the need to develop and expand interventions that address the causes of preterm birth and intrauterine growth restriction in particular. Much research is currently

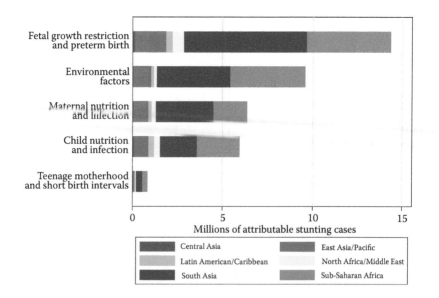

FIGURE 1.5 Number of stunting cases among children 2 years of age in 2011, attributable to risk factor clusters. (Reproduced from Danaei G, Andrews KG, Sudfeld CR et al., *PLoS Med* 2016; 13(11):e1002164. With permission.)

ongoing to understand the extent to which factors, such as hygiene and sanitation and aflatoxins, influence stunting risk. Additionally, adolescent pregnancy, common in many countries in which stunting is most prevalent, is getting considerable attention as a risk factor for stunting [30].

PUBLIC HEALTH AND ECONOMIC IMPORTANCE OF NUTRITION AND THE FIRST 1,000 DAYS

From an epidemiological perspective, the 1,000 days is a critical window in terms of mortality risk and other health outcomes. Over 6 million children under the age of 5 die each year, primarily in low- and middle-income countries [27]. The epidemiological burden of under-5 mortality is increasingly shifting to neonatal mortality: 42% of under-5 deaths occurred during the neonatal period in 2013, compared with 37% in 1990, suggesting that more attention needs to be paid to the factors that predispose children to early mortality [31].

The links between chronic undernutrition and child development have received greater attention over the past decade, building on early work that examined the effects of severe malnutrition in childhood on intelligence and school performance [29]. The first *Lancet* series on child development in developing countries identified several studies suggesting that term infants born with intrauterine growth restriction were associated with adverse developmental outcomes in the first several years of life, but identified a lack of studies that had outcomes followed up in later childhood [32]. More recently, meta-analyses have quantified the economic consequences of stunting and have estimated that early life growth faltering in low-income countries cost nearly 70 million years of educational attainment per birth cohort, resulting in total economic costs of over $616 billion at purchasing power parity-adjusted exchange rates [33].

CONCLUSION

The first 1,000 days represent a life window when growth rates and neuroplasticity are at their peak and where nutritional deficiencies can exert their most devastating impacts. Ensuring that the nutritional needs of women, infants, and young children are met during this period can help avert child mortality and lifelong disease burden, maximize growth, and enable children to reach their cognitive and developmental potential, particularly when combined with psychosocial stimulation. While these impacts represent a strong epidemiological rationale for focusing interventions on this window, it is important to remember that neither growth nor brain development stops at 1,000 days, and that ensuring good nutrition through the lifecycle is likely to have the greatest absolute benefit in terms of health and other outcomes.

REFERENCES

1. UNICEF/WHO/World Bank Group. *Levels and trends in child malnutrition: Key findings of the 2016 edition*. New York: UNICEF, 2016.
2. Black RE, Victora CG, Walker SP et al. Maternal and child undernutrition and overweight in low-income and middle-income countries. *Lancet* 2013; 382(9890):427–51.

3. Black RE, Allen LH, Bhutta ZA et al. Maternal and child undernutrition: Global and regional exposures and health consequences. *Lancet* 2008; 371(9608):243–60.

4. World Health Organization. The management of nutrition in major emergencies. Geneva: World Health Organization; 2000.

5. World Health Organization. WHA global nutrition targets 2025: Low birth weight policy brief. Geneva: World Health Organization; 2014.

6. Black RE. Global prevalence of small for gestational age births. In *Low-birthweight baby: Born too soon or too small*, Embleton ND, Katz J, Ziegler EE, editors. Basel: Nestle Nutrition Institute Karger; 2015.

7. Ezzati M, Lopez AD, Rodgers A et al. *Comparative quantification of health risks: The global and regional burden of disease attributable to selected major risk factors.* Geneva: World Health Organization; 2004.

8. Victora CG, de Onis M, Hallal PC et al. Worldwide timing of growth faltering: Revisiting implications for interventions. *Pediatrics* 2010; 125(3):e473–80.

9. Gernand AD, Schulze KJ, Stewart CP et al. Micronutrient deficiencies in pregnancy worldwide: Health effects and prevention. *Nat Rev Endocrinol* 2016; 12(5):274–89.

10. Balogun OO, da Silva Lopes K, Ota E et al. Vitamin supplementation for preventing miscarriage. *Cochrane Database Syst Rev* 2016; 5:CD004073.

11. Imdad A, Yakoob MY, Bhutta ZA. The effect of folic acid, protein energy and multiple micronutrient supplements in pregnancy on stillbirths. *BMC Public Health* 2011; 11(Suppl 3):S4.

12. Shamim AA, Schulze K, Merrill RD et al. First-trimester plasma tocopherols are associated with risk of miscarriage in rural Bangladesh. *Am J Clin Nutr* 2015; 101(2):294–301.

13. Otten JJ, Hellwig JP, Meyers LD. *Dietary reference intakes: The essential guide to nutrient requirements.* Washington, DC: National Academies Press; 2006.

14. Imdad A, Bhutta ZA. Maternal nutrition and birth outcomes: Effect of balanced protein-energy supplementation. *Paediatr Perinat Epidemiol* 2012; 26(Suppl 1):178–90.

15. Ota E, Hori H, Mori R, Tobe-Gai R et al. Antenatal dietary education and supplementation to increase energy and protein intake. *Cochrane Database Syst Rev* 2015; 6:CD000032.

16. Kramer MS. The epidemiology of adverse pregnancy outcomes: An overview. *J Nutr* 2003; 133(5 Suppl 2):S1592–6.

17. Haider BA, Olofin I, Wang M et al. Anaemia, prenatal iron use, and risk of adverse pregnancy outcomes: Systematic review and meta-analysis. *BMJ* 2013; 346:f3443.

18. Devakumar D, Fall CH, Sachdev HS et al. Maternal antenatal multiple micronutrient supplementation for long-term health benefits in children: A systematic review and meta-analysis. *BMC Med* 2016; 14(1):90.

19. West KP Jr, Shamim AA, Mehra S et al. Effect of maternal multiple micronutrient vs iron-folic acid supplementation on infant mortality and adverse birth outcomes in rural Bangladesh: The JiVitA-3 randomized trial. *JAMA* 2014; 312(24):2649–58.

20. World Health Organization. *WHO recommendations on antenatal care for a positive pregnancy experience.* Geneva: World Health Organization; 2016.

21. Liu L, Johnson HL, Cousens S et al. Global, regional, and national causes of child mortality: An updated systematic analysis for 2010 with time trends since 2000. *Lancet* 2012; 379(9832):2151–61.

22. Blencowe H, Cousens S, Oestergaard MZ et al. National, regional, and worldwide estimates of preterm birth rates in the year 2010 with time trends since 1990 for selected countries: A systematic analysis and implications. *Lancet* 2012; 379(9832):2162–72.

23. Blencowe H, Cousens S, Chou D et al. Born too soon: The global epidemiology of 15 million preterm births. *Reprod Health* 2013; 10(Suppl 1):S6.

24. Christian P, Lee SE, Angel MD et al. Risk of childhood undernutrition related to small-for-gestational age and preterm birth in low-and middle-income countries. *Int J Epidemiol* 2013; 42(5):1340–55.
25. Sundaram ME, Ali H, Mehra S et al. Early newborn ritual foods correlate with delayed breastfeeding initiation in rural Bangladesh. *Int Breastfeed J* 2016; 11:31.
26. Sundaram ME, Labrique AB, Mehra S et al. Early neonatal feeding is common and associated with subsequent breastfeeding behavior in rural Bangladesh. *J Nutr* 2013; 143(7):1161–7.
27. NEOVITA Study Group. Timing of initiation, patterns of breastfeeding, and infant survival: Prospective analysis of pooled data from three randomised trials. *Lancet Glob Health* 2016; 4(4):e266–75.
28. United Nations Children's Fund (UNICEF). *From the first hour of life: Making the case for improved infant and young child feeding everywhere.* New York: UNICEF; 2016.
29. Danaei G, Andrews KG, Sudfeld CR et al. Risk factors for childhood stunting in 137 developing countries: A comparative risk assessment analysis at global, regional, and country levels. *PLoS Med* 2016; 13(11):e1002164.
30. Prentice AM, Ward KA, Goldberg GR et al. Critical windows for nutritional interventions against stunting. *Am J Clin Nutr* 2013; 97(5):911–8.
31. Wang H, Liddell CA, Coates MM et al. Global, regional, and national levels of neonatal, infant, and under-5 mortality during 1990–2013: A systematic analysis for the global burden of disease study 2013. *Lancet* 2014; 384(9947):957–79.
32. Walker SP, Wachs TD, Gardner JM et al. Child development: Risk factors for adverse outcomes in developing countries. *Lancet* 2007; 369(9556):145–57.
33. Fink G, Peet E, Danaei G et al. Schooling and wage income losses due to early-childhood growth faltering in developing countries: National, regional, and global estimates. *Am J Clin Nutr* 2016; 104(1):104–12.

Section II

What Is Normal Growth?

2 World Health Organization Child Growth Standards

Mercedes de Onis

CONTENTS

INTRODUCTION

Adequate nutrition during the early years of life is of paramount importance for growth, development, and long-term health through adulthood. It is during infancy and early childhood that irreversible faltering in linear growth and cognitive deficits occur [1,2]. Poor nutrition during this critical period contributes to significant morbidity and mortality [3]. Similarly, the increasing prevalence of childhood obesity worldwide is associated with an increased risk of unfavorable health outcomes later in life and decreased longevity [4]. Apart from contributing positively to child survival, the quality of infant and young child feeding is fundamental to achieving optimal growth and development.

Pediatricians and other health professionals rely largely on the assessment of children's growth status to determine whether infant and young child nutrition is adequate. Growth charts are thus essential items in the pediatric toolkit for evaluating the degree to which physiological needs for growth and development are being met. However, the evaluation of child growth trajectories and the interventions designed to improve child health are highly dependent on the growth charts used.

In 2006, the World Health Organization (WHO) released new standards for assessing the growth and development of children from birth to 5 years of age [5,6]. The standards were the product of a detailed process initiated in the early 1990s, which involved various reviews of the uses of anthropometric references and alternative approaches to developing new tools to assess growth [7]. The WHO standards were developed to replace the National Center for Health Statistics (NCHS)/WHO international growth reference [8], the limitations of which have been described in detail elsewhere [9]. The purpose of this chapter is to describe the WHO Child Growth Standards, provide their background and rationale, detail how the charts were developed, and outline the main innovative aspects they offer. It also provides information on complementary charts for assessing the growth of fetuses, newborns by gestational age, preterm infants, and children over 5 years of age.

RATIONALE FOR DEVELOPING THE WORLD HEALTH ORGANIZATION CHILD GROWTH STANDARDS

The WHO standards date back to the early 1990s, when the WHO initiated a comprehensive review of the uses and interpretation of anthropometric references, and conducted an in-depth analysis of growth data from breastfed infants. This analysis showed that the growth pattern of healthy breastfed infants deviated to a significant extent from the NCHS/WHO international reference [10]. The review group concluded from these and other related findings that the NCHS/WHO reference did not adequately describe the physiological growth of children, and that its use to monitor the health and nutrition of individual children or to derive estimates of child malnutrition in populations was flawed. In particular, the reference was inadequate for assessing the growth pattern of healthy breastfed infants, because it was based on predominantly formula-fed infants, as are most national growth charts in use today. The group recommended the development of new standards, adopting a novel approach that would describe how children should grow when free of disease and when their care follows healthy practices such as breastfeeding and nonsmoking [7]. This approach would permit the development of a standard, as opposed to a reference merely describing how children grew in a particular place and time. Although a standard and a reference both serve as a basis for comparison, each enables a different interpretation. Since a standard defines how children should grow, deviations from the pattern it describes are evidence of abnormal growth. A reference, on the other hand, does not provide as sound a basis for such value judgments, although in practice references often are mistakenly used as standards.

Following a resolution from the World Health Assembly endorsing these recommendations (Resolution WHA47.5 on Infant and Young Child Nutrition, May 1994), the WHO Multicentre Growth Reference Study (MGRS) [11] was launched in 1997 to collect primary growth data that would allow the construction of new growth charts consistent with "best" health practices.

DESIGN OF THE WHO MULTICENTRE GROWTH REFERENCE STUDY

The goal of the MGRS was to describe the growth of healthy children. Implemented between 1997 and 2003, the MGRS was a population-based study conducted in six countries from diverse geographical regions: Brazil, Ghana, India, Norway, Oman, and the United States [11]. The study combined a longitudinal follow-up from birth to 24 months, with a cross-sectional component of children aged 18 to 71 months. In the longitudinal component, mothers and newborns were enrolled at birth and visited at home a total of 21 times (at weeks 1, 2, 4, and 6; monthly from 2 to 12 months; and bimonthly in the second year).

The study populations lived in socioeconomic conditions favorable to growth. The individual inclusion criteria were: no known health or environmental constraints to growth, mothers willing to follow MGRS feeding recommendations (i.e., exclusive or predominant breastfeeding for at least 4 months, the introduction of complementary foods by 6 months of age, and continued breastfeeding to at least 12 months of age); no maternal smoking before and after delivery; single term birth; and the absence of significant morbidity. Term low birth weight infants were not excluded. The eligibility criteria for the cross-sectional component were the same as those for the longitudinal component, with the exception of infant feeding practices. A minimum of 3 months of any type of breastfeeding was required for participants in the study's cross-sectional component. Rigorously standardized methods of data collection and procedures for data management across sites yielded exceptionally high-quality data. A full description of the MGRS and its implementation in the six study sites is found elsewhere [11].

The length of the children was strikingly similar among the six sites (Figure 2.1), with only about 3% of variability in length being due to intersite differences, as compared to 70% for individuals within sites [12]. The striking similarity in growth during early childhood across human populations means either a recent common origin as some suggest [13] or a strong selective advantage associated with the current pattern of growth and development across human environments.

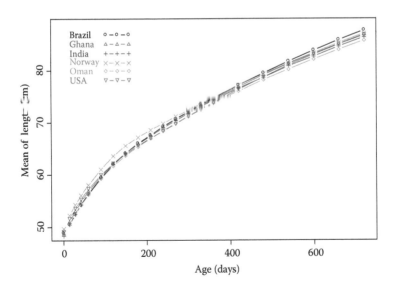

FIGURE 2.1 The mean length (centimeters) from birth through 2 years for each of the six study sites. (Reproduced from WHO Multicentre Growth Reference Study Group, *Acta Paediatr* 2006, (Suppl 450):56–65. With permission.)

CONSTRUCTION OF THE WHO CHILD GROWTH STANDARDS

Of 1,743 mother–child dyads enrolled in the MGRS longitudinal sample, 882 complied fully with the study's infant-feeding and nonsmoking criteria, and completed the follow-up period of 24 months. The remaining mother–child dyads did not comply with the study criteria ($n = 654$), dropped out of the follow-up ($n = 201$), or experienced morbidity which affected their growth ($n = 6$). The compliant sample ($n = 882$) was used to construct the WHO standards from birth to 2 years of age, combined with 6,669 children from the cross-sectional sample from age 2 to 5 years [14]. Data from all sites were pooled to construct the standards [12]. The generation of the standards followed state-of-the-art statistical methodologies that are described in detail elsewhere [6].

Weight-for-age, length/height-for-age, weight-for-length/height, and body mass index-for-age percentile and z-score values were generated for boys and girls aged 0 to 60 months and released in April 2006. The concordance between the smoothed curves and observed or empirical percentiles was excellent and free of bias at both the median and the edges, indicating that the resulting curves are a fair description of physiological growth of healthy children [6]. Detailed results of the MGRS study and the construction of the growth standards are available elsewhere [5,6]. The full set of tables and charts is presented on the WHO growth standards website (www.who.int/childgrowth/en), together with tools such as software and training materials that facilitate their application [15,16]. Figure 2.2 presents as an example the indicator of length/height-for-age, for boys aged 0 to 2 years (charts are available for different age groups). The length/height-based charts for the 0 to 5 years age group presents a

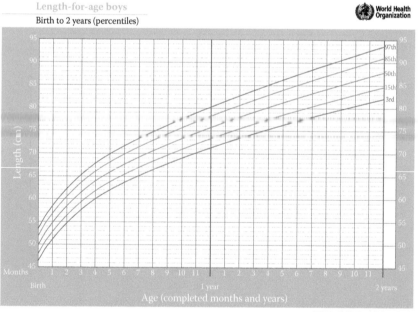

WHO Child Growth Standards

FIGURE 2.2 The length/height-for-age for boys aged 0 to 2 years (percentiles).

disjunction of 0.7 cm at 24 months, which corresponds to the change from measuring recumbent length to standing height.

Standards for other anthropometric variables (i.e., head circumference, mid-upper arm circumference, and triceps and subscapular skinfolds) were released in 2007 [17], and growth velocity standards for weight, length, and head circumference in 2009 [18,19]. The latter show striking differences compared to the U.S.-based reference values for weight and length gains most commonly used prior to the release of the WHO standards. The main differences relate to the spread of the distribution across the range of percentiles. Figure 2.3 presents, as an example, the

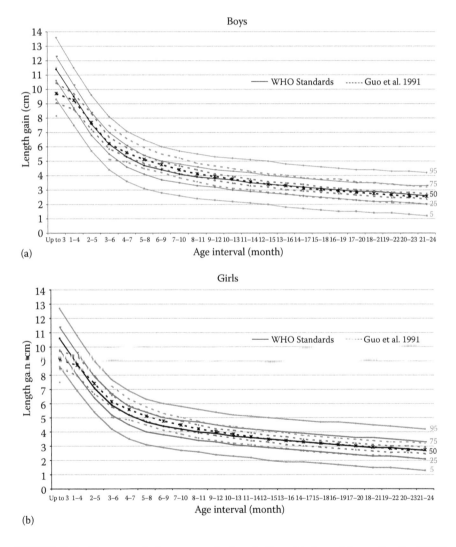

FIGURE 2.3 Three-month increments in length from birth to 24 months for WHO Standards and U.S.-based reference values for (a) boys and (b) girls. (Reproduced from de Onis M, Siyam A, Borghi E. et al., *Pediatrics* 2011, 128:e18–26. With permission.)

WHO three-month length increments for boys and girls from birth to 24 months compared to the U.S.-based reference values [19]. The use of growth velocity has considerable potential for the early identification of abnormal growth and treatment responses. However, apart from the inherent complexities of interpreting growth velocity, the dearth of reliable reference values has been a major impediment to gaining a better understanding of how to use growth velocities in a way that is helpful to clinicians. The WHO growth velocity standards fill this gap by providing a biologically robust tool, which reflects age-specific changes in the rate of growth for a varied selection of measurement intervals to suit pediatric follow-up routines in different settings.

The WHO standards do not include sexes-combined charts, and present separate charts for boys and girls for all anthropometric indicators. There are significant sex differences in the attained growth of boys and girls that justify the need to use separate charts. Figure 2.4 illustrates these differences for the indicator weight-for-age; the same applies to other growth indicators.

Last, windows of achievement for the six gross motor milestones collected in the MGRS (i.e., sitting without support, hands-and-knees crawling, standing with assistance, walking with assistance, standing alone, and walking alone) are also available in a published paper [20] and on the website (Figure 2.5).

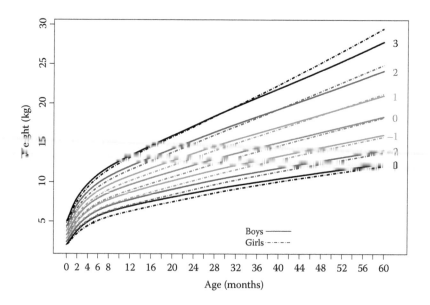

FIGURE 2.4   Attained weight-for-age for boys and girls.

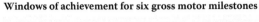

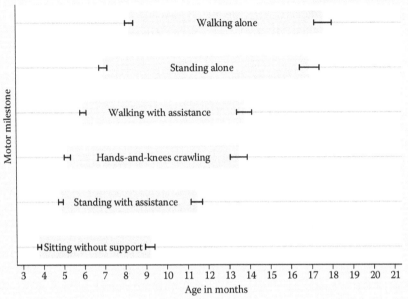

FIGURE 2.5 Windows of motor development milestone achievement in months. (Reproduced from WHO Multicentre Growth Reference Study Group, *Acta Paediatr* 2006, (Suppl 450):86–95. With permission.)

INNOVATIVE ASPECTS OF THE WHO CHILD GROWTH STANDARDS

The WHO standards were derived from children who were raised in environments that minimized constraints to growth, such as poor diets and infection. In addition, their mothers followed healthy practices, such as breastfeeding their children and not smoking during and after pregnancy. The standards depict normal human growth under optimal environmental conditions, and can be used to assess children everywhere, regardless of ethnicity, socioeconomic status, and type of feeding.

Another key characteristic of the new standards is that they explicitly identify breastfeeding as the biological norm, and establish the breastfed child as the normative model for growth and development. As an advocacy tool for the protection and promotion of breastfeeding, the new standards have the potential to significantly strengthen health policies and public support for breastfeeding.

Third, the pooled sample from the six participating countries allowed the development of a set of truly international growth standards, which underscore the fact that child populations grow similarly across the major regions of the world when their health and care needs are met. It also provides a tool that is timely and appropriate for the ethnic diversity seen within countries, and the evolution toward increasingly multiracial societies in the Americas and Europe, as elsewhere in the world.

Fourth, the wealth of data collected allowed not only the replacement of the current NCHS international references on attained growth (weight-for-age, length/height-for-age, and weight-for-length/height), but also the development of new standards for triceps and subscapular skinfolds, head and arm circumferences, and body mass index. These innovative references are particularly useful for monitoring the increasing epidemic of childhood obesity.

Fifth, the study's longitudinal nature also allowed the development of growth velocity standards. Pediatricians will not have to wait until children cross an attained growth threshold to make the diagnosis of undernutrition or overweight, since velocity standards will enable the early identification of children in the process of becoming undernourished or overweight.

Last, the development of accompanying motor development reference data has provided a unique link between physical growth and motor development. Although, in the past, the WHO issued guidelines concerning attained physical growth, it has not previously made recommendations for assessing motor development.

IMPLICATIONS OF ADOPTING THE WHO CHILD GROWTH STANDARDS FOR PUBLIC HEALTH AND CLINICAL PRACTICE

The scrutiny that the WHO standards have undergone is without precedent in the history of developing and applying growth assessment tools. Governments set up committees to scrutinize the new standards before deciding to adopt them, and professional groups conducted thorough examinations of the standards. The detailed evaluation allowed the assessment of the impact of the new standards and document their robustness and benefits for child health programs. Since their release in 2006, the WHO growth standards have been widely implemented globally [21] in over 130 countries by September 2014 (Figure 2.6). Reasons for adoption include: (1) the provision of a more reliable tool for assessing growth, which is consistent with the Global Strategy for Infant and Young Child Feeding; (2) the protection and promotion of breastfeeding; (3) the enablement of the monitoring of malnutrition's double burden, stunting and overweight; (4) the promotion of healthy growth and protection of the right of children to reach their full genetic potential; and (5) the harmonization of national growth assessment systems. In adopting the WHO growth standards, countries have harmonized best practices in child growth assessment, and established the breastfed infant as the norm against which compliance with children's right to achieve their full genetic growth potential is assessed.

FIGURE 2.6 The implementation of the WHO Child Growth Standards in more than 300 countries (by September 2014).

Detailed examination of the WHO growth standards by technical and scientific groups has provided a unique opportunity to validate their robustness and to improve understanding of their broad benefits:

- The WHO standards identify more children as being severely wasted [22]. Besides being more accurate in terms of the prediction of mortality risk [23–25], the use of the WHO standards results in shorter duration of treatment, higher rates of recovery, fewer deaths, and reduced loss to follow-up or need for inpatient care [26].
- The WHO standards confirm the dissimilar growth patterns between breastfed and formula-fed infants, and provide an improved tool for correctly assessing the adequacy of growth in breastfed infants [27–29]. They thereby considerably reduce the risk of unnecessary supplementation or breastfeeding cessation, which are major sources of morbidity and mortality in poor hygiene settings.
- In addition to confirming the importance of the first 2 years of life as a window of opportunity for promoting growth, the WHO standards demonstrate that intrauterine retardation in linear growth is more prevalent than previously thought [1], making a strong case for the need for interventions to start early in pregnancy and ideally before.
- Another important feature of the WHO standards is that they demonstrate that undernutrition during the first 6 months of life is a considerably more serious problem than previously detected [29,30], thereby reconciling the rates of growth faltering observed in young infants, and the prevalence of low birth weight and early abandonment of exclusive breastfeeding.

- The WHO standards also improve early detection of excess weight gain among infants and young children [31,32], showing that obesity often begins in early childhood, which is also when measures to tackle this global time bomb should be put in place.
- Last but not least, the WHO standards are an important means of ensuring the right of all children to be healthy, and to achieve their full physical and mental growth potential. They provide sound scientific evidence that, on average, young children everywhere experience similar growth patterns when their health and nutrition needs are met. For this reason, the WHO standards can be used to assess compliance with the UN Convention on the Rights of the Child, which recognizes the duties and obligations to children that cannot be met without attention to normal human development.

COMPLEMENTARY CHARTS FOR FETUSES, NEWBORNS BY GESTATIONAL AGE, PRETERM INFANTS, AND CHILDREN OLDER THAN 5 YEARS OF AGE

The WHO growth standards apply from birth to 5 years of age, and thus cannot be used to assess the growth of fetuses, newborns by gestational age, preterm infants, and children over 5 years of age. The recently completed International Fetal and Newborn Growth Consortium for the 21st Century (INTERGROWTH–21st) Project was designed to complement the WHO Child Growth Standards by providing comparable international standards for fetal growth from 9 weeks of gestation to birth and newborn size (weight, length, and head circumference) according to gestational age and sex [33].

Using a similar methodology and prescriptive approach to the MGRS [11], the eligibility criteria applied by INTERGROWTH–21st to select study populations and individual participants mirrored those used to construct the WHO Child Growth Standards (i.e., healthy populations seemingly free of disease, following current health recommendations, and living in environments highly unlikely to constrain growth) [34]. Implemented in eight countries several years after completion of the MGRS, the INTERGROWTH 21st project replicated the remarkable similarity in the linear growth of fetuses and newborns across study sites as that observed in the MGRS for children aged 0 to 60 months [12,35]. The results of the studies were also in strong agreement: The mean birth length for term newborns in INTERGROWTH–21st was 49.4 ± 1.9 cm, compared to 49.5 ± 1.9 cm in the MGRS [35]. Both studies consequently demonstrate that human growth potential is universal for healthy populations from conception to at least 5 years of age, when health, environmental, and care needs are met.

This view raises the inevitable question: How is the similarity in growth possible given the enrolled populations' distinct genetic backgrounds? It is true that tall parents tend to have tall children, and that short parents tend to have short children, but such expectations reflect interindividual rather than interpopulation variability. Moreover, recent studies of the genetics of height have identified about 200 genes associated with the genetic control of stature, explaining only approximately 10% of observed variability [36], which is much less than the 40% to 80% expected from earlier studies completed before the availability of genomic approaches. The small proportion of variability explained by that large number of genes likely reflects the influences on linear growth by nutrition and other care and environmental variables, in ways that remain not fully understood. These considerations lead to the expectation that most variation in growth exists among individuals rather than among populations, an expectation borne out by both the WHO MGRS and the INTERGROWTH–21st Project.

International standards for fetal growth, newborn-size, and preterm infants from the INTERGROWTH–21st Project are available for routine clinical practice and public health uses [37,38].

Contrary to the broad international consensus about the utility of the WHO Child Growth Standards for assessing the growth of preschool children, far less is known about the growth and nutritional status of school-aged children and adolescents. Reasons for this lack of knowledge include the rapid changes in somatic growth, problems in dealing with variations in maturation, and difficulties in separating normal variations from those associated with health risks. Consequently, there is far less agreement with regard to which reference to use to assess the growth of school-age children and adolescents. The release of the WHO Child Growth Standards for preschool children, and increasing public health concern over childhood obesity, stirred interest in developing appropriate growth curves for school-aged children and adolescents [39]. Some authors emphasize the use of contemporary, convenient (i.e., recent and logistically feasible) samples [40], while others feel it would be better to follow an approach analogous to the one used by the WHO in developing standards for preschool children, based on a prescriptive design [41].

A significant inherent problem when updating growth curves using contemporary samples is that the resulting weight-based curves, such as the body mass index (BMI), are markedly skewed to the right, thereby redefining overweight and obesity as "normal" [40,42]. This progressive redefinition of overweight and obesity cutoffs due to the secular increase in the prevalence of these conditions results in a substantial underestimation of the prevalence of childhood obesity [43].

In a recent Indian study [40], the 85th and 95th percentiles for BMI at 18 years are above 25 and 30, respectively. As the authors themselves acknowledge, using the 85th and 95th percentiles as cutoffs for defining overweight and obesity entails accepting higher BMI (overweight children) as "normal" at all ages. To overcome this flaw, the authors propose the use of the 75th percentile on the BMI curves as a cutoff for screening for overweight boys and girls. A key question is whether recommending a lower percentile, such as the 75th percentile, as the cutoff for defining overweight is the appropriate way forward. The central purpose of growth charts is to provide sensible cutoffs to screen for growth problems. Lowering the proposed cutoffs for defining childhood overweight as updated growth curves become increasingly skewed upward cannot be the solution. A better approach would be to construct growth curves using samples that have achieved expected linear growth, while still not being affected by excessive weight gain relative to linear growth.

The case made for using a national reference has traditionally been that it is more representative of a country's children than any other reference could be. However, given the child obesity epidemic, this no longer holds for either weight or BMI. As soon as a new reference is produced, it is out of date.

In order to complement the growth standards for children under 5 years of age, WHO developed a growth reference for school-aged children and adolescents, namely, the WHO Reference 2007 for School-Aged Children and Adolescents [44]. These curves are closely aligned with the WHO Child Growth Standards at 5 years and provide a suitable reference for the 5 to 19 years age group, for use in conjunction with the WHO Child Growth Standards from 0 to 5 years. The full set of tables and charts for height, weight, and BMI is available at www.who.int/growthref/en, and includes application tools such as software for clinicians and public health specialists [45].

CONCLUSION

The WHO Child Growth Standards were derived from children who were raised in environments that minimized constraints to growth, such as poor diets and infection. In addition, their mothers followed healthy practices, such as breastfeeding their children and not smoking during and after pregnancy. The standards depict normal human growth under optimal environmental conditions and can be used to assess children everywhere, regardless of their ethnicity, socioeconomic status, or type of feeding. They also demonstrate that healthy children from around the world, who are raised in healthy environments and follow recommended feeding practices, have strikingly similar patterns of growth. The International Pediatric Association has officially endorsed the use of the WHO standards, describing them as "an effective tool for detecting both undernutrition and overweight and obesity" [46].

Early recognition of growth problems, such as growth faltering and excessive weight gain relative to linear growth, should become standard clinical practice via

- Routine collection of accurate weight and height measurements to enable monitoring of childhood growth
- Interpretation of anthropometric indices, such as height-for-age and BMI-for-age, based on the WHO Child Growth Standards
- Early intervention, after changes on growth patterns (e.g., upward or downward crossing of percentiles) have been observed, to provide parents and caregivers appropriate guidance and support

NOTES

The author is a staff member of the World Health Organization. The author alone is responsible for the views expressed in this publication and they do not necessarily represent the decisions, policy, or views of the World Health Organization.

REFERENCES

1. Victora CG, de Onis M, Hallal PC et al. Worldwide timing of growth faltering: Revisiting implications for interventions using the World Health Organization growth standards. *Pediatrics* 2010; 125(3):e473–80.

2. Victora CG, Adair L, Fall C et al., for the Maternal and Child Undernutrition Study Group. Maternal and child undernutrition: Consequences for adult health and human capital. *Lancet* 2008; 371:340–57.
3. Caulfield LE, de Onis M, Blössner M et al. Undernutrition as an underlying cause of child deaths associated with diarrhea, pneumonia, malaria and measles. *Am J Clin Nutr* 2004; 80:193–8.
4. American Academy of Pediatrics Policy Statement. Prevention of pediatric overweight and obesity. *Pediatrics* 2003; 112:424–30.
5. de Onis M, Garza C, Onyango AW et al., editors. WHO Child Growth Standards. *Acta Paediatr Suppl* 2006; 450:1–101.
6. WHO Multicentre Growth Reference Study Group. *WHO Child Growth Standards: Length/height-for-age, weight-for-age, weight-for-length, weight-for-height and body mass index-for-age: Methods and development.* Geneva: World Health Organization; 2006.
7. Garza C, de Onis M, for the WHO Multicentre Growth Reference Study Group. Rationale for developing a new international growth reference. *Food Nutr Bull* 2004; 25(Suppl 1):S5–14.
8. Hamill PVV, Drizd TA, Johnson CL et al. Physical growth: National Center for Health Statistics percentiles. *Am J Clin Nutr* 1979; 32:607–29.
9. de Onis M, Yip R. The WHO growth chart: Historical considerations and current scientific issues. *Bibl Nutr Dieta* 1996; 53:74–89.
10. WHO Working Group on Infant Growth. An evaluation of infant growth: The use and interpretation of anthropometry in infants. *Bull World Health Organ* 1995; 73:165–74.
11. de Onis M, Garza C, Victora CG et al., editors. WHO Multicentre Growth Reference Study (MGRS): Rationale, planning and implementation. *Food Nutr Bull* 2004; 25(Suppl 1):S1–89.
12. WHO Multicentre Growth Reference Study Group. Assessment of differences in linear growth among populations in the WHO Multicentre Growth Reference Study. *Acta Paediatr* 2006; (Suppl 450):56–65.
13. Rosenberg NA, Pritchard JK, Weber JL et al. Genetic structure of human populations. *Science* 2002; 298:2381–5.
14. WHO Multicentre Growth Reference Study Group. Enrolment and baseline characteristics in the WHO Multicentre Growth Reference Study. *Acta Paediatr Suppl* 2006; 450:7–15.
15. WHO Anthro for personal computers, version 3.2.2. Software for assessing growth and development of the world's children. Geneva: World Health Organization, 2010. From: http://www.who.int/childgrowth/software/en/ (accessed February 3, 2017).
16. World Health Organization. Training course on child growth assessment. Geneva: World Health Organization, 2008. From: http://www.who.int/childgrowth/training/en/ (accessed February 3, 2017).
17. World Health Organization. *WHO Child Growth Standards: Head circumference-for-age, arm circumference-for-age, triceps skinfold-for-age and subscapular skinfold-for-age: Methods and development.* Geneva: World Health Organization; 2007.
18. World Health Organization. *WHO Child Growth Standards: Growth velocity based on weight, length and head circumference: Methods and development.* Geneva: World Health Organization; 2009.
19. de Onis M, Siyam A, Borghi E et al. Comparison of the World Health Organization growth velocity standards with existing US reference data. *Pediatrics* 2011; 128:e18–26.
20. WHO Multicentre Growth Reference Study Group. WHO Motor Development Study: Windows of achievement for six gross motor development milestones. *Acta Paediatr* 2006; Suppl 450:86–95.

21. de Onis M, Onyango A, Borghi E et al., for the WHO Multicentre Growth Reference Study Group. Worldwide implementation of the WHO Child Growth Standards. *Public Health Nutr* 2012; 15:1603–10.
22. Dale NM, Grais RF, Minetti A et al. Comparison of the new World Health Organization growth standards and the National Center for Health Statistics growth reference regarding mortality of malnourished children treated in a 2006 nutrition program in Niger. *Arch Pediatr Adolesc Med* 2009; 163:126–30.
23. Lapidus N, Luquero FJ, Gaboulaud V et al. Prognostic accuracy of WHO growth standards to predict mortality in a large-scale nutritional program in Niger. *PLoS Medicine* 2009; 6:e1000039.
24. Vesel L, Bahl R, Martines J et al., for the WHO Immunization-Linked Vitamin A Supplementation Study Group. Use of new World Health Organization child growth standards to assess how infant malnutrition relates to breastfeeding and mortality. *Bull World Health Organ* 2010; 88:39–48.
25. O'Neill S, Fitzgerald A, Briend A et al. Child mortality as predicted by nutritional status and recent weight velocity in children under two in rural Africa. *J Nutr* 2012; 142:520–5.
26. Isanaka S, Villamor E, Shepherd S et al. Assessing the impact of the introduction of the World Health Organization growth standards and weight-for-height z-score criterion on the response to treatment of severe acute malnutrition in children: Secondary data analysis. *Pediatrics* 2009; 123:e54–59.
27. Saha KK, Frongillo EA, Alam DS et al. Use of the new World Health Organization child growth standards to describe longitudinal growth of breastfed rural Bangladeshi infants and young children. *Food Nutr Bull* 2009; 30:137–44.
28. Bois C, Servolin J, Guillermot G. Usage comparé des courbes de l'organisation mondiale de la santé et des courbes françaises dans le suivi de la croissance pondérale des jeunes nourrissons. *Arch Pediatr* 2010; 17:1035–41.
29. de Onis M, Onyango AW, Borghi E et al., for the WHO Multicentre Growth Reference Study Group. Comparison of the WHO Child Growth Standards and the NCHS/WHO International Growth Reference: Implications for child health programmes. *Public Health Nutr* 2006; 9:942–7.
30. Kerac M, Blencowe H, Grijalva-Eternod C et al. Prevalence of wasting among under 6-month-old infants in developing countries and implications of new case definitions using WHO growth standards: A secondary data analysis. *Arch Dis Child* 2011; 96:1008–13.
31. van Dijk CE, Innis SM. Growth-curve standards and the assessment of early excess weight gain in infancy. *Pediatrics* 2009; 123:102–8.
32. Maalouf-Manassch Z, Metallinos-Katsaras E, Dewey KG. Obesity in preschool children is more prevalent and identified at a younger age when WHO growth charts are used compared with CDC charts. *J Nutr* 2011; 141:1154–8.
33. Villar J, Altman DG, Purwar M et al. The objectives, design, and implementation of the INTERGROWTH–21st Project. *BJOG* 2013; 120(Suppl 2):9–26.
34. Papageorghiou AT, Lambert A, Barros FC et al. The methodology of the INTERGROWTH–21st Project. *BJOG* 2013; 120(Suppl 2):1–142.
35. Villar J, Papageorghiou AT, Pang R et al. The likeness of fetal growth and newborn size across non-isolated populations in the INTERGROWTH–21st Project: The Fetal Growth Longitudinal Study and Newborn Cross-Sectional Study. *Lancet Diabetes Endocrinol* 2014; 2:781–92.
36. Lettre G. Recent progress in the study of the genetics of height. *Hum Genet* 2011; 129:465–72.

37. Papageorghiou AT, Ohuma EO, Altman DG et al. International standards for fetal growth based on serial ultrasound measurements: The Fetal Growth Longitudinal Study of the INTERGROWTH–21st Project. *Lancet* 2014; 384:869–79.
38. Villar J, Cheikh Ismail L, Victora CG et al. International standards for newborn weight, length, and head circumference by gestational age and sex: The Newborn Cross-Sectional Study of the INTERGROWTH–21st Project. *Lancet* 2014; 384:857–68.
39. Butte NF, Garza C, de Onis M. Evaluation of the feasibility of international growth standards for school-aged children and adolescents. *J Nutr* 2007; 137:153–7.
40. Khadilkar VV, Khadilkar AV, Cole TJ et al. Cross sectional growth curves for height, weight and body mass index for affluent Indian children, 2007. *Indian Pediatr* 2009; 46:477–89.
41. Butte NF, Garza C. Development of an international growth standard for preadolescent and adolescent children. *Food Nutr Bull* 2006; 27(Suppl):S169–326.
42. Fredriks AM, van Buuren S, Wit JM et al. Body mass index measurements in 1996–7 compared to 1980. *Arch Dis Child* 2000; 82:107–12.
43. de Onis M. The use of anthropometry in the prevention of childhood overweight and obesity. *Int J Obesity* 2004; 28:S81–85.
44. de Onis M, Onyango AW, Borghi E et al. Development of a WHO growth reference for school-aged children and adolescents. *Bull World Health Organ* 2007; 85:660–7.
45. *WHO AnthroPlus for personal computers manual: Software for assessing growth of the world's children and adolescents.* Geneva: World Health Organization; 2009. From: http://www.who.int/growthref/tools/en/ (accessed February 3, 2017).
46. International Pediatric Association. Endorsement of the New WHO Growth Standards for Infants and Young Children. Geneva: World Health Organization; 2006. From: http://www.who.int/childgrowth/Endorsement_IPA.pdf (accessed February 3, 2017).

Section III

Nutritional Requirements in the Life Stages

3 Nutrient Requirements and Recommendations during Pregnancy

Magali Leyvraz and Lynnette M. Neufeld

CONTENTS

INTRODUCTION

The importance of nutrition before and during pregnancy for women's health and fetal growth and development is well established, and the subject of other chapters in this book (see Chapters 4, 24, and 27).

The dietary requirements of pregnant women are higher to meet the needs of maternal tissue and plasma volume expansion, fetal growth, and preparation for lactation. To partially meet these needs, a number of normal cardiovascular, renal, endocrine, and metabolic changes during pregnancy improve the absorption and utilization of nutrients [1]. Even among healthy women with adequate nutritional status and reserves, these changes are insufficient to fully compensate for increased nutrient needs during pregnancy, and dietary requirements are therefore higher than those for nonpregnant women. Where women may not be able to meet their nutrient requirements from their daily dietary intake, a number of intervention programs exist, supported by recommendations from the World Health Organization (WHO) and national governments. The impact of such interventions is similarly covered in other chapters within this book.

In this chapter, we review nutrient requirements and recommended intakes during pregnancy, providing a brief biological justification for each. The first three sections focus on macronutrients, minerals, and vitamins, respectively. For each of these, we rely heavily on both published references (see Box 3.1) and, specifically, on Dietary Reference Intakes (DRIs; see Box 3.2). For each, we review the rationale, and provide an overview of current recommendations, including comments related to the strength of the evidence for these and any inconsistencies among diverse recommendation values. In each section, we make only brief reference to specific situations during pregnancy that influence requirements and recommendations, specifically adolescent pregnancy, obesity, gestational diabetes, multiple pregnancies, short interpregnancy interval, and vegetarian diets, as most of these are the topics of other chapters in this book. The fourth section briefly presents a number of food items or components to be avoided during pregnancy, specifically heavy metals, caffeine, and alcohol and other intoxicants. The final section provides some summary comments, as well as potential areas for further research.

BOX 3.1   SOURCES

Primary sources for information on requirements and recommendations for nutrition and dietary intake during pregnancy:

- Institute of Medicine (IOM)—http://iom.nationalacademies.org
- World Health Organization (WHO)—http://www.who.int/nutrition
- Food and Agriculture Organization (FAO)—http://www.fao.org/nutrition/en
- International Federation of Gynecology and Obstetrics (FIGO)—http://www.figo.org

BOX 3.2   DIETARY REFERENCE INTAKES

The Dietary Reference Intakes (DRIs) are a set of reference values for interpreting the adequacy of dietary nutrient intakes [2].

- The *Estimated Average Requirement* (EAR) is the daily intake level that is estimated to meet the requirements of half of all healthy individuals in a population.
- The *Recommended Dietary Allowance* (RDA) is the daily intake level that is sufficient to meet the nutrient requirements of nearly all (97%–98%) healthy individuals.
- The *Adequate Intake* (AI) is used when there is insufficient evidence to establish an EAR from which a RDA can be based, and is the recommended average daily intake level based on observed or experimentally determined estimates of nutrient intake among apparently healthy individuals, and is expected to meet or exceed the needs of most individuals in a population.
- The *Upper Limit* (UL) is the highest average daily intake of a nutrient that is likely to pose no risk of adverse health effects for nearly all persons in the general population.

REQUIREMENTS AND RECOMMENDATIONS FOR MACRONUTRIENTS DURING PREGNANCY

ENERGY

During pregnancy, basal metabolic rate and energy expenditure increase, and additional energy is required for storage in expanding maternal and fetal tissue as well as the placenta. Multiple studies have estimated the energy costs of pregnancy by measuring the increase in maternal basal metabolic rate and total energy expenditure. Assuming an average desirable gestational weight gain of 12 kg, it is estimated that between 76,048 and 76,652 additional kilocalories are required over the duration of a pregnancy, and that requirements progressively increase by trimester [1,3,4] (see Table 3.1).

Additional energy requirements are dependent on the expected weight gain of the mother, for which there are a number of determinants. Prepregnancy nutritional status (i.e., body mass index before pregnancy) is an important determinant of ideal weight gain. In 2009, the Institute of Medicine (IOM) updated weight gain recommendations for women based on prepregnancy body mass index (BMI; kg/m²). Underweight women (BMI <18.5) are recommended to gain between 12.7 and 18.1 kg; normal weight women (BMI 18.5 to 24.9) between 11.3 and 15.9 kg; overweight women (BMI 24.9 to 29.9) between 6.8 and 11.3 kg; and obese women (BMI ≥30.0) between 5.0 and 9.1 kg. It is recommended that women of short stature (<157 cm) gain weight at the lower range for their prepregnancy BMI [1,3].

TABLE 3.1

Additional Energy Requirements (kcal/day) during Pregnancy by Trimester (Assuming a Gestational Weight Gain of 12 kg)

	Reference		
	FAO, 2001	**Butte and King, 2005**	**IOM and NRC, 2009**
1st trimester	+69	+69	+0
2nd trimester	+266	+265	+340
3rd trimester	+496	+497	+452
Total pregnancy	+76,530	+76,652	+76,048

Sources: Food and Agriculture Organization (FAO), Human energy requirements: Report of a joint FAO/WHO/UNU Expert Consultation, Food and Nutrition Technical Report Series, Rome: Food and Agriculture Organization, 2001; Butte NF, King JC, *Public Health Nutr* 2005, 8(7A):1010–27; Institute of Medicine (IOM), National Research Council (NRC), *Weight gain during pregnancy: Reexamining the guidelines,* Rasmussen KM, Yaktine AL, eds., Washington, DC: National Academies Press, 2009.

For adolescents who are pregnant, energy requirements are further elevated by continued growth and development. Although specific energy and/or weight gain requirements have not been developed for pregnant adolescents, two studies suggest that weight gains in the upper range of those listed for each BMI group would be most appropriate for pregnant adolescents [5]. Based on limited evidence for multiple pregnancies (i.e., ≥2 fetuses), the IOM [1] suggests the following gestational weight gain recommendations for women with twins: between 16.8 and 24.5 kg for normal weight woman, between 14.1 and 22.7 kg for overweight women, and between 11.3 and 19.1 kg for obese women. Recommendations for underweight women with multiple pregnancies are still lacking.

FAT AND FATTY ACIDS

Fat is an important source of energy at all stages of life. Recommendations for fat intake are set as a proportion of total energy from fat and, at this time, there is no evidence to suggest that these should differ for pregnant women, compared to the general population. The IOM sets a recommended range of total energy from fat of 20% and 35% of daily total energy intake [2]. The International Federation of Gynecology and Obstetrics (FIGO) recommends slightly lower percentages, that is, between 15% and 30% [6].

Fatty acids can be categorized into two types: (1) nonessential fatty acids, which can be synthesized by the body, including saturated fatty acids and monounsaturated fatty acids; and (2) essential fatty acids, which cannot be synthesized by the body, and need to be provided by the diet, including polyunsaturated fatty acids, such as linoleic acid and α-linolenic acid. For saturated fatty acids and trans-fatty acids, recommendations for pregnant women are the same as for other adults; these nonessential fatty acids have been positively correlated with increased total and low-density lipoprotein (LDL) cholesterol levels and an increased risk of coronary heart diseases [2,7], so their intake should be limited. There are no recommended daily intakes for other nonessential fatty acids, as they can be synthesized by the body.

Over the past decade, interest has been rising in omega-3 fatty acids, specifically docosahexaenoic acid (DHA), a fatty acid that is partly made by the body from α-linolenic acid. As this fatty acid can be synthesized in the body, it is not considered to be an essential fatty acid; however, conversion of α-linolenic acid to DHA is limited. Increased prenatal DHA intake has been associated with several positive cognitive and motor development outcomes in children [8], and supplementation with 150 to 1,200 mg DHA per day may prolong gestation, increase birth weight, and reduce the risk of preterm delivery [9]. Recommendations therefore suggest that the intake of polyunsaturated fatty acids should be maintained or increased during pregnancy. More precisely, based on estimated mean intakes of healthy pregnant women, the IOM recommends 13 g linoleic acid per day and 1.4 g α-linolenic acid per day. Moreover, the Perinatal Lipid Metabolism Research Project of the European Commission recommends at least 200 mg DHA per day for pregnant women [10]. The Health Minister of Canada does not recommend any DHA supplementation; however, it recommends that pregnant women consume at least 150 g of cooked fish per week [11].

We found no studies or documents that provided varying recommendations for fat or fatty acid intakes based on maternal nutritional status prepregnancy, or for adolescents. The diet of vegetarian and vegan women is generally low in saturated fats but is also very low to null in DHA. FIGO recommends that pregnant women increase their intake of α-linoleic acid (ALA) to compensate for the low intakes of DHA [6].

PROTEIN

Due to adaptations in protein metabolism during pregnancy [12], the synthesis of protein in pregnant women is increased and protein is stored in the fetus, placenta, and maternal tissues. Similar to fat, recommendations for protein intake are expressed as a percent of total energy for protein: the IOM recommends 10% to 35% of total energy from protein.

Many studies have examined protein requirements during pregnancy, based on estimations of the total amount of protein deposited and nitrogen balance studies. The IOM recommends no additional intake of protein for the first half of the pregnancy, as the metabolic adaptations are thought to compensate for increased needs. During the second half of the pregnancy, the IOM set the estimated average requirement (EAR) at 0.88 g/kg/day and the RDA at 1.1 g/kg/day. In 2007, a WHO/ Food and Agriculture Organization (FAO)/United Nations University (UNU) Expert Consultation concluded, based on the estimated efficiency of protein utilization and an addition gestational weight gain of 13.8 kg, that during the first, second, and third trimesters of pregnancy, an additional 1 g, 9 g, and 31 g protein per day are required, respectively [13]. Despite this higher need, caution should be taken in increasing intakes beyond recommendations. High protein supplements may have deleterious effects, such as increased risk of small for gestational age, and both the *Cochrane*

Review and the WHO Expert Consultation recommend against their use, and against increasing protein intakes greatly above the requirements [13,14].

Based on one large effectiveness study and one nitrogen balance study, it is estimated that pregnant adolescent girls require slightly higher amounts of protein: 1.5 g/kg/day [13]. The WHO Expert Consultation suggests that the recommended intakes for women pregnant with twins should be 50 g additional per day during the second and third trimesters. However, no currently published trial has investigated this question, and recommendations for multiple fetus pregnancies have not been developed [13]. In 2006, the IOM concluded that vegetarian and vegan women are more likely to consume insufficient amounts of certain essential amino acids, and are therefore at risk of suboptimal protein utilization and synthesis. Although currently limited evidence exists, vegetarian and vegan pregnant women should carefully ensure sufficient intakes in all essential amino acids, again avoiding the use of high protein supplements [2].

CARBOHYDRATES

Carbohydrates are the primary source of energy for the cells of the pregnant women and the fetus. Glucose is the main source of energy for the brain, and the carbohydrate requirements of pregnant women are based on the estimated additional glucose needed for fetal brain development. Carbohydrates can be classified by their glycemic index (GI). Diets rich in low GI carbohydrates have been proposed to help prevent obesity-associated diseases, such as type 2 diabetes and coronary heart disease [15]. Studies with pregnant women have shown that a low GI diet can, among other things, prevent excessive gestational weight gain, improve glucose tolerance and, in the extent to which low GI foods are more likely to be rich in micronutrients, increase their intakes [16].

As for the general population, the IOM recommends that, for pregnant women, carbohydrates should provide 45% to 65% of the total energy. The IOM set the EAR for pregnant women at 135 g/day and the RDA at 185 g/d, based on the amount of carbohydrate on the EAR for the women, plus the amount of glucose needed for the fetal brain. Due to limited evidence, there are currently no specific recommendations for carbohydrate intake for obese women. However, low GI diets have been shown to prevent excessive weight gain, as well as increase weight loss after delivery [17]. In women with gestational diabetes, a low GI diet has been associated with lower insulin use and lower postprandial glucose levels [18].

REQUIREMENTS AND RECOMMENDATIONS FOR MINERAL INTAKES DURING PREGNANCY

IRON

Anemia remains one of the most common problems to affect women worldwide, and iron deficiency is among one of its many important causes. Anemia during pregnancy is associated with an increased risk of preterm birth, low birth weight, and reduced neonatal health, as well as reduced iron stores in the fetus [19]. The requirements in iron rise progressively during pregnancy, especially during the second and third tri-mesters when fetal growth is most rapid. Although iron absorption increases during pregnancy [20], iron intake is often insufficient, even in middle- and high-income populations [21]. The total amount of iron required for a normal pregnancy (for a women of 55 kg) is approximately 1,190 mg [20]. During the first trimester, the daily iron requirements increase by ~0.8 mg, during the second trimester they increase by between 4 and 5 mg, and during the third trimester by more than 6 mg [20]. Based on the estimated iron requirements of pregnancy, the IOM set the RDA for pregnant woman at 27 mg iron per day. The UL for iron is set at 45 mg iron per day.

As these levels of iron intake are not often reached by pregnant women, iron supplementation during pregnancy is implemented in countries where the prevalence of iron deficiency and anemia is high. Iron supplementation has been proven to be effective in preventing maternal iron deficiency and in reducing the risk of low birth weight, and the current WHO recommendations suggest 30 to 60 mg/day [22,23], together with folic acid (see "Folate" section, later). A short interpregnancy interval (i.e., <12 months) is associated with a higher risk of preterm birth, low birth weight, stillbirth, and neonatal death. A short interpregnancy interval is also associated with a higher risk of maternal iron deficiency and anemia; women could therefore benefit from higher iron intakes, although no specific recommendations have been devel-oped [24].

The iron requirements of adolescent girls are high due to growth and menstrua-tion and, during pregnancy, these are even further increased, putting pregnant ado-lescent girls at a higher risk of iron deficiency and low body iron stores. Studies have shown that the supplementation of pregnant adolescent girls with iron at current recommendations (60 mg iron daily) is efficacious at improving and maintaining iron status [25].

The prevalence of iron deficiency anemia is 2.4 to 4 times higher in twin preg-nancies than in singleton pregnancies. The suggested AI for twin pregnancies is 30 mg iron daily. Specific recommendations for iron during multiple pregnancies have not been developed. Obese pregnant women are at a higher risk of being iron deficient due to reduced iron absorption induced by the rise in hepcidin [26]. Some evidence suggests that this could have a negative impact on the iron transfer to, and the iron status of, the fetus. How this may translate into specific higher recommen-dations for iron during pregnancy in obese women has not been adequately studied [26]. Pregnant women with a vegetarian or vegan diet are at a higher risk of being iron deficient, because of the lower content and bioavailability of iron in vegetarian diets. The IOM recommends an iron intake that is 1.8 times higher for vegetarians, which corresponds to 48.6 mg iron per day instead of 27 mg per day for women with omnivorous diets [2].

ZINC

Zinc plays an important role in growth and development, and its effects during preg-nancy have been extensively studied. Maternal zinc deficiency during pregnancy is associated with reduced fetal growth, low birth weight, and preterm labor and preeclampsia [27]. A *Cochrane Review* [28] has concluded that there is moderate evidence that zinc supplementation (providing between 5 and 44 mg zinc per day) can reduce the risk of preterm birth, small for gestational age, and low birth weight. Findings for other pregnancy outcomes are mixed; possible reasons for the inconsistency of the findings are the lack of a reliable zinc status biomarker, small sample sizes, a variable start and duration of zinc supplementation, the presence of infections and other micronutrient deficiencies, and variable zinc absorption due to diet [29]. It is estimated that approximately 82% of the pregnant women in the world are zinc deficient [27].

Based on zinc absorption studies and estimated zinc accumulation in the fetus, the IOM estimated the EAR of pregnant women at 9.5 mg zinc per day and recommends an RDA of 11 mg zinc per day. The IOM set the UL for zinc at 40 mg per day. Zinc is essential for growth and is therefore needed in especially high amounts by pregnant adolescent girls. The IOM estimated the EAR at 10.5 mg zinc per day and recommends an RDA of 12 mg zinc per day for pregnant adolescent girls. The UL is 34 mg zinc per day [2].

Vegetarian pregnant women typically consume and absorb lower amounts of zinc than nonvegetarian women. In addition to a lower zinc content, zinc bioavailability is lower in vegetarian diets due to the high content of dietary fiber and phytate, inhibitors of zinc absorption. The IOM suggests that vegetarians should consume up to 50% more zinc daily to account for this reduced bioavailability [2]. It is suggested that zinc transfer to the fetus could be impaired in obese women; evidence is limited, however [30].

IODINE

Iodine is essential for the thyroid functioning of the mother and the fetus, and iodine deficiency during pregnancy has serious consequences, especially for the fetus, such as pregnancy loss, infant mortality, and irreversible cognitive impairments in the child [31].

The recommendations for iodine intake during pregnancy vary between 200 and 250 µg/day. The IOM recommends an RDA of 220 µg/day for pregnant women; the International Council for Control of Iodine Deficiency Disorders (ICCIDD) recommends 250 µg/day [32]; and the European Food Safety Authority (EFSA) recommends 200 µg/day [33]. The IOM advises on a UL of 1,100 µg/day. Obese women seem to be more likely to suffer from thyroid dysfunction during pregnancy [34]. However, the association between obesity and iodine requirements remains to be investigated.

CALCIUM

During pregnancy, calcium is needed for the growth and development of fetal bones. Although the mechanism is not as yet fully understood, low calcium intakes have been associated with hypertension during pregnancy [35]. Calcium supplementation (≥1 g per day) during pregnancy has been shown to reduce the risk of preeclampsia, preterm birth, and neonatal death and morbidity [36].

The IOM recommends 1 g calcium per day for pregnant women, based on bone mineral mass and on average intakes. Moreover, the WHO recommends the supplementation of 1.5 to 2 g calcium per day in populations where calcium intakes are low [23,36]. The IOM set an UL at 2,500 mg calcium per day.

Pregnant adolescent girls need sufficient calcium to meet the needs of their own bone growth, as well as the growth of the fetus's bones. The IOM recommends a daily intake of 1.3 g calcium for pregnant adolescent girls [2]. Bone metabolism differs between singleton and twin pregnancies, and requirements may be higher [37], but no specific recommendations for multiple pregnancies have been developed at this time. Calcium content and bioavailability is lower in vegetarian diets due to the high intakes in calcium absorption inhibitors (i.e., oxalic and phytic acids), and therefore recommended calcium intakes should be higher [2].

OTHER MINERALS

Many other essential minerals exist with mixed strength in the evidence base that underpins recommendations specifically for requirements during pregnancy. The recommended intakes of copper, chromium, fluoride, magnesium, manganese, molybdenum, phosphorus, selenium and sodium chloride for pregnant women are shown in Table 3.2.

REQUIREMENTS AND RECOMMENDATIONS FOR VITAMIN INTAKES DURING PREGNANCY

VITAMIN A

Vitamin A is essential for vision, gene expression, immune function, growth, reproduction, and embryonic development [38]. Maternal vitamin A deficiency increases the risk of maternal night blindness, maternal mortality, preterm birth, intrauterine growth restriction, low birth weight, and abnormal development of the fetus [6]. Vitamin A deficiency has also been shown to be associated with anemia and iron deficiency [38]

The requirements of pregnant women are estimated to be between 370 and 550 µg retinol activity equivalent (RAE) per day, and the recommended intake is 770 µg RAE/day [2]. The UL of vitamin A intake during pregnancy is 3,000 µg RAE/day [2]. WHO recommends antenatal vitamin A supplementation in regions where the prevalence of vitamin A deficiency is high [39], as a means to reduce maternal night blindness, maternal anemia, and probably maternal infection. However, a high intake of preformed vitamin A (retinol) has been shown to be teratogenic, especially in early pregnancy [40], and women in countries where vitamin A deficiency is not common should avoid taking vitamin A supplements.

Vitamin A requirements for pregnant adolescents are estimated to be 530 µg RAE/day; the recommended intake is set higher, at 750 µg RAE/day [2]. There is some suggestion that women with gestational diabetes may be at greater risk of vitamin A deficiency [41], although no specific recommendations to address this have been developed.

TABLE 3.2

Recommended Intakes of Select Minerals for Pregnant Women

Mineral	Recommendation	Strength of Evidence[a]
Copper	EAR: 800 µg/day RDA: 1,000 µg/day UL: 10,000 µg/day (8,000 µg/day for pregnant adolescents)	Moderate. Based on the results of nonpregnant women depletion/repletion studies and the estimated accumulation of copper in the fetus.
Chromium	AI: 30 µg/day (29 µg/day for pregnant adolescents)	Low. AI based on the average intake of adult women and on the estimated body weight of pregnant women.
Fluoride	AI: 3 mg/day UL: 10 mg/day	Low. AI set at the same level as for nonpregnant women.
Magnesium	EAR and RDA 19–30 years: 290 mg/day and 350 mg/day EAR and RDA 31–50 years: 300 mg/day and 360 mg/day EAR and RDA for adolescent girls: 335 mg/day and 400 mg/day UL: 350 mg/day	Moderate. Based on balance studies in nonpregnant women and estimated increase in lean mass in pregnant women.
Manganese	AI: 2 mg/day UL: 11 mg/day	Low. Based on the average intakes of nonpregnant women and modified for the average body weight of pregnant women.
Molybdenum	EAR: 40 µg/day RDA: 50 µg/day UL: 2,000 µg/day (1,700 µg per day for pregnant adolescents)	Moderate. Based on balance studies and body weight.
Phosphorus	EAR: 580 mg/day (1,055 mg/day for adolescent pregnancies) RDA: 700 mg/day (1,250 mg/day for adolescent pregnancies) UL: 3.5 g/day	Moderate. Based on studies of serum inorganic phosphate concentration studies.
Selenium	EAR: 49 µg/day RDA: 60 µg/day UL: 400 µg/day	Moderate. Based on the maximization plasma glutathione peroxidase activity and the saturation of fetal selenoprotein.
Sodium chloride	AI: 1.5 g and 2.3 g/day UL: 2.3 g and 3.6 g/day	Low. Based on the age-specific average intakes of nonpregnant women.

Source: Institute of Medicine (IOM), *Dietary reference intakes: The essential guide to nutrient requirements*, Otten JJ, Hellwig JP, Meyers LD, eds., Washington, DC: National Academy of Sciences, 2006.

[a] The strength of the evidence was assessed by the authors based to the extent possible on the GRADE system (https://www.essentialevidenceplus.com/).

Vitamin D

Vitamin D not only plays a key role in calcium and phosphorus homeostasis and bone health, but also in immune function, growth, development, and blood pressure. Vitamin D deficiency has been associated with, among others, increased risk of cardiovascular diseases, cancer, type 1 diabetes, and depression, also suggesting a prominent role in other processes [42]. During pregnancy, vitamin D deficiency has been associated with an increased risk of preeclampsia and gestational diabetes for the mother, and preterm birth, low birth weight, respiratory morbidity, rickets, and osteopenia for the child [6,43].

There is a considerable discrepancy among the existing recommendations for vitamin D intake during pregnancy. The IOM recommends an RDA of 600 International Unit (IU) vitamin D per day [44]; the FIGO recommends a daily intake of ≥600 IU/day; and the Endocrine Society recommends a daily intake of ≥600 IU/day, and suggests that 1,500 to 2,000 IU/day might be needed [45]. However, animal studies suggest that high intakes of vitamin D could be teratogenic [43], and the IOM recommends a UL of 4,000 IU per day [44]. Obese pregnant women and women who develop gestational diabetes are at an increased risk of vitamin D deficiency [43], as are vegetarians due to average lower intakes of vitamin D [46], but no specific recommendations exist.

Folate

Folate functions as a coenzyme, and is essential for the nucleic and amino acid metabolisms as well as cell division and tissue growth [47]. Insufficient folic acid intake during the periconceptional period is associated with an increased risk of neural tube defects (NTDs) and other congenital malformations [48].

The requirements of women during pregnancy are estimated to be 520 μg dietary folate equivalents (DFE)/day; the RDA is 600 μg DFE/day [2]. It is widely recommended that pregnant women take 400 μg DFE supplements per day [2,6,23]. Because the neural tube closes very early in gestation, often before women realize that they are pregnant, it is also recommended that women planning women, particularly those with a history of NTD, consume 400 μg of folic acid per day through supplements and/or fortified foods [6]. The IOM set the UL for folic acid intake during pregnancy at 1,000 μg DFE/day. It has been suggested that obese women and women with diabetes may benefit from folic acid supplementation during pregnancy [6], but this has not been translated into specific recommendations.

Vitamin B$_{12}$

Vitamin B$_{12}$ plays a key role in blood formation and neurological function. Vitamin B$_{12}$, along with folate, vitamin B$_6$, choline, and methionine, are involved in homocysteine metabolism and DNA methylation [2]. Vitamin B$_{12}$ deficiency during pregnancy is associated with increased risk of spontaneous abortion, preeclampsia, preterm delivery, low birth weight, and NTDs [49].

The IOM estimated that pregnant women's vitamin B_{12} requirement is 2.2 µg/day, based on the amount needed to maintain normal hematological status and serum vitamin B_{12} levels, and the amount of vitamin B_{12} deposited in the fetus, and assuming an absorption rate of 50%. The RDA for pregnant women is 2.6 µg vitamin B_{12} per day. An observational study in India showed an association between vitamin B_{12} deficiency and obesity, insulin resistance and gestational diabetes, suggesting that obese women might be at a higher risk of vitamin B_{12} deficiency [50]. As with all nutrients found primarily in animal source foods, vegetarian women are at a high risk of consuming insufficient amounts of vitamin B_{12}.

OTHER VITAMINS

Recommendations specific to pregnant women exist for a number of additional vitamins; however, the evidence that substantiates these recommendations varies considerably. The recommended intakes of vitamin B_6, riboflavin, thiamin, niacin, pantothenic acid, biotin, vitamin C, vitamin E, and vitamin K for pregnant women are shown in Table 3.3.

RECOMMENDATIONS FOR AVOIDANCE AND LIMITATION OF INTAKES DURING PREGNANCY

A number of substances consumed on their own, added to foods during processing, or entering the food chain through environmental contamination pose a health threat to pregnant women. Many of these are readily transported by the placenta and/or breast milk, with deleterious effects for the fetus/newborn.

TOXIC HEAVY METALS, CONTAMINANTS, AND FOOD ADDITIVES

Toxic heavy metals may enter the food chain naturally, for example, where ground water is high in arsenic, or through environmental contamination. In particular, exposure to and intake of toxic heavy metals should be avoided during pregnancy, as they can cause serious negative effects on fetal growth and development. Even low (<50 µg/L blood) exposure to arsenic has been shown to cause spontaneous abortion, stillbirth, low birth weight, and neonatal mortality [51]. Cadmium exposure has been associated with an increased risk of preterm birth and low birth weight [52]. Maximum levels during pregnancy have been defined for boron, lead, mercury, nickel, and vanadium (Table 3.4). Some studies suggest that the toxicity of heavy metals can be higher when micronutrient deficiencies are also present [53]. Strategies to limit the consumption of heavy metals during pregnancy have been implemented, including limiting the consumption of some types of fatty fish and shellfish prone to high levels of methylmercury, and the treatment of tube well water prior to consumption, where arsenic and other metals may be high.

TABLE 3.3
Recommended Intakes of Select Vitamins for Pregnant Women

Mineral	Recommendation	Strength of Evidence[a]
Vitamin B$_6$	EAR: 1.6 mg/day RDA: 1.9 mg/day UL: 100 mg/day (80 mg/day for adolescents)	Moderate. Based on plasma pyridoxal 5-phosphate level.
Riboflavin	EAR: 1.2 mg/day RDA: 1.4 mg/day	Moderate. Based on age-specific requirements of nonpregnant women estimated from riboflavin excretion and blood values and erythrocyte glutathione reductase activity coefficient with additional amount from estimated increased energy utilization and growth needs during pregnancy.
Thiamin	EAR: 1.2 mg/day RDA: 1.4 mg/day	Moderate. Based on age-specific requirements of nonpregnant women estimated from depletion-repletion studies with additional amount from estimated increased energy utilization and growth needs during pregnancy.
Niacin	EAR: 14 mg/day RDA: 18 mg/day UL: 35 mg/day (30 mg/day for adolescents)	Low. Based on age-specific requirements of nonpregnant women estimated from on urinary excretion of niacin metabolites with additional amount from estimated increased energy utilization and growth needs during pregnancy.
Pantothenic acid	AI: 6 mg/day	Low. Based on the mean intake of pregnant women.
Biotin	AI: 30 µg/day	Low. Extrapolated from concentration of biotin in human milk.
Vitamin C	RDA: 85 mg/day (85 mg/day for adolescents) UL: 2,000 mg/day (1,800 mg/day for adolescents)	Moderate. Based on age-specific requirements of nonpregnant women estimated from neutrophil concentration studies with additional amount for the transfer of vitamin C to the fetus.
Vitamin E	RDA: 15 mg/day UL: 1,000 mg/day (800 mg/day for adolescents)	Low. Based on age-specific requirements of nonpregnant women estimated from *in vitro* studies and plasma concentration.
Vitamin K	AI: 90 µg/day (75 µg/day for adolescents)	Low. Based on the median intake of adult women.

Source: Institute of Medicine (IOM), *Dietary reference intakes: The essential guide to nutrient requirements*, Otten JJ, Hellwig JP, Meyers LD, eds., Washington, DC: National Academy of Sciences, 2006.

[a] The strength of the evidence was assessed by the authors based to the extent possible on the GRADE system (https://www.essentialevidenceplus.com/).

TABLE 3.4

Maximum Levels of Heavy Metal Consumption during Pregnancy

Nutrient	Maximum Level	Reference
Boron	20 mg/day	[2,54]
Lead	5 µg/dL (blood lead level)	[55]
Mercury	1.3 µg methylmercury/kg/week	[56]
Nickel	1 mg/day	[2]
Vanadium	1.8 mg/day	[54]

During pregnancy, foodborne infections and contamination should be avoided by reducing the consumption of raw or other potentially contaminated foods, and by ensuring that the foods are stored, washed, and prepared well [6]. Animal studies have shown that aflatoxin consumption is teratogenic [6], and human studies have associated exposure to aflatoxin with low birth weight newborns [57].

We did not find studies that specifically reported the potential adverse effects of pregnancy outcomes from the consumption of food additives, nor did we find any specific recommendations to limit their consumption at this time.

CAFFEINE, ALCOHOL, AND OTHER INTOXICANTS

A *Cochrane Review* concluded that there is insufficient evidence to support that caffeine intake can have a negative or positive impact on pregnancy outcomes and fetal development [58]. However, the FIGO recommends limiting caffeine intake to 200 mg per day on the basis that high maternal caffeine consumption has been associated with fetal growth restriction.

The use of alcohol, tobacco, and other drugs should be avoided during pregnancy. Smoking during pregnancy is associated with spontaneous abortion, preterm birth, low birth weight, perinatal mortality, and impaired neurological and physical development of the child [59]. Alcohol consumption during pregnancy, even when low and occasional, has been associated with spontaneous abortion, fetus malformations, and behavioral and neurological impairments [59]. Moreover, alcohol can interfere with the absorption of key nutrients [59].

CONCLUSION AND FUTURE DIRECTIONS FOR RESEARCH

Dietary requirements and the corresponding recommended nutrient intakes during pregnancy are well established and based on evidence for energy, protein, and many nutrients, such as iron, folic acid, and vitamin A. For others, such as specific fatty acids, including DHA, types of carbohydrates, specifically related to the glycemic index, and a number of vitamins and minerals including zinc and vitamin D, there are important evidence gaps. For others, such as sodium, the evidence to provide pregnancy-specific guidance is unclear.

Dietary intake recommendations are based on the assumption of maintenance in healthy populations, and there is still much to learn about the implications of diverse health conditions on dietary nutrient needs during pregnancy, for example, underweight and obesity [1,2]. Similarly, the consequences of micronutrient deficiency for adverse pregnancy outcomes are well established, as is the potential of a number of interventions to improve pregnancy outcomes. Such interventions often provide supplements with a nutrient content at or above daily recommended intake amounts (e.g., iron folic acid supplementation during pregnancy). At the same time, there is some concern that high intakes of micronutrients may have deleterious effects, for example, vitamin A supplements in the absence of a deficiency, but the extent of other risks is not well established for other nutrients [2]. Improved methods of adequately assessing and balancing benefits and risks are needed. One such methodology, the Benefit-Risk Analysis for Foods (BRAFO) has been applied to a number of dietary interventions [60], but we found no studies at this time that use such methods to assess the balance of risks and benefits for nutrient intakes during pregnancy.

Similarly, there are considerable gaps in evidence related to additional nutrient requirements of specific subgroups, including adolescents and women with multiple pregnancies. Pregnant adolescents have increased nutrient requirements, not only for the fetus's growth, but also for their own growth [6]. Although some research has been done on the requirements of adolescent girls with respect to certain nutrients, such as energy, protein, and iron, evidence for setting specific recommendations is lacking for many nutrients (e.g., zinc and magnesium) [2]. At the same time, in many countries, women are getting pregnant at an older age, and the use of fertility treatments has become more common, resulting in a higher frequency of multiple pregnancies. However, specific recommendations for multiple pregnancies are lacking for most nutrients, such as proteins, iron, zinc, and calcium [2].

REFERENCES

1. Institute of Medicine (IOM), National Research Council (NRC) Weight gain during pregnancy: Reexamining the guidelines, Rasmussen KM, Yaktine AL, eds. Washington, DC: National Academies Press; 2009.
2. Institute of Medicine (IOM). *Dietary reference intakes: The essential guide to nutrient requirements*. Otten JJ, Hellwig JP, Meyers LD, eds. Washington, DC: National Academy of Sciences; 2006.
3. Food and Agriculture Organization (FAO). Human energy requirements: Report of a joint FAO/WHO/UNU Expert Consultation. Food and Nutrition Technical Report Series. Rome: Food and Agriculture Organization; 2001.
4. Butte NF, King JC. Energy requirements during pregnancy and lactation. *Public Health Nutr* 2005; 8(7A):1010–27.
5. Harper LM, Chang JJ, Macones GA. Adolescent pregnancy and gestational weight gain: Do the Institute of Medicine recommendations apply? *Am J Obstet Gynecol* 2011; 205(2):140.e1–8.
6. Hanson MA, Bardsley A, De-Regil LM et al. The International Federation of Gynecology and Obstetrics (FIGO) Recommendations on Adolescent, Preconception, and Maternal Nutrition: "Think nutrition first." 2015.

7. Smedts HP, Rakhshandehroo M, Verkleij-Hagoort AC et al. Maternal intake of fat, riboflavin and nicotinamide and the risk of having offspring with congenital heart defects. *Eur J Nutr* 2008; 47(7):357–65.

8. Carlson SE. Docosahexaenoic acid supplementation in pregnancy and lactation. *Am J Clin Nutr* 2009; 89(2):678S–84S.

9. Szajewska H, Horvath A, Koletzko B. Effect of n-3 long-chain polyunsaturated fatty acid supplementation of women with low-risk pregnancies on pregnancy outcomes and growth measures at birth: A meta-analysis of randomized controlled trials. *Am J Clin Nutr* 2006; 83(6):1337–44.

10. Koletzko B, Cetin I, Brenna JT, Perinatal Lipid Intake Working Group. Dietary fat intakes for pregnant and lactating women. *Br J Nutr* 2007; 98(5):873–7.

11. Health Canada. Prenatal nutrition guidelines for health professionals: Fish and omega-3 fatty acids. 2009.

12. Kalhan SC. Protein metabolism in pregnancy. *Am J Clin Nutr* 2000; 71(5 Suppl):1249S–55S.

13. World Health Organization (WHO). Protein and amino acid requirements in human nutrition. Report of a joint FAO/WHO/UNU Expert Consultation. WHO Technical Report Series, no 935. Geneva: World Health Organization; 2007.

14. Ota E, Hori H, Mori R et al. Antenatal dietary education and supplementation to increase energy and protein intake. *Cochrane Database Syst Rev* 2015; 6:CD000032.

15. Schwingshackl L, Hoffmann G. Long-term effects of low glycemic index/load vs. high glycemic index/load diets on parameters of obesity and obesity-associated risks: A systematic review and meta-analysis. *Nutr Metab Cardiovasc Dis* 2013; 23(8): 699–706.

16. McGowan CA, Walsh JM, Byrne J et al. The influence of a low glycemic index dietary intervention on maternal dietary intake, glycemic index and gestational weight gain during pregnancy: A randomized controlled trial. *Nutr J* 2013; 12(1):140.

17. Horan MK, McGowan CA, Gibney ER et al. Maternal diet and weight at 3 months post-partum following a pregnancy intervention with a low glycaemic index diet: Results from the ROLO randomised control trial. *Nutrients* 2014; 6(7):2946–55.

18. Viana LV, Gross JL, Azevedo MJ. Dietary intervention in patients with gestational diabetes mellitus: A systematic review and meta-analysis of randomized clinical trials on maternal and newborn outcomes. *Diabetes Care* 2014; 37(12):3345–55.

19. Allen LH. Anemia and iron deficiency: Effects on pregnancy outcome. *Am J Clin Nutr* 2000; 71(5 Suppl):1280S–4S.

20. Bothwell TH. Iron requirements in pregnancy and strategies to meet them. *Am J Clin Nutr* 2000; 72(1 Suppl):257S–64S.

21. Blumfield ML, Hure AJ, Macdonald-Wicks L et al. A systematic review and meta-analysis of micronutrient intakes during pregnancy in developed countries. *Nutr Rev* 2013; 71(2):118–32.

22. World Health Organization (WHO). *Guideline: Daily iron and folic acid supplementation in pregnant women*. Geneva: World Health Organization; 2012.

23. World Health Organization (WHO). *WHO recommendations on antenatal care for a positive pregnancy experience*. Geneva: World Health Organization; 2016.

24. Wendt A, Gibbs CM, Peters S et al. Impact of increasing inter-pregnancy interval on maternal and infant health. *Paediatr Perinat Epidemiol* 2012; 26 Suppl 1:239–58.

25. Angeles-Agdeppa I, Schultink W, Sastroamidjojo S et al. Weekly micronutrient supplementation to build iron stores in female Indonesian adolescents. *Am J Clin Nutr* 1997; 66(1):177–83.

26. Garcia-Valdes L, Campoy C, Hayes H et al. The impact of maternal obesity on iron status, placental transferrin receptor expression and hepcidin expression in human pregnancy. *Int J Obes* 2015; 39(4):571–8.

27. Caulfield LE, Zavaleta N, Shankar AH et al. Potential contribution of maternal zinc supplementation during pregnancy to maternal and child survival. *Am J Clin Nutr* 1998; 68(2 Suppl):499S–508S.
28. Ota E, Mori R, Middleton P et al. Zinc supplementation for improving pregnancy and infant outcome. *Cochrane Database Syst Rev* 2015; 2:CD000230.
29. Donangelo CM, King JC. Maternal zinc intakes and homeostatic adjustments during pregnancy and lactation. *Nutrients* 2012; 4(7):782–98.
30. Al-Saleh E, Nandakumaran M, Al-Harmi J et al. Maternal-fetal status of copper, iron, molybdenum, selenium, and zinc in obese pregnant women in late gestation. *Biol Trace Elem Res* 2006; 113(2):113–23.
31. De-Regil LM, Harding KB, Peña-Rosas JP et al. Iodine supplementation for women during the preconception, pregnancy and postpartum period (Protocol). *Cochrane Database Syst Rev* 2015; 6:CD011158.
32. World Health Organization (WHO)/United Nations Children's Fund (UNICEF)/ International Council for Control of Iodine Deficiency Disorders (ICCIDD). Assessment of iodine deficiency disorders and monitoring their elimination. A guide for programme managers. 3rd ed. Geneva: World Health Organization, United Nations Children's Fund, and International Council for Control of Iodine Deficiency Disorders; 2007.
33. European Food Safety Authority (EFSA). *Scientific opinion on dietary reference values for iodine.* European Food Safety Authority; 2014.
34. Gowachirapant S, Melse-Boonstra A, Winichagoon P et al. Overweight increases risk of first trimester hypothyroxinaemia in iodine-deficient pregnant women. *Matern Child Nutr* 2014; 10(1):61–71.
35. Hofmeyr GJ, Lawrie TA, Atallah AN et al. Calcium supplementation during pregnancy for preventing hypertensive disorders and related problems. *Cochrane Database Syst Rev* 2010; 8:CD001059.
36. World Health Organization (WHO). *Guideline: Calcium supplementation in pregnant women.* Geneva: World Health Organization; 2013.
37. Nakayama S, Yasui T, Suto M et al. Differences in bone metabolism between singleton pregnancy and twin pregnancy. *Bone* 2011; 49(3):513–9.
38. McCauley ME, van den Broek N, Dou L et al. Vitamin A supplementation during pregnancy for maternal and newborn outcomes. *Cochrane Database Syst Rev* 2015; 10:CD008666.
39. World Health Organization (WHO). *Safe vitamin A dosage during pregnancy and lactation. Recommendations and report of a consultation.* Geneva: World Health Organization; 1998.
40. Rothman KJ, Moore LL, Singer MR et al. Teratogenicity of high vitamin A intake. *N Engl J Med* 1995; 333(21):1369–73.
41. Resende FB, De Lira LQ, Grilo EC et al. Gestational diabetes: A risk of puerperal hypovitaminosis A? *An Acad Bras Cienc* 2015; 87(1):463-70.
42. Hossein-Nezhad A, Holick MF. Vitamin D for health: A global perspective. *Mayo Clin Proc* 2013; 88(7):720–55.
43. De-Regil LM, Palacios C, Ansary A et al. Vitamin D supplementation for women during pregnancy. *Cochrane Database Syst Rev* 2012; 2:CD008873.
44. Insitute of Medicine (IOM). *Dietary reference intakes for calcium and vitamin D.* Ross AC, Taylor CL, Yaktine AL, Del Valle HB, eds. Washington, DC: Insitute of Medicine; 2011.
45. Holick MF, Binkley NC, Bischoff-Ferrari HA et al. Evaluation, treatment, and prevention of vitamin D deficiency: An endocrine society clinical practice guideline. *J Clin Endocrinol Metab* 2011; 96(7):1911–30.
46. Reid MA, Marsh KA, Zeuschner CL et al. Meeting the nutrient reference values on a vegetarian diet. *Med J Aust* 2013; 199(4 Suppl):S33–40.

47. Lassi ZS, Salam RA, Haider BA et al. Folic acid supplementation during pregnancy for maternal health and pregnancy outcomes. *Cochrane Database Syst Rev* 2013; 3:CD006896.

48. De-Regil LM, Peña-Rosas JP, Fernandez-Gaxiola AC et al. Effects and safety of periconceptional oral folate supplementation for preventing birth defects. *Cochrane Database Syst Rev* 2015; 12:CD007950.

49. Finkelstein JL, Layden AJ, Stover PJ. Vitamin B_{12} and perinatal health. *Adv Nutr* 2015; 6(5):552–63.

50. Krishnaveni GV, Hill JC, Veena SR et al. Low plasma vitamin B_{12} in pregnancy is associated with gestational "diabesity" and later diabetes. *Diabetologia* 2009; 52(11):2350–8.

51. Quansah R, Armah FA, Essumang DK et al. Association of arsenic with adverse pregnancy outcomes/infant mortality: A systematic review and meta-analysis. *Environ Health Perspect* 2015; 123(5):412–21.

52. Ikeh-Tawari EP, Anetor JI, Charles-Davies MA. Cadmium level in pregnancy, influence on neonatal birth weight and possible amelioration by some essential trace elements. *Toxicol Int* 2013; 20(1):108–12.

53. Peraza MA, Ayala-Fierro F, Barber DS et al. Effects of micronutrients on metal toxicity. *Environ Health Perspect* 1998;106 Suppl 1:203–16.

54. European Food Safety Authority (EFSA). *Tolerable upper intake levels for vitamins and minerals.* Parma, Italy: European Food Safety Authority; 2006.

55. Gynecologists ACoOa. Comittee Opinion No. 533: Lead screening during pregnancy and lactation. *Obstet Gynecol* 2012; 120:416–20.

56. European Food Safety Authority (EFSA). *Scientific opinion on the risk for public health related to the presence of mercury and methylmercury in food.* Parma, Italy: European Food Safety Authority; 2012.

57. Shuaib FM, Ehiri J, Abdullahi A et al. Reproductive health effects of aflatoxins: A review of the literature. *Reprod Toxicol* 2010; 29(3):262–70.

58. Jahanfar S, Jaafar SH. Effects of restricted caffeine intake by mother on fetal, neonatal and pregnancy outcomes. *Cochrane Database Syst Rev* 2015; 6:CD006965.

59. Institute of Medicine (IOM), Committee on Nutritional Status During Pregnancy and Lactation. *Nutrition during pregnancy: Part I: Weight gain, Part II: Nutrient supplements.* Washington, DC: National Academy of Sciences; 1990.

60. Verhagen H, Andersen R, Antoine JM et al. Application of the BRAFO tiered approach for benefit-risk assessment to case studies on dietary interventions. *Food Chem Toxicol* 2012; 50 Suppl 4:S710–23.

4 Nutrient Requirements during Lactation

Victoria Hall Moran

CONTENTS

INTRODUCTION

Lactation is one of the most nutritionally demanding periods of a woman's life, when her nutritional requirements increase beyond both prepregnant and pregnant levels in order to support newborn and infant growth and development, and to maintain her own metabolic needs. Infants double their weight in the first 4 to 6 months postpartum, and the breast milk secreted in the first 4 months represents an amount of energy similar to the total energy cost of pregnancy [1]. Although the energy and nutrients stored during pregnancy are available to support milk production, there are undoubted increases in requirements for a broad range of nutrients during lactation. Most of these recommended intakes are based on our knowledge of the amount of milk produced during lactation, its energy and nutrient content, and the amount of maternal energy and nutrient reserves. Despite the increased nutrient demands of lactation, women are remarkably resilient in their ability to produce breast milk of sufficient quantity and quality to support the growth of their infant, even when the mother is deprived of nutrients herself. During lactation, the mammary gland exhibits metabolic priority for nutrients, often at the expense of maternal reserves.

Milk production may, therefore, have an impact on maternal body composition and nutritional status, and could impact negatively on any subsequent pregnancy, particularly if the interval between pregnancies is short. Pregnancies with short "recuperative intervals" (defined as the amount of time that the woman was not lactating prior to the next conception) may be particularly vulnerable to nutrient depletion. Yet reviews of the evidence, while highlighting weaknesses in the available studies, have found no clear evidence of a link between interpregnancy or recuperative interval and maternal anthropometric status, possibly due, in part, to changes in the hormonal regulation of nutrient partitioning between the mother and the fetus when a mother is malnourished [2,3].

Daily nutrient requirements during lactation are higher than requirements in pregnancy, and nutrient reference values are increased for a broad range of nutrients (Table 4.1). The exact nature of many of these requirements is not well understood, and it is difficult to conclude whether breastfeeding women need to make significant changes to their diet. Factors such as the duration and intensity of lactation (whether the infant is breastfed exclusively or only partially) are likely to influence nutritional requirements, as an exclusively breastfeeding woman will have greater energy and nutrient needs (with the exception of iron, attributed to the potential protective effect of lactational amenorrhea) than a woman who is partially breastfeeding [12]. Intensity of lactation has also been found to have an impact on the macronutrient content of breast milk. Recent research has found that the nutrient composition of the breast milk of exclusively breast feeding mothers is more calorific with a higher percentage fat content, and lower percentage protein and percentage carbohydrate content (percentages of total calorie content) than the milk of mixed feeding mothers [13].

In areas where micronutrient deficiency is prevalent, milk composition may be affected by marginal maternal deficiencies in fatty acids, iodine, and most vitamins. For most minerals, breast milk is more protected against variations in maternal nutrient intake and status by the mobilization of stores accrued during pregnancy (Table 4.2). Many minerals are transferred into milk by active transfer, rather than by passive diffusion, and this process compensates for variations in maternal mineral status.

ENERGY

The energy requirements of lactation are defined as "the level of energy intake from food that will balance the energy expenditure needed to maintain a body weight and body composition, a level of physical activity and breast milk production that are consistent with optimal health for the woman and her child" [14]. This definition implies that the energy required to produce an appropriate volume of milk must be added to the woman's usual energy requirement, and assumes that she resumes her usual level of physical activity soon after giving birth [14]. Although some studies suggest that metabolic adaptations and reduced physical activity act to conserve energy during lactation, it is generally believed that resting metabolic rate and thermogenesis are not significantly altered, and the savings gained from a sedentary lifestyle are unlikely to have much impact on availability of energy for lactation.

TABLE 4.1

United Kingdom, United States, and WHO/FAO Nutrient Reference Values during Lactation

	UK DRV [4]—RNI	U.S. DRI [5–9]—RDA	WHO/FAO—[10]
Energy	2270 kcal (0–6 m)		↑ 550
	↑[a] 330 kcal		
Protein	56 g/d (0–4 m)	71 g/d	↑ 19.4 g/d (0–6 m)
	↑ 11	↑ 25 g/d	↑ 12.5 g/d (6 m+)
	53 g/d (4 m and older)		
	↑8		
Vitamin A[b]	950 µg RE/d	1200 µg RE/d	850 µg RE/d
	↑ 350 µg RE/d	(age 14–18 y)	↑ 350 µg RE/d
		↑ 500 µg RE/d	
		1300 µg RE/d	
		(age 19–50 y)	
		↑ 600 µg RE/d	
Vitamin D	10 µg/d[c]	15 µg/d	5 µg/d
	↑ 10 µg/d	No increment	No increment
Vitamin E	None set	19 mg/d	None set
		↑ 4 mg/d	
Vitamin K	None set	75 µg/d (age 14–18 y)	55 µg/d
		90 µg/d (age 19–50 y)	No increment
		No increment	
Vitamin C	70 mg/d	115 mg/d (age 14–18 y)	70 mg/d
	↑ 30 mg/d	↑ 50 mg/d	↑ 25 mg/d
		120 mg/d (age 19–50 y)	
		↑ 45 mg/d	
Riboflavin	1.6 mg/d	1.6 mg/d	1.6 mg/d
	↑ 0.5 mg/d	↑ 0.6 mg/d (age 14–18 y)	↑ 0.5 mg/d
		↑ 0.5 mg/d (age 19–50 y)	
Vitamin B₁	1.0 mg	1.4 mg/d	1.5 mg/d
(Thiamin)	↑ 0.2 mg	↑ 0.4 mg/d (age 14–18 y)	↑ 0.1 mg/d
		↑ 0.3 mg/d (age 19–50 y)	
Vitamin B₃	16 mg/d (age 15–18 y)	17 mg/d	17 mg/d NE
(Niacin)[d]	15 mg/d (age 19–50 y)	↑ 3 mg/d	↑ 3 mg/d
	↑ 2 mg		
Vitamin B₆	1.2 mg	2.0 mg/d	2.0 mg/d
	No increment	↑ 0.8 mg/d (age 14–18 y)	↑ 0.7 mg/d
		↑ 0.7 mg/d (age 19–50 y)	
Folate	260 µg/d	500 µg/d	500 µg/d
	↑ 60 µg/d	↑ 100 µg/d	↑ 100 µg/d
Vitamin B₁₂	2.0 µg/d	2.8 µg/d	2.8 µg/d
(Cobalamin)	↑ 0.5 µg/d	↑ 0.4 µg/d	↑ 0.4 µg/d
Pantothenic	None set	7 mg/d (AI)	7 mg/d
acid		↑ 2 mg/d	↑ 2 mug/d

(Continued)

TABLE 4.1 (CONTINUED)
United Kingdom, United States, and WHO/FAO Nutrient Reference Values during Lactation

	UK DRV [4]—RNI	U.S. DRI [5–9]—RDA	WHO/FAO—[10]
Biotin	None set	35 µg/d (AI) ↑ 10 µg/d (age 14–18 y) ↑ 5 µg/d (age 19–50 y)	35 µg/d ↑ 5 µg/d
Choline	None set	550 mg/d (AI) ↑ 150 mg/d (age 14–18 y) ↑ 125 mg/d (age 19–50 y)	None set
Calcium	1350 mg/d (age 15–18 y) 1250 mg/d (age 19–50 y) ↑ 550 mg/d	1300 mg/d (age 14-18 y) 1000 mg/d (age 19-50 y) No increment	750 mg/d No increment (NPNL adolescents 1,000 mg/d particularly during the growth spurt)
Iron	14.8 mg/d No increment	10 mg/d (age 14–18 y) ↓ 5 mg/d 9 mg/d (age 19–50 y) ↓ 9 mg/d	10–30 mg/d depending on bioavailability of iron (a decrease of around 50% of NPNL values)
Phosphorus[e]	990 mg/d ↑ 440 mg/d	1250 mg/d (age 14–18 y) 700 mg/d (age 19–50 y) No increment	None set
Magnesium	350 mg/d (age 15–18 y) 320 mg/d (age 19–50 y) ↑ 50 mg/d	360 mg/d (age 14–18 y) 310 mg/d (age 19–30 y) 320 mg/d (age 31–50 y) No increment	270 mg/d ↑ 50 mg/d
Zinc	13 mg/d (0–4 months) ↑ 6 mg/d 9.5 mg/d (4+ months) ↑ 2.5 mg/d	13 mg/d (age 14–18 y) 12 mg/d (age 19–50) ↑ 4 mg/d	Depending on bioavailability of zinc: 5.8–19 mg/d (0–3 m) 5.3–17.5 mg/d (3–6 m) 4.3–14.4 (7–12 m) NPNL adolescents 4.3–14.4 mg/d NPNL women 3.0–9.8 mg/d
Copper	1.3 µg/d (age 15–18 y) 1.5 µg/d (age 19–50 y) ↑ 0.3 µg/d	1.3 µg/d ↑ 0.4 µg/d	None set
Selenium	75 µg/d ↑ 15 µg/d	70 µg/d ↑ 15 µg/d	35 µg/d (0–6 m) ↑ 9 µg/d 42 µg/d (7–12 m) ↑ 16 µg/d
Iodine	140 µg/d No increment	290 µg/d ↑ 140 µg/d	200 µg/d ↑ 50 µg/d
Potassium	3.5 g/d No increment	5.1 g/d (AI) ↑ 0.4 g/d	None set

(Continued)

TABLE 4.1 (CONTINUED)
United Kingdom, United States, and WHO/FAO Nutrient Reference Values during Lactation

	UK DRV [4]—RNI	U.S. DRI [5–9]—RDA	WHO/FAO—[10]
Manganese	None set	2.6 mg/d (AI) ↑ 1.0 mg/d (age 14–18 y) ↑ 0.8 mg/d (age 19–50)	None set
Molybdenum	None set	50 µg/d ↑ 7 µg/d (age 14–18 y) ↑ 5 µg/d (age 19–50)	None set

Abbreviations: AI, adequate intake; DRI, dietary reference intakes; DRV, dietary reference value; RAE, retinol activity equivalent; RDA, recommended dietary allowance; RE, retinol equivalent; RNI, recommended nutrient intake; NE, niacin equivalent; NPNL, nonpregnant, nonlactating.

[a] An upward-pointing arrow represents the amount of increase from nonpregnant, nonlactating levels. A downward-pointing arrow represents the amount of decrease from nonpregnant, nonlactating levels.

[b] As retinol activity equivalents (RAEs): 1 RAE = 1 µg retinol, 12 µg beta-carotene, 24 µg alpha-carotene, or 24 µg beta-cryptoxanthin. The RAE for dietary provitamin A carotenoids is twofold greater than the retinol equivalent, whereas the RAE for preformed vitamin A is the same as the retinol equivalent.

[c] The Scientific Advisory Committee on Nutrition (SACN) in the United Kingdom has recently proposed a Reference Nutrient Intake (RNI) for vitamin D of 10 µg/d for the UK population aged 4 years and over. This is the amount needed for 97.5% of the population to maintain a serum 25(OH)D concentration of 25 nmol/L when UVB sunshine exposure is minimal [11].

[d] As niacin equivalents: 1 mg of niacin = 60 mg tryptophan.

[e] RNI for phosphorus set to equal the RNI for calcium in millimoles (mmol). The increment for pregnancy therefore reflects the increment set for calcium.

Therefore, the mother must increase her energy intake in order to meet the additional demand of lactation.

The energy requirements for lactation are high. Dewey [12] calculated that the marginal energy costs for milk production for an exclusively breastfeeding woman are 595 kcal/day at 0 to 2 months postpartum and 670 kcal/day at 3 to 8 months postpartum. The energy needed to support this would be 440 to 515 kcal/day, allowing for 500 g of fat loss per month. The energy needs are lower for a partially breastfeeding woman, depending on the extent to which non–breast milk foods are consumed by her infant.

UK dietary reference values for energy during lactation increase by 330 kcal per day in the first 6 months [15], that is, a daily energy intake of 2,270 kcal. Thereafter, it is expected that the energy intake required to support breastfeeding will be modified by changes in maternal body composition and the breast milk intake of the infant. The World Health organization (WHO)/Food and Agriculture Organization of the United Nations (FAO) recommends an additional 505 kcal/day for well-nourished women with adequate gestational weight gain, while undernourished women and those with insufficient gestational weight gain should increase their calorific intake by 675 kcal/day during the first semester of lactation [14] (see Table 4.1).

TABLE 4.2

Nutrients in Human Breast Milk Categorized According to Whether They Are Likely to Be Affected by Maternal Dietary Intake

Nutrients Present in Breast Milk at Generally Stable Concentrations	Nutrients for Which Material Intake Is Likely to Affect Breast Milk Composition
	Polyunsaturated fatty acids (LA, ALA, ARA, DHA)
Total fat	Trans fatty acids
	Vitamin A (retinol and carotenes)
Carbohydrates (lactose)	Vitamin C
Calcium	Choline
	Thiamin
	Riboflavin
Iron	Vitamin B_6
Copper	Vitamin B_{12}
Zinc	Pantothenic acid
	Iodine
	Selenium
Folate	Vitamin D

There is conflicting evidence about whether women who breastfeed their infants lose more weight postnatally than women who do not breastfeed. Systematic reviews seeking to explore the relationship between breastfeeding and postpartum weight change have highlighted a high degree of heterogeneity among studies, particularly in relation to sample size, measurement time points, and in the classification of breastfeeding and postpartum weight change [16,17]. It has further been suggested that the lack of a consistent relation between breastfeeding and a reduction in postpartum weight retention may have resulted from a combination of a pattern of breastfeeding that would not be expected to modify weight loss appreciably, low statistical power, or the poor quality of information about the intensity and duration of breastfeeding (or all three factors), as well as a complex association with prepregnancy body mass index (BMI) [18]. Despite these limitations, Neville and colleagues [17] found that of the five studies that were considered to be of high methodological quality, four studies demonstrated a positive association between breastfeeding and weight change, and in a pooled analysis of eleven studies, He et al. [16] found that breastfeeding for 3 to ≤6 months had a negative influence on postpartum weight retention, but had little effect after 6 months of breastfeeding. Baker et al. [18] calculated that, if women exclusively breastfed for 6 months as recommended, postpartum weight retention could be eliminated by that time in women with gestational weight gain values of approximately 12 kg, and that the possibility of major weight gain (≥5 kg) could be reduced in all but the heaviest women. Moreover, longitudinal studies have suggested breastfeeding may provide long-term protection against postpartum weight retention among both normal weight [19] and obese women [20].

FATTY ACIDS

The major omega n-3 and n-6 long chain polyunsaturated fatty acid (LCPUFA) components of breast milk are docosahexaenoic acid (DHA), alpha-linoleic acid (ALA), and arachidonic acid (ARA). DHA and ARA in particular play a major role in infant neurological development. Research has indicated that the DHA content of breast milk may be linked to a child's optimal visual acuity development, neurodevelopment, and subsequent IQ [21]. A study of breast milk DHA levels across 28 countries found a strong positive relationship between breast milk DHA and cognitive performance, even when controlling for national wealth, investment in education and macronutrient intake [22]. Breast milk fatty acids may also contribute to the regulation of innate and adaptive immune responses later in childhood [23].

Although there is evidence that the fetus and infant are able to endogenously synthesize ARA and DHA, the synthesis is extremely low [24], making maternal LCPUFA supply critical during fetal and postnatal growth and development. The transfer of LCPUFA from mother to her infant during lactation is dependent largely upon maternal status, and variations in breast milk fatty acid status across populations are evident. Whereas the proportion of breast milk saturated and monounsaturated fatty acids have been found to be relatively constant across a large number of countries, some of the PUFAs, especially DHA, have shown high variability with particularly low levels in low marine food-consuming populations [25,26]. Maternal body mass index and age have also been found to affect fatty acid levels in human milk [27]. Very premature infants are particularly vulnerable to low DHA levels at birth, as LCPUFA accretion occurs primarily during the last trimester of pregnancy when maternal levels are high and growth and brain development are rapid. There is increasing evidence that optimizing LCPUFA provision postnatally may not only improve vision and neurodevelopment in very low birth weight and premature infants, but may also reduce the morbidity and mortality from diseases specific to premature infants, such as bronchopulmonary dysplasia, necrotizing enterocolitis, and retinopathy of prematurity [28].

Fish and seafood comprise the only food group that is a significant source of n-3 LCPUFA, which also offers a range of nutrients that are frequently underrepresented in habitual diets, including iodine, calcium, vitamin D, zinc, and iron [29]. Breast milk ALA levels tend to be better conserved than DHA, but DHA responds sensitively to dietary DHA and higher breast milk levels have been found in women consuming diets with high levels of fish intake [30]. LCPUFA levels in breast milk may be low when maternal fish intake is low and/or ALA intake is low and linoleic acid (LA) intake is high, as diets with very high levels of LA dramatically reduce conversion of ALA to DHA [31]. Food supplementation with DHA from midpregnancy and throughout lactation has been shown to improve the DHA status of the mother and the amount of DHA present in her milk [32], resulting in higher DHA plasma concentrations in infants [33] and improved child neurodevelopment [34].

PROTEIN

The protein content of milk varies according to the stage of lactation, with colostrum containing about 30 g/L and mature milk comprising 8 to 9 g/L, and whether the mother delivers at term: Preterm mothers' milk contains significantly more protein than that of mothers who deliver at term. The protein in breast milk is composed of lactalbumin, casein, lactoferrin, and IgA. It is estimated that women require an additional 11 g protein/day over the first 6 months to meet this need. In high-income countries, most women consume protein well in excess of this amount and would not need to alter their intake when breastfeeding. Vegetarian women should ensure a diet containing adequate combinations of plant proteins to provide all essential amino acids in sufficient quantities.

Human milk contains many proteins that have been shown to be bioactive, including lactoferrin, lysozyme, secretory IgA, haptocorrin, α-lactalbumin, and bile salt lipase. Their roles include enzymatic activities, enhancement of nutrient absorption, growth stimulation, modulation of the immune system, and defense against pathogens. Human milk proteins may be largely resistant to digestion in the gastrointestinal tract, be partially digested into bioactive peptides, or be completely digested and utilized as a source of amino acids, and it is still not known to what extent their bioactivities are exerted in breastfed infants [35].

CARBOHYDRATE

Lactose, the primary carbohydrate in human milk and the second-largest component of milk (70 g/L) after water, does not seem to be influenced by fluctuations in maternal diet. Human milk oligosaccharides (HMOs) represent a significant fraction of breast milk, reaching concentrations of up to 20 g/L. HMOs are a family of structurally diverse unconjugated glycans that serve as a prebiotic substrate enabling enrichment of *Bifidobacterium* species in the infant gut. Lactate and short-chain fatty acids (SCFAs), the end products of bacterial fermentation of HMOs, can then be absorbed and used as an energy source. There is increasing evidence to suggest that HMOs are important antiadhesive antimicrobials, which serve as soluble decoy receptors; prevent pathogen attachment to infant mucosal surfaces; and lower the risk for viral, bacterial, and protozoan parasite infections. HMOs may also modulate epithelial and immune cell responses, reduce excessive mucosal leukocyte infiltration and inflammation, lower the risk of necrotizing enterocolitis, and provide the infant with sialic acid as a potentially essential nutrient for brain development and cognition [36]. HMOs have been found in the circulation of breastfed infants, indicating good correlation between infant plasma HMO concentrations and levels in the corresponding mother's milk [37]. Differences in HMO composition in the mother's milk has been associated with infant growth and body composition [38], protection against postnatal HIV transmission [39], and reduced mortality during breastfeeding in HIV-exposed children [40]. Although some intervention studies have evaluated the effects of perinatal administration of prebiotics to pregnant women and to infants after birth, giving conflicting results depending on the specific strains tested, the conditions of use, and the population group, even less is known about the potential impact of prebiotic administration during lactation [41].

VITAMIN A

Vitamin A is needed for the growth and differentiation of cells and tissues and plays a key role in infant development. Vitamin A deficiency (VAD) is a major public health problem in low-income countries, and it is estimated that 140 million to 250 million children under the age of 5 years are affected globally. A lack of vitamin A is associated with a significant increased risk of illness and death from common childhood infections, such as measles and diarrhea, severe visual impairment and blindness. Breast milk is rich in vitamin A, derived mainly from maternal fat stores, and transported in the lipid fraction of milk as retinyl ester. Breastfed infants of women with adequate vitamin A status are generally protected from clinical VAD, as the amount of vitamin A transferred to the first 6 months of lactation is 60 times that which the fetus accumulates throughout gestation [42]. Breast milk vitamin A content is sensitive to maternal intake, and women with low intakes may be at risk of vitamin A deficiency, resulting from further depletion of their stores through breastfeeding. Although high doses of vitamin A should not be consumed during pregnancy to avoid teratogenesis, high-dose vitamin A supplementation of breast-feeding women with low serum retinol levels has been shown to be an effective way of ensuring adequate supplies for the infant through colostrum and breast milk, and preventing deficiency in the infant [43,44].

B VITAMINS

The B vitamins are a broad class of water-soluble vitamins, which play important roles in cell metabolism. The recommended dietary intakes for several B vitamins are increased during lactation (i.e., thiamin, riboflavin, niacin, vitamin B_6, vitamin B_{12}, and folate). Breast milk concentrations of thiamin, riboflavin, and vitamins B_6 and B_{12} are strongly dependent on maternal dietary intake and status, and low maternal intake/status can lead to the infant becoming deficient.

The two forms of thiamin present in breast milk are thiamin and thiamin monophosphate (approximate ratio 30:70). The risk of thiamin depletion is increased where diets are high in refined or polished unfortified grains, and low in animal source foods and legumes. Thiamin deficiency (beriberi) is uncommon in high-income countries, where deficiency is more often associated with chronic alcoholism, HIV/AIDS, or gastrointestinal conditions that impair vitamin absorption [45]. Infantile thiamin deficiency occurs mainly in infants who are breastfed by mothers with an inadequate intake of thiamin. There is evidence that infantile thiamin deficiency persists in parts of Southeast Asia, such as Laos, Cambodia, and Burma, where it may be a major cause of infant mortality [46]. Trials have shown that thiamin fortification of food consumed through pregnancy and early lactation improves breast milk thiamin concentrations, and has the potential to prevent infantile beriberi [47]. In 2003, several hundred Israeli infants were put at risk of thiamin deficiency after being fed a soy-based formula that was deficient in thiamin [48]. Approximately 20 infants were seriously affected, and 3 of them died. The long-term consequences of early thiamin deficiency included intellectual disability, motor abnormalities, severe epilepsy, and early kyphoscoliosis [49]. Some of the children with intellectual disability had been considered "asymptomatic," with no abnormal neurological symptoms during the thiamin deficiency period, suggesting that thiamin deficiency may exert its effects on the brain development of children without being detected [50].

Riboflavin is found mainly in animal source food and green vegetables, and so deficiency may be more common in populations whose diets are low in these food-stuffs. Marginal riboflavin status has also been reported to be prevalent in low-milk consumers in industrialized countries [51]. A low maternal dietary take of riboflavin has been associated with congenital limb deficiency, cleft lip, and palate deformities, and congenital heart defects in their offspring [52,53]. Riboflavin deficiency in moth-ers rapidly results in low breast milk concentrations of this vitamin, but supplemen-tation of both the mother and infant can improve the status of the infant [45].

The predominant form of vitamin B_6 in breast milk is pyridoxal (75%), and low breast milk concentrations have been associated with poor weight gain and linear growth, increased risk of seizure, and behavioral issues in early infancy, although evidence is limited [45]. The vitamin B_6 intake of mothers is a strong predictor of infant status [54]. Maternal vitamin B_6 supplementation results in a rapid increase in milk concentration of this vitamin. Associations between levels of vitamin B_6 in human breast milk and infant behavior have been suggested. A small study of Egyptian mothers reported that a low concentration of vitamin B_6 in breast milk (<415 nmol/L) was associated with less consolability, inappropriate buildup to a crying state, and poorer response to aversive stimuli in their infants [55]. A sub-sequent study, also carried out in Egyptian mother–infant pairs, reported a sig-nificant relationship between maternal dietary quality during lactation and infant drowsiness, and this was linked to certain B-vitamin intakes (particularly vitamin B_6, niacin, and riboflavin), and/or bioavailability of the nutrients [56]. A study in the United States of 25 lactating women whose vitamin B_6 intake was greater than the group median had infants whose habituation and autonomic stability scores on the Brazelton Neonatal Assessment Scale were positively correlated with their breast milk vitamin B_6 content [57]. All studies were small, however, and further work needs to be done to investigate whether these findings can be replicated in other contexts.

Neonates are thought to be at particular risk of vitamin B_{12} deficiency. Case studies have shown that breastfed infants born to vegetarian mothers or those with undiagnosed pernicious anemia may show clinical symptoms of vitamin B_{12} defi-ciency within months of birth. Such symptoms include severe growth failure (length, weight, and head circumference); cerebral atrophy; and muscular, behavioral, and developmental problems [58]. Studies that have investigated the association between maternal vitamin B_{12} supplementation and infant growth and cognitive function have, however, been inconsistent [59]. Whereas supplementation of vitamin B_{12}-depleted lactating women results in only modest increments in breast milk levels [45], vitamin B_{12} supplementation throughout pregnancy and early lactation has been found to significantly increase the status of mothers and infants [60], suggesting that supplementation must start early in pregnancy, or before, if it is to impact on child outcomes. However, despite the relatively high prevalence of vitamin B_{12} deficiency and its deleterious effects on pregnant women and their infants, there is still no con-sensus on the biochemical marker cutoff to correctly diagnose B_{12} deficiency, and the optimal dose of B_{12} to normalize B_{12} status of mother–infant pairs in a deficient population is not yet known [61].

The predominant form of folate in human milk is 5-methyl tetrahydrofolate. Folate deficiency and depletion is thought to be widespread in both industrialized and nonindustrialized countries. Unlike many of the other B vitamins, concentrations in human milk are maintained even when the mother is deficient in the vitamin, and there is no evidence that maternal supplementation with folate elicits beneficial effects on breast milk levels [62]. As a consequence of maintaining folate secretion in breast milk, women with low intakes will become more depleted as lactation progresses, and thus supplementation of the breastfeeding mother may be important for preserving maternal stores, which is particularly important for subsequent conception and pregnancy [45]. Steps to achieve folate sufficiency have included regulations for the mandatory fortification of wheat flour with folic acid. To date, 86 countries have legislation in place to mandate fortification of at least one industrially milled cereal grain, although in many cases these regulations have not been implemented [63]. Fortification of staple foodstuff with folic acid has reduced the number of births complicated by neural tube defects by up to 35% in the United States since folic acid fortification was mandated in 1998 [64].

VITAMIN C

Vitamin C is an essential water-soluble antioxidant vitamin and, in its role as a cofactor for mixed function oxidases, it participates in the synthesis of various macromolecules, including collagen and carnitine. Breast milk contains 30 to 80 mg/L vitamin C, and to compensate for the excretion of vitamin C into breast milk, dietary recommendations advocate an increased intake of vitamin C during lactation. High-dose vitamin C supplementation (>200 mg/day) to the mother has not been found to result in higher breast milk concentrations in well-nourished women, but rather resulted in increased urinary excretion of vitamin C [65]. However, in women with low vitamin C status, breast milk concentrations can be doubled or tripled by taking vitamin C supplements, thereby improving the nutrient status of the breastfed infant [66]. No cases of infantile scurvy, characterized by hemorrhage around the connective tissue and bone, impaired bone development, listlessness, loss of appetite, and weight loss, have been reported in breastfed infants, even when maternal vitamin C intake is low. A higher breast milk vitamin C concentration, however, has been associated with a reduced risk of atopy in the infant [67]. After weaning, the vitamin C intake of infants declines significantly, but the potentially negative influences of low ascorbic acid intake in early life related to its antioxidant properties have yet to be evaluated.

VITAMIN D

Through its action in regulating calcium and phosphate absorption, vitamin D plays a vital role in the growth and development of bones. Infants, in particular, are at risk of vitamin D deficiency because of their relatively large vitamin D needs related to the high rate of skeletal growth [10]. Breast milk contains relatively small amounts of vitamin D that may not be sufficient to prevent vitamin D deficiency in exclusively breastfed infants if sunlight exposure is limited [68]. Since several groups, such as the American Academy of Pediatrics, recommend that infants younger than 6 months should be kept out of direct sunlight and protected with clothing and hats, opportunities for vitamin D synthesis from sunlight may be further restricted. Pregnant women and newborn infants are recognized as populations at an increased risk of poor vitamin D status. Globally, 54% of pregnant women and 75% of newborns suffer from vitamin D deficiency (25(OH)D <50 nmol/L), and 18% of pregnant women and 29% of newborns suffer from severe vitamin D deficiency (25(OH)D concentration <25 nmol/L) [69].

During lactation, the efficiency of maternal dietary calcium absorption is increased to ensure adequate calcium content of breast milk. The metabolism of 25(OH)D to 1,25(OH)2D is enhanced to meet this demand. However, because circulating concentrations of 1,25(OH)2D are 500 to 1000 times less than 25(OH)D, the increased metabolism probably does not significantly alter the daily requirement for vitamin D [70]. Dietary guidelines recommend that pregnant and breastfeeding women should take 5 to 15 µg (200–600 IU) vitamin D per day (Table 4.1). Public health guidance in the United Kingdom highlights those at greatest risk of vitamin D deficiency, including women who are obese; have limited skin exposure to sunlight; or who are of South Asian, African, Caribbean, or Middle Eastern descent [71]. As there are few dietary sources of vitamin D available to meet this recommended target, vitamin D supplements should be taken. Supplementation with 10 µg/d (400 IU/day), however, has been shown to have only a modest effect of maternal 25(OH)D levels, leading for calls to increase levels of supplementation in lactating women to at least 37.5 to 50 µg/d (1500–2000 IU/d) or 100 to 150 µg/d (4000–6000 IU/d) if mothers choose not to give their infant a vitamin D supplement [70]. Studies have indicated that supplementation of healthy breastfeeding mothers with doses of 100 to 160 µg/d vitamin D (4000–6400 IU/d) provides the breastfeeding infant with adequate levels of vitamin D, despite both mother and infant being limited in sunlight exposure [72,73]. Maternal high-dose supplementation remains controversial, however, and research has shown that supplements administered directly to the infant during lactation can easily achieve vitamin D sufficiency. Several groups, such as the European Society for Pediatric Gastroenterology, Hepatology and Nutrition (ESPGHAN) and Health Canada, advocate that breastfed infants should receive an oral supplementation of 10 µg/d (400 IU/day) of vitamin D [74].

CALCIUM

Breastfeeding women typically lose calcium at a rate of 280 to 400 mg/day through their breast milk [75]. To meet this increased demand, the mother must mobilize calcium from her own skeletal reserves. Physiologic adaptations, such as upregulation of intestinal calcium absorption and bone resorption, provide much of the calcium in breast milk (Figure 4.1). Once breastfeeding has ceased, bone density is fully restored over the subsequent 6 to 12 months [76]. Although the total duration of breastfeeding has not been found to have an adverse impact on bone mineral density in later life, short interpregnancy intervals (1 to 12 months) are associated with an increased risk of osteoporosis [77]. Several investigations of the effect of maternal dietary calcium intake on breast milk levels indicate that they may be independent.

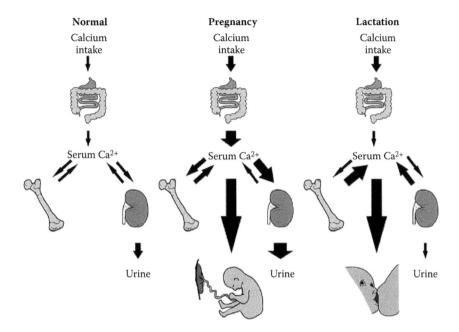

Normal	Pregnancy	Lactation
Calcium intake	Calcium intake	Calcium intake

FIGURE 4.1 The contrast of calcium homeostasis in human pregnancy and lactation, compared with nonpregnant, nonlactating levels. The thickness of arrows indicates a relative increase or decrease with respect to the normal nonpregnant state. The serum (total) calcium is decreased during pregnancy, while the ionized calcium remains normal during both pregnancy and lactation. (Reproduced from Kovacs CS, *Endocrinol Metab Clin North Am* 2011, 40(4):795–826. With permission.)

A randomized controlled trial of pregnant and lactating women in The Gambia showed that, despite having low dietary intakes of calcium (300–400 mg/d) and low breast milk concentrations, calcium supplements had no significant benefit in terms of breast milk concentration [78].

The calcium requirements of adolescents are elevated as a result of intensive bone and muscular development, and most adolescents worldwide fail to achieve recommended calcium intakes [79]. There is some concern, therefore, that adolescent mothers with low calcium intakes may not achieve normal peak bone mass as a consequence of lactation-induced bone loss. Although there is evidence that adolescents experience bone mineral density loss during lactation, this effect has been shown to be transient, with subsequent repletion of bone mineral density once breastfeeding is ceased [80]. Evidence remains limited, however, and adolescent mothers are advised to maintain or increase their dietary intake to recommended levels (1300 mg/d) while breastfeeding to ensure that their growth needs are met and that peak bone mass is achieved [81].

IRON

Iron deficiency is the most common micronutrient deficiency globally, and young children are at particular risk due to high iron requirements to support their rapid growth. Controversy exists regarding the adequacy of exclusive breastfeeding for 6 months in maintaining optimum iron status in term babies. The ESPGHAN Committee on Nutrition recently reviewed the evidence for iron requirements in early childhood and concluded that there is no convincing evidence that iron supplements should be provided to normal-birth weight, exclusively breastfed infants during the first 6 months of life in populations with a low prevalence of iron deficiency anemia (IDA) among 6 month olds [82]. After this time, many infants exhaust their iron stores and require a secondary dietary iron supply in the form of iron-rich complementary foods. In low-resource settings, however, where newborn iron stores may be suboptimal and there is reduced access to iron-rich foods, it has been suggested that breastfeeding mothers may need to be supplemented with iron to ensure the hematological status of the infant is not compromised, particularly in low-birth-weight infants [83], but evidence is limited. Several studies have found no correlation between the levels of iron in breast milk and maternal iron status, and some have demonstrated that exclusively breastfed infants of both anemic and nonanemic mothers in resource-poor areas do not develop iron deficiency or iron deficiency anemia by 6 months of age [84]. The WHO recommends delayed umbilical cord clamping (not earlier than one minute after birth) for the prevention of iron deficiency anemia in infants [85].

IODINE

The iodine concentration of human breast milk varies markedly as a function of the iodine intake of the population, and depends on the iodine levels in local soil, seafood consumption, and the availability and use of iodized salt. An average consumption of 750 mL breast milk per day would provide an intake of iodine of about 60 µg/day in Europe and 120 µg/day in the United States, which, in some areas, falls short of the WHO recommendation of 90 µg iodine/day from birth [10]. For women with marginal iodine status, the demands of lactation can precipitate clinical and biochemical symptoms including increased thyroid volume, altered thyroid hormone levels, and impaired mental function [86,87].

Thyroidal iodine turnover rate is more rapid in infants [88] so adequate breast milk iodine levels are particularly important for neurodevelopment in breastfed infants. A 33% increase in iodine intake is needed to accommodate the changes in maternal thyroid metabolism to support lactation and to supply sufficient iodine for milk to meet the needs for growth and development of the infant [10]. The levels recommended to ensure that pregnant and lactating women do not suffer from iodine deficiency postpartum are listed in Table 4.1.

Although substantial progress has been made over the recent decades through programs of universal salt iodization, iodine deficiency remains a significant health problem worldwide, and affects both industrialized and developing nations [89]. Evidence suggests that pregnant and lactating women may be at risk of iodine deficiency where universal salt iodization is not fully implemented. The WHO and UNICEF (United Nations Children's Fund) recommend iodine supplementation for pregnant and lactating women in countries where less than 20% of households have access to iodized salt, until the salt iodization program is scaled up. Countries in which household access to iodized salt ranges from 20% and 90% should make efforts to accelerate salt iodization programs or assess the feasibility of increasing iodine intake in the form of a supplement or other iodine-fortified foods by the most susceptible groups [90].

ZINC

Zinc is vital for infant development, and an adequate supply of zinc is essential for mammary gland function and for milk synthesis and secretion. The level of zinc in colostrum is 17 times higher than in maternal blood but declines rapidly in breast milk during the first 3 months postpartum [91]. A number of studies of lactating women with marginal zinc status have revealed that homeostatic mechanisms can compensate for low maternal dietary zinc intakes. The proportion of dietary zinc absorbed in such women has been shown to increase by over 70%, compared to nonlactating women or preconception values [92]. Nevertheless, the adequacy of breast milk zinc for optimal infant growth and development is likely to depend on the amount of infant zinc reserves at birth; more research is needed to elucidate this [93,94].

COPPER AND SELENIUM

Copper and selenium are essential trace minerals, and the requirements for both are increased during lactation. Copper has a number of biochemical roles, most notably as a cofactor for enzymes that regulate various physiologic pathways such as energy production, iron metabolism, connective tissue maturation, and neurotransmission. Selenium is a component of selenoproteins that act as antioxidant enzymes and regulate thyroid hormone metabolism and immune function. The copper content of human milk is highest during early lactation, and declines as lactation progresses, averaging approximately 250 µg/L in the first 6 months. Maternal intake of copper has little effect on breast milk concentrations or infant status [95]. The selenium content of breast milk also decreases over the course of lactation but, unlike copper, is influenced by maternal status.

CONCLUSION

Although dietary requirements for many nutrients are increased during lactation, the requisite levels of intake are likely to be met within the normal diets of well-nourished women. Other factors, such as the mother's nutritional stores, reductions in physical activity, and metabolic adaptation, all combine to provide a buffer to protect the breastfed infant against suboptimal nutrient intake. There is evidence, however, that the secretion of certain nutrients into breast milk is rapidly or substantially reduced by maternal depletion. Such nutrients include vitamin A, thiamin, riboflavin, vitamin B_6, vitamin B_{12}, vitamin D, selenium, and iodine. The concentrations of nutrients less sensitive to maternal depletion (folate, calcium, iron, copper, and zinc) may be adequate and maintained through lactation, but low maternal intakes are likely to deplete mothers' reserves as lactation progresses.

The nutritional status of pregnant women and women of childbearing age, particularly in resource-poor settings, is therefore of critical importance. Multiple-micronutrient supplementation of mothers during pregnancy and lactation may be an important intervention to improve nutrient concentrations in breast milk [96]. The magnitude of the effect of supplementation of pregnant and breastfeeding mothers on their breast milk composition, however, has been highly variable, perhaps due in part to methodological issues such as differences in the dose of the supplement and the timing of sample collection, as many nutrients exhibit considerable kinetic variation in their secretion into milk following consumption. In addition, education is important in order to change cultural taboos and practices that impair maternal and infant status, most notably maternal perinatal and/or postpartum food avoidance. Optimal infant and young child feeding practices, which include the early introduction of breastfeeding, exclusive and continued breastfeeding, and good hygiene and sanitation (including the use of clean water and the introduction of safe and appropriate complementary feeding from 6 months of age), should all be promoted to ensure that the health-promoting properties of breast milk are provided to the infant to sustain optimal health.

REFERENCES

1. Picciano MF. Pregnancy and lactation: Physiological adjustments, nutritional requirements and the role of dietary supplements. *J Nutr* 2003; 133(6):S1997–2002.
2. Dewey KG, Cohen RJ. Does birth spacing affect maternal or child nutritional status? A systematic literature review. *Matern Child Nutr* 2007; 3(3):151–73.
3. Conde-Agudelo A, Rosas-Bermudez A, Castaño F et al. Effects of birth spacing on maternal, perinatal, infant, and child health: A systematic review of causal mechanisms. *Stud Fam Plann* 2012; 43(2):93–114.
4. Committee on Medical Aspects of Food Policy (COMA). *Dietary reference values for food energy and nutrients for the United Kingdom.* London: TSO; 1991.
5. Institute of Medicine, Food and Nutrition Board. *Dietary reference intakes for calcium, phosphorus, magnesium, vitamin D and fluoride.* Washington, DC: National Academy Press; 1997.
6. Institute of Medicine, Food and Nutrition Board. *Dietary reference intakes for thiamin, riboflavin, niacin, vitamin B_6, folate, vitamin B_{12}, pantothenic acid, biotin, and chlorine.* Washington, DC: National Academy Press; 1998.
7. Institute of Medicine, Food and Nutrition Board. *Dietary reference intakes for vitamin C, vitamin E, selenium, and carotenoids.* Washington, DC: National Academy Press; 2000.
8. Institute of Medicine, Food and Nutrition Board. *Dietary reference intakes for vitamin A, vitamin K, arsenic, boron, chromium, copper, iodine, iron, manganese, molybdenum, nickel, silicon, vanadium, and zinc.* Washington, DC: National Academy Press; 2001.
9. Institute of Medicine, Food and Nutrition Board. *Dietary reference intakes for water, potassium, sodium, chloride, and sulphate.* Washington, DC: National Academy Press; 2004.
10. World Health Organization. *Vitamin and mineral requirements in human nutrition.* Geneva: World Health Organization; 2005.
11. Scientific Advisory Committee on Nutrition (SACN). Vitamin D and health report [Internet]. Available at: https://www.gov.uk/government/publications/sacn-vitamin-d-and-health-report (accessed February 14, 2017).

12. Dewey KG. Impact of breastfeeding on maternal nutritional status. *Adv Exp Med Biol* 2004; 554:91–100.
13. Prentice P, Ong KK, Schoemaker MH et al. Breast milk nutrient content and infancy growth. *Acta Paediatr* 2016; 105(6):641–7.
14. Food and Agriculture Organization of the United Nations (FAO). Human energy requirements: Report of a Joint FAO/WHO/UNU Expert Consultation, Rome, 17–24 October 2001. Rome: FAO; 2001.
15. Scientific Advisory Committee on Nutrition (SACN). *Dietary reference values for energy.* London: TSO; 2012.
16. He X, Zhu M, Hu C et al. Breast-feeding and postpartum weight retention: A systematic review and meta-analysis. *Public Health Nutr* 2015; 18(18):3308–16.
17. Neville CE, McKinley MC, Holmes VA et al. The relationship between breastfeeding and postpartum weight change: A systematic review and critical evaluation. *Int J Obes (Lond)* 2014; 38(4):577–90.
18. Baker JL, Gamborg M, Heitmann BL et al. Breastfeeding reduces postpartum weight retention. *Am J Clin Nutr* 2008; 88(6):1543–51.
19. da Silva Mda C, Oliveira Assis AM, Pinheiro SM et al. Breastfeeding and maternal weight changes during 24 months post-partum: A cohort study. *Matern Child Nutr* 2015; 11(4):780–91.
20. Sharma AJ, Dee DL, Harden SM. Adherence to breastfeeding guidelines and maternal weight 6 years after delivery. *Pediatrics* 2014; 134(Suppl 1):S42–49.
21. Lauritzen L, Brambilla P, Mazzocchi A et al. DHA effects in brain development and function. *Nutrients* 2016;8(1):E6.
22. Lassek WD, Gaulin SJ. Maternal milk DHA content predicts cognitive performance in a sample of 28 nations. *Matern Child Nutr* 2015; 11(4):773–9.
23. Prentice AM, van der Merwe L. Impact of fatty acid status on immune function of children in low-income countries. *Matern Child Nutr* 2011; 7(Suppl 2):89–98.
24. Uauy R, Mena P, Rojas C. Essential fatty acids in early life: Structural and functional role. *Proc Nutr Soc* 2000; 59(1):3–15.
25. Yuhas R, Pramuk K, Lien EL. Human milk fatty acid composition from nine countries varies most in DHA. *Lipids* 2006; 41(9):851–8.
26. Brenna JT, Varamini B, Jensen RG et al. Docosahexaenoic and arachidonic acid concentrations in human breast milk worldwide. *Am J Clin Nutr* 2007; 85(6):1457–64.
27. Grote V, Verduci E, Scaglioni S et al. Breast milk composition and infant nutrient intakes during the first 12 months of life. *Eur J Clin Nutr* 2016; 70(2):250–6
28. Harris WS, Baack ML. Beyond building better brains: Bridging the docosahexaenoic acid (DHA) gap of prematurity. *J Perinatol* 2015; 35(1):1–7.
29. Hunt W, McManus A. Seafood and omega 3s for maternal and child mental health. In *Maternal and infant nutrition and nurture: Controversies and challenges*, 2nd ed., Hall Moran V, editor. London: Quay Books; 2013.
30. Quinn EA, Kuzawa CW. A dose-response relationship between fish consumption and human milk DHA content among Filipino women in Cebu City, Philippines. *Acta Paediatr* 2012; 101(10):e439–45.
31. Briend A, Dewey KG, Reinhart GA. Fatty acid status in early life in low-income countries: Overview of the situation, policy and research priorities. *Matern Child Nutr* 2011; 7(Suppl 2):141–8.
32. van Goor SA, Dijck-Brouwer D, Hadders-Algra M et al. Human milk arachidonic acid and docosahexaenoic acid contents increase following supplementation during pregnancy and lactation. *Prostaglandins Leukot Essent Fatty Acids* 2009; 80(1):65–69.
33. Jensen CL, Maude M, Anderson RE et al. Effect of docosahexaenoic acid supplementation of lactating women on the fatty acid composition of breast milk lipids and maternal and infant plasma phospholipids. *Am J Clin Nutr* 2000; 71(1 Suppl):S292–9.

34. Helland IB, Smith L, Saarem K et al. Maternal supplementation with very-long-chain n-3 fatty acids during pregnancy and lactation augments children's IQ at 4 years of age. *Pediatrics* 2003; 111(1):e39–44.

35. Lönnerdal B. Bioactive proteins in breast milk. *J Paediatr Child Health* 2013; 49(Suppl 1):1–7.

36. Bode L. Human milk oligosaccharides: Every baby needs a sugar mama. *Glycobiology* 2012; 22(9):1147–62.

37. Goehring KC, Kennedy AD, Prieto PA et al. Direct evidence for the presence of human milk oligosaccharides in the circulation of breastfed infants. *PLoS One* 2014; 9(7):1–11.

38. Alderete TL, Autran C, Brekke BE et al. Associations between human milk oligosaccharides and infant body composition in the first 6 mo of life. *Am J Clin Nutr* 2015; 102(6):1381–8.

39. Bode L, Kuhn L, Kim H et al. Human milk oligosaccharide concentration and risk of postnatal transmission of HIV through breastfeeding. *Am J Clin Nutr* 2012; 96(4):831–9.

40. Kuhn L, Kim HY, Hsiao L et al. Oligosaccharide Composition of breast milk influences survival of uninfected children born to HIV-infected mothers in Lusaka, Zambia. *J Nutr* 2015; 145(1):66–72.

41. Thum C, Cookson AL, Otter DE et al. Can nutritional modulation of maternal intestinal microbiota influence the development of the infant gastrointestinal tract? *J Nutr* 2012; 142(11):1921–8.

42. Gluckman P, Hanson M, Seng CY et al. *Nutrition and lifestyle for pregnancy and breastfeeding.* London: Oxford University Press; 2014.

43. Sommer A, Davidson FR. Assessment and control of vitamin A deficiency: The Annecy Accords. *J Nutr* 2002; 132(9 Suppl):S2845–50.

44. Grilo EC, Lima MS, Cunha LR. Effect of maternal vitamin A supplementation on retinol concentration in colostrum. *J Pediatr* (Rio J) 2015; 91(1):81–86.

45. Allen L. Maternal nutrient metabolism and requirements in pregnancy and lactation. In *Present knowledge in nutrition*, 10th ed., Erdman JW, MacDonald IA, Zeisel SH, editors. Oxford: Wiley-Blackwell; 2012.

46. Barennes H, Sengkhamyong K, René JP et al. Beriberi (thiamine deficiency) and high infant mortality in northern Laos. *PLoS Negl Trop Dis* 2015; 9(3):e0003581.

47. Whitfield KC, Karakochuk CD, Kroeun H et al. Perinatal consumption of thiamine-fortified fish sauce in rural Cambodia: A randomized clinical trial. *JAMA Pediatr* 2016: 170(10):e162065.

48. Fattal-Valevski A, Kesler A, Sela DA et al. Outbreak of life-threatening thiamine deficiency in infants in Israel caused by a defective soy-based formula. *Pediatrics* 2005; 115(2):e233–8.

49. Mimouni-Bloch A, Goldberg-Stern H, Strausberg R et al. Thiamine deficiency in infancy: Long-term follow-up. *Pediatr Neurol* 2014; 51(3):311–6.

50. Fattal-Valevski A, Azouri-Fattal I, Greenstein YJ et al. Delayed language development due to infantile thiamine deficiency. *Dev Med Child Neurol* 2009; 51(8):629–34.

51. Powers HJ, Hill MH, Mushtaq S et al. Correcting a marginal riboflavin deficiency improves hematologic status in young women in the United Kingdom (RIBOFEM). *Am J Clin Nutr* 2011; 93(6):1274–84.

52. Smedts HP, Rakhshandehroo M, Verkleij-Hagoort AC et al. Maternal intake of fat, riboflavin and nicotinamide and the risk of having offspring with congenital heart defects. *Eur J Nutr* 2008; 47(7):357–65.

53. Robitaille J, Carmichael SL, Shaw GM et al. Maternal nutrient intake and risks for transverse and longitudinal limb deficiencies: Data from the National Birth Defects Prevention Study, 1997–2003. *Birth Defects Res A Clin Mol Teratol* 2009; 85(9):773–9.

54. Kang-Yoon SA, Kirksey A, Giacoia G et al. Vitamin B_6 status of breast-fed neonates: Influence of pyridoxine supplementation on mothers and neonates. *Am J Clin Nutr* 1992; 56(3):548–58.
55. McCullough AL, Kirksey A, Wachs TD et al. Vitamin B_6 status of Egyptian mothers: Relation to infant behavior and maternal-infant interactions. *Am J Clin Nutr* 1990; 51(6):1067–74.
56. Rahmanifar A, Kirksey A, Wachs TD et al. Diet during lactation associated with infant behavior and caregiver-infant interaction in a semirural Egyptian village. *J Nutr* 1993; 123(2):164–75.
57. Ooylan LM, Hart S, Porter KB et al. Vitamin B_6 content of breast milk and neonatal behavioral functioning. *J Am Diet Assoc* 2002; 102(10):1433–8.
58. Dror DK, Allen LH. Effect of vitamin B_{12} deficiency on neurodevelopment in infants: Current knowledge and possible mechanisms. *Nutr Rev* 2008; 66(5):250–5.
59. Srinivasan K, Thomas T, Kapanee AR et al. Effects of maternal vitamin B_{12} supplementation on early infant neurocognitive outcomes: A randomized controlled clinical trial. *Matern Child Nutr* 2016; (epub ahead of print). doi: 10.1111/mcn.12325.
60. Duggan C, Srinivasan K, Thomas T et al. Vitamin B_{12} supplementation during pregnancy and early lactation increases maternal, breast milk, and infant measures of vitamin B_{12} status. *J Nutr* 2014; 144(5):758–64.
61. Siddiqua TJ, Allen LH, Raqib R et al. Vitamin B_{12} deficiency in pregnancy and lactation: Is there a need for pre and post-natal supplementation? *J Nutr Disord Ther* 2014; 4:142.
62. Mackey AD, Picciano MF. Maternal folate status during extended lactation and the effect of supplemental folic acid. *Am J Clin Nutr* 1999; 69(2):285–92.
63. Food fortification initiative: Global progress [Internet]. 2016 May. Available at: http://www.ffinetwork.org/global_progress/ (accessed February 14, 2017).
64. Williams J, Mai CT, Mulinare J et al. Updated estimates of neural tube defects prevented by mandatory folic acid fortification: United States, 1995–2011. *MMWR Morb Mortal Wkly Rep* 2015; 64(1):1–5.
65. Byerley LO, Kirksey A. Effects of different levels of vitamin C intake on the vitamin C concentration in human milk and the vitamin C intakes of breast-fed infants. *Am J Clin Nutr* 1985; 41(4):665–71.
66. Daneel-Otterbech S, Davidsson L, Hurrell R. Ascorbic acid supplementation and regular consumption of fresh orange juice increase the ascorbic acid content of human milk: Studies in European and African lactating women. *Am J Clin Nutr* 2005; 81(5):1088–93.
67. Hoppu U, Rinne M, Salo-Väänänen P et al. Vitamin C in breast milk may reduce the risk of atopy in the infant. *Eur J Clin Nutr* 2005; 59(1):123–8.
68. Hoogenboezem T, Degenhart HJ, de Muinck Keizer-Schrama SM et al. Vitamin D metabolism in breast-fed infants and their mothers. *Pediatr Res* 1989; 25(6):623–8.
69. Saraf R, Morton SM, Camargo CA Jr et al. Global summary of maternal and newborn vitamin D status: A systematic review. *Matern Child Nutr* 2015; 12(4):647–68.
70. Holick MF, Binkley NC, Bischoff-Ferrari HA et al. Evaluation, treatment, and prevention of vitamin D deficiency: An Endocrine Society clinical practice guideline. *J Clin Endocrinol Metab* 2011; 96(7):1911–30.
71. National Institute for Health and Care Excellence (NICE). Vitamin D: Increasing supplement use in at-risk groups. Public health guidance #56 [Internet]. Available at: https://www.nice.org.uk/guidance/ph56?unlid=7843398020162310553 (accessed February 14, 2017).
72. Hollis BW, Wagner CL. Vitamin D requirements during lactation: High-dose maternal supplementation as therapy to prevent hypovitaminosis D for both the mother and the nursing infant. *Am J Clin Nutr* 2004; 80(6):S1752–8.

73. Wagner CL, Hulsey TC, Fanning D et al. High-dose vitamin D3 supplementation in a cohort of breastfeeding mothers and their infants: A 6-month follow-up pilot study. *Breastfeed Med* 2006; 1(2):59–70.
74. Braegger C, Campoy C, Colomb V et al. Vitamin D in the healthy European paediatric population. *J Pediatr Gastroenterol Nutr* 2013; 56(6):692–701.
75. Kovacs CS. Calcium and bone metabolism disorders during pregnancy and lactation. *Endocrinol Metab Clin North Am* 2011; 40(4):795–826.
76. Polatti F, Capuzzo E, Viazzo F et al. Bone mineral changes during and after lactation. *Obstet Gynecol* 1999; 94(1):52–56.
77. Sahin Ersoy G, Giray B, Subas S et al. Interpregnancy interval as a risk factor for post-menopausal osteoporosis. *Maturitas* 2015; 82(2):236–40.
78. Jarjou LM, Prentice AM, Sawo Y et al. Randomized, placebo-controlled, calcium supplementation study in pregnant Gambian women: Effects on breast-milk calcium concentrations and infant birth weight, growth, and bone mineral accretion in the first year of life. *Am J Clin Nutr* 2006; 83(3):657–66.
79. Mesias M, Seiquer I, Navarro MP. Calcium nutrition in adolescence. *Crit Rev Food Sci Nutr* 2011; 51(3):195–209.
80. Bezerra FF, Mendonça LM, Lobato EC et al. Bone mass is recovered from lactation to postweaning in adolescent mothers with low calcium intakes. *Am J Clin Nutr* 2004; 80(5):1322–6.
81. Institute of Medicine. *Dietary reference intakes for calcium and vitamin D*. Washington, DC: National Academies Press; 2011.
82. Domellof M, Braegger C, Campoy C et al. Iron requirements of infants and toddlers. *J Pediatr Gastroenterol Nutr* 2014; 58(1):119–29.
83. Kramer MS, Kakuma R. Optimal duration of exclusive breastfeeding. *Cochrane Database Syst Rev* 2012; 8:CD003517.
84. Raj S, Faridi M, Rusia U et al. A prospective study of iron status in exclusively breast-fed term infants up to 6 months of age. *Int Breastfeed J* 2008; 3:3.
85. World Health Organization. *Guideline: Delayed umbilical cord clamping for improved maternal and infant health and nutrition outcomes*. Geneva: World Health Organization; 2014.
86. Dorea JG. Iodine nutrition and breast feeding. *J Trace Elem Med Biol* 2002; 16(4):207–20.
87. Zimmermann M, Delange F. Iodine supplementation of pregnant women in Europe: A review and recommendations. *Eur J Clin Nutr* 2004; 58(7):979–84.
88. Zimmermann MB. Iodine deficiency. *Endocr Rev* 2009; 30(4):376–408.
89. Pearce EN, Andersson M, Zimmermann MB. Global iodine nutrition. Where do we stand in 2013? *Thyroid* 2013; 23(5):523–8.
90. World Health Organization (WHO)/United Nations Children's Fund (UNICEF). *Reaching optimal iodine nutrition in pregnant and lactating women and young children*. Joint statement by WHO and UNICEF. Geneva: WHO; 2007.
91. Almeida AA, Lopes CM, Silva AM et al. Trace elements in human milk: Correlation with blood levels, inter-element correlations and changes in concentration during the first month of lactation. *J Trace Elem Med Biol* 2008; 22(3):196–205.
92. Sian L, Krebs NF, Westcott JE et al. Zinc homeostasis during lactation in a population with a low zinc intake. *Am J Clin Nutr* 2002; 75(1):99–103.
93. Nissensohn M, Sánchez-Villegas A, Fuentes Lugo D et al. Effect of zinc intake on growth in infants: A meta-analysis. *Crit Rev Food Sci Nutr* 2016; 56(3):350–63.
94. Islam MM, Brown KH. Zinc transferred through breast milk does not differ between appropriate- and small-for-gestational-age, predominantly breast-fed Bangladeshi infants. *J Nutr* 2014; 144(5):771–6.

95. Domellöf M, Hernell O, Dewey KG et al. Factors influencing concentrations of iron, zinc, and copper in human milk. In *Protecting infants through human milk: Advancing the scientific evidence*, Pickering LK, Morrow AL, Ruiz-Palacios GM et al. editors. New York: Kluwer Academic; 2004.
96. Shahab-Ferdows S, Hampel D, Islam M et al. Effects and kinetics of maternal supplementation on vitamin concentrations in human milk. *FASEB J* 2015; 29(1):133.5.

5 Nutrient Requirements of Infants and Young Children

Minghua Tang, Kathryn G. Dewey, and Nancy F. Krebs

CONTENTS

NUTRITION CONSIDERATIONS FOR YOUNG INFANTS (0–6 MONTHS)

Infancy is the period of the life cycle with the highest growth rate, and thus meeting energy and nutrient needs is critical to ensure optimal development. Infants have comparatively high needs for most nutrients relative to body size. Moreover, compared with older children, very young infants have immature digestive, absorptive, and renal capacities. Therefore, human milk, with low renal solute load and compositional factors that facilitate nutrient absorption, is the ideal sole source of nutrition for healthy, term infants from birth through approximately 6 months of age. Besides promoting optimal growth, human milk contains many bioactive components, such as anti-infectious and anti-inflammatory agents, growth factors, and prebiotics [1].

Mature human milk remains relatively stable in composition during lactation, although longitudinal declines occur in protein content and some micronutrients. The exact mechanisms for such physiologic changes are not yet characterized but, teleologically, such changes are concurrent with the gradually reduced rate of infant growth, and serve to spare the demands on maternal nutrition. Other aspects of milk composition are more variable, for example, lipid content, which varies over the course of both a feed and a day.

The overall macronutrient content of human milk is well characterized, but recent research emphasizes numerous distinctive features within this traditional framework

75

that contribute to its uniquely beneficial aspects for the young infant. The main carbohydrate in human milk is lactose; mature, term human milk has an average lactose concentration of ~70 g/L, which gradually increases over time as protein concentration declines. Besides lactose, multiple oligosaccharides are found in human milk, and exert potent prebiotic effects and protective immunomodulatory functions, such as mediating leukocyte adhesion and activation [2]. Recent investigations have highlighted the importance of stage of lactation, maternal phenotype, and environmental factors with regard to diversity in the oligosaccharide profile of human milk [3]. These differences, in turn, have been associated with infant outcomes. For example, an increase in lacto-N-fucopentaose is associated with a decrease of fat mass in 6-month-old breastfed infants [4]. Total protein in mature, term human milk is approximately 10 g/L and is whey protein-dominant. Of this, α-lactalbumin, which is high in tryptophan and cysteine, accounts for 41% of whey and 28% of total protein in human milk [5]. In addition to containing high quantities of essential amino acids, α-lactalbumin is found to have protective effects against common pathogens such as *Escherichia coli* [6]. Other whey proteins, such as lactoferrin and secretory IgA, are also present in human milk, and exert unique functions, including facilitating iron absorption and boosting the innate immune system, respectively [7,8]. Lipids in human milk are high in palmitic and oleic acids [1], and also contain essential fatty acids such as long-chain polyunsaturated fatty acids (LCPUFA), which are important components of cell membranes and of cell-signaling pathways, and make up a large proportion of brain tissue. The highest variability of human milk composition is found with regard to the lipid composition, which is dependent on a number of maternal factors, such as diet, age, body mass index (BMI), and the stage of lactation. In particular, the fatty acid composition of human milk, including the proportions of omega-6 and omega-3 fatty acids, reflects maternal diet and fatty acid status. The ratio of omega-6 to omega-3 fatty acids in human milk, which has been observed to differ between normal weight and obese women, may influence the rate and quality of infant adipogenesis [9]. Overall, it is virtually impossible to define a "standard" lipid composition of human milk. The quantity of lipid also varies; a range of 10 to 50 g/L in human milk has been reported [10]. Correspondingly, this variability in the lipid content of human milk results in variable energy density, which has been proposed to contribute to infants' satiety and self-regulation of energy intake. Generally, lipids are assumed to provide approximately 50% of the energy content of human milk.

The micronutrient content of human milk may also vary by maternal diet and nutritional status; this is true especially for water-soluble vitamins, with the exception of folate. Recent reports suggest that concentrations may vary substantially [11]. Concentrations of fat-soluble vitamins also can be affected by maternal intake. Vitamin D is generally low in human milk relative to infants' needs, especially for mothers with limited sun exposure [12], for whom prenatal transfer to the fetus would likely also have been modest. However, a recent trial of high dose vitamin D supplementation to mothers demonstrated its capacity to sufficiently increase the level to meet estimated infant needs [13]. In contrast to vitamins, the concentrations of most macro- and trace minerals in human milk are independent of maternal dietary intake; iodine and selenium are exceptions. The concentration of iron is consistently low (0.3–0.5 mg/L), which fosters healthy bacterial colonization of the newborn's intestine, and thus is one of the nonspecific protective factors of human milk. In contrast to iron, the zinc content of human milk (~4 mg/L) is initially quite high relative to the infant's requirements, but sharply and steadily declines over the early months postpartum; by 6 months, typical concentration is <1.0 mg/L. Calcium concentrations follow a similar but less dramatic pattern, with a postpartum decline of ~15% from birth to 6 months. Term infants are born with an endowment of iron that is typically adequate for the first approximately 6 months of life [14], which also coincides with the time when the amount of zinc provided by milk alone also becomes marginally adequate [15,16]. These considerations thus impact the timing and quality of complementary feeding.

When human milk is not available, an alternative human milk substitute, ideally one which is modified specifically for infants, should be recommended [17]. This poses an important challenge in low-resource settings, where the expense of commercial infant formulas is typically prohibitive and the availability of clean water is often problematic. Unmodified bovine milk is unsuitable for young human infants, primarily because of the high protein content, which is 3 times the concentration of human milk, as well as the overall high mineral content. These factors result in a high renal solute load and can lead to dehydration due to the young infant's limited ability to concentrate the urine. Commercial infant formulas, with a lower and modified protein content, are considered a safe alternative when human milk is unavailable and if clean water and sanitation can be ensured. Most standard commercial infant formulas are produced with bovine milk concentrates and bovine whey protein extracts, and tend to have a higher protein content than human milk. Despite higher total protein content, infant formulas still have less than half of the α-lactalbumin found in human milk. Efforts are being made to promote a lower protein infant formula while increasing α-lactalbumin and certain essential amino acids, such as tryptophan, which is rich in human milk. Attention to the impact of intact bovine milk protein on atopic disease has increased in recent years, with results from randomized trials supporting a lower rate of atopic disease from formulas with partially or extensively hydrolyzed protein [18]. Although limited components discovered in human milk, including prebiotics (e.g., oligosaccharides), are also added to some infant formulas, it is impossible to completely reproduce the unique features of human milk (e.g., secretory IgA) or to replicate the enormous public health benefits that derive from breastfeeding. Indeed, the challenge to health care providers and policy makers is to actively protect, promote, and support breastfeeding in all settings. This includes initiation of nursing within the first 30 to 60 minutes after birth (the "golden hour") and exclusive breastfeeding through the critical early months of life when enteric colonization is established and influences the developing innate immune response.

NUTRITION CONSIDERATIONS FOR OLDER INFANTS AND TODDLERS (6–24 MONTHS)

Complementary feeding, the inclusion of other foods and liquids besides human milk or infant formula, generally starts around 6 months of age and continues to 24 months of age. This ~18-month interval accounts for a substantial portion of the "1,000 days," and represents a critical window for disease prevention and long-term health. The best time to introduce complementary foods to infants is still being debated [19], particularly in high-resource settings in which concern about ingestion of contaminated foods and liquids is low. Very few randomized controlled trials have been conducted to specifically address this question, but the importance of context and considerations of risks versus benefits clearly merit discussion. Regardless of the setting, the most recent systematic review identified less morbidity from gastrointestinal infection in infants who were exclusively breastfed for 6 months, compared to those who were partially breastfed by 3 to 4 months [19]. No deficits in growth or effects on risk of allergy development were identified with 6 months of exclusive breastfeeding. The review also found that, in low-resource settings, where newborn iron stores may be low, exclusive breastfeeding for 6 months (without iron supplementation) may be associated with compromised hematologic status [20]. Data for the optimal timing of initiating complementary feeding in infants who are exclusively formula-fed are even more limited, but generally suggest no difference in growth or micronutrient status when comparing introduction of complementary foods at 3 to 4 months versus 6 months. Exclusive breastfeeding without complementary foods beyond 6 months increases the risk of micronutrient deficiencies (especially iron and zinc) and of atopic disease. Complementary foods are generally introduced gradually to accommodate the infant's oral-motor skill development and acceptance of new flavors and food consistencies. However, the evidence base for strict adherence to a prescribed order of food introduction is limited, since the gastrointestinal tract is considered to be essentially mature by approximately 5 months of age.

Breastfed and formula-fed infants have different dependence on complementary foods to meet nutrient needs. The most notable difference is for micronutrients, due to the generous fortification levels in all standard infant formulas, which result in a minimal risk of micronutrient deficiencies and make the choice of specific complementary foods less critical. In contrast, exclusively breastfed infants will need either to consume iron- and zinc-rich complementary foods (e.g., meats or fortified foods) or to receive supplements to supply these two micronutrients. Without additional iron intake starting at approximately 6 months, the risk for iron deficiency progressively increases. Iron deficiency is the most common nutritional deficiency in breastfed infants [21]. Zinc deficiency is also very common in older infants and toddlers. Many traditional complementary foods are low in both iron and zinc, as described later, and bioavailability from cereal-based foods is limited. Additionally, in low-resource settings, other factors, such as chronic inflammation and environmental enteric dysfunction, may interfere with iron absorption due to the stimulation of hepcidin, which blocks the release of iron from the basal surface of the enterocyte into the circulation. Similarly, zinc absorption may also be impaired in older infants who bear a high inflammatory burden [16].

The order of complementary food introduction in Western settings is often not specified but, traditionally, the order is fortified cereal, fruits, vegetables, and meats. The recommendation for iron-fortified infant cereal as the first food recognizes the need for iron. However, absorption from the iron fortificant in the cereal is quite low; furthermore, many infant cereals are not routinely fortified with zinc. On the other hand, animal-flesh sources of complementary foods are generally considered good sources of both iron and zinc with a more favorable bioavailability. From an evolutionary perspective, red meat (premasticated) and other animal source foods may have been more readily available to infants as complementary foods when humans relied on hunting and fishing as their primary means of getting food, and before the agricultural revolution. This dietary pattern, an early reliance on inclusion of animal source foods, is most consistent with the nutrients in human milk that become limiting for the older infant and toddler. In contrast, the modern day complementary diet is largely cereal based; not only does this generally provide lower concentrations of iron and zinc, but it also means that such foods also contain many inhibitors of absorption, such as phytates and polyphenols [22].

Consumption of meat also increases protein intake, for which caution has been raised with regard to potential accelerated weight gain and risk of obesity later in life. As discussed later, high protein intake from infant formula and dairy foods has been associated with early rapid weight and fat gain during infancy, and with higher BMI at school age [23–25]. Consumption of meat, on the other hand, may not have the same weight-gain-accelerating effect in infants and young children, possibly because of the different amino acid profile between meat and dairy protein, although research is limited. Some current guidelines for complementary feeding do not recommend against the use of meats as early complementary foods, but there is often no emphasis on introducing them as an important source of protein and micronutrients, especially for breastfed infants. The World Health Organization (WHO) has been more explicit about this, particularly in low-resource settings, where fortified complementary foods are often not available: "Meat, poultry, fish or eggs should be eaten daily, or as often as possible" [26]. The new recommendations for complementary feeding by the European Society for Paediatric Gastroenterology, Hepatology, and Nutrition (ESPGHAN) Committee on Nutrition also emphasized the importance of red meat, especially between 6 and 12 months, primarily due to its high iron and zinc content [19].

With regard to promoting positive feeding behaviors in older infants and toddlers, "responsive feeding" is recommended. Caregivers are encouraged to follow infant cues of hunger and satiety, rather than being overly prescriptive in urging the infant to consume "expected" volumes, which can result in either the infant's resistance to eating or excessive intakes. Both restrictive and permissive feeding styles can lead to poor regulation of energy intake and result in overconsumption of energy [27], which increases the risk of being overweight. Breastfed infants have also been found to have better satiety regulation than formula-fed infants, and formula-fed infants tend to consume more energy in general [28]. In a study of cereal versus meat-based complementary feeding for 9- to 10-month-old breastfed infants, strikingly different macronutrient distributions and total energy intakes from complementary foods in the two groups resulted in the adjustment of breast milk intake accordingly, and virtually identical total daily energy intake [29]. When introducing new foods to infants, initial refusal of new foods is common. Studies have demonstrated that repeated exposure improves the acceptance of new foods in infants, including pureed meats [30]. Offering a variety of healthy foods is also important to provide adequate nutrition and build a healthy eating pattern. Breastfed infants have been shown to be more likely to accept a new food than formula-fed infants [31], possibly due to exposure to a greater variety of flavors via human milk, which may increase the acceptance of foods with similar flavors during the complementary feeding period [32].

CONTEMPORARY HEALTH ISSUES RELATED TO COMPLEMENTARY FEEDING PRACTICES

The following subsections will address atopic disease and the risk of obesity, health problems of great public health significance in resource-rich environments and of emerging relevance in settings undergoing the nutrition transition. The last section highlights current understanding of the relationship between nutrients, feeding choices, and neurodevelopment.

COMPLEMENTARY FEEDING AND ALLERGY PREVENTION

In Westernized settings, complementary foods have traditionally been introduced one food and one food group at a time with the rationale that allergies/intolerance can be more readily identified. The evidence base for this practice is limited, however, and the pattern is not generally followed in low-resource settings, where family foods are typically offered to older infants. Food allergy occurs in up to 12% of American children and adults, and rates have been steadily increasing over the last two decades.

Infants are at the highest risk of developing food allergies, due in part to immature gastrointestinal function and the developing immune system, as well as to genetic predisposition. Contrary to past recommendations to delay introduction of allergenic foods such as peanuts and eggs, infants at risk of developing food allergies should not delay the consumption of such foods. The American Academy of Allergy, Asthma & Immunology recommends introducing highly allergenic complementary foods between 4 and 6 months of age once a few typical foods have been fed and tolerated. Emerging evidence suggests that delaying the introduction of allergenic foods may actually increase the risk of allergic reactions [18]. For example, the "Learning Early About Peanut" allergy (LEAP) study found that infants with a high risk of peanut allergies consuming a peanut-butter mush between 4 and 11 months of age were 80% less likely to develop a peanut allergy by age 5, compared with high-risk infants who were not exposed to peanuts early in life [33].

OBESITY RISK AND PROTEIN INTAKE

Young children who are obese have a higher likelihood of becoming obese adults [34]. Evidence-based consensus holds that the first 2 years of life are critical in obesity "programming," and that obesity prevention should be implemented early in life [35]. Given the current childhood obesity rates in the United States and other Westernized settings [36], a better understanding of potentially modifiable risk factors underpinning excessive weight and adiposity gain early in life is urgently needed. Multiple observational studies [37–39] have reported greater weight gain in formula-fed infants compared with breastfed infants. As described earlier, standard bovine-milk-based formula has a higher protein content than human milk and the difference in protein intakes has been considered a key contributor to the greater weight gain in formula-fed infants. This speculation has been strongly supported by a large-scale randomized controlled trial [23] by the European Childhood Obesity Group to examine how protein quantity relates to infant growth, using isocaloric infant formula with high- or low-protein content from birth to 24 months. The results showed that linear growth (length-for-age z-scores, LAZ) did not differ between groups during the intervention, whereas the weight-for-length z-scores (WLZ) of infants in the high-protein formula group were on average 0.20-z higher than those of the low-protein formula group at 24 months. The high-protein formula group also had a higher adiposity at 24 months [24]. Overall, the current literature suggests that consuming more protein, particularly dairy protein, increases WLZ and adiposity early in life, and may contribute to later obesity development.

The physiologic mechanisms underlying the relation between protein quantity and increased obesity risk are not clear. Insulin-like growth factor I (IGF-1) has been proposed to be the key mediator [23], since high protein intake is associated with high circulating IGF-1. However, findings are inconclusive [29,40]. Another potential mechanism may be driven by protein source, for example, through branched chain amino acids (BCAA), which tend to be higher in milk-derived protein compared to meats. Emerging metabolomics studies report associations of adult obesity and insulin resistance with BCAA and their degradation products, short-chain acylcarnitines [41]. Another proposed critical variable is the specific amino acid glutamate, which is considered a key regulator of appetite and energy intake [42]. Thus, current research findings reflect the complexity of interactions between protein intake (both quantity and quality), weight gain, adiposity, and later obesity risk.

Although the mechanism is not clear, emerging consensus for infant feeding recommendations in Westernized settings is to reduce protein intake to no more than 15% of total energy [19]. A caution regarding such a recommendation is that it does not distinguish between sources of protein (e.g., meats versus dairy versus plant-based sources), or between breastfed and formula-fed infants. This proposed guidance does not necessarily have equal relevance to settings in which breastfeeding predominates. As described earlier, meat can be an important dietary component to avoid micronutrient deficiencies in breastfed infants, and there is no evidence for its contribution to excessive weight gain. However, research that directly compares protein sources is limited and an area of active investigation.

NEUROCOGNITIVE DEVELOPMENT AND EARLY NUTRITION

Critical brain development occurs during the first 24 months of postnatal life. Brain and peripheral nervous tissues are especially sensitive to nutrient inadequacies, and suboptimal nutrition during this critical phase may lead to irreversible deficits in neurodevelopment. The developing brain clearly demonstrates plasticity, but the effects of nutritional insults, and their persistence, depend on the timing, duration, and severity of deficiencies and other risk factors. Before complementary foods are introduced, human milk is considered the ideal source of nutrition for brain development. Multiple studies have demonstrated that breastfed infants have better cognitive function than formula-fed infants, likely due to the difference in nutrient composition between human milk and infant formula, and the potentially greater mother–infant interactions with breastfeeding [43]. For the older infant, many of the micronutrients that are critical for brain development are found primarily, or in some cases exclusively, in animal source foods, especially flesh foods. Key nutrients to be discussed in this section in relation to brain development include LCPUFA, iron, zinc, iodine, choline, and vitamin B_{12}, which are either highly variable or commonly present the risk of deficiency.

Sixty percent of brain matter is lipid, including long-chain polyunsaturated fatty acids, with docosahexaenoic acid (DHA) and arachidonic acid (AA) being the most abundant fatty acids in the brain, the content of which continues to increase in the brain from birth to approximately 2 years of age [44]. DHA is essential for neurogenesis, as it is a structural component of cell membrane phospholipids, especially in the central nervous system [45]. Synthesis of DHA from α-linolenic acid occurs at a relatively low rate in young infants, and the primary source of DHA should be the diet. DHA content in human milk is dependent on maternal diet and is highly variable. When complementary feeding starts, omega-3-rich seafood, such as salmon, is a good source of dietary DHA, but this is not a typical food choice. Studies in Westernized settings investigating the effect of adding DHA to infants' diet on neurodevelopment early in life have not demonstrated a clear benefit in term infants [46,47]; stronger positive effects of DHA-enriched formulas have been observed for preterm infants. Potential confounding factors include sex, dose, duration, and cognitive assessment methods. One long-term follow-up study found that the benefits of LCPUFA may not be evident until 3 to 5 years of age [48].

The three most relevant micronutrient deficiencies to brain development are iron, zinc, and iodine. Iron is involved in numerous cellular processes in the brain, including myelination, and also serves as a cofactor for serotonin and catecholamines biosynthesis. As discussed earlier, human milk is uniformly low in iron and infants rely on prenatal iron stores, the transfer of the erythrocytes from delayed cord clamping at birth, and iron-rich complementary foods or supplementation. Despite the fortification of infant and many toddler foods in the United States, a significant number of children still develop iron deficiency or iron deficiency anemia, partially because of poor absorption from iron-fortified foods [49]. Multiple observational studies comparing infants and toddlers with and without iron deficiency have shown that iron deficiency, even without anemia, is associated with significantly lower scores of mental development, such as information processing and social-emotional development, in otherwise healthy term infants [50]. These impacts can be long term. A 19-year follow-up of term infants with iron deficiency in late infancy showed persisting motor differences, more grade repetition, anxiety/depression, social problems, and inattention in adolescence [51]. These adverse effects can be prevented if iron is supplemented before iron deficiency becomes severe or chronic [52]. Maternal diabetes and obesity are associated with low iron status in the offspring, and may be an important contributor to development of iron deficiency in Westernized settings where diabetes and obesity rates are high. In low-resource settings, chronic inflammation likely stimulates hepcidin, an anti-inflammatory hormone that inhibits iron absorption. Zinc is essential for neurogenesis, neuronal migration, and synaptogenesis, and plays an important role in neurotransmission and subsequent neuropsychological behavior. Zinc deficiency early in life may result in poorer learning, memory, and attention. After about 6 months of age, infant and toddler reliance on zinc-rich complementary foods to meet their relatively high needs places them at risk of developing at least mild zinc deficiency, particularly in settings where the zinc in foods may be poorly absorbed. Controlled trials of zinc supplementation on developmental outcomes, however, have had more mixed results. The role of iodine in brain development is through thyroid hormone synthesis. Iodine is routinely added to salt and widely available in both high- and low-resource settings; thus, iodine deficiency is not as common as iron and zinc deficiencies. However, iodine is one of the nutrients in human milk that reflects maternal status. Low iodine concentrations in human milk have been observed in regions with a high prevalence of goiter, and prevalence of cretinism has been reported in 5% to 15% of breastfed infants in such regions.

Choline is also considered a critical nutrient for brain and neurodevelopment. Choline is the precursor of the neurotransmitter acetylcholine and is also the major source of methyl groups in the diet. Human milk is rich in choline, and infant formulas are also typically fortified with choline. Animal source foods, such as liver and eggs, are excellent sources of dietary choline. Choline deficiencies are not common in infants and young children with a balanced diet, but the prevalence of deficiency is unknown in low-resource settings where access to animal source foods is limited.

Vitamin B_{12} is an enzyme cofactor in the formation of myelination. Deficiency of B_{12} can cause brain myelination impairment and affect neural development in infants. Because the vitamin B_{12} content in breast milk is substantially affected by maternal status, B_{12} deficiency can be caused by maternal lack of animal source foods intake, including meat and dairy. In addition, infants born to vegan mothers or mothers who have limited access to animal source foods are at risk for inadequate B_{12} stores at birth [11].

The human brain is composed of multiple regions, each with unique developmental processes. The hippocampus, for example, which is critical for special memory and navigation, begins its peak growth at approximately 32 weeks gestation and growth continues for at least 18 months. However, the frontal cortex, which controls complex processing behaviors (e.g., attention), has a much more protracted developmental trajectory. Thus, understanding the timing of the developmental trajectories is critical for assessing the impact of nutrient deficiencies. If a certain deficiency occurs during the time when the need for that nutrient is high, impairment is more likely to occur. The timing in five key neurodevelopmental domains and their relation with nutrient needs have recently been summarized [53].

CONCLUSION

The biology of the young infant is influenced by fetal exposures, but the postnatal period exemplifies an additional window of critical programming potential with long-term impact. The robust but dynamic composition of human milk provides nutritional and immunologic advantages to the young infant, while mitigating nutritional demands for the mother over time. The macronutrient profile of human milk includes a distinctive protein and amino acid composition, which provides immunologic benefits and optimizes growth and brain development; an array of oligosaccharides that modify infectious disease susceptibility; and lipid composition that enhances neurodevelopment and may influence the quality of infant weight gain. Likewise, the micronutrient composition of human milk is relatively robust and is dynamically tailored to meet the infant's changing nutritional needs through the critical early postnatal months. While human milk substitutes continue to be modified to better align with the beneficial matrix of human milk, many gaps remain.

The importance of high-quality complementary foods for the older infant and toddler is evident, as choices have clear impact on the nutritional adequacy of the complete diet. Attaining optimal intakes of several critical micronutrients is strongly dependent on high-quality complementary foods, which should include animal source foods and/or fortified products. Dietary exposures during the complementary feeding period, which comprises two-thirds of the 1,000 day window, are likely to affect long-term outcomes, including the risk of atopic disease, later obesity and metabolic health, and, perhaps most important, optimal neurodevelopment. As understanding of these critical biological underpinnings continues to emerge, the need for rigorous research as well as targeted programs and policies to support this most vulnerable time of the life cycle cannot be overstated.

REFERENCES

1. Ballard O, Morrow AL. Human milk composition: Nutrients and bioactive factors. *Pediatr Clin North Am* 2013; 60(1):49–74.
2. Bode L. Human milk oligosaccharides: Prebiotics and beyond. *Nutr Rev* 2009; 67(Suppl 2): S183–91.
3. Xu G, Davis JC, Goonatilleke E et al. Absolute quantitation of human milk oligosaccharides reveals phenotypic variations during lactation. *J Nutr* 2017; 147(1):117–24.
4. Alderete TL, Autran C, Brekke BE et al. Associations between human milk oligosaccharides and infant body composition in the first 6 mo of life. *Am J Clin Nutr* 2015; 102(6):1381–8.
5. Lien EL. Infant formulas with increased concentrations of alpha-lactalbumin. *Am J Clin Nutr* 2003; 77(6):S1555–8.
6. Lonnerdal B, Lien EL. Nutritional and physiologic significance of alpha-lactalbumin in infants. *Nutr Rev* 2003; 61(9):295–305.
7. Lonnerdal B. Nutritional and physiologic significance of human milk proteins. *Am J Clin Nutr* 2003; 77(6):S1537–43.
8. Lonnerdal B. Infant formula and infant nutrition: Bioactive proteins of human milk and implications for composition of infant formulas. *Am J Clin Nutr* 2014; 99(3):S712–7.
9. Rudolph MC, Young BE, Lemas DJ, Hernandez TL, Barbour LA, Friedman JE, Krebs NF, MacLean PS. Early infant adipose deposition is positively associated with human milk omega-6 to omega-3 fatty acid ratio independent of maternal BMI. *Int J Obes* 2017; 41:510–517.
10. Saarela T, Kokkonen J, Koivisto M. Macronutrient and energy contents of human milk fractions during the first six months of lactation. *Acta Paediatr* 2005; 94(9):1176–81.
11. Allen LH. B vitamins in breast milk: Relative importance of maternal status and intake, and effects on infant status and function. *Adv Nutr* 2012; 3(3):362–9.
12. Dawodu A, Zalla L, Woo JG et al. Heightened attention to supplementation is needed to improve the vitamin D status of breastfeeding mothers and infants when sunshine exposure is restricted. *Matern Child Nutr* 2014; 10(3):383–97.

13. Hollis BW, Wagner CL, Howard CR et al. Maternal versus infant vitamin D supplementation during lactation: A randomized controlled trial. *Pediatrics* 2015; 136(4): 625–34.
14. Dewey KG, Chaparro CM. Session 4: Mineral metabolism and body composition iron status of breast-fed infants. *Proc Nutr Soc* 2007; 66(3):412–22.
15. Qasem WA, Friel JK. An overview of iron in term breast-fed infants. *Clin Med Insights Pediatr* 2015; 9:79–84.
16. Krebs NF, Miller LV, Hambidge KM. Zinc deficiency in infants and children: A review of its complex and synergistic interactions. *Paediatr Int Child Health* 2014; 34(4):279–88.
17. Informal Working Group on Feeding of Nonbreastfed Children. Feeding of nonbreastfed children from 6 to 24 months of age: Conclusions of an informal meeting on infant and young child feeding organized by the World Health Organization, Geneva, March 8–10, 2004. *Food Nutr Bull* 2004; 25(4):403–6.
18. Fleischer DM, Spergel JM, Assa'ad AH et al. Primary prevention of allergic disease through nutritional interventions. *J Allergy Clin Immunol Pract* 2013; 1(1):29–36.
19. Fewtrell M, Bronsky J, Campoy C et al. Complementary feeding: A position paper by the European Society for Paediatric Gastroenterology, Hepatology, and Nutrition (ESPGHAN) Committee on Nutrition. *J Pediatr Gastroenterol Nutr* 2017; 64(1):119–32.
20. Kramer MS, Kakuma R. Optimal duration of exclusive breastfeeding. *Cochrane Database Syst Rev* 2012; 15(8):CD003517.
21. Baker RD, Greer FR, and Committee on Nutrition American Academy of Pediatrics. Diagnosis and prevention of iron deficiency and iron-deficiency anemia in infants and young children (0–3 years of age). *Pediatrics* 2010; 126(5):1040–50.
22. Dewey KG. The challenge of meeting nutrient needs of infants and young children during the period of complementary feeding: An evolutionary perspective. *J Nutr* 2013; 143(12):2050–4.
23. Koletzko B, von Kries R, Closa R et al. for the European Childhood Obesity Trial Study Group. Lower protein in infant formula is associated with lower weight up to age 2 y: A randomized clinical trial. *Am J Clin Nutr* 2009; 89(6):1836–45.
24. Escribano J, Luque V, Ferre N et al. for the European Childhood Obesity Trial Study Group. Effect of protein intake and weight gain velocity on body fat mass at 6 months of age: The EU Childhood Obesity Programme. *Int J Obes (Lond)* 2012; 36(4):548–53.
25. Weber M, Grote V, Closa-Monasterolo R et al. for the European Childhood Obesity Trial Study Group. Lower protein content in infant formula reduces BMI and obesity risk at school age: Follow-up of a randomized trial. *Am J Clin Nutr* 2014; 99(5):1041–51.
26. Pan American Health Organization (PAHO)/World Health Organization (WHO). *Guiding Principles for Complementary Feeding of the Breastfed Child.* Washington, DC: Pan American Health Organization/World Health Organization, 2003.
27. Anzman SL, Rollins BY, Birch LL. Parental influence on children's early eating environments and obesity risk: Implications for prevention. *Int J Obes (Lond)* 2010; 34(7):1116–24.
28. Dewey KG, Heinig MJ, Nommsen LA et al. Growth of breast-fed and formula-fed infants from 0 to 18 months: The DARLING Study. *Pediatrics* 1992; 89(6 Pt 1):1035–41.
29. Tang M, Krebs NF. High protein intake from meat as complementary food increases growth but not adiposity in breastfed infants: A randomized trial. *Am J Clin Nutr* 2014; 100(5):1322–8.
30. Krebs NF, Westcott JE, Butler N et al. Meat as a first complementary food for breastfed infants: Feasibility and impact on zinc intake and status. *J Pediatr Gastroenterol Nutr* 2006; 42(2):207–14.
31. Sullivan SA, Birch LL. Infant dietary experience and acceptance of solid foods. *Pediatrics* 1994; 93(2):271–7.

32. Mennella JA, Jagnow CP, Beauchamp GK. Prenatal and postnatal flavor learning by human infants. *Pediatrics* 2001; 107(6):E88.

33. Du Toit G, Roberts G, Sayre PH et al. Randomized trial of peanut consumption in infants at risk for peanut allergy. *N Engl J Med* 2015; 372(9):803–13.

34. Guo SS, Chumlea WC. Tracking of body mass index in children in relation to overweight in adulthood. *Am J Clin Nutr* 1999; 70(1 Part 2):S145–8.

35. Dattilo AM, Birch L, Krebs NF et al. Need for early interventions in the prevention of pediatric overweight: A review and upcoming directions. *J Obes* 2012; 2012:123023.

36. Ogden CL, Carroll MD, Kit BK et al. Prevalence of childhood and adult obesity in the United States, 2011–2012. *JAMA* 2014; 311(8):806–14.

37. Dewey KG. Growth characteristics of breast-fed compared to formula-fed infants. *Biol Neonate* 1998; 74(2):94–105.

38. Victora CG, Morris SS, Barros FC et al. Breast-feeding and growth in Brazilian infants. *Am J Clin Nutr* 1998; 67(3):452–8.

39. Krebs NF, Westcott JE, Culbertson DL et al. Comparison of complementary feeding strategies to meet zinc requirements of older breastfed infants. *Am J Clin Nutr* 2012; 96(1):30–35.

40. Chellakooty M, Juul A, Boisen KA et al. A prospective study of serum insulin-like growth factor I (IGF-I) and IGF-binding protein-3 in 942 healthy infants: Associations with birth weight, gender, growth velocity, and breastfeeding. *J Clin Endocrinol Metab* 2006; 91(3):820–6.

41. Newgard CB, An J, Bain JR et al. A branched-chain amino acid-related metabolic signature that differentiates obese and lean humans and contributes to insulin resistance. *Cell Metab* 2009; 9(4):311–26.

42. Ventura AK, Beauchamp GK, Mennella JA. Infant regulation of intake: The effect of free glutamate content in infant formulas. *Am J Clin Nutr* 2012; 95(4):875–81.

43. Victora CG, Bahl R, Barros AJ et al. and the Lancet Breastfeeding Series Group. Breastfeeding in the 21st century: Epidemiology, mechanisms, and lifelong effect. *Lancet* 2016; 387(10017):475–90.

44. Martinez M. Tissue levels of polyunsaturated fatty acids during early human development. *J Pediatr* 1992; 120(4):S129–38.

45. Innis SM. Dietary omega 3 fatty acids and the developing brain. *Brain Res* 2008; 1237:35–43.

46. Simmer K, Patole SK, Rao SC. Long-chain polyunsaturated fatty acid supplementation in infants born at term. *Cochrane Database Syst Rev* 2011; 12.CD000376.

47. Qawasmi A, Landeros-Weisenberger A, Leckman JF et al. Meta-analysis of long-chain polyunsaturated fatty acid supplementation of formula and infant cognition. *Pediatrics* 2012; 129(6):1141–9.

48. Colombo J, Carlson SE, Cheatham CL et al. Long-term effects of LCPUFA supplementation on childhood cognitive outcomes. *Am J Clin Nutr* 2013; 98(2):403–12.

49. Krebs NF, Sherlock LG, Westcott J et al. Effects of different complementary feeding regimens on iron status and enteric microbiota in breastfed infants. *J Pediatr* 2013; 163(2):416–23.

50. Lozoff B, Beard J, Connor J et al. Long-lasting neural and behavioral effects of iron deficiency in infancy. *Nutr Rev* 2006; 64(5 Pt 2):S34–43.

51. Lozoff B, Jimenez E, Smith JB. Double burden of iron deficiency in infancy and low socioeconomic status: A longitudinal analysis of cognitive test scores to age 19 years. *Arch Pediatr Adolesc Med* 2006; 160(11):1108–13.

52. Lozoff B, Georgieff MK. Iron deficiency and brain development. *Semin Pediatr Neurol* 2006; 13(3):158–65.

53. Prado EL, Dewey KG. Nutrition and brain development in early life. *Nutr Rev* 2014; 72(4):267–84.

Section IV

Endocrinology in the Regulation of Growth

6 Hormonal Regulation during Lactation and Human Milk Production

William R. Crowley

CONTENTS

THE HORMONES OF LACTATION: ACTIONS
AND PATTERNS OF SECRETION

Lactation, the physiological condition during which milk is produced for the purpose of providing nutrition to the offspring, features numerous anatomical, cellular, physiological, and behavioral adaptations, which affect the central nervous system as well as the mammary gland. Traditionally, lactation comprise of the sequential phases of lactogenesis I, the onset of milk synthesis without secretion during late pregnancy; lactogenesis II, the onset of active milk secretion shortly after parturition; and lactogenesis III (galactopoiesis), the maintenance of lactation [1–3]. Milk ejection, or letdown, refers to the process by which milk is removed from the mammary gland during lactation by the nursing offspring [1,2,4]. All aspects of lactation are influenced by multiple, interacting endocrine and paracrine messengers, including ovarian and adrenal steroids and insulin [1,2]. However, prolactin (PRL), released from the anterior pituitary gland and critical for milk synthesis and secretion, and oxytocin (OT), released from the posterior pituitary and required for milk ejection, are arguably the most important hormonal regulators of lactation, and their patterns of secretion, physiological actions and neuroendocrine regulation will be the focus of this review. Table 6.1 shows some actions of these and other hormones, which are exerted within the mammary gland and important for lactation.

The release of both PRL and OT during lactation is elicited through neuroendocrine reflex arcs that are initiated by suckling and, likely, other sensory stimuli

TABLE 6.1
Actions of Hormones in Lactogenesis

Hormone	Effects in Mammary Gland
Estradiol	Stimulation of ductal development[a]
	Inhibition of milk secretion[a]
	Stimulation of PRL gene expression
	Stimulation of OT gene expression[a]
Progesterone	Stimulation of ductal development[a]
	Inhibition of milk secretion[a]
	Stimulation of lactose synthesis[a]
	Inhibition of OT gene expression[a]
Cortisol	Stimulation of casein gene expression
	Stimulation of PRL receptor expression
Prolactin	Stimulation of milk proteins gene expression
	Stimulation of milk lipid synthetic enzymes gene expression
	Stimulation of progestin receptor gene expression
Oxytocin	Stimulation of myoepithelial cell contractions

Notes: OT, oxytocin; PRL, prolactin.
[a] Effects exerted during pregnancy.

from the offspring, and that engage complex ascending pathways into the neuroendo-crine hypothalamus. These processes have been investigated in highly sophisticated animal models for many years and are the subject of detailed reviews [5–10], yet our understanding of these processes remains incomplete. This chapter provides an overview of our current knowledge of the mechanisms that mediate these neuro-endocrine reflexes. Although the vast majority of the experimental data have been obtained from studies in animals, it seems highly likely that the regulation of lacta-tion in humans will follow similar principles.

ACTIONS OF PROLACTIN (PRL)

PRL participates in multiple aspects of lactation, both in the mammary gland and in the central nervous system [2,3,9,11]. PRL is secreted from the lactotrophs (or mam-motrophs) of the anterior pituitary gland, and also from a population that secretes both PRL and growth hormone (somatomammotrophs) [8]. The chemistry of PRL has been thoroughly reviewed elsewhere [12]. Briefly, native PRL is a protein of approximately 23,000 molecular weight, but PRL also circulates in a number of isoforms, for example, as phosphorylated or glycosylated monomers, as well as in multimeric species, modifications that can influence biological activity [8].

The actions of PRL in the mammary gland to promote lactogenesis have likewise been reviewed extensively and are reasonably well understood [1,2,9,12]. The PRL receptor is a member of the cytokine superfamily. The binding of PRL occurs in a stepwise fashion across two PRL receptor monomers, which then promotes their dimerization and initiates signal transduction, mediated predominantly via Janus kinase-signal transducer and activator of transcription proteins, which are coupled to gene expression [9,12]. Mammary gland genes that are regulated by PRL encode most milk proteins, as well as enzymes that are involved in synthesis of other milk constituents (e.g., lipids) (Table 6.1). A current concept is that PRL, along with cor-tisol, supports the onset of lactogenesis II when progesterone, which inhibits milk secretion, falls after parturition and also is essential in maintaining lactation [2,3]. PRL also gains access to the brain and exerts multiple actions that support lactation, such as the initiation of maternal behavior, the stimulation of food intake, and, pos-sibly, facilitatory effects on OT secretion [11].

ACTIONS OF OXYTOCIN (OT)

Although its actions are largely limited to the stimulation of milk ejection, OT, a nine-amino acid peptide released from the brain, is no less essential for a successful lactation, in as much as milk ejections are totally absent in mice bearing a deletion of the OT gene [13]. Stimulation of OT receptors increases the contractions of myoepithelial cells that are present on the surface of the alveoli and ducts of the mammary gland, with the increase in intramammary pressure thereby forcing milk to the nipple [4]. OT receptor structure, signal transduction pathways, and effects within the mammary gland to affect milk letdown have been authoritatively reviewed by Arrowsmith and Wray [14] and Burbach et al. [6]. Acting within the central nervous system as a central neurotransmitter, OT is importantly involved in the induction and regulation of maternal behavior [1,15].

Differential Patterns of PRL and OT Release in Response to Suckling

Animal Studies

Although both PRL and OT are released in a reflexive way by the sensory stimulation from the suckling offspring, the patterns of release of these two hormones are quite distinct. When female rats are suckled by their offspring after a period of separation (the separation–reunion paradigm), plasma concentrations of PRL begin to increase almost immediately, reach peak levels by 10 to 15 minutes, and exhibit irregular pulses, with peaks and troughs superimposed on the elevated baseline [4,16,17]. In contrast, the release of OT from the neurohypophysis may be delayed for a considerable period of time in lactating rats [16,18], although this latency is not seen in other species, including humans. OT release in response to suckling is pulsatile in all species examined, including humans, but in a pattern distinct from that described for PRL. Thus, the blood levels of OT during suckling are not elevated from baseline, but remain generally low and exhibit intermittent, abrupt increases of release that occur at approximately 5 to 10 minute intervals, last only several minutes, and then return to baseline [10,17,18]. During each secretory episode, a bolus of OT is delivered to the mammary gland, resulting in the increase in intramammary pressure and milk ejection [4,6,16]. Milk is thus available to the offspring only intermittently as a result of this pulsatile pattern of OT release.

Studies in Humans

Although not studied as extensively, the release of PRL and OT in lactating women exhibits a number of similarities to common animal models, such as the rat. For example, most studies show that PRL levels increase quickly and remain elevated throughout the period of suckling in women [1,19–21], and there is evidence of pulsatility as well [20,22]. Of considerable interest, the release of PRL in lactating women appears to be specifically linked to the suckling stimulus, as PRL is not increased prior to the onset of nursing (e.g., from visual or olfactory cues), or in response to tactile, non-suckling stimulation of the breast [1,20,21].

In nursing women, OT release in response to suckling begins quickly and exhibits a pulsatile pattern very much like that seen in animal models, that is, transient increases of OT release at intervals, followed by a return to baseline [20–23]. An important experimental consideration is that blood sampling must be sufficiently frequent in order to detect such pulsatility in women, which may account for the failure of some studies, which used more infrequent blood sampling protocols to observe OT pulses [1]. In contrast to PRL, OT is often released in lactating women prior to suckling and in response to tactile stimulation [19–21].

The neuroendocrine controls over lactation are prepared during pregnancy, particularly in late pregnancy, under the influence of ovarian hormones [7]. As noted in Table 6.1, estradiol and progesterone exert important effects during pregnancy to ready the mammary gland anatomically and physiologically, yet both also inhibit milk secretion, most likely to conserve the supply until needed following parturition. Indeed, the fall in progesterone following delivery of the placenta has been identified as a key event in the initiation of lactogenesis II in experimental animals and humans, with PRL and cortisol exerting stimulatory influences at this time [1–3].

The fall in progesterone is also important for the upregulation of OT gene expression in the hypothalamus [7]. Estradiol, in addition to stimulating PRL gene expression [8], at the same time inhibits milk production, an effect that also has a clinical impact on the choice of contraception during lactation.

NEUROENDOCRINE CONTROL OF PRL AND OT SECRETION

The general consistencies in the patterns of PRL and OT release across animal models and humans support the view that the neuroendocrine mechanisms identified in animal studies are likely to be very similar in humans, so that the wealth of mechanistic data obtained from animal models can likely provide insights into the physiological processes that occur in lactating women. However, it is important to keep in mind the obvious caveat that definitive mechanistic studies in lactating women are not going to be forthcoming. The next sections review the neural mechanisms that mediate suckling-induced PRL and OT release, with descriptions of the separate neuroendocrine control systems that regulate each hormone and the mechanisms that mediate their release in response to suckling during lactation.

THE SUCKLING-ACTIVATED AFFERENT PATHWAY TO THE NEUROENDOCRINE HYPOTHALAMUS

It has long been a goal of neuroendocrinologists to identify the neural pathways activated by suckling, which ultimately generate the distinct patterns of PRL and OT secretion described earlier. The following discussion briefly summarizes the results of numerous investigations; for more detail, the reader is referred to recent detailed summaries of this literature [5–7,10]. Sensory information from mechanoreceptors in the mammary glands is conveyed by the mammary nerves into the dorsal horn of the spinal cord and ascends ipsilaterally in the spino-cervical tract into the caudal medulla. From this level, most suckling-activated fibers cross the midline, and eventually reach the border between the mesencephalon and diencephalon. However, along the way, numerous connections are made with neurons in various nuclei of the brainstem, which, in turn, also contribute their processes to the ascending system. The exact routes taken by the nerve fibers of the afferent pathway from the mesencephalic–diencephalic border into the hypothalamic OT and PRL regulatory centers are not known with much certainty, at present.

NEUROENDOCRINE REGULATION OF PRL SECRETION IN LACTATION

The secretion of protein hormones from the anterior pituitary gland, a true endocrine secretory tissue, is governed by the central nervous system, but not through direct neural connections. Rather, clusters of neurons in the hypothalamus and basal forebrain express chemical messengers, which are released from nerve terminals in response to action potentials into the primary capillary plexus in the median eminence at the base of the hypothalamus. These hypophysiotropic hormones (also known as releasing factors or hormones) travel through the long portal veins into the anterior pituitary, and exert either stimulatory or inhibitory effects on specific anterior pituitary cell types. In the case of PRL, inhibitory and excitatory hypophy-siotropic hormones exist [7,8].

TUBEROINFUNDIBULAR DOPAMINE

The predominant mode of hypothalamic regulation over PRL secretion is tonically inhibitory and is accomplished by dopamine (DA), in this case acting as a hypophy-siotropic hormone rather than a neurotransmitter. The hypothalamic DA systems that act in this manner have been described extensively ([11,24]; see Figure 4 in Freeman et al. [8]). The major DA cell group regulating PRL is located in the arcuate nucleus of the medial-basal hypothalamus and is referred to as the tuberoinfundibu-lar dopamine (TIDA) system. The axons of these cells project to the external zones of the median eminence in proximity to the portal capillaries.

The administration of pharmacological agents that inhibit DA synthesis, deplete dopamine, or block the D_2 subtype of DA receptor, which is the predominant subtype present on lactotrophs, rapidly increases PRL secretion in animals; conversely, DA receptor agonists, especially those specific to the D_2 receptor, decrease PRL release in response to physiologic stimuli, including suckling [8]. That the TIDA system exerts these effects in humans is strongly supported by abundant evidence that D_2 antagonists used clinically (e.g., typical and atypical neuroleptics for treatment of psychosis and bipolar disorder) routinely elevate PRL, and several, such as metaclopramide and dom-peridone, have also been investigated for clinical utility as "galactogues," that is, stimu-lants of milk production in women experiencing difficulties in breast feeding [1,25–27]. Conversely, D_2 agonists, such as bromocriptine and cabergoline, are commonly used to lower PRL hypersecretion from prolactinomas of the anterior pituitary gland [25,26].

There is strong evidence from animal studies that this tonic inhibitory control exerted by DA over the lactotroph is removed as an obligatory step before PRL can be secreted in response to physiologic stimuli, including suckling during lactation [8,11]. For example, suckling or a suckling-mimetic stimulus, such as mammary nerve stimulation, reduces the release of DA from the TIDA system into the por-tal vasculature, as measured with various experimental techniques, including direct measurement of DA in portal blood [7,8,24,28]. Moreover, the evidence indicates that such a reduction in DA release itself constitutes a stimulus for an increase in PRL secretion; this has been demonstrated not only by the previously mentioned pharmacological studies with DA antagonists, but also in *in vitro* preparations, for example, in cultured anterior pituitary cell preparations, in which an increase in PRL release is observed following the removal of DA after a period of exposure [28].

THYROTROPIN-RELEASING HORMONE

PRL secretion in response to suckling also requires a stimulatory hypophysiotropic influence, and the tripeptide thyrotropin-releasing hormone (TRH), which also stimu-lates thyroid-stimulating hormone, appears to fulfill this role, at least in some species. TRH-immunopositive neurons are found in several hypothalamic regions, most nota-bly in the parvicellular (small cell) regions of the paraventricular nucleus of the hypo-thalamus [29]. TRH is highly effective in physiological concentrations in stimulating

PRL secretion from anterior pituitary cells and PRL-secreting pituitary tumor cell lines [8]. Administration of TRH to lactating women also results in an increase of PRL secretion [1,30], strongly suggestive of a physiological role of this peptide in humans. Further, the suckling-mimetic stimulus of mammary nerve stimulation produces a significant increase in TRH concentrations in hypophyseal portal blood, and inhibition of TRH action via immunoneutralization prevents suckling-induced PRL release in rats [7,8]. Moreover, DA withdrawal, as described earlier during suckling, enhances the ability of TRH to stimulate PRL secretion [4,28]. Hence, our current concept is that suckling activates a dynamic interplay of hypophysiotropic regulation that involves DA withdrawal from and TRH stimulation of the lactotroph as the critical hypophysiotropic mechanism governing PRL secretion in response to suckling.

"Upstream" Regulation over Prolactin: Central Neurotransmitters

This dual hypophysiotropic control represents the final output of the central nervous system in mediating suckling-induced PRL release, but what neural systems are activated by suckling, connect with these neuroendocrine systems, and effect the withdrawal of DA and activation of TRH release? Pharmacological approaches have provided evidence in animals and humans that strongly suggests important stimulatory actions of serotonin (5-hydroxytryptamine, or 5-HT) and opioid peptides.

Pharmacological studies indicate a generally excitatory action of 5-HT on PRL secretion in different physiological conditions [8,31], an effect also observed in humans [32]. Pharmacological disruption of serotonergic transmission inhibits suckling-induced PRL release in animals, implying an obligatory role for this system [7,31]. Serotonin neurons originating in the midbrain innervate the hypothalamic nuclei containing the TIDA system and TRH neurons, and these serotonergic cells are likely activated by the suckling stimulus as a component of the suckling-activated ascending pathway [7,31]. Moreover, evidence suggests that, when activated, the serotonin system concomitantly inhibits TIDA activity and stimulates TRH neurons [7,8,31], thereby setting in motion the critical hypophysiotropic mechanism for suckling-induced PRL release discussed earlier.

Opioid drugs, such as morphine, and the endogenous peptides, β-endorphin, the enkephalins, and the dynorphins, the receptors of which are stimulated by opioid drugs, also stimulate PRL secretion in animals and humans [1,7,8,31,33]. Suggesting an obligatory role of opioid peptides during lactation, administration of the opioid receptor antagonist naloxone can inhibit suckling-induced PRL release in rats [31,34]. On the other hand, naloxone treatment failed to inhibit PRL release in lactating women evoked by stimulation of the nipples by breast pump [35], so that an obligatory role in the mechanism of suckling-induced PRL release in humans is much less certain.

Interactions of opioids with the TIDA system have been addressed in many animal studies and, as might be expected, opioid peptides and agonists decrease TIDA neuronal activity, thereby diminishing DA release [34]. An opioid–serotonin interaction also seems plausible, and it seems reasonable to propose that one or more endogenous opioid systems contribute to the physiological process of suckling-induced PRL release via (1) suppression of the TIDA inhibitory system, and (2) enhancement of excitatory serotonergic activity [7,31]. Figure 6.1 presents an overall mechanism for suckling-induced PRL secretion that incorporates these mechanisms.

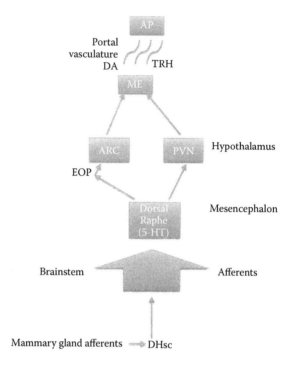

FIGURE 6.1 Basic mechanisms of suckling-induced prolactin secretion during lactation. Serotonin (5-HT) neurons in the mesencephalon are activated by the suckling stimulus and reduce the tonic inhibitory influence over PRL secretion exerted by the hypophysiotropic tuberoinfundibular dopamine (TIDA) system present in the arcuate nucleus of the hypothalamus (ARC), likely through a link with endogenous opioid peptides (EOP). Concomitantly, the 5-HT system activates the release of the stimulatory hypophysiotropic hormone thyrotropin-releasing hormone (TRH) neurons present in the hypothalamic paraventricular nucleus (PVN). AP, anterior pituitary; DHsc, dorsal horn of the spinal cord; ME, median eminence.

NEUROENDOCRINE REGULATION OF OT SECRETION IN LACTATION

The major focus of investigations on OT secretion during lactation has been on the complex neurophysiological mechanisms responsible for generating the episodic, pulsatile pattern of OT release described earlier, a pattern which occurs only during lactation and parturition, and not in response to other physiological stimuli, and which is also different from that seen with PRL (see earlier). Because OT release results from OT neuronal activity, it is critical to understand the process by which OT neurons largely remain at baseline levels of activity, despite the continual presence of the suckling stimulus, but are interrupted at varying intervals by a dramatic increase in neuronal activity that lasts only a few seconds, resulting in a brief pulse of OT release into the systemic vasculature, followed by a return to baseline. Out of necessity, these investigations have been performed in animal models, and there is a great paucity of information on the regulation of OT secretion in humans.

ANATOMY OF THE OT NEUROENDOCRINE SYSTEM

The neuroanatomical organization of the OT neuroendocrine system is well established ([5]; see also Figure 3 in Armstrong [36] and Figure 5 in Crowley and Armstrong [18]). Briefly, OT is synthesized by large (magnocellular) neurons in the hypothalamic paraventricular (PVN) and supraoptic (SON) nuclei, and in several nearby "accessory" clusters of neurons, and is transported in vesicles down the axons that project into the neurohypophysis, in reality an outgrowth of the brain rather than a true endocrine organ. OT is released into the circulation via the classic process of calcium-dependent neuronal exocytosis, in response to invasion of the nerve terminals by action potentials that are generated in the cell bodies [10].

THE MILK EJECTION REFLEX: NEURONAL MECHANISMS

Underlying the episodic pattern of systemic OT release is the intermittent bursting behavior of magnocellular OT neurons in response to suckling ("milk ejection bursts"), which has been extensively studied in sophisticated electrophysiological experiments in lactating animals [5,6,10,37]. As observed in the separation–reunion paradigm in rats, the onset of suckling typically does not immediately change the irregular, baseline pattern of OT neuronal firing. However, after a variable latent period, and approximately 15 to 20 seconds before each milk ejection, a brief and explosive increase in the firing rate of the OT neurons occurs, lasting 2 to 4 seconds, followed by a short period of inhibition and then a return to irregular baseline activity (see Figure 1 in Crowley and Armstrong [18] and Figure 8 in Wakerley [10]). These milk ejection bursts recur at intervals of 5 to 10 minutes throughout the period of suckling. Further, during these events, it appears that virtually the entire magnocellular OT population on both sides of the brain fires bursts of action potentials in a coordinated manner in order to deliver each bolus pulse of OT to the systemic circulation and from there to the mammary gland [10,37]. Given the similar pattern of pulsatile OT secretion in lactating women, there is every reason to believe that this milk ejection burst mechanism also occurs in humans.

However, the mechanisms by which (1) the continuous suckling stimulus generates intermittent bursting behavior of magnocellular OT neurons, and (2) this bursting activity becomes coordinated and synchronized across the entire population in diverse hypothalamic nuclei both remain to be determined. A common view is that suckling initially activates mechanisms that promote local excitatory interactions among OT neurons within each PVN and SON, progressively leading to more coordination among these cells in their firing patterns. As increasing numbers of OT neurons are recruited into this coordinated firing, eventually the entire population explodes with a burst of action potentials that last for a few seconds, followed by a brief period of inhibition, before the process begins anew [5,7,10]. However, to accomplish this, there must not only be excitatory interactions among OT neurons within a nucleus, but also intercommunications between the multiple neurosecretory nuclei to mediate activation of the entire population. Some of the mechanisms that may participate in this complex process are discussed next, and the reader is referred to several recent, more detailed reviews [5–7,10].

NEUROTRANSMITTERS: GLUTAMATE, NOREPINEPHRINE, AND γ-AMINOBUTYRIC ACID

Multidisciplinary studies indicate that glutamate (Glu) and norepinephrine (NE) are critical excitatory neurotransmitters for suckling-induced OT release [7]. OT neurons receive a dense innervation from local glutamatergic neurons adjacent to the magnocellular nuclei [38] and from noradrenergic neurons in the lower brainstem [18]; both systems are components of the suckling-activated ascending afferent pathway [7,18]. Agonists at the AMPA receptor for Glu and α-1 receptor for NE stimulate OT release, while antagonists at these respective receptors inhibit suckling-induced OT release [39–41], thus providing strong evidence that these systems are critically involved in the neural mechanism mediating the suckling-induced OT neuroendocrine reflex. Further, intrahypothalamic Glu systems appear to be particularly important in generating the burst-type of firing pattern in OT neurons during lactation [38]. On the other hand, γ-aminobutyric acid (GABA), also released from local hypothalamic neurons near the magnocellular nuclei, inhibits OT neuronal firing, and has been proposed to be important in shaping the burst pattern of OT neuronal activity [42].

INTRANUCLEAR RELEASE OF OT

An important neurochemical signal for the generation of the milk ejection burst in OT neurons involves an action of OT itself. In addition to being released into the systemic circulation, OT is also released within the magnocellular nuclei, mainly from dendrites, in response to suckling, and such "intranuclear release" is critical in promoting the milk ejection bursts of neuronal firing [5,7,10,43,44]. For example, OT activates OT neurons but, more specifically, increases the frequency and amplitude of milk ejection bursts in response to suckling [45]; further, because central administration of an OT receptor antagonist inhibits both intranuclear and systemic OT release, it is proposed that OT acts in a positive feedback manner to increase its own intranuclear release, which, in turn, ultimately leads to the milk ejection burst and release into the systemic circulation [44].

These and other experiments have led to the hypothesis [5,10,43] that, prior to the onset of milk ejection bursts, suckling may initially stimulate the release of OT within the magnocellular nuclei, most likely in response to NE and Glu [7,39]. As suckling continues, the intranuclear release of OT progressively increases and, acting on OT receptors on the same neuron and/or on adjacent OT neurons, intranuclear OT exerts multiple neurochemical effects that progressively excite and coordinate the firing of OT neurons, until the entire population unites in firing the brief milk ejection burst [5,43].

SYNCHRONIZATION MECHANISMS

Whereas local mechanisms such as those described earlier can account for the coordinated activation of OT neurons within a single PVN or SON (i.e., intranuclear synchronization), how can we explain the synchronized activation of the entire OT neurosecretory population during the milk ejection burst, involving OT neurons residing in multiple nuclei and on both sides of the brain (internuclear synchronization)? This question cannot be answered definitively at present, but there is anatomical and electrophysiological evidence for local interconnections between, for example, the PVN and SON [38], which could propagate excitation from one magnocellular nucleus to the other ipsilaterally. In addition, interconnections exist between the two SONs and between the two PVNs [5,7], which could account for the contralateral spread of excitation. It may also be the case that these local interconnections among the magnocellular nuclei could relay synchronizing signals that originate in more distal structures. For example, investigators have identified neurons in the ventromedial medulla that are activated by suckling and that project bilaterally to both the PVN and SON [46]. Other regions put forth as synchronization "centers" include the mammillary bodies [47]. Figure 6.2 presents a mechanism for suckling-induced OT secretion that incorporates some of these interactions.

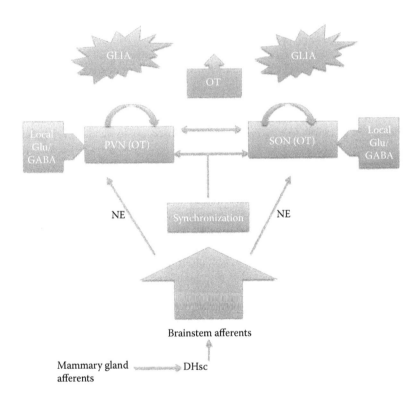

FIGURE 6.2 Basic mechanisms of suckling-induced oxytocin secretion during lactation. Ascending systems from the brainstem activated by suckling include noradrenergic neurons (NE), which stimulate activity of OT neurons in the paraventricular (PVN) and supraoptic (SON) nuclei of the hypothalamus. Intrahypothalamic glutamate (Glu) and γ-aminobutyric acid (GABA) neurons also provide important excitatory and inhibitory inputs, respectively, and OT released within the magnocellular nuclei exerts critical facilitatory actions on OT neuronal firing and coordination of firing. Arrows represent local and extrinsic mechanisms for synchronization of the entire OT population to produce a brief milk ejection burst of firing. Withdrawal of glial elements may also promote an excitatory neurochemical microenvironment. DHsc, dorsal horn of the spinal cord; OT, oxytocin.

Glial Plasticity in the Magnocellular Nuclei and Neurohypophysis

An additional mechanism that may facilitate the milk ejection burst involves a remodeling of the anatomical relationships between astroglial cells and OT neurons in the magnocellular nuclei [5,48,49]. To summarize this work briefly, during lactation, and dependent upon the continual presence of the suckling stimulus, astroglial processes surround magnocellular OT neurons retract, thereby permitting close somatic appositions between OT neurons, and also increasing the incidence of "double synapses" (i.e., presynaptic nerve terminals that contact multiple OT neurons postsynaptically). Interestingly, many of these double synapses are glutamatergic and noradrenergic [49], which provide the major excitatory drives to OT neurons in response to suckling (see earlier). Intranuclear release of OT may also promote this plasticity in the glial-neuronal interactions in the magnocellular nuclei [49]. While the precise mechanisms and physiological significance of these morphological changes remain to be conclusively determined, it is attractive to hypothesize that they could enhance the actions of excitatory neurotransmitters to create a stimulatory neurochemical microenvironment affecting multiple OT neurons, which might then facilitate coordinated activity among OT neurons within each magnocellular nucleus. OT-containing nerve terminals in the neurohypophysis are also enclosed by glial cells, which separate them from the capillaries; these glia are also withdrawn from the neurosecretory endings during parturition and lactation, increasing access of released OT to the capillaries [50].

ENDOCRINE DYSFUNCTION AND LACTATION

Although detailed information on pathophysiologic mechanisms is generally not available, lactation may be compromised or complicated in women with endocrine disease, perhaps as a result of disruptions in intermediary metabolism, and the supply of nutrients to the mammary gland [1]. For example, both hypo- and hyperthyroidism can impair lactation, as can all forms of diabetes mellitus. On the other hand, there is suggestive evidence in humans that the metabolic adaptations of lactation can lower insulin requirements and even serve a metabolically "protective" function in women with diabetes mellitus [1]. Polycystic ovarian syndrome, a complex, poorly understood neuroendocrine disorder, also complicates pregnancy and lactation, perhaps via alterations in steroid receptor function [1]. As previously noted, both estradiol and progesterone can inhibit milk production, which is likely to be important during pregnancy, but can have clinical significance in the context of choice of contraceptives during lactation. Generally, combined oral contraceptives have been avoided from concerns over the estrogenic component, in favor of progestin-only contraceptives, which are usually initiated only after lactation has been established to avoid any inhibition by progestin.

REFERENCES

1. Lawrence RA, Lawrence RM. *Breastfeeding: A guide for the medical professional.* 6th ed. Philadelphia: Elsevier Mosby; 2005.
2. Neville MC. Lactation and its hormonal control. In *Physiology of reproduction*, 3rd ed., Plant TM, Zeleznik AJ, editors. New York: Elsevier; 2006.

3. Neville MC, Morton J. Physiology and endocrine changes underlying human lactogenesis II. *J Nutr* 2001; 131(11):S3005–8.
4. Grosvenor CE, Mena F. Regulating mechanisms for oxytocin and prolactin secretion during lactation. In *Neuroendocrine perspectives*, Muller EE, MacLeod RM, editors. New York: Elsevier; 1982.
5. Brown CH, Bains JS, Ludwig M et al. 2013. Physiological regulation of magnocellular neurosecretory cell activity: Integration of intrinsic, local and afferent mechanisms. *J Neuroendocrinol* 2013; 25(8):678–710.
6. Burbach JP, Young LJ, Russell JA. Oxytocin: Synthesis, secretion, and reproductive functions. In *Physiology of reproduction*, 3rd ed., Plant TM, Zeleznik AJ, editors. New York: Elsevier; 2006.
7. Crowley WR. Neuroendocrine regulation of lactation and milk production. *Compr Physiol* 2015; 5(1):255–91.
8. Freeman ME, Kanyicska B, Lerant A et al. Prolactin: Structure, function, and regulation of secretion. *Physiol Rev* 2000; 80(4):1523–631.
9. Horseman ND, Gregerson KA. Prolactin actions. *J Mol Endocrinol* 52:R95–106.
10. Wakerley JB. Milk ejection and its control. In *Physiology of reproduction*, 4th ed., Plant TM, Zeleznik AJ, editors. New York: Elsevier; 2006.
11. Grattan DR. The hypothalamo-prolactin axis. *J Endocrinol* 2015; 226(2):T101–22.
12. Bole-Feysot C, Goffin V, Edery M et al. Prolactin (PRL) and its receptor: Actions, signal transduction pathways and phenotypes observed in PRL receptor knockout mice. *Endocr Rev* 1998 19(3):225–68.
13. Nishimori K, Young LJ, Guo Q et al. Oxytocin is required for nursing but is not essential for parturition or reproductive behavior. *Proc Natl Acad Sci USA* 1996; 93(21):11699–704.
14. Arrowsmith S, Wray S. Oxytocin: Its mechanism of action and receptor signaling in the myometrium. *J Neuroendocrinol* 2014; 26(6):356–69.
15. Olza-Fernandez I, Marin Gabriel MA, Gil-Sanchez A et al. Neuroendocrinology of childbirth and mother-child attachment: The basis of an etiopathogenic model of perinatal neurobiological disorders. *Front Neuroendocrinol* 2014; 35(4):459–72.
16. Crowley WR, Parker SL, Armstrong WE et al. Neurotransmitter and neurohormonal regulation of oxytocin in lactation. *Ann NY Acad Sci* 1992; 652:286–302.
17. Higuchi T, Honda T, Fukuoka T et al. Pulsatile secretion of prolactin and oxytocin during nursing in the lactating rat. *Endocrinol Jpn* 1983; 30(3):353–9.
18. Crowley WR, Armstrong WE. Neurochemical regulation of oxytocin secretion during lactation. *Endocr Rev* 1992; 13(1):33–65.
19. Howie PW, McNeilly AS, McArdle T et al. The relationship between suckling-induced prolactin responses and lactogenesis. *J Clin Endocrinol Metab* 1980; 50(4):670–3.
20. McNeilly AS, Robinson IC, Houston MJ et al. Release of oxytocin and prolactin in response to suckling. *Br Med J (Clin Res Ed)* 1983; 286(6361):257–9.
21. Yokoyama Y, Ueda T, Irahara M et al. Release of oxytocin and prolactin during breast message and suckling in puerperal women. *Eur J Obstet Gynecol Reprod Biol* 1994; 53(1):17–20.
22. Tay CC, Glasier AF, McNeilly AS. Twenty-four hour patterns of prolactin secretion during lactation and the relationship to suckling and the resumption of fertility in breast-feeding women. *Hum Reprod* 1996; 11(5):950–5.
23. Johnston JM, Amico JA. A prospective longitudinal study of the release of oxytocin and prolactin in response to infant suckling in long-term lactation. *J Clin Endocrinol Metab* 1986; 62(4):653–7.
24. Ben-Jonathan N, Hnasko R. Dopamine as a prolactin (PRL) inhibitor. *Endocr Rev* 2001; 22(6):724–63.
25. Peuskens J, Pani L, Detraux J et al. The effects of novel and newly approved antipsychotics on serum prolactin levels: A comprehensive review. *CNS Drugs* 2014; 28(5):421–53.

26. Voicu V, Medvedovici A, Ranetti AE et al. Drug-induced hypo- and hyperprolactinemia: Mechanisms, clinical and therapeutic consequences. *Expert Opin Drug Metab Toxicol* 2013; 9(8):955–68.

27. Zuppa AA, Sindico P, Orchi C et al. Safety and efficacy of galactogogues: Substances that induce, maintain, and increase breast milk production. *J Pharm Pharm Sci* 2010; 13(2):162–74.

28. Martinez de la Escalera G, Weiner RI. Dissociation of dopamine from its receptor as a signal in the pleiotropic hypothalamic regulation of prolactin secretion. *Endocr Rev* 1992; 13(2):241–55.

29. Lechan RM, Jackson IM. Immunohistochemical localization of thyrotropin-releasing hormone in the rat hypothalamus and pituitary. *Endocrinology* 1982; 111(1):55–65.

30. Tyson JE, Perez A, Zanartu J. Human lactational response to oral thyrotropin releasing hormone. *J Clin Endocrinol Metab* 1976; 43(4):760–8.

31. Crowley WR. Role of endogenous opioid neuropeptides in the physiological regulation of luteinizing hormone and prolactin secretion. In *Peptide hormones: Effects and mechanisms of action*, Negro-Vilar A, Conn PM, editors. Boca Raton: CRC Press; 1988.

32. Coker F, Taylor D. Antidepressant-induced hyperprolactinemia: Incidence, mechanisms and management. *CNS Drugs* 2010; 24(7):563–74.

33. Vuong C, Van Uum SH, O'Dell LE et al. The effects of opioids and opioid analogs on animal and human endocrine systems. *Endocr Rev* 2010; 31(1):98–132.

34. Arbogast LA, Voogt JL. Endogenous opioid peptides contribute to suckling-induced prolactin release by suppressing tyrosine hydroxylase activity and messenger ribonucleic acid levels in tuberoinfundibular dopaminergic neurons. *Endocrinology* 1998; 139(6):2857–62.

35. Cholst IN, Wardlaw SL, Newman CB et al. Prolactin response to breast stimulation in lactating women is not mediated by endogenous opioids. *Am J Obstet Gynecol* 1984; 150(5 Pt 1):558–61.

36. Armstrong WE. Hypothalamic supraoptic and paraventricular nuclei. In *The rat nervous system*, 2nd ed., Paxinos G, editor. New York: Academic Press; 1994.

37. Richard P, Moos F, Freund-Mercier MJ. Bursting activity in oxytocin cells. In *Pulsatility in neuroendocrine systems*, Leng G, editor. Boca Raton: CRC Press; 1998.

38. Israel JM, Le Masson G, Theodosis DT et al. Glutamatergic input governs periodicity and synchronization of bursting activity in oxytocin neurons in hypothalamic organotypic cultures. *Eur J Neurosci* 2003; 17(12):2619–29.

39. Bealer SL, Crowley WR. Noradrenergic control of central oxytocin release during lactation in rats. *Am J Physiol* 1998; 274(3 Pt 1):E453–8.

40. Crowley WR, Shyr SW, Kacsoh B et al. Evidence for stimulatory noradrenergic and inhibitory dopaminergic regulation of oxytocin release in the lactating rat. *Endocrinology* 1987; 121(1):14–20.

41. Parker SL, Crowley WR. Stimulation of oxytocin release in the lactating rat by central interaction of alpha 1-adrenergic and alpha-amino-3-hydroxy-5-methyisoxazole-4-propionic acid-sensitive excitatory amino acid mechanisms. *Endocrinology* 1993; 133(6):2855–60.

42. Moos FC. GABA-induced facilitation of the periodic bursting activity of oxytocin 43. neurons in suckled rats. *J Physiol* 1995; 488(Pt 1):103–14.

43. Leng G, Caquineau C, Ludwig M. Priming in oxytocin cells and gonadotrophs. *Neurochem Res* 2008; 33(4):668–77.

44. Neumann I, Koehler E, Landgraf R et al. An oxytocin receptor antagonist infused into the supraoptic nucleus attenuates intranuclear and peripheral release of oxytocin during suckling in conscious rats. *Endocrinology* 1994; 134(1):141–8.

45. Freund-Mercier MJ, Richard P. Electrophysiological evidence for facilitatory control of oxytocin neurons by oxytocin during suckling in the rat. *J Physiol* 1984; 352:447–66.
46. Moos F, Marganiec A, Fontanaud P et al. Synchronization of oxytocin neurons in suckled rats: Possible role of bilateral innervation of hypothalamic supraoptic nuclei by single medullary neurons. *Eur J Neurosci* 2004; 20(1):66–78.
47. Wang YF, Negoro H, Higuchi T. Lesions of hypothalamic mammillary body desynchronize milk-ejection bursts of rat bilateral supraoptic oxytocin neurons. *J Neuroendocrinol* 2013; 25(1):67–75.
48. Montagnese C, Poulain DA, Vincent JD et al. Synaptic and neuronal-glial plasticity in the adult oxytocinergic system in response to physiological stimuli. *Brain Res Bull* 1998; 20(6):681–92.
49. Theodosis DT. Oxytocin-secreting neurons: A physiological model of morphological neuronal and glial plasticity in the adult hypothalamus. *Front Neuroendocrinol* 2002; 23(1):101–35.
50. Tweedle CD, Hatton GI. Morphological adaptability at neurosecretory axonal endings on the neurovascular contact zone of the rat neurohypophysis. *Neuroscience* 1987; 20(1):241–6.

7 Thyroid Function and Growth
The Mechanisms of Iodine

Jessica Farebrother and Fabian Rohner

CONTENTS

CONSEQUENCES OF SUBOPTIMAL IODINE STATUS DURING THE FIRST 1,000 DAYS

Iodine deficiency used to be highly prevalent in many parts of the world until salt iodization programs and other prevention strategies were implemented and scaled up, beginning in the 1980s in most countries [1]. Great progress toward reducing iodine deficiency and its consequences has been achieved since; however, large numbers of people continue to be affected [2,3].

The Biology of the First 1,000 Days

In particular, the fetus and the neonate are important at-risk groups, both with regard to physiological needs and supply from the mother, and because of the profound negative effects that iodine deficiency can have during perinatal development [4,5]. During pregnancy and lactation, maternal iodine requirements are increased to supply the fetus, neonate, and, later, infant, and iodine stores (if available at the onset of pregnancy) can become rapidly depleted if the pregnant or lactating woman does not consume sufficient dietary iodine [6–9]. Iodine requirements during these life stages are thus higher than for the general population.

Figure 7.1 summarizes the consequences of iodine deficiency during the first 1,000 days, along with a judgment of the quality of evidence for each effect and each early life stage. All effects presented are child-related and do not include adverse effects on pregnant or lactating women. Upward arrows indicate an improvement of the situation. For example, an upward arrow indicates less child cretinism if iodine is provided during pregnancy. These effects are usually stronger when intervening in populations that are severely iodine deficient, but, to some extent, they also apply in situations of moderate deficiency.

In summary, there is strong evidence as to the negative consequences of suboptimal iodine status during the first 1,000 days, such as reduced cognition, cretinism, and lower birth weight, as outlined in several recent reviews. Whereas Bougma et al. [10] concluded that iodine deficiency has an important association with the mental development in children under 5 years, Zimmermann [11] concluded that iodine supplementation at varying levels before or during early pregnancy among women living in moderate to severely iodine-deficient areas eliminates cretinism, increases cognitive development in young children, increases birth weight, and reduces infant mortality.

Outcome	Life Stage of Intervention			Quality of Evidence
	Pregnancy	Lactation	Infancy/Childhood	
Cretinism	↗ 1	NA	NA	Low —→
				Medium →
Cognition	↗	○	↗	High ⇒
Birth weight	↗	NA	NA	Insufficient literature ∩
Growth	∩	○	○	
Perinatal/infant mortality	↗	↗ 2	NA	No effect →
				Positive effect ↗
Abortion/stillbirth	○	○	○	Negative effect ↘

FIGURE 7.1 The effect of iodine interventions on known consequences of iodine deficiency during the 1,000 day window, by outcome and quality of evidence. Quality of evidence rating criteria: insufficient literature, only individual studies that are somewhat unclear or conflicting; low, small number of studies, but mostly consistent findings; medium, existing meta-analyses or systematic reviews, but somewhat conflicting findings; high, existing meta-analyses or systematic reviews, with mostly consistent findings. Outcomes reported for cretinism [10], cognition [10–12], birth weight [10,12], growth [13], perinatal and infant mortality [10], and abortion/stillbirth [14]. NA, not applicable (intervention–outcome combination not applicable for this life stage). [1]Graded medium level despite only a few studies available. Studies showed consistent and clear improvements and thus, for ethical reasons, intervention studies to further corroborate findings are no longer feasible. [2]Iodine supplements were given to neonate directly, not via maternal breast milk.

Similarly, Gunnarsdottir and Dahl [12] suggest that an improved prenatal iodine status is associated with improvements in cognitive function for infants and toddlers to 18 months. In contrast, there is little evidence on the effect of iodine status, or iodine supplementation or fortification on infant and child growth outcomes [13]. However, as seen in the reviews discussed above, in the case of iodine there are clear positive effects of correcting iodine status on infant morbidities, likely due to the underlying mechanisms of action of iodine in the body. Iodine plays an essential role in the synthesis of thyroid hormones; these hormones are required for many metabolic reactions essential to growth and development such as protein synthesis, and bone turnover and regulation, which will be discussed in more detail later in the chapter.

Given the paucity of available evidence about the effect of iodine deficiency on growth as an outcome measure, the objective of this chapter is to elucidate the underlying mechanisms linking iodine nutrition to growth during the life stages of the first 1,000 days.

IODINE METABOLISM AND THE THYROID HORMONES

The only known function of iodine in the human body is in the synthesis of thyroid hormones, of which it is a key component. This makes iodine an essential nutrient. The thyroid hormones T4 and T3 are hereafter collectively referred to as "thyroid hormones" (TH). The functions of TH in the human body are many-fold; those directly involved with somatic growth and development are discussed later in the section "Iodine Metabolism and Thyroid Hormones during the First 1,000 Days."

The thyroid prohormone 3,5,3′,5′-L-tetraiodothyronine (thyroxine, T4) and its biologically active counterpart 3,5,3′-L-triiodothyronine (T3) are small, biphenolic compounds produced by the thyroid gland in a process that is rate-limited by iodine availability. Very little T3 is secreted by the thyroid gland itself, as T3 is principally formed in the peripheral tissues by the deiodination of circulating T4 [15]. Iodine comprises 65% and 59% of T4 and T3, respectively [16]. They are structurally identical, with the exception of one less iodine at the 5′ position on the outer ring of T3.

Figure 7.2 provides a visual overview of the following sections of this chapter and will be referred to frequently.

Iodine Uptake from the Gut

After ingestion, iodine uptake into the bloodstream is facilitated by the sodium/iodine (Na/I) symporter (NIS), found on the apical surface of enterocytes in the stomach and duodenum [17], as shown in Figure 7.2. Uptake is autoregulated; with increasing concentrations of iodine in the gut, a regulatory mechanism is initiated that downregulates the genetic expression and, thus, production and activity of the NIS [17]. Iodine circulates in the blood in three forms: (1) inorganic iodide; (2) as TH bound to carrier proteins; and (3) to a very small extent, as part of free TH [15]. Thyroid hormones are released by the thyroid gland, which is a highly vascularized organ with a unique structure of thyroid cells surrounding a colloid. The main constituent of the colloid is thyroglobulin (Tg), a thyroid-specific, large molecular weight glycoprotein that provides the structure for thyroid hormone synthesis and iodine storage [5].

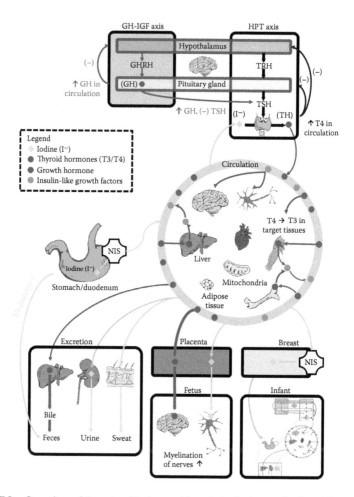

FIGURE 7.2 Overview of the role of iodine and its role in the human body, with a particular focus on the first 1,000 days. Iodine taken up via the stomach/duodenum is used for the production of TH by the thyroid gland. TH are secreted into the blood circulation, where they join free circulating iodine. TH act on almost all cells and tissues in the body to influence cellular metabolism and growth mechanisms as described later in the chapter. They do this in conjunction with GH and IGFs, which promote growth and cell survival for almost every cell in the body. During pregnancy, TH and free circulating iodine move from the maternal circulation through the placenta to the fetus where they promote growth and development, in particular an accelerated myelination of nerves in the brain and central nervous system. After birth, free circulating iodine, provided by the mother via breast milk, is critical for the continued development of neonate and infant. Any superfluous TH and free circulating iodine not removed by the placenta or mammary gland are eliminated via the liver or kidney respectively. Some free iodine is also lost in sweat. Feedback mechanisms maintain homeostasis in blood hormone levels. GH, growth hormone; GHRH, growth hormone releasing hormone; HPT-axis, hypothalamus-pituitary-thyroid axis; I⁻, Iodine; IGFs, insulin-like growth factors; NIS, sodium/iodine (Na/I) symporter; T3, 3,5,3′-L-triiodothyronine; T4, thyroid pro-hormone 3,5,3′,5′-L-tetraiodothyronine (thyroxine); TH, thyroid hormones; TRHs, thyroid-releasing hormones; TSHs, thyroid-stimulating hormones. (Reproduced with permission from Sabine Douxchamps.)

METABOLISM OF IODINE: THE FORMATION
AND CELLULAR UPTAKE OF THYROID HORMONES

Iodine metabolism and TH synthesis are regulated via intricate interactions between the hypothalamus, pituitary, and thyroid glands, which together form the hypothalamus-pituitary-thyroid axis (HPT) (see Figure 7.2).

The synthesis of TH starts in the hypothalamus, where thyrotropin-releasing hormones (TRHs) stimulate both the synthesis and release of thyroid-stimulating hormones (TSHs). TSHs have the same α-subunit structure as other glycoprotein hormones synthesized by the pituitary. Its β-subunit provides receptor-binding specificity and binds to TSH receptors at the thyroid cell surface. This initiates a cascade of reactions in the thyroid, resulting in the synthesis of TH, which are subsequently released into the bloodstream. Upon release into the circulation, TH are bound noncovalently to carrier proteins, mainly to thyroxine-binding globulin (75%) but also albumin (10%) and transthyretin (15%) [16,18]. TH are taken up from the blood into target tissues by active transport to a concentration of approximately 10 times that of the circulation [15]. Circulating TH exert a negative feedback effect to control the release of TSH from the pituitary gland and on the activity of the TRH-stimulating neurons in the hypothalamus. In this way, similar to the autoregulation of iodine uptake, TH regulate their own synthesis (Figure 7.2).

Once inside the cell, TH bind to a nuclear TH receptor to elicit a response. TH receptors are found throughout the body, including the liver, kidney, heart, skeletal muscle, brain, pituitary gland, adipose tissue [16], and bone [19], but not in the adult brain, spleen, testes, uterus, or thyroid gland itself. Additionally, TH can act via nongenomic mechanisms [20], or other indirect mechanisms through their influence on other endocrine systems such as the growth hormone and insulin-like growth factor axis (IGF-axis; see Figure 7.2).

IODINE EXCRETION

Circulating inorganic iodide is removed from the bloodstream for excretion by the kidney [16]. Renal uptake occurs by passive diffusion and is relatively constant at 85% to 90% of daily iodine intake under conditions of sufficiency [21]. There is no mechanism by which the body can reduce renal excretion to retain iodine [21]. Iodine circulating as TH is excreted hepatically through conjugation via sulfotransferase or glucuronyl-transferase. The conjugates are eliminated in the bile, and the iodine is excreted in feces along with any other iodine not absorbed from the gut [15,22]. Iodine is also lost in sweat, which may be an important factor to consider in hot environments, since these losses can contribute to a depletion of iodine stores [23]. This may be particularly pronounced in infants due to the high surface-area-to-body-mass ratio (see Figure 7.2).

IODINE METABOLISM AND THYROID HORMONES
DURING THE FIRST 1,000 DAYS

In addition to renal and hepatic iodine excretion, both the placenta and mammary gland remove iodine and TH from the mother for the support of growth and development of the fetus and infant, as shown in Figure 7.2. In particular, the transfer of maternal TH is important in the early stages of pregnancy until the fetal thyroid is functional and can make use of the transplacental transfer of iodine, which is the only source of iodine for the fetus during gestation, for fetal TH production. Similarly, during exclusive breastfeeding as recommended by the World Health Organization (WHO) until 6 months of age [24], the iodine in breast milk must sustain the needs of the infant (Figure 7.2), despite the wide variations in breast milk iodine concentration corresponding to the mother's intake [25].

Pregnancy and the Fetus

Pregnancy induces several major changes in thyroid function and iodine metabolism. Maternal requirements are increased in pregnancy due to an increased iodine demand, and a higher than usual iodine clearance rate, as described later. The concentration of TH *in utero* regulates fetal growth, development, and viability via a number of factors, as discussed later. Formation of the fetal thyroid does not occur until about week 12 of gestation, and it is not capable of iodine organification until around week 20 [26]. At this point, the fetal thyroid can produce and secrete TH under the control of the HPT axis and, although it is fully functional at birth [27], during gestation the fetus is reliant on maternal TH and, later in pregnancy, iodine from the maternal circulation. At the start of gestation, therefore, maternal T4 crosses the placenta in small amounts [27]. To reflect this, in the first trimester maternal T4 production is increased by about 50% to ensure that the fetus has adequate T4 for local deiodination to the active TH, T3, for correct cerebral development [28]. An increase in circulating estrogen inhibits the breakdown of thyroid-binding globulin, maintaining higher levels of total circulating TH [26,29]. Transiently during pregnancy, there is a trend toward a reduction in free circulating TH, which stimulates a rise in TSH that in turn stimulates the synthesis of TH to maintain homeostasis in levels of unbound TH in the bloodstream [26,30]. Additional iodine is needed for the increase in TH synthesis; to compensate for this, iodine uptake and the use of maternal iodine stores are increased. Additionally, human chorionic gonadotropin, which shares the same α subunit as TSH and is produced in the first days of pregnancy, can also bind to TSH receptors on the maternal thyroid cells and stimulate TH synthesis [30]. Once the fetal thyroid function is established and is capable of organification of iodine from week 20, fetal iodine supply is met entirely from maternal intake [30].

From early pregnancy, there is a 30% to 50% increase in maternal glomerular filtration rate and a corresponding increase in renal blood flow, leading to a greater loss of iodine that persists until the end of the pregnancy [26,30]. To compensate, the thyroid gland increases iodine uptake [31]. In iodine-replete regions, women will typically have between 10 and 20 mg of iodine stored in their thyroid gland, and if iodine intake remains sufficient during gestation, the increased demands can be met [26]. However, if the mother's own thyroidal iodine stores are depleted at the start of gestation and her intake during pregnancy does not compensate for the higher need, then the risks of the effects of iodine deficiency on both mother and child are increased [31].

THE NEONATE AND THE INFANT

Though infants are born with a functioning thyroid gland, they have only scant iodine stores at birth. The average iodine content of the thyroid gland of a neonate is 50 to 100 μg [32], compared to 15 to 20 mg in iodine-sufficient adults [16]. Yet, in terms of iodine requirements per kilogram body weight, infants have the highest relative requirements for iodine of any life stage group [33]. There are significant changes in thyroid physiology and circulating TH concentrations after birth. There is a distinctly high turnover of T4 in infants relative to that of adults: Turnover is estimated to be about 5 to 6 μg/kg/day for infants under 3 years, compared to 1.5 μg/kg/day in adults [34].

   In the neonate and young infant, excluding infants who receive formula milk (which is usually iodized), the main source of external iodine is breast milk. Breastfeeding is recommended until 2 years of age [24]. The milk must therefore contain adequate levels of iodine to ensure continued growth and development until the infant can obtain adequate iodine from food sources and iodized salt. A number of factors can influence breast milk iodine concentration: maternal iodine status, recent maternal iodine intake, duration of lactation, and maternal fluid intake [35]. In iodine-sufficient populations where the mother has had adequate intakes of iodine during pregnancy, breast milk iodine concentrations are considered adequate to meet the needs of neonates and young infants [29]. The mammary gland is able to concentrate iodine at levels 20 to 50 times higher than in plasma [36] due to the upregulated expression of NIS in breast alveoli during lactation [37]. About 20% of iodine in breast milk is present as organic iodine, including a small amount of TH, which are considered insufficient for neonatal and infant TH requirements, particularly because, via oral ingestion, the TH are likely to be destroyed during digestion [38]. The remaining 80% of the iodine present in breast milk is in the form of free iodide from the maternal bloodstream. This should cover the deficit in iodine requirement, provided that the mother's dietary iodine intake is adequate [29], and in which case, breast milk is considered sufficient to fulfill the iodine requirements of the breastfed infant.

IODINE REQUIREMENTS DURING THE FIRST 1,000 DAYS

To reflect the increased nutritional needs to support normal fetal, neonatal, and infant growth and development, the daily recommended iodine intakes for pregnant and lactating women are increased. The iodine requirements for these groups, women of reproductive age, and infants are summarized in Table 7.1 [39,40].

TABLE 7.1

Iodine Intake Recommendations for Individuals during the First 1,000-Day Window

World Health Organization		Institute of Medicine	
Population Group	RNI (µg/day)	Population Group	RDA (µg/day)
Women and children >12 years	150	Women and children ≥14 years	150
Children 0–5 years	90	Infants 0–12 months[a]	110–130
		Children 1–8 years	90
Pregnant	250	Pregnant	220
Lactating	250	Lactating	290

Source: Data from Institute of Medicine, Academy of Sciences, USA, *Dietary reference intakes for vitamin A, vitamin K, arsenic, boron, chromium, copper, iodine, iron, manganese, molybdenum, nickel, silicon, vanadium and zinc*, Washington DC: National Academy Press, 2001; and World Health Organization, United Nations Children's Fund, International Council for the Control of Iodine Deficiency Disorders, *Assessment of iodine deficiency disorders and monitoring their elimination: A guide for programme managers*, 3rd ed., Geneva: World Health Organization, 2007.

Notes: RNI, Recommended Nutrient Intake; RDA, Recommended Dietary Allowance.

[a] Adequate intake.

IODINE, THYROID HORMONES, AND GROWTH DURING THE FIRST 1,000 DAYS

An adequate and continued supply of iodine is needed throughout life for the synthesis of TH, yet, as outlined earlier, the first 1,000 days are of particular importance. Iodine supplementation of mothers during pregnancy, and directly to infants, has been shown to improve infant survival [10,41] and, if undertaken before or during pregnancy, maternal iodine supplementation [10,42,43], or an adequate iodine status [44,45], has been positively associated with birth weight [10,42–45] and infant growth at 6 months [46]. Furthermore, recent analyses by Krämer et al. (2016) suggest a positive association between the absence of iodized salt availability at household level and low birth weight [47]. Additionally, they observed significant associations between household iodized salt availability and growth indicators (e.g., stunting, wasting, and underweight); however, the association was only positive for infants over 5 months of age, probably due to breastfeeding practices, although some benefit via maternal milk is proposed [47]. In toddlers, Neumann and Harrison (1994) reported that household availability of iodized salt use in Kenya was related to improvements in height [48] and, in toddlers and older infants, the use of iodized salt has been associated with increased weight-for-age z-scores and mid-upper-arm circumference in some countries in Asia [49].

During fetal development and the neonatal period, TH are responsible for a multitude of effects that include accelerated myelination in the brain, and improved central nervous system cell migration, differentiation, and maturation [50,51] (see Figure 7.2). TH also promote the growth and maturation of the peripheral tissues and skeleton, and they raise the basal metabolic rate [16], which provides energy for growth (see Figure 7.2). Indeed, TH may be the best surrogate biochemical markers for healthy fetal development [28,52].

The actions of TH in target tissues are mediated by thyroid hormone receptors that regulate the transcription of target genes, which can induce pathways that stimulate or inhibit protein synthesis [16]. However, the numerous effects of TH integral to human growth are not limited to TH and TH receptor interactions: Nongenomic functions of TH have also been identified. Such effects do not require gene transcription or protein synthesis, and can have a rapid onset [20]. Furthermore, and importantly, TH can also indirectly affect growth via other hormonal axes, the most consequent of which involves human growth hormone (GH). Both TH and GH are essential for normal growth and development [53]: Thyroid function and growth mechanisms are intertwined in a complex relationship; neither can be considered without the other, and iodine plays a central and fundamental role.

The impact of TH on growth mechanisms is discussed next. A detailed and fully comprehensive discussion on growth mechanisms above those directly implicated with thyroid function is beyond the scope herein. For further in-depth reading, additional publications are suggested at the end of this chapter.

EFFECTS OF THYROID HORMONES ON THE REPRODUCTIVE SYSTEM

Although not directly a growth mechanism, the ability to reproduce successfully is the primary determinant of the viability of a new life. Normal reproductive physiology in both women and men is dependent on having normal levels of TH, and evidence points to the association between reproductive complications or failure and TH levels. This association is mainly based upon the interrelationship between the HPT axis and the hypothalamic-pituitary-gonadal axis, which controls the release of sex steroid hormones, as both axes influence each other.

Abnormal amounts of TH can cause problems with sperm morphology and motility and induce erectile dysfunction [54]. In women, reproductive problems due to incorrect TH levels may impact menstruation, oocyte quality, and endometrial thickness [54]. In both sexes, TH abnormalities cause changes in sex hormone-binding globulin and sex steroids.

EFFECTS OF THYROID HORMONES ON MUSCLE, CELLULAR ENERGY TURNOVER, AND GLUCOSE METABOLISM

Skeletal muscle is a principal target of TH signaling, and TH regulate the expression of a broad range of genes with key roles in skeletal muscle development, homeostasis, function, and metabolism [55]. TH transporter proteins and deiodination enzymes that are required for TH binding and the conversion of the inactive T4 to the active metabolite T3 are expressed in skeletal muscle tissue [55], and provide the means to control TH uptake and activation. Starting with early embryonic development of the trunk and limbs, the development of fetal skeletal muscle is dependent upon TH, a process that continues postnatally with the transition from fetal muscle fiber phenotypes to adult phenotypes [55]. Furthermore, muscle is one of the major tissues involved in glucose uptake, and TH can control glucose uptake both from the gut and also by skeletal muscle, thereby influencing the overall glucose homeostasis of the body. This is particularly important since the fetus is highly reliant upon glucose, not only for energy but also as a precursor for biochemical reactions promoting tissue growth.

TH are unique in their ability to affect the resting metabolic rate [20], primarily through their actions on skeletal muscle both at rest and while active. The sodium/potassium (Na^+/K^+) adenosine triphosphatase (ATP) pump, responsible for the maintenance of the resting cellular membrane potential, is a direct target gene for T3. The ability of TH to regulate energy utilization is closely linked to effects on the function of mitochondria, which provide about 90% of intracellular energy in the form of ATP [56], and actions can also be via nongenomic influences of TH directly on mitochondria [20]. TH can also increase the number of mitochondria in a cell [57], thereby increasing the capacity to generate ATP. These factors influence fetal growth and metabolism, and hypothyroid fetuses will obtain less ATP, and therefore have less energy available for growth of nonessential tissues [27].

Last, TH have an important role to play in the survival of the neonate at birth. After delivery, the neonate must expend ATP to maintain the extrauterine body temperature, which requires the generation of more heat than while in the protective environment of the uterus. Activation of this thermogenesis in brown adipose tissue is reliant on T3 [27,55].

EFFECTS OF THYROID HORMONES ON THE CARDIOVASCULAR SYSTEM

TH are critical regulators of cardiac development during fetal and postnatal life. Cardiac functions, such as heart rate, cardiac output, and contraction force, are linked to TH. In the fetus, TH are responsible for the maturation of the cardiovascular system adaptation of the heart at birth, when the ventricles switch from pumping in parallel to pumping in series [27].

EFFECTS OF THYROID HORMONES ON SKELETAL BONE DEVELOPMENT AND MAINTENANCE

The dependence of bone maturation, growth, and development on thyroid function is well recognized [19,58], and the association between myxedematous cretinism and short stature is confirmed [59]. A euthyroid state is essential for normal skeletal development. Thyroid hormone receptors (THRs) have been identified in bone cells (osteoblasts, osteoclasts) and cartilage cells (chondrocytes) [19,58], and TH have been shown to accelerate osteoblastic differentiation [60]. In response to TH, bone cells stop proliferating and develop differentiated functions including production of growth factors, cytokines, prostaglandins, and structural proteins [60]. Osteoblasts are stimulated directly by T3 and, indirectly, via the action of

T3 on growth factors, such as insulin-like growth factor-1 (see next section) [61]. Furthermore, TSH receptors, expressed predominantly on thyroid cells, have also been identified in bone, suggesting that TSH may have a direct effect on bone and cartilage itself [62].

EFFECTS OF THYROID HORMONES ON GROWTH HORMONES AND INSULIN-LIKE GROWTH FACTORS

The human growth hormone (GH) is the most abundant hormone secreted from the anterior pituitary. It is secreted in response to a stimulus on the anterior pituitary by growth hormone releasing hormone, which itself is secreted from the hypothalamus. GH is essential for normal growth and development, and exerts its effects on almost all tissues in the body. TH promotes GH synthesis and secretion from the pituitary, and has a permissive effect on the anabolic and metabolic effects of GH. In turn, GH also affects TH activity: GH depresses the secretion of TSH from the pituitary, which in turn will dampen the cascade of reactions initiated by TSH to produce TH.

The main function of GH is to promote the synthesis and secretion of IGFs, which mediate the effects of GH. IGFs cause cell growth and multiplication by increasing the uptake of amino acids into the cell, thereby promoting protein synthesis. GH acts on muscle, cartilage, bone, and other tissues and, indirectly, on the liver, to promote IGF secretion. IGFs circulate in the plasma in complexes with structurally related binding proteins, called IGF-binding proteins. IGF-binding protein-3 is the most common and has the highest binding affinity for IGF-1, binding approximately 95% of the growth factor [63].

Via IGFs, GH promotes accelerated protein growth, an increase in cell size, an increase in cell mitosis, and a corresponding increase in cell number. It promotes the specific differentiation of certain types of cells, for example, bone growth cells and early muscle cells. In infants and children, in whom the bone has not yet reached its adult length, GH increases the growth rate of the skeleton and skeletal muscles.

GH stimulates the increased deposition of proteins around the body, yet the most visible effect is on somatic growth, the stimulation of growth of the skeletal frame. This impact on skeletal growth occurs as a result of multiple effects on bone, including (1) an increased deposition of protein by chondrocytic and osteogenic cells that cause bone growth; (2) an increased rate of reproduction of these cells; and (3) the specific effect of converting chondrocytes into osteogenic cells, thus causing specific deposition of new bone.

Indeed, the growth-promoting actions of GH and IGF-1 are critical for growth during early life: IGFs are widely expressed in fetal tissues and have major influences on fetal growth [27,53], and in neonates and infants, where a deficiency of GH is established by 6 months of age, growth failure may result in stunted growth of up to 3 or 4 standard deviations below the mean [53].

An adequate iodine status and euthyroid state can thereby promote growth, through the promotion of the secretion of GH. The secretion of GH, IGF-1, and IGFBP-3 is dependent on thyroid function, both directly and indirectly via the effects of TH on pituitary secretion [63].

CONCLUSION

Iodine is an essential micronutrient for optimal health, being the principal component of thyroid hormones. Thyroid hormones are implicated in nearly all cells in the body, and control a vast array of biochemical reactions. During the first 1,000 days, iodine and thyroid hormones are indispensable for the viability of the developing fetus and for perinatal survival, including brain formation and cognitive development, and thereafter for the growth and ongoing development of the child.

Whereas the biological mechanisms are well characterized for iodine to have a clear impact on growth processes and growth outcomes, measurable effects of thyroid hormones on growth are principally seen in situations of severe or moderate deficiency. In situations of mild iodine deficiency, where the body's regulatory mechanisms are better prepared to maintain homeostasis, we may be unlikely to see a measureable impact on growth. That said, there is a distinct lack of literature on the effects of iodine status, or supplementation or fortification strategies, on infant and child growth outcomes [13]. The available data are mainly cross-sectional, which are limited in scope since they make assumptions that current iodine status is reflective of past status, and may not control for socioeconomic confounders [63]. The little evidence from iodine intervention strategies on growth is of low quality: There are few randomized controlled trials, and those conducted have generally been underpowered to study growth outcomes [13,64,65]. Furthermore, growth as an outcome is difficult to measure in single nutrient nutritional studies, since the factors impacting growth are vast and spread beyond the reach of a single nutrient or even nutrition alone, and studies need to be of sufficient duration to detect a measureable difference. There is a clear need for well-designed, long-term controlled trials with adequate sample sizes to appropriately assess the impact of iodine status on growth outcomes.

SUGGESTED ADDITIONAL READING

Bassett JH, Williams GR. Role of thyroid hormones in skeletal development and bone maintenance. *Endocr Rev* 2016; 37(2):135–87.

Cheng SY, Leonard JL, Davis PJ. Molecular aspects of thyroid hormone actions. *Endocr Rev* 2010; 31(2):139–70.

Davis PJ, Goglia F, Leonard JL. Nongenomic actions of thyroid hormone. *Nat Rev Endocrinol* 2016; 12(2):111–21.

Forhead AJ, Fowden AL. Thyroid hormones in fetal growth and prepartum maturation. *J Endocrinol* 2014; 221:R87–103.

REFERENCES

1. Andersson M, Karumbunathan V, Zimmermann MB. Global iodine status in 2011 and trends over the past decade. *J Nutr* 2012; 142:744–50.
2. Pearce EN, Andersson M, Zimmermann MB. Global iodine nutrition: Where do we stand in 2013? *Thyroid* 2013; 23:523–8.
3. Lazarus JH. Iodine status in Europe in 2014. *Eur Thyroid J* 2014; 3:3–6.
4. Zimmermann MB, Gizak M, Abbott K et al. Iodine deficiency in pregnant women in Europe. *Lancet Diabetes Endocrinol* 2015; 3:672–4.

5. Rohner F, Zimmermann M, Jooste P et al. Biomarkers of nutrition for development: Iodine review. *J Nutr* 2014; 144:S1322–42.

6. Zimmermann MB, Boelaert K. Iodine deficiency and thyroid disorders. *Lancet Diabetes Endocrinol* 2015; 3:286–95.

7. Zimmermann MB, Jooste PL, Pandav CS. Iodine-deficiency disorders. *Lancet* 2008; 372:1251–62.

8. Zimmermann MB. Iodine deficiency in pregnancy and the effects of maternal iodine supplementation on the offspring: A review. *Am J Clin Nutr* 2009; 89:S668–72.

9. Glinoer D. Iodine nutrition requirements during pregnancy. *Thyroid* 2006; 16:947–8.

10. Bougma K, Aboud FE, Harding KB et al. Iodine and mental development of children 5 years old and under: A systematic review and meta-analysis. *Nutrients* 2013; 5:1384–1416.

11. Zimmermann MB. The effects of iodine deficiency in pregnancy and infancy. *Paediatr Perinat Epidemiol* 2012; 26(Suppl 1):108–17.

12. Gunnarsdottir I, Dahl L. Iodine intake in human nutrition: A systematic literature review. *Food Nutr Res* 2012; 56.

13. Farebrother J, Naude CE, Nicol L et al. Iodised salt and iodine supplements for prenatal and postnatal growth: A rapid scoping of existing systematic reviews. *Nutr J* 2015; 14:89.

14. Monahan M, Boelaert K, Jolly K et al. Costs and benefits of iodine supplementation for pregnant women in a mildly to moderately iodine-deficient population: A modelling analysis. *Lancet Diabetes Endocrinol* 2015; 3(9):715–22.

15. Zoeller RT, Tan SW, Tyl RW. General background on the hypothalamic-pituitary-thyroid (HPT) axis. *Crit Rev Toxicol* 2007; 37:11–53.

16. Zimmermann MB. Iodine and iodine deficiency disorders. In *Present knowledge in nutrition*, 10th ed., Erdman JWJ, Macdonald IA, Zeisel SH, editors. Ames, IA: John Wiley & Sons, Inc.; 2012.

17. Nicola JP, Basquin C, Portulano C et al. The Na+/I-symporter mediates active iodide uptake in the intestine. *Am J Physiol Cell Physiol* 2009; 296:C654–62.

18. Schussler GC. The thyroxine-binding proteins. *Thyroid* 2000; 10:141–9.

19. Bassett JHD, Williams GR. Role of thyroid hormones in skeletal development and bone maintenance. *Endocrine Rev* 2016; 37:135–87.

20. Davis PJ, Goglia F, Leonard JL. Nongenomic actions of thyroid hormone. *Nat Rev Endocrinol* 2016; 12:111–21.

21. Hurrell RF. Bioavailability of iodine. *Eur J Clin Nutr* 1997; 51(Suppl 1):09–12.

22. Wu S, Green WL, Huang W et al. Alternative pathways of thyroid hormone metabolism. *Thyroid* 2005; 15:943–58.

23. European Food Safety Authority (EFSA). EFSA Panel on Dietetic Products Nutrition and Allergies: Scientific opinion on dietary reference values for iodine. *EFSA Journal* 2014, 12.

24. United Nations Children's Fund (UNICEF). *Infant and young child feeding programming guide*. New York: UNICEF; 2011.

25. Dorea JG. Iodine nutrition and breast feeding. *J Trace Elem Med Biol* 2002; 16:207–20.

26. Yarrington C, Pearce EN. Iodine and pregnancy. *J Thyroid Res* 2011; 2011:934104.

27. Forhead AJ, Fowden AL. Thyroid hormones in fetal growth and prepartum maturation. *J Endocrinol* 2014; 221:R87–103.

28. Morreale de Escobar G, Obregon MJ, Escobar del Rey F. Role of thyroid hormone during early brain development. *Eur J Endocrinol* 2004;151(Suppl 3):U25–37.

29. Leung AM, Pearce EN, Braverman LE. Iodine nutrition in pregnancy and lactation. *Endocrinol Metab Clin North Am* 2011; 40:765–77.

30. Glinoer D. The regulation of thyroid function during normal pregnancy: Importance of the iodine nutrition status. *Best Pract Res Clin Endocrinol Metab* 2004; 18:133–52.

31. Glinoer D. Pregnancy and iodine. *Thyroid* 2001; 11:471–81.

32. Delange F, Ermans A. Iodine deficiency. In *Werner and Ingbar's The thyroid: A fundamental and clinical text*, 9th ed., Braverman LE, Utiger RD, editors. Philadelphia PA: Lippincott Williams & Wilkins; 1996.

33. Andersson M, Aeberli I, Wust N et al. The Swiss iodized salt program provides adequate iodine for school children and pregnant women, but weaning infants not receiving iodine-containing complementary foods as well as their mothers are iodine deficient. *J Clin Endocrinol Metab* 2010; 95:5217–24.

34. Brown RS. Disorders of the thyroid gland in infancy, childhood and adolescence. March 21, 2012. http://www.thyroidmanager.org/chapter/disorders-of-the-thyroid-gland-in-infancy-childhood-and-adolescence/ (accessed February 4, 2017).

35. Dold S, Baumgartner J, Zeder C et al. Optimization of a new mass spectrometry method for measurement of breast milk iodine concentrations and an assessment of the effect of analytic method and timing of within-feed sample collection on breast milk iodine concentrations. *Thyroid* 2016; 26:287–95.

36. Azizi F, Smyth P. Breastfeeding and maternal and infant iodine nutrition. *Clin Endocrinol (Oxf)* 2009; 70:803–9.

37. Tazebay UH, Wapnir IL, Levy O et al. The mammary gland iodide transporter is expressed during lactation and in breast cancer. *Nature Medicine* 2000; 6:871–8.

38. Semba RD, Delange F. Iodine in human milk: Perspectives for infant health. *Nutr Rev* 2001; 59:269–78.

39. Institute of Medicine, Academy of Sciences, USA. *Dietary reference intakes for vitamin A, vitamin K, arsenic, boron, chromium, copper, iodine, iron, manganese, molybdenum, nickel, silicon, vanadium and zinc.* Washington DC: National Academy Press; 2001.

40. World Health Organization, United Nations Children's Fund, International Council for the Control of Iodine Deficiency Disorders. *Assessment of iodine deficiency disorders and monitoring their elimination: A guide for programme managers*, 3rd ed. Geneva: World Health Organization; 2007.

41. Cobra C, Muhilal, Rusmil K et al. Infant survival is improved by oral iodine supplementation. *J Nutr* 1997; 127:574–8.

42. Zhuang C, Wang D. A tentative investigation on iodine deficiency of pregnant women. *Shanghai Med J* 1998; 21:198–200.

43. Chaouki ML, Benmiloud M. Prevention of iodine deficiency disorders by oral administration of lipiodol during pregnancy. *Eur J Endocrinol* 1994; 130:547–51.

44. Alvarez-Pedrerol M, Guxens M, Mendez M et al. Iodine levels and thyroid hormones in healthy pregnant women and birth weight of their offspring. *Eur J Endocrinol* 2009; 160:423–9.

45. Rydbeck F, Rahman A, Grander M et al. Maternal urinary iodine concentration up to 1.0 mg/L is positively associated with birth weight, length, and head circumference of male offspring. *J Nutr* 2014; 144:1438–44.

46. Huang Z, Zhu Q, Liu L et al. Effects of iodine deficiency on infants' cognitive development. *Acta Univ Med Tongji* 1995; 24:445–8.

47. Kramer M, Kupka R, Subramanian SV et al. Association between household unavailability of iodized salt and child growth: Evidence from 89 demographic and health surveys. *Am J Clin Nutr* 2016; 104(4):1093–1100.

48. Neumann CG, Harrison GG. Onset and evolution of stunting in infants and children. Examples from the Human Nutrition Collaborative Research Support Program. Kenya and Egypt studies. *Eur J Clin Nutr* 1994; 48(Suppl 1):S90–102.

49. Mason JB, Deitchler M, Gilman A et al. Iodine fortification is related to increased weight-for-age and birthweight in children in Asia. *Food Nutr Bull* 2002; 23:292–308.

50. Bernal J. Thyroid hormones and brain development. In *Hormones, brain and behaviour*, 2nd ed., Pfaff DW, Arnold AP, Fahrbach SE, Etgen AM, Rubin RT, editors. New York: Academic Press; 2009.

51. Morreale de Escobar G, Obregon MJ, del Rey FE. Maternal thyroid hormones early in pregnancy and fetal brain development. *Best Pract Res Clin Endocrinol Metab* 2004; 18:225–48.

52. Zimmermann MB. The role of iodine in human growth and development. *Semin Cell Dev Biol* 2011; 22:645–52.

53. Robson H, Siebler T, Shalet SM et al. Interactions between GH, IGF-1, glucocorticoids and thyroid hormones during skeletal growth. *Pediatr Res* 2002; 52:137–47.

54. Dittrich R, Beckmann MW, Oppelt PG et al. Thyroid hormone receptors and reproduction. *J Reprod Immunol* 2011; 90:58–66.

55. Salvatore D, Simonides WS, Dentice M et al. Thyroid hormones and skeletal muscle: New insights and potential implications. *Nat Rev Endocrinol* 2014; 10:206–14.

56. Weitzel JM, Iwen KA. Coordination of mitochondrial biogenesis by thyroid hormone. *Mol Cell Endocrinol* 2011; 342:1–7.

57. Cheng SY, Leonard JL, Davis PJ. Molecular aspects of thyroid hormone actions. *Endocr Rev* 2010; 31:139–70.

58. Capelo LP, Beber EH, Huang SA et al. Deiodinase-mediated thyroid hormone inactivation minimizes thyroid hormone signaling in the early development of fetal skeleton. *Bone* 2008; 43:921–30.

59. Delange F. The disorders induced by iodine deficiency. *Thyroid* 1994; 4:107–28.

60. Klaushofer K, Varga F, Glantschnig H et al. The regulatory role of thyroid hormones in bone cell growth and differentiation. *J Nutr* 1995; 125:S1996–2003.

61. Galliford TM, Murphy E, Williams AJ et al. Effects of thyroid status on bone metabolism: A primary role for thyroid stimulating hormone or thyroid hormone? *Minerva Endocrinol* 2005; 30:237–46.

62. Endo T, Kobayashi T. Excess TSH causes abnormal skeletal development in young mice with hypothyroidism via suppressive effects on the growth plate. *Am J Physiol Endocrinol Metab* 2013; 305:E660–6.

63. Zimmermann MB, Jooste PL, Mabapa NS et al. Treatment of iodine deficiency in school-age children increases insulin-like growth factor (IGF)-I and IGF binding protein-3 concentrations and improves somatic growth. *J Clin Endocrinol Metab* 2007; 92:437–42.

64. Farebrother J, Naude CE, Nicol L et al. Systematic review of the effects of iodised salt and iodine supplements on prenatal and postnatal growth: Study protocol. *BMJ Open* 2015; 5:e007238.

65. Farebrother J, Naude CE, Nicol L et al. Effects of iodized salt and iodine supplements on prenatal and postnatal growth: A systematic review. (Submitted.)

Section V

Adverse Pregnancy and Birth Outcomes (Pathophysiology and Consequences)

8 Neural Tube Defects
Mechanisms of Folate

Anne M. Molloy

CONTENTS

INTRODUCTION

The year 2016 marks the 25th anniversary of the seminal Medical Research Council (MRC) randomized trial that conclusively demonstrated a marked protection against the recurrence of a neural tube defect (NTD) affected pregnancy among previously affected mothers by the simple prophylactic treatment of folic acid supplements in the periconceptional period [1]. This trial and a contemporaneous Hungarian trial [2] that assessed the efficacy of multivitamins containing folic acid on the first occurrence of an NTD were the culmination of a decade of prior research, primarily led by the work of Professor Smithells in Leeds, United Kingdom. Smithells's work strongly suggested that women entering pregnancy with deficiency or inadequate status of specific micronutrients might be predisposed to these congenital defects, and that multivitamin supplements, taken before and during the first trimester of pregnancy, might be protective. The work resolved well-established epidemiologic data on the prevalence and patterns of occurrence of NTDs suggesting that maternal nutrition was an important factor. In the 25 years since these seminal trials, many countries worldwide have introduced mandatory folic acid food fortification programs to reduce NTD occurrence; other countries have limited their official involvement to public health recommendations and promoting increased awareness of the benefits of folic acid. The fortification programs have had more success, but the worldwide prevalence of NTDs is still documented at more than 300,000 births per year, with great variation in prevalence from region to region and an unclear extent of underreporting [3]. Moreover, efforts to understand the molecular nature of the effect of folic acid in preventing NTDs have had little success, despite intense research.

CLINICAL FEATURES AND EPIDEMIOLOGY

NTDs are the largest group of anomalies of the central nervous system and are among the most significant congenital causes of morbidity and mortality in infants worldwide [3–5]. They occur when the embryonic neural plate fails to fuse and form a closed structure, known as the neural tube. Primary neural tube closure is an early developmental event that produces the prototype of the spinal column and brain in the fetus and is completed in the human embryo between days 21 and 28 of gestation. The fusion of the neural plate is thought to progress at multiple sites simultaneously, and NTDs occur when the process malfunctions in any respect [6–8]. Incomplete closure gives rise to lesions of different types and severity, depending on the portion of the neural tube that fails to close (Figure 8.1). The most severe defect, where there

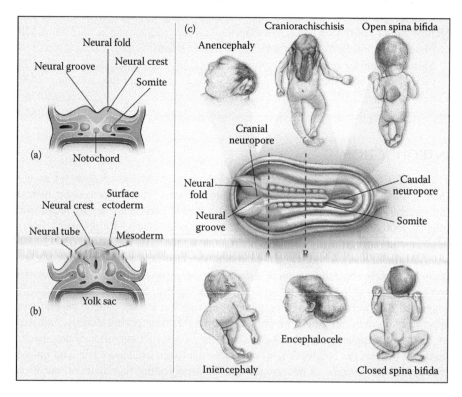

FIGURE 8.1　Neural tube defects: The left panel shows cross-sectional diagrams of embryonic neural tube development, (a) during neural tube closure and (b) after closure. Panel (c) shows the clinical features of the main types of neural tube defects. The central diagram shows a dorsal view of the embryo with dotted lines referring to the cross-sections in the left panel. (Reproduced from Botto LD, Moore CA, Khoury MJ et al., *N Engl J Med* 1999, 341(20):1509–19. With permission.)

is a complete failure to commence fusion, is craniorachischisis, a rare form of NTD occurring in less than 5% of cases.

The two most common forms of lesion are anencephaly (affecting the cranium and involving some 40% of all NTD cases) and spina bifida (affecting the spinal cord and including several open spinal dysraphisms such as myelomeningocele and myelocele), which comprise about 50% of all NTD cases. Other rarer forms, including encephalocele and iniencephalocele, make up the remaining proportion of cases [3]. Severely affected cases may include combinations of these lesions. As might be expected, the position and the extent of the lesion are the main factors in determining the medical management and long-term prognosis. Anencephaly is a lethal disorder. Spina bifida is not intrinsically lethal, but the position and extent of the lesion dictates severity. Most surviving individuals require intensive medical intervention and lifelong care. There is some evidence that at least part of the damage evident at birth in children born with open spina bifida is caused by infiltration of amniotic fluid over the course of the pregnancy. Newly developed methods to repair the lesion *in utero* offer the possibility of improved quality of life for children with myelomeningocele [9]. Several other forms of NTD result from malformations occurring somewhat later in development and are termed secondary or postneurulation defects. They are generally not considered to have the same causative background as primary neural tube defects, and it is not known whether folic acid can protect against their occurrence.

Worldwide, the birth prevalence of NTDs is reported to be between 6 and 60 cases per 10,000 births [5], but there are wide geographic variations in prevalence. In low-income countries, NTDs have been estimated to account for up to 29% of neonatal deaths due to visible congenital abnormalities [10]. A recent systematic study found prevalence rates between 0.3 and 199.4 per 10,000 births across different regions, but reported great difficulty in obtaining accurate estimates, particularly in low-income countries [3].

ENVIRONMENTAL AND GENETIC INFLUENCES

It has been accepted for many years that NTDs have a multifactorial etiology, containing both environmental and genetic influences, and probably requiring an interaction of both elements for the defect to occur. The influence of environmental factors is shown in several ways. In addition to variations in the underlying rates across geographic and ethnic groups, changing trends over time have been well documented. Variations with season, social class, and food availability (e.g., increased incidence during periods of economic and wartime nutritional deprivation) have been described [10–12]. Of particular note is a significant decline in the prevalence of NTDs over some 40 years of monitoring in developed countries, most noticeably in areas with historically high rates, and occurring mainly prior to the introduction of food fortification practices [13–15]. Although the reason for this decline is not understood, improved nutrition is probably an important consideration. Better prenatal detection and increased use of screening programs to terminate affected pregnancies have also contributed significantly to the decline in the number of liveborn cases in most developed countries.

The evidence of a genetic component is strongly demonstrated in familial recurrence patterns, which are at least 10 times the general occurrence risk [11,16]. There are also higher rates in females (particularly for anencephaly) and there are well-recognized Mendelian syndromes that feature an NTD as part of the defect [7,11]. The underlying genetic component can also be observed in prevalence differences among some ethnic groups compared with others. For example, data from the United States indicate that Hispanic women have a higher susceptibility than non-Hispanic whites, with black non-Hispanic women having the lowest risk [17]. Studies from other parts of the world also show lower observed rates among African populations and highest in some Chinese provinces [3]. Finally, the importance of genetic factors is clearly demonstrated by the generation of more than 240 laboratory mouse strains exhibiting NTDs [8,18]. Despite these known environmental and genetic influences, it must be said that most NTDs are isolated events.

FOLIC ACID AS A PROTECTIVE AGENT

EVIDENCE FROM INTERVENTION TRIALS

The past 25 years has seen extensive literature on the role of folic acid in the prevention of NTDs, but the possible involvement of micronutrients and particularly folic acid in protection against these devastating defects had been suggested since the 1960s (reviewed by Molloy et al. [19]). Smithells et al. [20] suggested the possibility of a micronutrient deficiency in women who had previously had an NTD-affected pregnancy. In a series of intervention trials conducted in the early 1970s, they tested the hypothesis that preconceptional vitamin supplementation would prevent these defects [21]. The amount of folic acid delivered to women in the treatment group was 360 µg/day in multivitamin supplement tablets (Pregnavite Forte F). The control group consisted of women who chose not to take the multivitamins or who were already pregnant at the time of the intervention. The results were striking and showed highly protective effects in the treatment group, but because the intervention was not randomized, the level of evidence was considered low. There followed a heated debate on the need for a more scientifically rigorous randomized controlled trial [22–24]. This culminated in two seminal trials. The UK MRC trial [1] tested the risk of NTD recurrence in women who had previously delivered an affected infant. Because of the known familial aggregation of the defect, these women were at high risk and one might say that they were implicitly considered to be genetically susceptible. The second trial, conducted in Hungary [2], tested the first occurrence in a pregnancy and therefore was aimed at an unbiased population of women. Both trials demonstrated conclusively that supplements containing folic acid were highly protective in preventing NTDs. Several features of the trials are worth noting. The MRC trial used a high folic acid dose (4 mg/day), and the research design was to investigate the effects of folic acid alone, a combination of other vitamins plus folic acid, a combination of other vitamins without folic acid, and a placebo control. There were a total of 1,195 completed pregnancies in the trial, with 27 recorded NTDs, of which 6 were in the groups that took folic acid and 21 were in the two other groups. The calculated relative risk for taking folic acid was 0.28 (95% CI: 0.12, 0.71) indicating that 72% of recurrences could be prevented by folic acid. An important point guiding future research was that the multivitamin combination, which had no protective effect on its own, included vitamin B_6 and B_2 but did not include vitamin B_{12}. The Hungarian trial had just two treatment arms: one with multivitamins that contained folic acid at a lower dose of 800 µg/day and a multimineral preparation considered to function as a virtual placebo. There were no NTDs from 2,104 pregnancies in the multivitamin-treated group but 6 from 2,052 pregnancies in the multimineral group. The study demonstrated a similarly striking effect to the MRC trial but was notably different in design because it showed that multivitamins containing folic acid could prevent NTDs, but it did not show that folic acid was the active agent. Interestingly vitamin B_{12} was also included in the list of multivitamins in this trial.

In relation to the dose of folic acid that would be protective, several smaller randomized intervention trials [4], a large ($n = 247,831$) nonrandomized trial in China [25], and the trials carried out by Smithells's group in Leeds all demonstrated significant protection with doses as low as 400 µg/day. The efficacy of 400 µg/day was also supported by several observational studies, most of which found a significant protective effect of maternal periconceptional use of vitamin supplements containing between 400 and 800 µg/day folic acid. The totality of evidence after the randomized trials created a dilemma for public health authorities. Since NTDs are rare events in reproduction, most women would be regarded as at low risk for having an NTD-affected pregnancy (see earlier discussion of prevalence rates). A woman's genetic susceptibility becomes evident after she has an NTD-affected pregnancy and she is then regarded as high-risk for having another affected pregnancy. In developing public health guidelines for primary prevention of NTDs, recommending that high-risk women take 4 mg/day folic acid was a straightforward decision for public health authorities. However, experts agreed that the folic acid doses in both the MRC and Hungarian trials were too high to recommend for use in low-risk women. The additional information from the other studies above was considered in the public health guidelines for primary prevention of NTDs in low-risk women. Within 5 years of the randomized trials, guidelines were published by many countries worldwide and by the World Health Organization (WHO) [26–28]. For low-risk women, the guidelines state that all women capable of becoming pregnant (the wording varies in this respect) should consume 400 µg/day folic acid, in addition to their usual dietary intake of folate, which in most developed countries approximates 200 µg/day.

In recent years, there have been calls to recommend higher doses for specific subgroups of women who may have a moderately higher risk of having an NTD-affected pregnancy [29]. These include women who are obese or have diabetes, both of which conditions are associated with approximately double the risk of having an NTD-affected pregnancy compared to the general population of women [30–32]. The intention of these revisions is to improve health and ensure maximum protection against NTDs, but the evidence for changing the original recommendations is low, since it is not clear that these risk factors are folate responsive. It must also be said that no dose-finding studies have ever been carried out in relation to folic acid and NTDs (such studies could not now be ethically approved), and so the minimal requirement for folic acid to prevent NTDs is not known but, as discussed later, is likely to be dependent on the underlying folate status of individual women as they enter pregnancy.

Primary Prevention Pitfalls and Opportunities

The goal of primary prevention is to prevent the development of an NTD *in utero*, either as a first occurrence or as a recurrence. In the case of NTDs, although the finding that folic acid was protective was hailed as an important medical advance, a major obstacle to successful primary prevention has been the very early timing of neural tube closure in pregnancy, when many women are unaware that they are pregnant. Thus for optimum efficacy, folic acid supplements need to be taken prior to becoming pregnant. Even after 25 years of knowledge by public health authorities, campaigns to encourage women to take folic acid supplements remain unsuccessful [33–36]. Moreover, numerous studies show sociodemographic differences with respect to a woman's knowledge and use of folic acid, with those in lower socioeconomic groups, who were historically known to be most at risk of having an NTD-affected pregnancy, being least likely to know the benefits of folic acid supplementation or take folic acid supplements prior to becoming pregnant.

Because many women do not plan their pregnancies and do not understand the benefit of periconceptional folic acid supplement use, fortification of flour or grain products that are consumed by all sectors of the population was initially established in the United States and Canada, and is now mandated in approximately 80 countries worldwide. These programs of mandatory fortification have been successful in reducing the prevalence of NTDs [17,37]. Fortification is not mandated in most countries in Europe, Africa, or Asia [27,38], but a number of countries allow voluntary food fortification with folic acid, including many European states and New Zealand. Although this has been effective in raising the general folate status of the populations involved, as documented in several studies across Europe [39,40], the lack of reduction in the prevalence of NTDs across Europe over the past 2 decades strongly suggests that such commerce-driven strategies are not adequately effective in improving the folate status of women of child-bearing age [41]. A recent systematic review concluded that spina bifida is significantly more common in areas of the world without mandatory fortification, regardless of population type [38].

At this point, after nearly 20 years of mandatory fortification in the United States and Canada, the decline in NTD rates in the United States has leveled off at a total prevalence for anencephaly and spina bifida of approximately 7 per 10,000 births [17]. Other observational data from several population-based studies give some indication there is a limit to the reduction achievable by folic acid fortification [42,43] and that the threshold of NTD occurrence below which folic acid is not effective is in the region of 6 or 7 per 10,000 pregnancies [37,43]. This is in line with data from the MRC trial, indicating that not all NTDs are folic acid responsive.

MECHANISM OF FOLIC ACID PROTECTIVE EFFECT

Maternal Blood Folate Status and Neural Tube Defects

In parallel with the intervention studies carried out over some 20 years from 1976, other biological evidence was also accumulating that implicated low maternal folate status as an important factor. Several authors reported lower total folate concentrations in serum or red blood cells of women during or after an NTD-affected pregnancy, although other studies showed no effect, and a limitation of all was low sample size, as expected since an NTD-affected pregnancy is a rare event. For this reason, Kirke et al. in Dublin, Ireland, a traditionally high-risk area, collected blood samples at an average gestation of 15 weeks from over 56,000 women over a 4-year period to assemble a large enough sample of bloods from women who were undergoing an NTD-affected pregnancy (reviewed by Molloy et al. [19]). Then, in a nested case control study from this large pregnancy cohort, consisting of 84 women who were carrying NTD-affected fetuses and 247 nonaffected women, they found significantly lower concentrations of plasma and red blood cell (RBC) folate in NTD-affected women [44]. In a follow-up risk analysis of the data, the group reported an approximate eightfold higher risk for NTDs among women with early pregnancy RBC folate concentrations below 340 nmol/l, compared with those who had RBC folate above 906 nmol/l [45]. Although it has never been possible to replicate this sample size with RBC folate measurements in early pregnancy, the hypothesis was recently tested in an *in silico* Bayesian analysis by amalgamating data from two large Chinese population studies; one, a folic acid dose response study in which RBC folate was measured after different folic acid supplement interventions [46] and the other a folic acid intervention trial in which the prevalence of NTDs was monitored in an ethnically identical population of women [25]. Amazingly, the Bayesian calculations resulted in risks that were almost identical to the Dublin study [47]. This strengthened the evidence that women should aim to enter pregnancy with an RBC folate concentration above 906 nmol/L in order to be optimally protected against an NTD. This work has recently led to new WHO guidelines on optimal folate status for protection against NTDs [48]. Two points are noteworthy. First, was the observation in both studies of increased risk among women with low folate status but who would not be described as clinically folate deficient. Second, was the presence of a continuous relationship between folate status and risk reduction, apparently even beyond RBC concentrations of 906 nmol/l. The findings argue against a mechanism of simple folate deficiency and support a mechanism whereby an inadequate nutritional status associated an underlying genetic susceptibility leading inadvertently to an NTD, as would be predicted for a condition caused by a low-penetrance gene-nutrient interaction.

Molecular and Genetic Considerations

The complex processes involved during embryonic neural tube closure have been the subjects of ongoing research for decades. It is known that the progression of closure is highly orchestrated at the molecular level and involves multiple processes, including cell proliferation and differentiation plus cell apoptotic events, all of which involve complex genetic expression and regulatory activities [7,8,18]. Despite these advances, there is little understanding of what developmental mechanisms are folate sensitive. The unexplained effect of folic acid on neural tube closure is all the more extraordinary when one considers that other important processes occurring during early embryonic life, such as the development of the lip and palate, appear not to be so unequivocally sensitive to folic acid treatment.

Folic acid is not a biologically active vitamin. It is the stable, commercially available precursor of the B-vitamin folate but it can be readily converted within the enteral cells and liver to the biologically active molecule, tetrahydrofolate (THF). THF plays a central role in one-carbon metabolism of the cell. One-carbon units (more precisely formyl, methylene, and methyl groups) are essential components of nucleotides and an extensive variety of methylated cell components that are critical to cell signaling, gene expression, and other functions that ensure optimum performance of the living organism. THF accepts one-carbon units from small, simple molecules, such as the amino acids serine and glycine and the organic acid formate. These THF derivatives can then donate one-carbon units to other molecules (Figure 8.2). The two most important elements of folate metabolism are as follows. First, THF donates one-carbon groups in the *de novo* synthesis of purines and the pyrimidine, thymidine. This means that folate metabolism is needed for DNA synthesis and is an essential nutritional factor in situations that involve extensive and rapid cellular proliferation, as is the case in early fetal development. Second, the derivative methyl-THF is used to regenerate the amino acid methionine from its precursor homocysteine (Figure 8.2). Of note, the enzyme responsible for this reaction (methionine synthase) requires cobalamin (vitamin B_{12}). In a sequence of enzymatic steps, termed the methylation cycle, the newly synthesized methionine is used to produce S-adenosylmethionine, the universal donor of methyl groups for numerous biological reactions, including the methylation of DNA, an important mechanism of gene silencing, and the methylation of histone proteins that confer epigenetic control over gene expression. Transfer of the methyl group to its final destination, results in a molecule of S-adenosylhomocysteine, which is converted to homocysteine and recycled back to methionine. In this way, the provision of methyl groups, derived ultimately from folate metabolism, is a crucial factor involved in the choreography of gene activation and silencing activities required by the developing cell. Such control would be expected to play an important role in the molecular stages through which the developing embryo must progress during the fashioning of the neural tube.

Because folate metabolism is involved in so many fundamental cellular processes, it is easy to see the difficulty in identifying any single reaction or series of steps that might be the key to understanding the mechanism that fails and causes an NTD. Nevertheless, in the wake of the randomized trials, the likely involvement of a gene–folate interaction in the pathogenesis of NTDs prompted researchers to study common genetic variation in candidate genes that are linked to folate metabolism as the most productive way to approach the problem of identifying a folate-responsive mechanism. Such an approach was not trivial, when one considers the dozens of relevant gene products such as membrane receptors, folate transporters, and handling proteins, as well as enzymes involved in the interconversion of THF derivatives and the movement of one-carbon units, and so forth. However, in 1995 an important polymorphism (677C>T) in the folate pathway gene methylenetetrahydrofolate reductase (*MTHFR*) was discovered [49]. MTHFR is an important checkpoint enzyme in the folate metabolic pathway because it catalyzes the essentially irreversible conversion of 5,10-methyleneTHF to 5-methylTHF and thereby channels one-carbon units away from purine and pyrimidine synthesis and into the provision of methyl groups for S-adenosylmethionine mediated methylation reactions (Figure 8.2). Individuals who

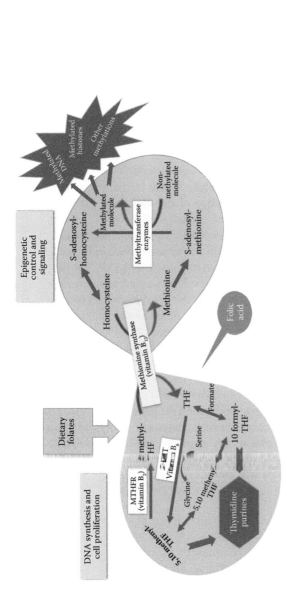

FIGURE 8.2 The main features of folate metabolism in two main cyclic pathways: A DNA synthesis cycle and a methylation cycle. The parent biologically active vitamin is tetrahydrofolate (THF). Dietary folates usually enter cells as the 5-methyl derivative of THF (5-methylTHF). The methyl group is transferred to homocysteine in the reaction catalyzed by the vitamin B₁₂-dependent enzyme, methionine synthase, producing methionine. The other product is THF. THF can accept one-carbon units from formate, thereby producing 10-formylTHF, or from serine through the enzyme serine hydroxymethyltransferase (SHMT) to produce 5,10-methyleneTHF. 10-FormylTHF donates one-carbon units to be inserted at the C2 and C8 position in the *de novo* synthesis of purines. 5,10-methyleneTHF is used in the conversion of deoxyuridine monophosphate to deoxythymidine monophosphate, during *de novo* synthesis of pyrimidines. 5,10-methyleneTHF can also be converted to 5-methylTHF via the vitamin B₂ requiring enzyme methylenetetrahydrofolate reductase (MTHFR). This provides a supply of methyl groups from the folate pool to remethylate homocysteine and maintain a functioning methylation cycle. Methionine is converted to S-adenosylmethionine, the universal donor of methyl groups for biological methylations. The secondary product is S-adenosylhomocysteine, which is a potent competitive inhibitor of most methyltransferase reactions and must be recycled back to methionine through homocysteine in order to maintain the system. Folic acid is not biologically active but can be directly reduced to THF.

have the rarer TT genotype tend to have lower serum and RBC folate concentrations, suggesting an increased requirement for folate [50]. The polymorphism was the first to be examined in relation to NTDs and remains more thoroughly explored than other folate enzymes in relation to NTDs, showing an approximate twofold risk in mothers and in cases themselves, calculated to be about 12% of the population attributable risk [51,52]. This relatively moderate effect emphasized the need to find other factors. Candidate polymorphism analysis very quickly led to the broader task of candidate gene analysis, where all of the common variability in a gene is evaluated for association with risk. However, despite a considerable body of research on candidate genes in folate pathways, other than the MTHFR gene, which confers at most a moderately increased risk, there has been little success in uncovering a genetic cause for NTDs that is based on common genetic variability in folate candidate genes [19,53,54].

A major limitation of candidate gene analysis is that the list of candidates that can be produced is limited by our knowledge of the molecular pathways. More unfocused methods, such as genome-wide associated studies (GWAS) apply a broad unbiased approach to searching for common genetic variation. Such an analysis has not yet been applied to the study of NTDs. Other approaches, such as exome sequencing, aim to find rare and *de novo* mutations that may confer risk [55]. The lack of success in discovering risk factors at the genetic level has also prompted researchers to consider the possibility that changes in epigenetic patterns rather than genetic factors may be an important feature of neural tube closure. Since folate metabolism is essential for DNA and histone methylations, it might be expected that low maternal folate status could cause an inadequate supply of methyl groups leading to changes at the chromatin level and altered transcriptional activity during mammalian development [56]. A recent study found altered methylation patterns in a study of ten spina bifida cases and six controls [57]. These types of studies are still in their infancy but give an example of the directions being taken to understand the basis of folic acid responsive NTDs. The tools and expertise to carry out such analysis of the human genome and methylome are beginning to be more widely available, and it will soon be possible to explore a broader architecture of the genome and epigenome in relation to these defects. Whether such explorations will provide a breakthrough in our understanding of the genetic factors causing susceptibility to NTDs is unknown.

OTHER INTERACTING NUTRIENTS

Finally, it is important to include some comment on the role of vitamin B_{12}, because of the requirement for cobalamin in providing folate derived methyl groups for methionine synthesis. There are no trials demonstrating that vitamin B_{12} can prevent NTDs, but several studies have found evidence of low or altered vitamin B_{12} status in maternal blood or amniotic fluid during and after an NTD-affected pregnancy [58,59]. Furthermore, genes in vitamin B_{12} metabolizing proteins have emerged as significant risk factors for NTDs [60]. No evidence of a role for vitamin B_6 has been found, although several steps in folate metabolism require pyridoxal phosphate. Smithells's group reported low blood riboflavin and vitamin C concentrations in women during NTD-affected pregnancies, but as noted earlier, none of these had an effect in the randomized trials. Anecdotally, the Hungarian trial, which included vitamin B_{12} in the treatment group, reported no cases of NTD in that group. This could be due to chance, but it is also possible that folic acid plus vitamin B_{12} reduced the risk to an even lower level than that seen in the MRC study. It is also possible that the importance of this vitamin to NTD prevention may be much greater in low-income countries where vitamin B_{12} deficiency is widespread.

CONCLUSION

There are many unsolved questions in relation to the mechanism of folic acid protection against NTDs, despite a quarter century of intense research. The multifactorial nature of the defect is probably the biggest obstacle to research efforts. Intervention trials indicate that folate is the most important environmental factor, but the contribution of the interacting nutrient vitamin B_{12} cannot be ignored. It is important to note that a proportion of NTDs are not responsive to folic acid. Certain toxins such as fumonisin, hyperthermia during early pregnancy, use of antifolate medicines such as some antiepileptic drugs, obesity, and diabetes are all maternal risk factors for having an NTD-affected birth. It is not clear if increasing folic acid supplement use can moderate the increased risk that is associated with conditions such as obesity and diabetes. The ongoing surge of progress in genomic and epigenomic research will soon give us much deeper knowledge of gene structure and control, but until this is linked to the equally complex tapestry of our nutritional environment, a true understanding of the nature of NTDs is unlikely to emerge.

REFERENCES

1. Medical Research Council (MRC). Prevention of neural tube defects: Results of the medical research council vitamin study. MRC Vitamin Study Research Group. *Lancet* 1991; 338(8760):131–7.
2. Czeizel AE, Dudas I. Prevention of the first occurrence of neural-tube defects by periconceptional vitamin supplementation. *N Engl J Med* 1992; 327(26):1832–5.
3. Zaganjor I, Sekkarie A, Tsang BL et al. Describing the prevalence of neural tube defects worldwide: A systematic literature review. *PLoS One* 2016; 11(4):e0151586.
4. Botto LD, Moore CA, Khoury MJ et al. Neural-tube defects. *N Engl J Med* 1999; 341(20):1509–19.
5. Christianson AL, Howson CP, Modell B. *Global report on birth defects: The hidden toll of dying and disabled children.* White Plains, NY: March of Dimes Birth Defects Foundation; 2006.
6. Greene ND, Copp AJ. Neural tube defects. *Annu Rev Neurosci* 2014; 37:221–42.
7. Au KS, Ashley-Koch A, Northrup H. Epidemiologic and genetic aspects of spina bifida and other neural tube defects. *Dev Disabil Res Rev* 2010; 16(1):6–15.
8. Harris MJ, Juriloff DM. An update to the list of mouse mutants with neural tube closure defects and advances toward a complete genetic perspective of neural tube closure. *Birth Defects Res A Clin Mol Teratol* 2010; 88(8):653–69.
9. Adzick NS, Thom EA, Spong CY et al. A randomized trial of prenatal versus postnatal repair of myelomeningocele. *N Engl J Med* 2011; 364(11):993–1004.
10. Blencowe H, Cousens S, Modell B et al. Folic acid to reduce neonatal mortality from neural tube disorders. *Int J Epidemiol* 2010; 39(Suppl 1):i110–21.

11. Elwood JM, Little J, Elwood JH. *Epidemiology and control of neural tube defects.* Oxford: Oxford University Press; 1992.
12. Neugebauer R, Hoek HW, Susser E. Prenatal exposure to wartime famine and development of antisocial personality disorder in early adulthood. *JAMA* 1999; 282(5):455–62.
13. Murphy M, Seagroatt V, Hey K et al. Neural tube defects 1974–94: Down but not out. *Arch Dis Child Fetal Neonatal Ed* 1996; 75(2):F133–4.
14. Besser LM, Williams LJ, Cragan JD. Interpreting changes in the epidemiology of anencephaly and spina bifida following folic acid fortification of the U.S. grain supply in the setting of long-term trends, Atlanta, Georgia, 1968–2003. *Birth Defects Res A Clin Mol Teratol* 2007; 79(11):730–6.
15. Kallen B, Lofkvist E. Time trends of spina bifida in Sweden 1947–81. *J Epidemiol Community Health* 1984; 38(2):103–7.
16. Deak KL, Siegel DG, George TM et al. Further evidence for a maternal genetic effect and a sex-influenced effect contributing to risk for human neural tube defects. *Birth Defects Res A Clin Mol Teratol* 2008; 82(10):662–9.
17. Williams J, Mai CT, Mulinare J et al. Updated estimates of neural tube defects prevented by mandatory folic acid fortification: United States, 1995–2011. *Morb Mortal Wkly Rep* 2015; 64(1):1–5.
18. Copp AJ, Greene ND. Neural tube defects: Disorders of neurulation and related embryonic processes. *Wiley Interdiscip Rev Dev Biol* 2013; 2(2):213–27.
19. Molloy AM, Brody LC, Mills JL et al. The search for genetic polymorphisms in the homocysteine/folate pathway that contribute to the etiology of human neural tube defects. *Birth Defects Res A Clin Mol Teratol* 2009; 85(4):285–94.
20. Smithells RW, Sheppard S, Schorah CJ. Vitamin deficiencies and neural tube defects. *Arch Dis Child* 1976; 51(12):944–50.
21. Smithells RW, Sheppard S, Schorah CJ et al. Possible prevention of neural-tube defects by periconceptional vitamin supplementation. *Lancet* 1980; 1(8164):339–40.
22. Schorah C. Commentary: From controversy and procrastination to primary prevention. *Int J Epidemiol* 2011; 40(5):1156–8.
23. Abel J. Neural tube defects: British vitamin trial starts slowly. *Nature* 1984; 309(5970):661.
24. Wald NJ, Polani PE. Neural-tube defects and vitamins: The need for a randomized clinical trial. *Br J Obstet Gynaecol* 1984; 91(6):516–23.
25. Berry RJ, Li Z, Erickson JD et al. Prevention of neural-tube defects with folic acid in China. China-U.S. Collaborative Project for Neural Tube Defect Prevention. *N Engl J Med* 1999; 341(20):1485–90.
26. Centers for Disease Control. Use of folic acid for prevention of spina bifida and other neural tube defects: 1983–1991. *Morb Mortal Wkly Rep* 1991; 40(30):513–6.
27. Chitayat D, Matsui D, Amitai Y et al. Folic acid supplementation for pregnant women and those planning pregnancy: 2015 update. *J Clin Pharmacol* 2016; 56(2):170–5.
28. Gomes S, Lopes C, Pinto E. Folate and folic acid in the periconceptional period: Recommendations from official health organizations in thirty-six countries worldwide and WHO. *Public Health Nutr* 2016; 19(1):176–89.
29. Wilson RD, Wilson RD, Audibert F et al. Pre-conception folic acid and multivitamin supplementation for the primary and secondary prevention of neural tube defects and other folic acid-sensitive congenital anomalies. *J Obstet Gynaecol Can* 2015; 37(6):534–52.
30. Watkins ML, Rasmussen SA, Honein MA et al. Maternal obesity and risk for birth defects. *Pediatrics* 2003; 111(5 Pt 2):1152–8.
31. Shaw GM, Velie EM, Schaffer D. Risk of neural tube defect-affected pregnancies among obese women. *JAMA* 1996; 275(14):1093–6.

32. Hendricks KA, Nuno OM, Suarez L et al. Effects of hyperinsulinemia and obesity on risk of neural tube defects among Mexican Americans. *Epidemiology* 2001;12(6):630–5.
33. Arth A, Tinker S, Moore C et al. Supplement use and other characteristics among pregnant women with a previous pregnancy affected by a neural tube defect: United States, 1997–2009. *Morb Mortal Wkly Rep* 2015; 64(1):6–9.
34. Nelson CR, Leon JA, Evans J. The relationship between awareness and supplementation: Which Canadian women know about folic acid and how does that translate into use? *Can J Public Health* 2014; 105(1):e40–6.
35. Zeng Z, Yuan P, Wang Y et al. Folic acid awareness and intake among women in areas with high prevalence of neural tube defects in China: A cross-sectional study. *Public Health Nutr* 2011; 14(7):1142–7.
36. Peake JN, Copp AJ, Shawe J. Knowledge and periconceptional use of folic acid for the prevention of neural tube defects in ethnic communities in the United Kingdom: Systematic review and meta-analysis. *Birth Defects Res A Clin Mol Teratol* 2013; 97(7):444–51.
37. De Wals P, Tairou F, Van Allen MI et al. Reduction in neural-tube defects after folic acid fortification in Canada. *N Engl J Med* 2007; 357(2):135–42.
38. Atta CA, Fiest KM, Frolkis AD et al. Global birth prevalence of spina bifida by folic acid fortification status: A systematic review and meta-analysis. *Am J Public Health* 2016; 106(1):e24–34.
39. Hopkins SM, Gibney MJ, Nugent AP et al. Impact of voluntary fortification and supplement use on dietary intakes and biomarker status of folate and vitamin B_{12} in Irish adults. *Am J Clin Nutr* 2015; 101(6):1163–72.
40. Martiniak Y, Heuer T, Hoffmann I. Intake of dietary folate and folic acid in Germany based on different scenarios for food fortification with folic acid. *Eur J Nutr* 2015; 54(7):1045–54.
41. Khoshnood B, Loane M, de Walle H et al. Long-term trends in prevalence of neural tube defects in Europe: Population-based study. *BMJ* 2015; 351:h5949.
42. Mosley BS, Cleves MA, Siega-Riz AM et al. Neural tube defects and maternal folate intake among pregnancies conceived after folic acid fortification in the United States. *Am J Epidemiol* 2009; 169(1):9–17.
43. Heseker HB, Mason JB, Selhub J et al. Not all cases of neural-tube defect can be prevented by increasing the intake of folic acid. *Br J Nutr* 2009; 102(2):173–80.
44. Kirke PN, Molloy AM, Daly LE et al. Maternal plasma folate and vitamin B12 are independent risk factors for neural tube defects. *Q J Med* 1993; 86(11):703–8.
45. Daly LE, Kirke PN, Molloy A et al. Folate levels and neural tube defects: Implications for prevention. *JAMA* 1995; 274(21):1698–702.
46. Hao L, Yang QH, Li Z et al. Folate status and homocysteine response to folic acid doses and withdrawal among young Chinese women in a large-scale randomized double-blind trial. *Am J Clin Nutr* 2008; 88(2):448–57.
47. Crider KS, Devine O, Hao L et al. Population red blood cell folate concentrations for prevention of neural tube defects: Bayesian model. *BMJ* 2014; 349:g4554.
48. World Health Organization (WHO). *WHO guideline: Optimal serum and red blood cell folate concentrations in women of reproductive age for prevention of neural tube defects.* Geneva: World Health Organization; 2015.
49. Frosst P, Blom HJ, Milos R et al. A candidate genetic risk factor for vascular disease: A common mutation in methylenetetrahydrofolate reductase. *Nat Gen* 1995; 10(1):111–3.
50. Molloy AM, Daly S, Mills JL et al. Thermolabile variant of 5,10-methylenetetrahydrofolate reductase associated with low red-cell folates: Implications for folate intake recommendations. *Lancet* 1997; 349(9065):1591–3.

51. Shields DC, Kirke PN, Mills JL et al. The "thermolabile" variant of methylenetetra-hydrofolate reductase and neural tube defects: An evaluation of genetic risk and the relative importance of the genotypes of the embryo and the mother. *Am J Hum Genet* 1999; 64(4):1045–55.

52. Botto LD, Yang Q. 5,10-Methylenetetrahydrofolate reductase gene variants and con-genital anomalies: A HuGE review. *Am J Epidemiol* 2000; 151(9):862–77.

53. Boyles AL, Hammock P, Speer MC. Candidate gene analysis in human neural tube defects. *Am J Med Genet C Semin Med Genet* 2005; 135C(1):9–23.

54. Pangilinan F, Molloy AM, Mills JL et al. Evaluation of common genetic variants in 82 candidate genes as risk factors for neural tube defects. *BMC Med Genet* 2012; 13:62.

55. Krupp DR, Soldano KL, Garrett ME et al. Missing genetic risk in neural tube defects: Can exome sequencing yield an insight? *Birth Defects Res A Clin Mol Teratol* 2014; 100(8):642–6.

56. Smith ZD, Chan MM, Humm KC et al. DNA methylation dynamics of the human pre-implantation embryo. *Nature* 2014; 511(7511):611–5.

57. Rochtus A, Winand R, Laenen G et al. Methylome analysis for spina bifida shows SOX18 hypomethylation as a risk factor with evidence for a complex (epi)genetic inter-play to affect neural tube development. *Clin Epigenetics* 2016; 8:108.

58. Molloy AM, Kirke PN, Troendle JF et al. Maternal vitamin B_{12} status and risk of neural tube defects in a population with high neural tube defect prevalence and no folic acid fortification. *Pediatrics* 2009; 123(3):917–23.

59. Ray JG, Blom HJ. Vitamin B_{12} insufficiency and the risk of fetal neural tube defects. *QJM* 2003; 96(4):289–95.

60. Pangilinan F, Mitchell A, VanderMeer J et al. Transcobalamin II receptor polymor-phisms are associated with increased risk for neural tube defects. *J Med Genet* 2010; 47(10):677–85.

9 Cleft Lip and Palate

Eman Allam, Ahmed Ghoneima,
and Katherine Kula

CONTENTS

NORMAL EMBRYOLOGICAL FACE STRUCTURE DEVELOPMENT

Early in the fourth week intrauterine, five primordial swellings, consisting of neural crest derived mesenchyme, appear around the stomodeum and are mainly responsible for face development (Figure 9.1). The process starts with the appearance of the nasal placodes on the inferior aspect of the frontonasal process. Proliferation of the ectomesenchyme on both sides of each placode results in the formation of the lateral and medial nasal processes, and the formation of the depression representing the primitive nostril, or "nasal pit." As the two maxillary processes enlarge, they fuse medially with the medial nasal processes to form the upper lip. The failure of one or both of the maxillary processes to fuse with the medial nasal processes leads to a unilateral or bilateral cleft of the lip, respectively [1 4]

Development of the palate begins at the end of the fifth week. The primary palate is then formed by the fusion of the medial nasal processes to form the intermaxillary segment. This segment is responsible for the formation of the premaxilla, which is the triangular bony piece that will support the maxillary incisor teeth. The secondary palate is the primordia of the hard and soft palate posterior to the premaxilla, and is formed from the maxillary processes of the first branchial arches [3,4]. Two lateral palatine processes develop from the internal aspects of both maxillary processes, initially in a vertical position on each side of the developing tongue. With the development and growth of the mandible, the tongue moves inferiorly, allowing the palatine processes to move horizontally and grow toward each other. By the end of the eighth week, fusion occurs between the two palatine processes or shelves, and also with the primary palate and nasal septum. The process of palate formation is completed by the twelfth week. Bone develops in the anterior region, forming the

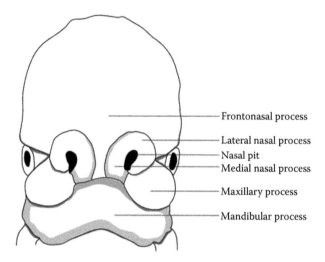

FIGURE 9.1   Structures contributing to the face formation.

hard palate, and the posterior part develops as the muscular soft palate. Defective fusion of the palatine processes results in a cleft palate [1–4].

EPIDEMIOLOGY OF CLEFT LIP AND PALATE (CLP)

Incidence reports estimate about 1 per 2,500 births for cleft lip and palate (CLP), and about 1 per 1,500 births for cleft palate (CP) alone. Cleft lip (CL) and CP occur together in about 45% of cases, with about 30% being only CP and about 25% being isolated CL. CLP occurs more frequently in males, while CP is more predominant in females. Approximately 50% of CL patients also have CP, which is considered to be a secondary effect resulting from the abnormal fusion of facial prominences that precedes palate development [5,6]. The majority of CLP (about 70%) are classified as nonsyndromic, while the rest are considered to be syndromic ones, where cleft-ing occurs as part of a range of chromosomal anomalies, or other teratogenic or Mendelian syndromes. Unilateral clefts are more prevalent than bilateral clefts, and left-sided unilateral clefts are more common than right-sided clefts [6,7].

The prevalence of CLP in general shows large variations among different popu-lations, with high rates reported in Asians, moderate rates among Caucasians, and low rates among African populations. Reports indicate a risk of 14 per 10,000 births in Asians, 10 per 10,000 births in Caucasians, and 4 per 10,000 births in African Americans. In addition to genetic association, the wide variation in reported data was suggested to be due, in part, to methodological problems, such as selection bias and underreporting associated with using different data sources [8–10]. Data from some regions depends on investigating hospital records and birth certificates, rather than large population data registries. Some reports only included live births, whereas others included stillborn and aborted fetuses. Data was sometimes based on iso-lated clefts, whereas other sources included clefts associated with other congenital abnormalities or syndromes. There were also differences in the classification system used in each report. It is thus considered to be difficult to draw definite conclusions from these figures; effective comparisons as well as accurate descriptive epidemiological data are unlikely to be possible with the existence of all these variations [10,11].

CLP ETIOLOGY AND ASSOCIATED RISK FACTORS

CLP is etiologically heterogeneous, with complex genetic and environmental interactions. It is essential to distinguish between isolated nonsyndromic cases and those clefts associated with particular syndromes, as both are considered etiologically distinct. More than 350 specific genetic syndromes are identified as being associated with CLP [12,13]. Advances in segregation analysis, genetic linkage, and association studies, as well as twin studies, have made it possible to identify major genetic factors related to clefts [12–14]. Three classes of genes are suggested to play a role in the susceptibility to clefting: (1) genes involved in the process of palate development, specifically the transforming growth factors alpha and beta (TGFα, TGFβ 2, TGFβ 3); (2) genes identified in transgenic animal models with clefts, such as the Msx1, Msx2, Gabrb3, and Ap-2 genes; and (3) genes involved in certain biological activities that are linked to CLP pathogenesis, such as xenobiotic metabolism (CYP1A1, glutathione S-transferase μ1 [GSTM1], N-acetyltransferase [NAT2]), and nutrient metabolism (methylene tetrahydrofolate reductase [MTHFR], retinoic acid receptor [RAR-α], and folate receptor [FOLR1]). Candidate gene studies on CLP cases associated with Van der Woude (VWS) syndrome have also provided strong evidence for the involvement of the interferon regulatory factor 6 (IRF6) gene in cleft development. VWS is an autosomal dominant disorder that is considered to be the most important model for isolated CLP [15–17].

Several environmental factors have also been implicated in the CLP etiology, such as maternal smoking and alcohol consumption, dietary vitamins and folate deficiency, medications during pregnancy, infections, and other environmental triggers. These environmental components are suggested to be modulators for the genetic elements [18–20]. The most extensively studied teratogen linked to CLP is maternal tobacco and cigarette smoking. Studies have indicated that maternal smoking is a central covariate in clefting, and that it doubles the frequency of cleft development when compared to nonsmoking mothers [19,20]. The genes glutathione S-transferase-θ1 (GSTT1) and nitric oxide synthase 3 (NOS3), which are both major players in metabolic detoxification pathways, were confirmed to be directly linked to maternal smoking and the susceptibility to the development of clefts [3].

Maternal alcohol consumption has also been suggested as a risk factor; however, most reports indicate that evidence linking alcohol to cleft development is yet to be confirmed [21]. Other teratogens that are linked to increased risk of clefts via ingestion during pregnancy include anticonvulsant medications, specifically phenytoin, corticosteroids, thalidomide, and benzodiazepines, or pesticides, such as dioxin. Maternal use of phenytoin has been associated with a tenfold increase in the incidence of CL [21,22]. Some nutritional factors, such as folate, play a critical role in neural tube formation. Folate deficiency was recognized as being potentially responsible for several congenital malformations, including CLP. Several investigators suggested that folate deficiency is associated with an increased risk of CLP development through a common mechanism that interferes with normal embryonic development. Other reports supported the beneficial effects of periconceptional folic acid supplementation in reducing this risk [23,24].

CLASSIFICATION OF CLP

Although many systems have been offered to classify CLP, none of them has been considered to be either comprehensive or universally accepted. The simple numerical method of classification proposed by Veau is widely used, and categorizes clefts into four groups: I, clefts of the soft palate; II, clefts of the soft and hard palate; III, complete unilateral clefts of the lip and palate; and IV, complete bilateral clefts of the lip and palate. More complex classification systems were developed later, but Veau classification has remained popular because it is simple and easy to use [25,26].

Several embryologically based and more comprehensive classification systems and codes have also been suggested in the literature, and later on have been subjected to further modifications and revisions, albeit without general agreement. They mainly divide the groups into clefts of the lip, alveolus, and palate, where the palate is identified as either the primary or secondary palate. Failure to apply a standardized, comparable, and internationally accepted classification has always presented a barrier to collecting accurate epidemiological data, and to efforts to plan proper preventive and public health services for cleft patients [6,27,28].

A CL can be either unilateral or bilateral, and can be isolated or associated with CP. CP cases demonstrate a wide range of severity. The defect can be as minimal as a "bifid uvula," in which the uvula is affected by a median cleft without any clinical consequences, it can involve the soft palate alone, or the hard and soft palates (Figures 9.2 and 9.3). CLP may be associated with other major congenital anomalies, such as malformations of the upper and lower limbs or the vertebral column, mental retardation, or congenital heart disease [29–32]. In addition, clefting occurs in a subset of more than 350 syndromes. Examples of the syndromes and chromosomal abnormalities most frequently associated with CLP include trisomy 13, trisomy 18, Meckel syndrome, Treacher Collins syndrome, VWS, Pierre Robin syndrome, Crouzon syndrome, Apert

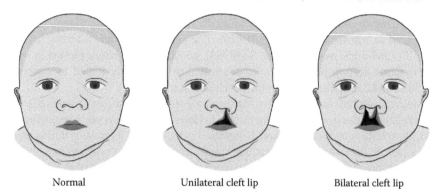

          Normal   Unilateral cleft lip   Bilateral cleft lip

FIGURE 9.2   (From left) A normal lip, unilateral, and bilateral cleft lip.

| Normal | Cleft palate | Unilateral cleft lip and palate | Bilateral cleft lip and palate |

FIGURE 9.3   The ventral view of the palate in "normal," and cleft lip and palate-affected individuals.

syndrome, Stickler syndrome, fetal alcohol syndrome, Turner syndrome, Down syndrome, cri-du-chat syndrome, amniotic band anomalad, velocardiofacial syndrome, Wolf-Hirschhorn syndrome, and Fryns syndrome [29–32].

DIAGNOSIS AND PREVENTION

Advances in molecular technology, and the identification of genes' environmental interactions associated with cleft development, has allowed for hypothesis testing for prevention. Preventive measures include efforts toward avoiding maternal smoking and alcohol exposure, as well as anticonvulsants and other risk-related medications [33]. Genetic counseling is likely to be significant in terms of primary prevention. Reports have indicated that the molecular identification of mutations in certain genes, such as Msx1 and VWS, might deliver valuable information on the expected risk. Maternal supplementation, in particular with folic acid and other important micronutrients, such as zinc and vitamin B_6, is recommended for prevention of clefts and other birth anomalies in newborns [8,28,33].

The prenatal detection of CLP, as well as any other genetic abnormalities, provides parents with a chance to plan adequate counseling and neonatal care. Counseling should provide information on the possible associated risks, anticipated treatment plans and surgical protocols and the overall prognosis of each case [34,35]. Antenatal ultrasound is currently considered to be a routine procedure of prenatal care and is reported to be an efficient diagnostic tool for the detection of CLP. Diagnosis of facial cleft anomalies can be identified as early as the 12th week intrauterine. Abnormal ultrasound findings that indicate possible clefting should be considered to be a forewarning of the possibility of other serious associated defects, which could be ruled out via karyotyping and amniocentesis. Reports have indicated that about 25% of neonates diagnosed with CLP have other associated malformations, syndromes, or trisomies [35–37].

TREATMENT AND MANAGEMENT

CLP is considered to be a debilitating condition associated with significant feeding, hearing, speech, and psychological problems. Affected children suffer from multiple aesthetic and functional impairments. Feeding difficulties due to problems with the oral seal, for example, start very early in life and can greatly affect the nutritional status of the child [38]. Other complications, such as hearing problems and nasal regurgitation, and speech and dental problems, may lead to negative social and psychological burdens for both the child and the parents. Management of a CLP case is challenging and ideally requires a multidisciplinary approach. The process starts from birth and may continue into adulthood, and involves cooperation between different health professionals including, but not limited to, a pediatrician, plastic surgeon, otolaryngologist, pediatric dentist, psychologist, speech pathologist, social worker, and orthodontist [39–41].

Surgical management is generally required, and it involves multiple primary and secondary surgeries, which may start in infancy and continue throughout childhood. Surgeries reconstruct the lip and palate, and improve esthetic appearance and functions, including swallowing, speech, and breathing. The specific nature and timing of the surgical procedures vary considerably, depending on the treatment protocol set by the therapeutic team and the severity of the defect. Primary lip closure is usually indicated early during the first months after birth, followed by a palatoplasty, the procedure used for palatal repair, later in life [42–45]. Autogenous bone grafts, prosthetic appliances, and orthognathic procedures may be required at certain points to ensure better functional and esthetic results according to the treatment plan. Social and psychological care is also indicated, to assist the parents in dealing with the condition. Other forms of treatment such as dental, hearing, and speech therapy, should be provided throughout the whole treatment process. Genetic counseling is highly recommended. Reports have indicated that the risk of clefting, in isolated CLP cases, in a sibling of an affected child is 3% to 5%; this increases to about 10% to 20% if any first-degree relative or other family member is also affected. The risk is significantly higher in syndromic cases [42–45].

CONCLUSION

CLP is a common human birth defect of complex etiology including multiple genetic and environmental factors. Complete understanding of causes and associated risk factors requires extensive research through epidemiological, animal models, and genetics studies. It is only possible for a CLP patient to receive the essential care and support that help ensure better living conditions with the support of a multidisciplinary approach from a dedicated team of professionals from various fields.

REFERENCES

1. Avery JK, Steele PF. Development of the face and palate. In *Essentials of oral histology and embryology: A clinical approach*, 3rd ed. Maryland Heights, MO: Mosby; 2006.
2. Gorlin, RJ, Cohen MM, Hennekam RC. Orofacial clefting syndromes: General aspects. In *Gorlin's syndromes of the head and neck*, Hennekam RCM, Krantz ID, Allanson JE, editors. London: Oxford University Press; 2010.
3. Dixon MJ, Marazita ML, Beaty TH et al. Cleft lip and palate: Understanding genetic and environmental influences. *Nat Rev Genet* 2011; 12(3):167–78.
4. Kirschner RE, LaRossa D. Cleft lip and palate. *Otolaryngol Clin North Am* 2000; 33(6):1191–215.

5. Schutte BC, Murray JC. The many faces and factors of orofacial clefts. *Hum Mol Genet* 1999; 8(10):1853–9.
6. Thornton JB, Nimer S, Howard PS. The incidence, classification, etiology, and embryology of oral clefts. *Semin Orthod* 1996; 2(3):162–8.
7. Gundlach KH, Maus C. Epidemiological studies on the frequency of clefts in Europe and world-wide. *J Craniomaxillofac Surg* 2006; 34(Suppl 2):1–2.
8. Agbenorku P. Orofacial clefts: A worldwide review of the problem. ISRN Plastic Surgery 2013 [Internet]. Available at: http://dx.doi.org/10.5402/2013/348465 (accessed February 14, 2017).
9. Vanderas AP. Incidence of cleft lip, cleft palate, and cleft lip and palate among races: A review. *Cleft Palate* 1987; 24(3):216–25.
10. Derijcke A, Eerens A, Carels C. The incidence of oral clefts: A review. *Br J Oral Maxillofac Surg* 1996; 34(6):488–94.
11. Mossey PA, Modell B. Epidemiology of oral clefts 2012: An international perspective. *Front Oral Biol* 2012; 16:1–18.
12. Murray JC. Gene/environment causes of cleft lip and/or palate. *Clin Genet* 2002; 61(4):248–56.
13. Murray JC, Schutte BC. Cleft palate: Players, pathways, and pursuits. *J Clin Invest* 2004; 113(12):1676–8.
14. Spritz RA. The genetics and epigenetics of orofacial clefts. *Curr Opin Pediatr* 2001; 13(6):556–60.
15. Cobourne MT. The complex genetics of cleft lip and palate. *Eur J Orthod* 2004; 26(1):7–16.
16. Jezewski PA, Vieira AR, Nishimura C et al. Complete sequencing shows a role for MSX1 in non-syndromic cleft lip and palate. *J Med Genet* 2003; 40(6):399–407.
17. Rahimov F, Jugessur A, Murray JC. Genetics of nonsyndromic orofacial clefts. *Cleft Palate Craniofac J* 2012; 49(1):73–91.
18. Jugessur A, Murray JC. Orofacial clefting: Recent insights into a complex trait. *Current Opin Genet Dev* 2005; 15(3):270–8.
19. Murray JC. Face facts: Genes, environment, and clefts. *Am J Hum Genet* 1995; 57(2):227–32.
20. Honein MA, Rasmussen SA, Reefhuis J et al. Maternal smoking and environmental tobacco smoke exposure and the risk of orofacial clefts. *Epidemiology* 2007; 18(2):226–33.
21. Kohli SS, Kohli VS. A comprehensive review of the genetic basis of cleft lip and palate. *J Oral Maxillofac Pathol* 2012; 16(1):64–72.
22. Chen H. Cleft lip and/or cleft palate. In *Atlas of genetic diagnosis and counseling.* Totowa, NJ: Humana Press; 2006.
23. Wehby GL, Murray JC. Folic acid and orofacial clefts: A review of the evidence. *Oral Dis* 2010; 16(1):11–19.
24. Krapels IP, Vermeij-Keers C, Muller M et al. Nutrition and genes in the development of orofacial clefting. *Nutr Rev* 2006; 64(6):280–8.
25. Eppley BL, van Aalst JA, Robey A et al. The spectrum of orofacial clefting. *Plast Reconstr Surg* 2005; 115(7):e101–4.
26. Yadav SP, Sunil K, Sajid K et al. Cleft lip and palate. *Br J Mat Sci Tech* 2015; 1:11–14.
27. Converse JM, Hogan VM, McCarthy JG. Cleft lip and palate: Introduction. In *Reconstructive plastic surgery*, Converse JM, editor. Saunders, 1977.
28. Coleman JR Jr, Sykes JM. The embryology, classification, epidemiology, and genetics of facial clefting. *Facial Plastic Surg Clin North Am* 2001; 9(1):1–13.
29. Milerad J, Larson O, Hagberg C et al. Associated malformations in infants with cleft lip and palate: A prospective, population-based study. *Pediatrics* 1997; 100(2 Pt 1):180–6.

30. Kirby ML, Bockman DE. Neural crest and normal development: A new perspective. *The Anat Rec* 1984; 209(1):1–6.
31. Shprintzen RJ, Siegel-Sadewitz VL, Amato J et al. Anomalies associated with cleft lip, cleft palate, or both. *Am J Med Genet* 1985; 20(4):585–95.
32. Hagberg C, Larson O, Milerad J. Incidence of cleft lip and palate and risks of additional malformations. *Cleft Palate Craniofac J* 1998; 35(1):40–45.
33. Mossey PA, Little J, Munger RG et al. Cleft lip and palate. *Lancet* 2009; 374(9703):1773–85.
34. Davalbhakta A, Hall P. The impact of antenatal diagnosis on the effectiveness and timing of counselling for cleft lip and palate. *Br J Plast Surg* 2000; 53(4):298–301.
35. Clementi M, Tenconi R, Bianchi F et al. Evaluation of prenatal diagnosis of cleft lip with or without cleft palate and cleft palate by ultrasound: Experience from 20 European registries. *Prenat Diagn* 2000; 20(11):870–5.
36. Lee W, Kirk JS, Shaheen KW et al. Fetal cleft lip and palate detection by three-dimensional ultrasonography. *Ultrasound Obstet Gynecol* 2000; 16(4):314–20.
37. Sohan K, Freer M, Mercer N et al. Prenatal detection of facial clefts. *Fetal Diagn Ther* 2001; 16(4):196–9.
38. Wehby GL, Cassell CH. The impact of orofacial clefts on quality of life and healthcare use and costs. *Oral Dis* 2010(1); 16:3–10.
39. Tolarova MM, Cervenka J. Classification and birth prevalence of orofacial clefts. *Am J Med Genet* 1998; 75(2):126–37.
40. Weinberg SM, Neiswanger K, Martin RA et al. The Pittsburgh Oral-Facial Cleft study: Expanding the cleft phenotype. Background and justification. *Cleft Palate Craniofac J* 2006; 43(1):7–20.
41. Prahl-Andersen B. Dental treatment of predental and infant patients with clefts and craniofacial anomalies. *Cleft Palate Craniofac J* 2000; 37(6):528–32.
42. Millard DR Jr, Latham RA. Improved primary surgical and dental treatment of clefts. *Plast Reconstr Surg* 1990; 86(5):856–71.
43. Reisberg DJ. Dental and prosthodontic care for patients with cleft or craniofacial conditions. *Cleft Palate Craniofac J* 2000; 37(6):534–7.
44. Harville EW, Wilcox AJ, Lie RT et al. Epidemiology of cleft palate alone and cleft palate with accompanying defects. *Eur J Epidemiol* 2007; 22(6):389–95.
45. Habel A, Sell D, Mars M. Management of cleft lip and palate. *Arch Dis Child* 1996; 74(4):360–6.

10 Obesity and Gestational Diabetes

Maria Farren and Michael J. Turner

CONTENTS

INTRODUCTION

Maternal obesity in early pregnancy is an important risk factor for gestational diabetes mellitus (GDM) [1]. Epidemiological studies show that both conditions are prevalent and at an increasing rate worldwide, and they are associated with increased fetomaternal complications. They carry lifelong consequences for both the mother and her offspring, and thus place heavy demands on health care resources [1–3]. Authors of a meta-analysis concluded that GDM was increased twofold in women with mild obesity, fourfold in women with moderate obesity, and eightfold in women with severe obesity [1], as compared with women with a normal body mass index (BMI). There is little consensus internationally about the diagnosis and subsequent management of GDM, which results in wide variations in obstetric care in countries throughout the world [4,5]. Whereas much of the debate around GDM is focused on the screening tests used and the diagnostic criteria involved, until recently little attention has been given to the importance of preanalytical glucose sample handling and the subsequent effect this has on diagnostic accuracy [6]. This has the potential to revolutionize our approach to GDM in the future.

DIAGNOSIS OF GESTATIONAL DIABETES MELLITUS

GDM may be defined as any degree of glucose intolerance with onset or first recognition during pregnancy [7]. This includes women with type 2 diabetes mellitus (T2DM) who present for the first time during pregnancy. Other experts define women with GDM as those with onset or first diagnosis during pregnancy who return to normoglycemia postnatally [8,9]. The original laboratory criteria used for diagnosis of GDM were based on oral glucose tolerance test (OGTT) values, which determined the risk of developing T2DM in a nonpregnant population and were not based on any increased pregnancy risk to the woman or her offspring [10].

The findings of the Hyperglycaemia and Adverse Pregnancy Outcome (HAPO) study, which was a large, international multicenter study of 25,000 women, led to recommendations that revised the diagnostic thresholds for diagnosis of GDM. In 2010, the World Health Organization (WHO) convened an international group of experts to review recommendations for the diagnosis of GDM in light of the findings of the HAPO study [9]. This panel recognized that the existing recommendations published in 1999 were not evidence based. A systematic review was undertaken, focused on studies that looked at short-term pregnancy outcomes and perinatal outcomes. Potential long-term benefits were not evaluated. The review considered the effects on pregnancy, when both the International Association of Diabetes and Pregnancy Study Groups (IADPSG) and the WHO criteria (which had different diagnostic cutoffs) were used. GDM was associated with an increased risk ratio (RR) of large for gestational age (LGA) (RR 1.73 of IADPSG, RR 1.53 for WHO) and with an increased risk of preeclampsia (RR 1.71 for IADPSG, RR 1.69 for WHO). The evidence for an increased risk of caesarean section (CS) and shoulder dystocia with GDM was weaker. It was predicted that adopting the IADPSG criteria would reduce LGA and preeclampsia by 0.32% and 0.12%, respectively, and would have no effect on CS rates.

Following the convening of this expert panel three important recommendations were made by the WHO:

1. Hyperglycemia detected at any time during pregnancy must be classified as either:
   a. Diabetes mellitus in pregnancy (DMIP)
   b. Gestational diabetes mellitus
2. DMIP should be diagnosed by the 2006 WHO criteria for diabetes if one or more of the following criteria are met:
   a. Fasting plasma glucose ≥11.1 mmol/L (200 mg/dL)
   b. Three-hour plasma glucose ≥11.1 mmol/L (200 mg/dL) following a 75 g oral glucose load
   c. Random plasma glucose ≥11.1 mmol/L (200 mg/dL) in the presence of diabetes symptoms
3. GDM diagnosed in pregnancy should be based on the following:
   a. Fasting plasma glucose of 5.1–6.9 mmol/L (92–125 mg/dL)
   b. One-hour post 75 g oral glucose load ≥10.0 mmol/L (180 mg/dL)
   c. Two-hour post 75 g oral glucose load 8.5–11.0 mmol/L (153–199 mg/dL)

In the interest of consensus, WHO decided to adopt the IADPSG criteria for the diagnosis of GDM. Screening for GDM in pregnancy, however, continues to vary worldwide [5]. In Europe, women are selectively screened based on a host of risk factors including BMI, a family history of diabetes mellitus (DM), ethnicity, previous history of GDM, macrosomia (≥4.5 kg) or stillbirth, or metabolic issues such as polycystic ovarian syndrome (PCOS) [11]. There is, however, variation in the test used for diagnosis. The IADPSG recommends a one-step 75 g 2-hour oral glucose tolerance test (OGTT) between 24 and 28 weeks gestation, with additional consideration given to screening women who are at high risk of T2DM based on a fasting or random glucose at the first antenatal visit. In the United States, Canada, and Australia, however, the American College of Obstetrics and Gynecology recommends a universal approach, where all women are screened using a 50 g one-hour oral glucose challenge test (OGCT) between 24 and 28 weeks gestation [12].

The result of such disparity is reflected in the wide variety of reported prevalence of GDM. Within the United States, the prevalence varies from 1% to 26% [3]. A large review of European centers identified 32 different estimates of prevalence of GDM. In over half the studies ($n = 17$), the prevalence was between 2% and 6%, with the remainder varying between 7% and 25% [5]. Regional variation was observed in that the Northern and Atlantic seaboard regions of Europe mostly had a prevalence rate below 4%, while many areas of Southern Europe and the Mediterranean seaboard exceeded 6%. Of note, most of the previously reported European studies used a 100 g OGTT.

The type of OGTT used is only one factor accounting for the wide variety in the prevalence of GDM. The criteria used, the gestational age at which women are screened, the ethnicity of the population, the obesity levels of the population screened, and even whether the screening is universal or selective further complicates the issue of diagnosis. Moreover, many women with an abnormal OGTT do not have a postnatal OGTT to determine if they remain glucose intolerant, making it difficult to determine whether they have GDM or T2DM [13].

The way in which the samples of maternal glucose are handled as a subject until now has received scant attention. The importance of glycolysis postsampling in the accurate measurement of plasma glucose level is well established [14]. Blood glucose levels fall proportionate to time, in particular when the time from sampling to analysis exceeds 90 minutes [15,16]. The HAPO study attempted to standardize research conditions across all 15 centers with special attention being paid to how the samples were handled and analyzing maternal glucose centrally [17]. Subsequently, in 2011 the American Diabetes Association (ADA) recommended stricter guidelines for laboratory standards when diagnosing diabetes mellitus. It is recommended that, when using fluoride tubes to prevent glycolysis, "one should place the sample tube immediately in an ice-water slurry, and plasma should be separated from the cells within 30 mins" [18]. Although most laboratory scientists understand this, there is a paucity of information on the implementation of these strict conditions when screening women within the maternity services internationally. The impact of implementing such conditions was demonstrated in a prospective observational study in our hospital in 2014 [6]. Obese women were recruited after accurate measurement of BMI. One group of 24 women had a fasting glucose test in early pregnancy, while a second group of 24 women had an oral glucose tolerance test between 24 and 28 weeks gestation. In both groups paired glucose samples were taken. One sample was handled under customary conditions (i.e., not placed on ice for transfer), while the other sample was handled under strict research conditions (i.e., on ice with immediate analysis). The current WHO diagnostic criteria were used [9]. In the first cohort of women who had fasting glucose in early pregnancy, the result was abnormal in 29% of samples analyzed under hospital conditions, as compared to 67% of samples analyzed under strict, fast-tracked research conditions ($P < 0.01$). In the second cohort of women, who had an OGTT in the second trimester, the fasting glucose was abnormal in 17% of samples analyzed under customary conditions, as compared to 54% of samples under research conditions ($P < 0.001$). These results highlight an important aspect of the diagnosis of GDM. Unless strict laboratory standards are implemented, we could potentially be underdiagnosing GDM. This finding may explain the poor reproducibility of OGTTs reported in the literature [4,19,20].

THE HYPERGLYCEMIA AND ADVERSE PREGNANCY OUTCOME (HAPO) STUDY AND GESTATIONAL DIABETES MELLITUS

The HAPO study examined the relationship between mild hyperglycemia in the third trimester using a 75 g OGTT and pregnancy outcomes in a multicenter, multicultural, heterogeneous, ethnically diverse cohort of 25,000 women. The primary outcomes were birth weight (BW) >90th percentile, primary CS, neonatal hypoglycemia and cord C-peptide, which reflects fetal hyperinsulinemia, >90th percentile. A number of observations can be made about this study. There was a wide variation in the primary outcome and the incidence of GDM across the 15 field centers. The incidence of GDM ranged from 9% to 25%. The cause for such variation remains uncertain but is possibly attributable to obesity levels among the study population. The primary outcome of CS was 18%, with a variation of 8% to 24%. There was no distinction between elective and emergency CS, and there was no analysis by parity. Neonatal hypoglycemia ranged from 0.3% to 6% between centers. However, this was not based on the measurement of neonatal glucose in all cases but was based on recording in medical charts.

The IADPSG Consensus Panel, established after publication of the HAPO study, recommended that OGTTs be performed between 24 and 28 weeks gestation [21]. The HAPO study was intended to exclude women with overt DM [22]. However, the panel also noted that women with preexisting or overt T2DM might not be diagnosed until they actually become pregnant and this issue become more prominent as obesity rates continue to rise. It further recommended early testing of populations with a high prevalence of T2DM.

When establishing increased risk with mild hyperglycemia, the IADPSG concluded that the predefined value for this odds ratio (OR) at the threshold relative to the mean should be 1.75 for the fasting, 1-hour and 2-hour plasma glucose of the cohort. Within the HAPO cohort, approximately 2% were unblinded due to severe hyperglycemia [23]. When the remaining cohort was considered, the diagnosis of GDM was made on the fasting glucose in 8% of cases, the 1-hour glucose in an additional 5%, and the 2-hour glucose in 2% [21]. A further prospective observational study carried out in our hospital examined the role of the fasting, 1-hour and 2-hour tests in the OGTT [24]. Paired samples were taken from 155 women. One sample of the pair was handled under normal hospital conditions (i.e., not on ice), whereas a second sample was immediately put on ice and underwent fast-tracked analysis. Again, the rates of diagnosis were higher in the samples collected under research conditions, but it was also concluded that, based on the IADPSG criteria for diagnosis, GDM could be diagnosed on the fasting sample in 86% of cases, and on the 1-hour sample in the remaining 14% of cases. The 2-hour sample was not required for the diagnosis of GDM. This could potentially eliminate the need for the 2-hour OGTT, which would make the test more time efficient and tolerable for women and hospital staff alike.

The subsequent response following HAPO remains controversial. The OR of 1.75 was chosen arbitrarily and the reproducibility of IADPSG criteria is unknown [19]. When we consider the HAPO findings, we must also consider what we now know about preanalytical sample handling. In the future, it may not be necessary for women to undergo a 2-hour OGTT. It also may not be necessary for women with GDM to receive the intensive fetal and maternal monitoring previously received by women with GM, and which should continue to be received by women with T1DM and T2DM.

CATEGORIZATION OF MATERNAL OBESITY

We have already considered in detail the challenges of accurately diagnosing GDM, but must also consider the categorization of maternal obesity [25]. A diagnosis of obesity is usually made based on the WHO categorization of BMI, which is a surrogate measure for adiposity and does not provide information on the distribution of adiposity. It also does not reflect the possibility that there is variation in adiposity in different ethnic groups at the same BMI measurement [26]. This may explain why GDM is more common in certain ethnic groups.

Most epidemiological studies base the calculation of BMI on weight and height that is self-reported, which leads to 22% of women being assigned to the wrong BMI category and the diagnosis of obesity being missed in 5% [27]. A Canadian study of 2,667 women demonstrated the pitfalls of self-reported height and weight. Women in this study were being screened for T2DM outside of pregnancy, and self-reported both height and weight. This led to an exaggerated risk in obese women because women who were mildly obese reported themselves in the overweight and not the obese category [28].

The impact of accurate diagnosis of obesity is particularly important in countries where obesity is an indication for selective screening for GDM. Many studies use prepregnancy self-reported weight, which is unreliable [29]. Other studies use the weight and BMI calculated at the first antenatal visit. This calls into question the optimum timing of the calculation of BMI. Ideally, recording of BMI should take place before 18 weeks gestation [4]. Therefore, if the first antenatal visit takes place after 18 weeks, this measurement may not be accurate.

There is little information on the prevalence of obesity in pregnancy, unlike the well-documented rise in obesity in the nonpregnant adult population in developed countries [30]. In our hospital, based on accurate measurement of height and weight, 17% of women booking for antenatal care are obese and 1 in 50 are morbidly obese with a BMI ≥ 39.9 kg/m^2 [31].

In an analysis in the United States between 2004 and 2006, which only included seven states, 21% of women were obese with a prevalence of GDM of 4% [32].

The prevalence of GDM in normal-weight women was approximately 2%, but was 5% in overweight women, approximately 6% in women with mild obesity, and 12% in women with moderate or severe obesity. However, it is worth noting that these rates were based on self-reported prepregnancy weight. It is not clear what population was screened and how. However, this does highlight the close relationship between increasing obesity and GDM.

In a study of white European women in our hospital, 547 women were selectively screened using the 100 g OGTT [33]. Compared with overweight women, women with mild obesity were not more likely to have an abnormal OGTT, but women with moderate and severe obesity were more likely to have an abnormal result ($P = 0.008$). Nearly one in four women in this study with moderate or severe obesity had an abnormal OGTT. The risk of abnormal OGTT increased at the 90th percentile for BMI, which was 33 kg/m^2 and not the standard cutoff for obesity of >29.9.0 kg/m^2.

In another study of women with moderate to severe obesity, 100 women were offered screening before 20 weeks gestation, and if results were normal, the OGTT was repeated at 28 weeks gestation [34]. Prior to 20 weeks gestation, 88 women were screened, of whom approximately 21% had an abnormal OGTT. Of these 88 women, 11% had an abnormal early OGTT, and 10% had an abnormal OGTT at 28 weeks. This suggests that obese women diagnosed with GDM may have impaired glucose in early pregnancy and calls into question the optimum gestation at which we should be screening. Screening should be performed early and, if negative, should be repeated at between 24 and 28 weeks gestation. This is to ensure that GDM is not missed or diagnosis delayed. The clinical outcomes of treating GDM in early pregnancy are not known, and may depend on BMI levels and gestational weight gain before the diagnosis [35].

MATERNAL OBESITY AND GESTATIONAL DIABETES MELLITUS

The relationship between maternal BMI and hyperglycemia is poorly characterized because BMI has not been calculated accurately in early pregnancy and the definition of GDM has varied over time and between studies. There is uncertainty around the role of inflammatory biomarkers in the interplay between obesity and GDM. It is difficult to ascertain if inflammation in obese women contributes to the development of GDM or whether abnormal inflammatory biomarkers reflect an epidemiological association [36]. There is also uncertainty around the role of hyperglycemia in programming intrauterine fetal growth development and whether other metabolic abnormalities such as hypertriglyceridemia may be of greater importance [37].

It was previously thought that increasing obesity levels may lead to an increase in the rate of fetal macrosomia. However, there is no evidence at present to suggest an increasing fetal macrosomia rate in developed countries, despite rising rates of obesity and GDM [38].

THE HAPO STUDY AND MATERNAL OBESITY

A secondary analysis of the HAPO study specifically focused on the association between maternal obesity, GDM, and adverse pregnancy outcome. This found that both GDM and obesity were independently associated with adverse outcomes such as CS, birth weight, and birth injury [39]. However, it is worth noting that the diagnosis of obesity was not made until the time of the OGTT, at between 24 and 32 weeks gestation, and that the categorization for obesity was based on BMI ≥33 kg/m², the 90th percentile for BMI in the study above and not the WHO standard of 29.9 kg/m² [7]. To allow for weight gain during pregnancy, the recategorization of BMI at OGTT was based on prepregnancy BMI and gestational age at the time of testing. This, however, assumed weight gain in pregnancy to be linear and did not consider the wide variation in timing of the OGTT.

When we consider the outcomes, the OR for the different statistical models found that the associations with GDM were consistently stronger than the associations with obesity at the OGTT. There was no association between obesity alone with shoulder dystocia/birth injury despite obesity alone having an OR of 1.7 (95% CI: 1.5, 2.0) for BW >90th percentile. This was shown in a previous meta-analysis that showed no relationship between obesity and shoulder dystocia [30].

Following the findings of HAPO, obese women were recommended to follow the Institute of Medicine guidelines with regard to gestational weight gain. This included all obese women, both those with and without GDM. However, it is worth noting that obese women already gain less weight in pregnancy than nonobese women. Therefore, weight management interventions during pregnancy in obese women with GDM may not lead to improved clinical outcomes [29].

THE RISKS AND BENEFITS OF SCREENING FOR GESTATIONAL DIABETES MELLITUS

The debate continues about the risks and benefits of screening for GDM and the optimum management once the diagnosis is made [22,40]. While there is consensus that women with a BMI ≥29.9 kg/m² should be screened, in certain ethnic groups, the diagnosis of obesity should perhaps be made at a lower BMI [33].

One of the disadvantages of screening, however, is that it may lead to further testing, ultrasound examinations, and medical interventions, such as the induction of labor and CS. We must also be mindful that ultrasound examination of fetal growth is more technically challenging in the obese woman [41]. This can lead to potential overdiagnosis of fetal macrosomia and subsequent unnecessary intervention. For example, induction of labor in the obese woman is more likely to be unsuccessful, resulting in an increase in emergency CS. It is also important to note that obese women are more likely to need CS, even in the absence of GDM [42]. The increase in CS rates in obese women with GDM may become a self-fulfilling prophecy, and result in more adverse outcomes due the risk associated with this cohort.

A meta-analysis in 2009 quantified the risk of developing T2DM in women diagnosed with GDM [43]. This included 20 studies involving 675,455 women. Compared to women who did not have GDM, women with GDM had an increased risk for T2DM (RR of 7.4) [43]. The diagnosis of GDM was not standardized, and the duration of the follow-up varied. However, the authors concluded that GDM was a low-cost screening test of T2DM. This raises the possibility that initiating cardioprotective interventions in obese women with GDM may reduce their risk of cardiovascular disease, and its complications in later life. The ADA now recommends annual testing for women in the prediabetes range and testing every 3 years for women with GDM [11].

CONCLUSION

Based on the implementation of the HAPO findings and subsequent research, the number of women diagnosed with GDM will double, and one in three pregnancies in obese women may be complicated with GDM. This may further increase should the strict handling of glucose samples be implemented. The strength of the relationship between maternal obesity and GDM remains uncertain. What interventions, either prepregnancy or in the antenatal period, may lead to improved clinical outcomes for the woman and her baby also remain to be determined.

REFERENCES

1. Chu SY, Callaghan WM, Kim SY et al. Maternal obesity and risk of gestational diabetes mellitus. *Diabetes Care* 2007; 30:2070–6.
2 Sathyapalan T, Mellor D, Atkin SL. Obesity and gestational diabetes. *Semin Fetal Neonatal Med* 2010; 15:89–93.
3. Hartling L, Dryden DM, Guthrie A et al. Benefits and harms of treating gestational diabetes mellitus: A systematic review and meta-analysis for the U.S. Preventive Services Task Force and the National Institute Office of Medical Applications of Research. *Ann Int Med* 2013; 159:123–30.
4. O'Higgins A, Dunne F, Lee B et al. A national survey of the implementation of guidelines for gestational diabetes mellitus. *Ir Med J* 2014; 107:231–3.
5. Buckley BS, Harreiter J, Damm P et al. for the DALI Core Investigator Group. Gestational diabetes mellitus in Europe: Prevalence, current screening practice and barriers to screening. A review. *Diabet Med* 2012; 29(7):844–54.
6. Daly N, Stapleton M, O'Kelly R et al. The role of preanalytical glycolysis in the diagnosis of gestational diabetes mellitus in obese women. *Am J Obstet Gynecol* 2015; 213(1):84.e1–5.
7. Metzger BE, Couston DR. Proceedings of the Fourth International Workshop-Conference on Gestational Diabetes Mellitus. *Diabetes Care* 1998; 21:B1–167
8. National Collaborating Centre for Women's and Children's Health (UK). *Diabetes in pregnancy management of diabetes and its complications from preconception to the postnatal period.* NICE Guideline. London: RCOG Press; 2008.
9. World Health Organization. *Diagnostic criteria and classification of hyperglycemia first detected in pregnancy.* Geneva: World Health Organization; 2013.
10. O'Sullivan J, Mahan C. Criteria for the oral glucose tolerance test in pregnancy. *Diabetes* 1964; 13:278–85.
11. American Diabetes Association. Standards of medical care in diabetes 2014. *Diabetes Care* 2014; 37 Suppl 1:S14–80.
12. American College of Obstetrics and Gynecology (ACOG). Practice Bulletin No. 137. Gestational diabetes mellitus. *Obstet Gynecol* 2013; 122:406–16.
13. Almario CV, Ecker T, Moroz LA et al. Obstetricians seldom provide postpartum diabetes screening for women with gestational diabetes. *Am J Obstet Gynecol* 2008; 33:676–82.

14. Gupta S, Kaur H. Inhibition of glycolysis for glucose estimation in plasma: Recent guidelines and their implications. *Indian J Clin Biochem* 2014; 29(2):262–4.
15. Chan AY, Swaminathan R, Cockram CS. Effectiveness of sodium fluoride as a preservative of glucose in blood. *Clin Chem* 1989; 35(2):315–7.
16. Gambino R, Bruns DE. Stabilization of glucose in blood samples: Out with the old, in with the new. *Clin Chem Lab Med* 2013; 51(10):1883–5.
17. HAPO Study Cooperative Research Group, Nesbitt GS, Smye M, Sheridan B et al. Integration of local and central laboratory functions in a worldwide multicentre study: Experience from the hyperglycemia and adverse pregnancy outcome (HAPO) study. *Clin Trials* 2006; 3(4):397–407.
18. Sacks DB, Arnold M, Bakris GL et al. Guidelines and recommendations for laboratory analysis in the diagnosis and management of diabetes mellitus. *Diabetes Care* 2011; 34(6):e61–99.
19. Langer O, Umans JG, Miodovnik M. Perspectives on the proposed gestational diabetes mellitus diagnostic criteria. *Obstet Gynecol* 2013; 121(1):177–82.
20. Catalano PM, Avallone DA, Drago NM et al. Reproducibility of the oral glucose tolerance test in pregnant women. *Am J Obstet Gynecol* 1993; 169(4):874–81.
21. International Association of Diabetes and Pregnancy Study Groups. Recommendations on the diagnosis and classification of hyperglycemia in pregnancy. *Diabetes Care* 2010; 33(3):676–82.
22. The HAPO Study Cooperative Research Group. Hyperglycemia and Adverse Pregnancy Outcome (HAPO) study. Associations with Neonatal Anthropometrics. *Diabetes* 2009; 58(2):453–9.
23. Coustan DR, Lowe LP, Metzger BE et al. The Hyperglycemia and Adverse Pregnancy Outcome (HAPO) study: Paving the way for new diagnostic criteria for gestational diabetes mellitus. *Am J Obstet Gynecol* 2010; 202(6):654.e1–6.
24. Daly N, Flynn I, Carroll C et al. Comparison of citrate-fluoride-EDTA with fluoride-EDTA additives to stabilize plasma glucose measurements in women being screened during pregnancy with an oral glucose tolerance test: A prospective observational study. *Clin Chem* 2016; 62(6):886–7.
25. Turner MJ. The measurement of maternal obesity: Can we do better? *Clin Obes* 2011; 1(2–3):127–9.
26. Farah N, Murphy M, Ramphul M et al. Comparison in maternal body composition between Caucasian Irish and Indian women. *J Obstset Gynaecol* 2011; 31(6):483–5.
27. Fattah C, Farah N, O'Toole F et al. Body mass index (BMI) in women booking for antenatal care: Comparison between self-reported and digital measurements. *Eur J Obstet Gynecol Reprod Biol* 2009; 144(1):32–34.
28. Shields M, Connor Gorber S, Tremblay MS. Associations between obesity and morbidity: Effects of measurement methods. *Obes Rev* 2008; 9:501–2.
29. O'Dwyer V, O'Toole F, Darcy S et al. Maternal obesity and gestational weight gain. *J Obstet Gynaecol* 2013; 33(7):671–4.
30. Heslehurst N, Simpson H, Ells LJ et al. The impact of maternal BMI status on pregnancy outcomes with immediate short-term obstetric response implications: A meta-analysis. *Obes Rev* 2008; 9:635–83.
31. McKeating A, Maguire PJ, Daly N et al. Trends in maternal obesity in a large university hospital 2009–2013. *Acta Obstet Gynecol Scand* 2015; 94(9):969–75.
32. Kim SY, England L, Wilson HG et al. Percentage of gestational diabetes mellitus attributable to overweight and obesity. *Am J Public Health* 2010; 100(6):1047–52.
33. Farah N, McGoldrick A, Fattah C et al. Body mass index (BMI) and glucose intolerance during pregnancy in white European women. *J Reprod Infertil* 2012; 24:673–85.

34. O'Dwyer V, Farah N, Hogan J et al. Timing of screening for gestational diabetes mellitus in women with moderate and severe obesity. *Acta Obstet Gynecol Scand* 2012; 91(4):447–51.
35. Leary J, Pettitt DJ, Jovanovic L. Gestational diabetes guidelines in a HAPO world. *Best Pract Res Clin Endocrinol Metabol* 2010; 24(4):673–85.
36. Wolf M, Sandler L, Hsu K et al. First trimester c-reactive protein and subsequent gestational diabetes. *Diabetes Care* 2003; 26:819–24.
37. Whyte K, Kelly H, O'Dwyer V et al. Offspring birth weight and maternal fasting lipids in women screened for gestational diabetes mellitus (GDM). *Eur J Obstet Gynecol Reprod Biol* 2013; 170:67–70.
38. Farren M, Turner MJ. Can fetal macrosomia be predicted or prevented? In *Textbook of diabetes and pregnancy*, 3rd ed., Hod M, Jovanovic LG, Di Renzo GC, De Leiva A, Langer O, editors. London: CRC Press; 2016.
39. Catalano PM, McIntyre HD, Cruickshank JK et al. The hyperglycemia and adverse pregnancy outcome study: Associations of GDM and obesity with pregnancy outcomes. *Diabetes Care* 2012; 35:780–6.
40. Landon MB, Gabbe SG. Gestational diabetes mellitus. *Obstet Gynecol* 2011; 118:1379–93.
41. Paladini D. Sonography in obese and overweight pregnant women: Clinical, medicolegal and technical issues. *Ultrasound Obstet Gynecol* 2009; 33(6):720–9.
42. O'Dwyer V, Farah N, Fattah C et al. The risk of caesarean section in obese women analysed by parity. *Eur J Obstet Gynecol Reprod Biol* 2011; 158(1):28–32.
43. Bellamy L, Casas JP, Hingorani AD et al. Type 2 diabetes mellitus after gestational diabetes: A systematic review and meta-analysis. *Lancet* 2009; 373(9677):1773–9.

11 Nutrition and Hypertensive Disorders of Pregnancy

Antonia W. Shand and Tim J. Green

CONTENTS

INTRODUCTION

Hypertensive disorders of pregnancy (HDP) affect 4% to 9% of women and are among the leading causes of maternal and perinatal morbidity and mortality worldwide [1]. In low-income countries where maternal mortality is high, HDP-related maternal death ranges from 9% of deaths in Asia and Africa to 26% in South America. In high-income countries where maternal mortality rates are lower and effective diagnosis and treatment are available, HDP still account for 16% of maternal deaths [2]. The burden of maternal morbidity, perinatal morbidity and mortality, and costs to health care systems are even higher. There has been considerable interest in identifying modifiable risk factors, such as nutritional and lifestyle factors that might reduce the burden of HDP. Nutrition interventions, such as dietary modification and micronutrient supplementation, have been explored in relation to HDP risk and, given that obesity is associated with an increased risk of HDP, lifestyle interventions have also been investigated. In this chapter, following an overview of HDP and its pathogenesis, we examine the evidence relating to nutrition and HDP, and try to elucidate some of the mechanisms by which nutrition may influence HDP.

DEFINITION, TREATMENT, AND OUTCOME

HDP include preeclampsia (PE), gestational hypertension, PE superimposed on preexisting (chronic) hypertension, and chronic hypertension. Although there are several classification systems, PE generally refers to the onset of hypertension with either proteinuria and/or end organ dysfunction in a previously normotensive woman [3]. PE can be categorized as early (<34 weeks gestation) or late (≥34 weeks gestation) onset, depending on when it first develops. Gestational hypertension is the development of hypertension in a pregnant woman after 20 weeks gestation, without the presence of protein in the urine or other signs of PE. Hypertension in pregnancy is defined as a systolic blood pressure of ≥140 mmHg and/or diastolic blood pressure of ≥90 mmHg. Overall, HDP affects 3.6% to 9.1% of all pregnancies, with PE affecting 1.4% to 5% of pregnancies; however, early-onset PE affects only 0.3% to 0.7% of pregnancies [1,4].

Maternal morbidity from PE may be significant, including stroke, renal failure, eclampsia (seizures), or pulmonary edema [4]. In addition, perinatal and infant morbidity may result from iatrogenic or spontaneous preterm birth (<37 weeks gestation), and may lead to chronic lung disease, cerebral palsy, and neurodevelopmental disability [5]. An emerging concept is that PE may have several subtypes by gestation, resulting in varying degrees of intrauterine growth restriction, maternal hypertension, and maternal organ system involvement [6].

Treatment for PE is delivery of the fetus. When PE occurs preterm (<37 weeks gestation), the risks and benefits of delivery in terms of fetal, neonatal, and maternal outcomes must be evaluated [7]. Optimal antenatal treatment may include the provision of antenatal steroids for fetal lung maturation, transfer to an appropriate hospital for birth, magnesium sulfate for fetal and/or maternal neuroprotection, and expectant care or delivery, depending on the severity of PE and gestational age. Antihypertensive treatment is recommended for severe hypertension (BP ≥160/110) to reduce the risk of stroke. However, guidelines for less severe hypertension vary in terms of the medications recommended and the blood pressure targets utilized. It is recommended that women with gestational hypertension be observed for features of PE <37 weeks, with the consideration of delivery at term, despite a low risk of adverse outcomes [3,7].

PATHOPHYSIOLOGY OF HYPERTENSIVE DISORDERS OF PREGNANCY

PE is a multisystem syndrome with multifactorial causation. Reduced uterine perfusion and/or certain disease states (e.g., metabolic, immune, and cardiovascular) are associated with the maternal and placental abnormalities. In addition, there may be a genetic component predisposing women to the development of PE and influencing its severity [8]. It is proposed that PE is a two-stage disorder (see Figure 11.1),

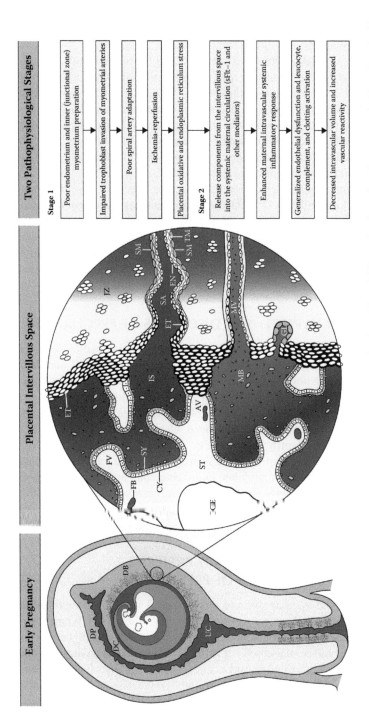

FIGURE 11.1 Possible pathophysiological processes in preeclampsia. AV, anchoring villus; COE, coelomic cavity; CY, cytotrophoblast; DB, decidua basalis; DC, decidua capsularis; DP, decidua parietalis; EN, endothelium; ET, extravillous trophoblast; FB, fetal blood vessel; FV, floating villus; GL, gland; IS, intervillous space; JZ, junctional zone; myometrium; MB, maternal blood, leaving the intervillous space with various components such as antiangiogenic factors; MV, maternal vein; SA, spiral artery; SM, smooth muscle; ST, stroma; SY, syncytiotrophoblast; TM, tunica media; UC, uterine cavity; sFlt-1, soluble form of the vascular endothelial growth factor receptor. (Reproduced from Steegers EAP, von Dadelszen P, Duvekot JJ, Prijnenborg R, Pre-eclampsia, *Lancet* 2010; 376: 631–44. With permission.)

with abnormal placentation leading to a maternal inflammatory response [9]. First, defective trophoblast invasion during early implantation may contribute to the release of vasoactive agents and subsequent remodeling of the uterine spiral arteries. This ultimately presents as defective uteroplacental blood circulation and placental ischemia. The second stage is a maternal systemic inflammatory response and maternal endothelial dysfunction. This results in the manifestation of clinical features, such as hypertension, proteinuria, cerebrovascular effects, and hepatic dysfunction, which may occur sometime later in pregnancy [8]. Interestingly, women with HDP are at greater risk of of cardiovascular disease in later life, the incidence of which is associated with the severity and gestation of onset in pregnancy, and the severity of intrauterine growth restriction [10].

The placental features of PE include a smaller placenta, excessive syncytial knots, infarcts covering more than 10% of the placental area, decidual vasculopathy including unconverted spiral arterioles, fibrinoid necrosis, atheroma, thrombosis of the uteroplacental vessels, the absence of intermediate villi (distal villous hypoplasia), and placental erythroblastosis [11]. These changes are thought to occur as a result of reduced perfusion or hypoxia of the maternal–fetal interface leading to oxidative stress, which can be detected long before signs of PE appear [9,11].

RISK FACTORS FOR HYPERTENSIVE DISORDERS OF PREGNANCY

Maternal, placental, and fetal mechanisms associated with inflammation, immune responses, and metabolism may have a role in the development of HDP, and these may be influenced by nutrition. Risk factors for HDP include obesity, diabetes, previous hypertensive disorders, polycystic ovarian syndrome, multiple pregnancy, first pregnancy, later gestation, ethnicity, extremes of maternal age, renal disease, the presence of antiphospholipid antibodies, prolonged interpregnancy interval, assisted reproductive technologies, a family history of PE, and a new partner [7].

There is a substantial amount of evidence to suggest that many foods, nutrients, and nonnutrient food components modulate both acute and chronic inflammation. However, the mechanisms underlying the association between maternal nutrition and HDP are unknown, with genetic, inflammatory, hormonal, metabolic, and vascular changes proposed [8,12].

OBSERVATIONAL STUDIES OF DIETARY PATTERNS
AND HYPERTENSIVE DISORDERS OF PREGNANCY

A systematic review and meta-analysis of observational studies has reported an association between HDP and several dietary factors [13] (Table 11.1). Women who had low calcium intake from their diet, lower magnesium intake, and higher energy intake were more likely to be diagnosed with HDP. The few studies on diet included in the meta-analysis suggest a trend toward a beneficial effect of diets high in vegetables and fruit on PE risk. In a prospective cohort study, prepregnancy fruit intake also increased the chance of an uncomplicated pregnancy outcome [14]. In a population-based study of Australian women, a protective dose response was observed between HDP risk and prepregnancy consumption of a Mediterranean-style dietary pattern,

TABLE 11.1

Summary of Current Evidence Assessing the Impact of Nutrient Supplementation, Diet, and/or Exercise in the Prevention of Preeclampsia

Intervention	Type of Study	Reference	Number of Women	Population	Relative Risk (95% CI)
Diet and nutritional counseling intervention	Meta-analysis of RCTs	Allen et al. 2014 [15]	2,695	Obese and normal weight pregnant women (n = 2,085 with gestational diabetes mellitus)	0.67 (0.53, 0.85)[a]
Diet and exercise intervention	Meta-analysis of RCTs	Allen et al. 2014 [15]	1,438	Obese and normal weight pregnant women	0.93 (0.66, 1.32)
Dietary and lifestyle intervention	RCT	Dodd et al. 2014 [16]	2,212	Women with BMI ≥ 25 kg/m^2	1.03 (0.71, 1.47)
Diet and activity intervention	RCT	Poston et al. 2015 [17]	1,555	Women with BMI ≥ 30 kg/m^2	1.00 (0.59, 1.69)
Calcium supplementation	Meta-analysis of RCTs	Hofmeyr et al. 2014 [18]	15,730	Pregnant women	0.45 (0.31, 0.65)[a]
Calcium supplementation	Meta-analysis of RCTs	Hofmeyr et al. 2014 [18]	10,678	Women with low calcium diets	0.36 (0.20, 0.65)[a]
Calcium supplementation	Meta-analysis of RCTs	Hofmeyr et al. 2014 [18]	5,022	Women with adequate calcium diets	0.62 (0.32, 1.20)
Vitamin D supplementation	Meta-analysis of RCTs	De-Regil et al. 2016 [19]	219	Pregnant women	0.52 (0.25, 1.05)
Vitamin D and calcium supplementation	Meta-analysis of RCTs	De-Regil et al. 2016 [19]	1,114	Pregnant women	0.50 (0.32, 0.80)[a]
Vitamin C supplementation	Meta-analysis of RCTs	Rumbold et al. 2015 [20]	21,956	Pregnant women	0.92 (0.80, 1.05)

(Continued)

TABLE 11.1 (CONTINUED)
Summary of Current Evidence Assessing the Impact of Nutrient Supplementation, Diet, and/or Exercise in the Prevention of Preeclampsia

Intervention	Type of Study	Reference	Number of Women	Population	Relative Risk (95% CI)
Vitamin E supplementation	Meta-analysis of RCTs	Rumbold et al. 2015 [21]	20,878	Pregnant women	0.91 (0.79, 1.06)
Salt (sodium)	Meta-analysis of RCTs	Duley et al. 2005 [22]	603	Hypertensive and normotensive pregnant women	1.11 (0.46, 2.66)
Zinc[b]	Meta-analysis of RCTs	Ota et al. 2015 [23]	2,975	Pregnant women	0.83 (0.64, 1.08)
Magnesium supplementation	Meta-analysis of RCTs	Makrides et al. 2014 [24]	1,040	Pregnant women	0.87 (0.58, 1.32)
Marine oil supplementation	Meta-analysis of RCTs	Allen et al. 2014 [15]	2,304	Pregnant women, most with previous pregnancy induced hypertension	0.93 (0.66, 1.32)
Marine oil supplementation	RCT	Zhou et al. 2012 [25]	2,399	Low-risk pregnant women with singleton pregnancies	0.87 (0.60, 1.25)

Note: RCT, randomized controlled trial.
[a] Relative risk statistically significant.
[b] Gestational hypertension or preeclampsia.

characterized by vegetables, legumes, nuts, tofu, rice, pasta, rye bread, red wine, and fish [26]. The authors found a 42% lower risk of HDP in women in the highest quartile of Mediterranean-style diet intake compared to women in the lowest quartile. This was similar to findings from a Norwegian study, which found that women with a diet high in vegetables, plant foods, and vegetable oils had a lower risk of PE [27]. This may be attributable to specific flavonoids that affect inflammation [28]. It has been proposed that unresolved chronic inflammation is a cause of a range of chronic diseases, and that diet may have anti-inflammatory and proinflammatory components. However, the mechanisms by which diet influences HDP risk has not been well studied.

OBESITY

Obesity, defined as a body mass index greater than 30 kg/m^2, has been identified as a major risk factor for the development of HDP. Obesity not only increases the risk of entering pregnancy with high blood pressure, but it also independently increases the risk of PE. Obesity rates are rising not only in Western countries such as the United States, where 50% of pregnant women are overweight or obese (BMI ≥25 kg/m^2), but also in middle-income countries, such as India and China. Rates of PE have risen in tandem with rising rates of obesity. The degree of obesity and risk of PE is a graded relationship, where with a normal BMI (18.5–24.9 kg/m^2) the risk is 3%, rising to 13% in the superobese (BMI ≥50 kg/m^2) [29].

Obesity is a risk factor for a number of other adverse pregnancy outcomes, including gestational diabetes, macrosomia, birth trauma, preterm birth, fetal anomalies, and stillbirth. Excessive weight gain during pregnancy has also been associated with an increased risk of PE [30]. This is, in part, why the Institute of Medicine recommends weight gain in pregnancy based on prepregnancy weight, with obese and overweight women recommended to gain less weight than normal weight women [31].

The mechanisms by which obesity leads to adverse pregnancy outcomes as well as chronic diseases are likely to be similar. Obesity can lead to a proinflammatory state with elevated cytokine levels, possibly making these women more susceptible to PE. Body mass index is also associated with the activation of inflammatory pathways in the placenta. It has been suggested that increased related metabolic factors alter uteroplacental vascular remodeling leading to placental ischemia, which in turn leads to exaggerated release of soluble placental factors, such as anti-angiogenic sFlt-1 and proinflammatory TNF-α and AT1-AA. Obesity can lead to an exaggerated response to placental factors, ultimately leading to reduced renal excretory function and hypertension. Obesity also increases the risk of type 2 diabetes mellitus and gestational diabetes, both of which are associated with HDP; women with poor glycemic control are at even higher risk [32]. Hyperglycemia may lead to oxidative stress and inflammation via nitric oxide (NO) pathways, thereby reducing NO concentrations and causing impairments in vasodilation [33].

A number of clinical trials have examined the effect of dietary and lifestyle interventions to promote appropriate weight gain in both obese and normal weight pregnant women, with and without gestational diabetes mellitus. Overall, the use of mainly dietary interventions, including energy-restricted or cholesterol-lowering diets, healthy dietary advice provided by a trained nutritionist, and the use of food diaries, was found to be associated with a 33% reduced risk of PE compared to controls (RR 0.67, 95% CI: 0.53, 0.85; 6 studies, n = 2,695 women). However, there was no reduction in the risk of PE with mixed interventions that included diet, physical activity, and lifestyle (RR 0.93, 95% CI: 0.66, 1.32; 6 studies, n = 1,438 women) [15]. Subsequent to this meta-analysis, in the largest study to date, Dodd et al. [16] randomized 2,252 women with a BMI ≥25 kg/m^2 between 10 and 20 weeks of gestation to either dietary and lifestyle intervention or standard care. In this trial, they found no difference in rates of hypertension (RR 1.05, 95% CI: 0.81, 1.38) or PE (RR 1.03, 95% CI: 0.71, 1.47) between the intervention and control groups. However, it should be noted that the intervention did not result in a lower gestational weight gain compared to the control (9.39 kg versus 9.44 kg). Likewise, the UPBEAT trial randomized 1,555 obese pregnant women with a mean BMI of 36 kg/m^2 to diet and activity or usual care, and found no difference in the rates of hypertensive disorders (RR 1.00, 95% CI: 0.59, 1.69) [17]. Thus, the effects of intervention to prevent excess weight gain on pregnancy outcomes such as HDP still remain to be determined. Currently, the World Health Organization (WHO) recommends that all women receive counseling about healthy eating and keeping physically active during pregnancy to stay healthy and to prevent excessive weight gain during pregnancy [34].

CALCIUM

The idea that calcium could reduce HDP came from Mayan Indians, who have a low incidence of PE and a high calcium intake due to their practice of soaking corn, a dietary staple, in lime (calcium oxide or calcium hydroxide) prior to cooking. A recent meta-analysis reported that women with the lowest quintile of dietary calcium intake were more likely to be diagnosed with HDP than women with the highest quintile of calcium intake (gestational hypertension, adjusted OR 0.63, 95% CI: 0.41, 0.97 and, overall, HDP adjusted OR 0.76, 95% CI: 0.57, 1.01; 5 studies, n = 2,203 women) [13]. A subsequent meta analysis reported a significant reduction in the risk of PE with calcium supplementation (RR 0.45, 95% CI: 0.31, 0.65; 13 studies, n = 15,730 women). In particular, the effect was greatest for women with low calcium diets (RR 0.36, 95% CI: 0.20, 0.65; 8 studies, n = 10,678 women) and women at high risk of PE (RR 0.22, 95% CI: 0.12, 0.42; 5 studies, n = 587 women) [18]. Calcium supplementation was also associated with a lower rate of gestational hypertension, with or without proteinuria (RR 0.65, 95% CI: 0.53, 0.81; 12 studies, n = 15,470 women). Again, a larger effect size was seen in areas with low background dietary calcium and in women already at a high risk of PE. Based largely on these meta-analyses, the WHO has issued guidelines recommending that women in areas where calcium intake is low take 1.5 to 2.0 g of oral elemental calcium daily to reduce the risk of PE. Calcium intake is typically low in areas where the main dietary source of calcium is grains. The WHO guidelines also emphasize the need for pregnant women at risk of gestational hypertension to take calcium supplements [34].

The mechanism for how calcium influences vessel contractility and how supplementation decreases HDP and PE is unknown. Low dietary calcium intake may lead to lower serum calcium and, in response, increased parathyroid hormone that may lead to higher levels of intracellular free calcium. Increased parathyroid hormone has been shown to be associated with raised blood pressure and has vasoconstrictive properties. Higher intracellular free calcium levels result in smooth muscle contractility and vasoconstriction. Other pathways, including the renin-angiotensin system, or antiangiogenic factors may play a role in PE [35].

VITAMIN D

Vitamin D deficiency, assessed by measuring 25-hydroxyvitamin D (25OHD) concentration, is common among pregnant women, with a prevalence of over 80% in some countries [36]. Because humans normally synthesize vitamin D through their skin by the action of ultraviolet (UV) light, populations of women who live at extremes of latitudes or "cover up" for religious or cultural regions tend to be at greatest risk. Low 25OHD during pregnancy has been associated with an increased risk of HDP in some but not all observational studies [37]. There have been at least five meta-analyses of the association of maternal vitamin D deficiency and risk of PE producing very different odds ratios, ranging from 0.78 to 2.11 [38]. The meta-analyses differed in their inclusion criteria, used different cutoffs for 25OHD to define deficiency, and often reported significant between-study heterogeneity. Several more studies have been published on the association between 25OHD and PE since the latest meta-analyses, and a number of studies are ongoing.

Vitamin D deficiency impairs calcium absorption and may increase HDP risk through similar mechanisms to having a low calcium diet. Vitamin D deficiency may also lead to immune dysfunction, abnormal angiogenesis, excessive inflammation, and hypertension, conditions that are involved in the pathogenesis of PE. In PE, high proinflammatory and low anti-inflammatory cytokine concentrations have been reported compared to normal pregnancies. Vitamin D receptors are located in the placenta, and it is possible that vitamin D deficiency promotes a cytokine profile that leads to placental dysfunction and PE [39].

Unfortunately, there have been few randomized trials evaluating vitamin D supplementation and PE risk, and those conducted have often used a combination of calcium and vitamin D. The most recent meta-analysis includes data from two trials involving 219 women, and suggests that women who received vitamin D supplements may have a lower risk of PE than those receiving no intervention or placebo (8.9% versus 15.5%; RR 0.52, 95% CI: 0.25, 1.05) [19]. However, the studies were very small, and the rate of PE in these studies was very high, limiting the generalizability of the findings. The meta-analysis also reported that women who received a combination of vitamin D and calcium supplementation had a reduced risk of PE compared to controls (5% versus 9%, RR 0.50, 95% CI: 0.32, 0.80; 3 studies, n = 1,114 women) but an increased risk of preterm birth (RR 1.57, 95% CI: 1.02, 2.43) [19]. Conversely, a reduction of preterm birth has been reported in the meta-analysis of calcium supplementation with or without vitamin D [34]. These findings suggest the need to determine the independent effects of vitamin D and calcium on HDP and other pregnancy outcomes.

At present, there is insufficient evidence to recommend vitamin D supplementation for PE prevention. Whether vitamin D supplementation will reduce the risk of HDP requires well-designed randomized control trials. Careful consideration needs to be given to study design, dose of vitamin D, and population.

ANTIOXIDANT NUTRIENTS

Vitamins E and C, as well as selenium, have antioxidant roles in the body. As oxidative stress is a known cause of endothelial dysfunction, it has been proposed that lower antioxidant levels may predispose women to PE. A recent study found an association between midpregnancy antioxidant nutrients and early onset PE [40]. This included the ratio of α-tocopherol:cholesterol, retinol, and lutein; however, the sources, such as diet or supplement use, were not measured.

Although earlier studies showed promise, the most recent meta-analyses have shown no effect of vitamin C or vitamin E supplementation on the risk of PE. There was no difference in the rate of PE in women with vitamin C supplementation (with or without other vitamins, including vitamin E) compared to placebo or no control (RR 0.92, 95% CI: 0.80, 1.05; 16 studies, n = 21,956 women) [20]. In the meta-analysis, there was no difference in the risk of PE in women with vitamin E supplementation (with or without other vitamins, including vitamin C) compared to placebo or no control (RR 0.91, 95% CI: 0.79, 1.06; 14 studies, n = 20,878 women) [21]. Therefore, vitamin C and E supplementation is not recommended in pregnancy.

An association with selenium deficiency and PE has been found [41]; however, there is limited data to support the benefits of selenium supplementation in preventing HDP. One randomized study, which included 230 primiparous women, found no difference in PE rates in women treated with selenium supplementation compared to controls. However, given the small number of subjects, the trial was underpowered for this outcome [42].

FOLATE

A number of studies have reported an association between folate deficiency and observed biomarkers in women at risk of PE. However, there is limited data from prospective studies as to whether folate supplementation reduces the risk of HDP [43]. A recent prospective cohort study found an association between folate supplementation or multivitamins containing folic acid and a reduction in PE in high-risk women (OR 0.17, 95% CI: 0.03, 0.95) [44]. However, a meta-analysis of observational studies and randomized studies found no difference in outcomes with folic acid or multivitamin containing folic acid [43]. A large multicountry randomized trial conducted among high-risk women (n = 3,656) comparing 4000 μg/d folate versus placebo has just been completed and is expected to be published in the next year [45].

Folate may exert a protective effect through its role in modifying DNA and influence on DNA methylation. Elevated homocysteine concentrations, which can be lowered by taking folate and certain other B vitamins, have been associated with higher rates of PE in numerous studies. Homocysteine might contribute to the placental vasoconstriction and ischemia of PE by increasing collagen deposition in arteries leading to vascular thickening, promoting apoptosis in umbilical vein endothelial cells and smooth muscle cells, increasing proinflammatory cytokines within blood vessels, and downregulating glutathione peroxidase, thus reducing nitric oxide, a major vasodilator.

SALT (SODIUM)

Dietary salt is a well-established risk factor for hypertensive disorders, and a reduction in salt intake can reduce blood pressure in some people [46]. Excess salt can cause the kidneys to retain water, increasing plasma volume, and thereby increasing blood pressure. Two trials, both conducted in the Netherlands, compared dietary advice to reduce salt intake versus the continuation of a normal diet: one study included women with hypertension (diastolic blood pressure >85 mmHg), and the other only included healthy women. Neither study, nor the meta-analysis of the two studies, showed a significant benefit of sodium restriction on PE risk (RR 1.11, 95% CI: 0.46, 2.66; 2 studies, $n = 603$ women) [22]. At present, salt restriction is not recommended in pregnancy, even for women at high risk of PE. The role of sodium restriction in the general population for the prevention of cardiovascular disease also remains uncertain [47].

ZINC

Zinc is a trace metal with a wide range of metabolic functions. Observational studies of serum zinc levels and PE have found conflicting information, with some studies finding an association between low serum zinc levels and HDP, and other studies reporting no association [48]. The mechanisms for this association between zinc and PE are not entirely understood; however, it is thought that zinc may alleviate oxidative stress by increasing antioxidants, or by serving as a cofactor or substrate for the activation of antioxidant enzymes. However, a meta-analysis of 21 randomized studies found no difference in any maternal outcomes including HDP, with zinc supplementation in pregnancy compared to controls, although there was a small reduction in the rate of preterm birth (RR 0.86, 95% CI: 0.76, 0.97; 16 studies, $n = 7,637$ women) [23]. Therefore, zinc supplementation is not recommended for PE prevention.

MAGNESIUM

Magnesium is an essential mineral that is required for nucleic acid and protein synthesis, body temperature control, and maintenance of nerve and muscle cell electrical potentials. Magnesium occurs naturally in many foods, and studies have shown that many women, especially those from disadvantaged populations, have intakes that are lower than recommended. Intravenous magnesium sulfate supplementation is the first-line anticonvulsant of choice in the prevention and treatment of eclampsia and has been used for this purpose for many years [49]. However, the mechanism of action, optimal dosing regimen, and pharmacokinetic properties remain unknown.

The use of intravenous magnesium sulfate gave rise to the idea that magnesium sulfate supplementation in pregnancy might reduce the risk of HDP. In a systematic review and meta-analysis of studies, oral magnesium supplementation in pregnancy compared to no magnesium or placebo was associated with no significant difference in PE (RR 0.87; 95% CI: 0.58, 1.32; 3 studies, n = 1,042 women), perinatal mortality, or small-for-gestational age fetuses. Therefore, oral magnesium supplementation is not recommended in pregnancy [24].

MARINE OILS

Greenland Inuit have lower rates of HDP than other populations, ascribed to their intake of marine oil, which contains high amounts of long-chain omega-3 fatty acids, eicosapentaenoic acid (EPA), and docosahexaenoic acid (DHA) [49]. Omega-3 fatty acids have been shown to give rise to cytokines, which have anti-inflammatory properties. In contrast, omega-6 fatty acids, such as arachidonic acid, which predominate in most Western diets, give rise to cytokines that have proinflammatory effects and promote vasoconstriction. PE and gestational hypertension are associated with vasoconstriction and endothelial damage. Both DHA and EPA compete with the thromboxane A2 precursor, arachidonic acid, and downregulate these responses.

In the most recent systematic review of studies of marine oils and HDP, there was no difference in either the rate of high blood pressure during pregnancy or the incidence of PE in women treated with marine oils compared to controls (RR 0.93, 95% CI: 0.66, 1.32; 6 studies, n = 2,304 women) [15]. Most of these studies were completed in women with previous pregnancy complications, usually gestational hypertension. One exception was a study that randomized 2,399 low-risk pregnant women with singleton pregnancies to DHA-enriched fish oil (800 mg) or placebo from midpregnancy. However, they also found no difference in the rate of gestational hypertension (RR 0.97, 95% CI: 0.74, 1.27) or PE (RR 0.87, 95% CI: 0.60, 1.25) in the women treated with fish oil compared to placebo [50].

CONCLUSION AND FURTHER RESEARCH

The mechanisms by which diet and/or dietary supplementation influence HDP are not well understood. Currently, it appears that calcium supplementation in women with a low calcium diet reduces the risk of HDP. A diet high in vegetables and fruit also appears to increase the odds of an uncomplicated pregnancy outcome; however, analyzing diet separately to other lifestyle considerations is difficult. Prepregnancy weight loss and glycemic control in women with diabetes mellitus have evidence for benefit. Further studies are needed to assess the effect of vitamin D, with or without calcium, and folate supplementation in pregnancy. Further studies would benefit from standardized definitions of HDP, and describing risk factors for HDP, including maternal and family history, blood pressure, body mass index, uterine artery Doppler studies, and biomarkers. Ideally, future intervention studies should enroll women either prepregnancy or very early in pregnancy, when placentation occurs; however, this comes with logistic issues.

REFERENCES

1. Roberts CL, Algert CS, Morris JM et al. Increased planned delivery contributes to declining rates of pregnancy hypertension in Australia: A population-based record linkage study. *BMJ Open* 2015; 5:e009313.
2. Khan KS, Wojdyla D, Say L et al. WHO analysis of causes of maternal death: A systematic review. *Lancet* 2006; 367:1066–74.
3. Lowe SA, Bowyer L, Lust K et al. The SOMANZ guidelines for the management of hypertensive disorders of pregnancy 2014. *Aust N Z J Obstet Gynaecol* 2014; 55:11–16.
4. Lisonkova S, Sabr Y, Mayer C et al. Maternal morbidity associated with early-onset and late-onset preeclampsia. *Obstet Gynecol* 2014; 124:771–81.
5. Ancel P-Y, Goffinet F, Kuhn P et al. Survival and morbidity of preterm children born at 22 through 34 weeks' gestation in France in 2011: Results of the EPIPAGE-2 cohort study. *JAMA Pediatr* 2015; 169:230–8.
6. Redman CW, Sargent IL, Staff AC. IFPA senior award lecture: Making sense of pre-eclampsia—Two placental causes of preeclampsia? *Placenta* 2014; 35:S20–25.
7. Mol BW, Roberts CT, Thangaratinam S et al. Pre-eclampsia. *Lancet* 2016; 387: 999–1011.
8. Ali SM, Khalil RA. Genetic, immune and vasoactive factors in the vascular dysfunction associated with hypertension in pregnancy. *Expert Opin Ther Targets* 2015; 8222:1–21.
9. Fisher SJ. Why is placentation abnormal in preeclampsia? *Am J Obstet Gynecol* 2015; 213:S115–22.
10. Staff AC, Redman CWG, Williams D et al. Pregnancy and long-term maternal cardiovascular health. *Hypertension* 2015; 67:251–60.
11. Devisme L, Merlot B, Ego A et al. A case-control study of placental lesions associated with pre-eclampsia. *Int J Gynecol Obstet* 2013; 120:165–8.
12. Vannuccini S, Clifton VL, Fraser IS et al. Infertility and reproductive disorders: Impact of hormonal and inflammatory mechanisms on pregnancy outcome. *Hum Reprod Update* 2016; 22:104–15.
13. Schoenaker D, Soedamah-Muthu SS, Mishra GD. The association between dietary factors and gestational hypertension and pre-eclampsia: A systematic review and meta-analysis of observational studies. *BMC Med* 2014; 12:157.
14. Chappell LC, Seed PT, Myers J et al. Exploration and confirmation of factors associated with uncomplicated pregnancy in nulliparous women: Prospective cohort study. *BMJ* 2013; 347:f6398.
15. Allen R, Rogozinska E, Sivarajasingam P et al. Effect of diet- and lifestyle-based metabolic risk-modifying interventions on preeclampsia: A meta-analysis. *Acta Obstet Gynecol Scand* 2014; 93:973–85.
16. Dodd JM, Turnbull D, McPhee AJ et al. Antenatal lifestyle advice for women who are overweight or obese: LIMIT randomised trial. *BMJ* 2014; 348:g1285.
17. Poston L, Bell R, Croker H et al. Effect of a behavioural intervention in obese pregnant women (the UPBEAT study): A multicentre, randomised controlled trial. *Lancet Diabetes Endocrinol* 2015; 3:767–77.
18. Hofmeyr JG, Lawrie TA, Atallah AN et al. Calcium supplementation during pregnancy for preventing hypertensive disorders and related problems. *Cochrane Database Syst Rev* 2014; 6:CD001059.
19. De-Regil LM, Palacios C, Ansary A et al. Vitamin D supplementation for women during pregnancy. *Cochrane Database Syst Rev* 2016; 2:CD008873.
20. Rumbold A, Ota E, Nagata C et al. Vitamin C supplementation in pregnancy. *Cochrane Database Syst Rev* 2015; 9:CD004072.
21. Rumbold A, Ota E, Hori H et al. Vitamin E supplementation in pregnancy. *Cochrane Database Syst Rev* 2015; 9:CD004069.

22. Duley L, Henderson-Smart D, Meher S. Altered dietary salt for preventing pre-eclampsia, and its complications. *Cochrane Database Syst Rev* 2005; 4:Cd005548.

23. Ota E, Mori R, Middleton P et al. Zinc supplementation for improving pregnancy and infant outcome. *Cochrane Database Syst Rev* 2015; 2:CD000230.

24. Makrides M, Crosby DD, Bain E et al. Magnesium supplementation in pregnancy. *Cochrane Database Syst Rev* 2014; 4: CD000937.

25. Zhou S, Yelland L, McPhee A et al. Fish-oil supplementation in pregnancy does not reduce the risk of gestational diabetes or preeclampsia. *Am J Clin Nutr* 2012; 95:1378–84.

26. Schoenaker D, Soedamah-Muthu S, Callaway L et al. Prepregnancy dietary patterns and risk of developing hypertensive disorders of pregnancy: Results from the Australian Longitudinal Study on Women's Health. *Am J Clin Nutr* 2015; 102:94–101.

27. Brantsaeter AL, Haugen M, Samuelsen SO et al. A dietary pattern characterized by high intake of vegetables, fruits, and vegetable oils is associated with reduced risk of preeclampsia in nulliparous pregnant Norwegian women. *J Nutr* 2009; 139:1162–8.

28. Minihane AM, Vinoy S, Russell WR et al. Low-grade inflammation, diet composition and health: Current research evidence and its translation. *Br J Nutr* 2015; 114:999–1012.

29. Mbah AK, Kornosky JL, Kristensen S et al. Super-obesity and risk for early and late pre-eclampsia. *BJOG* 2010; 117:997–1003.

30. Marchi J, Berg M, Dencker A et al. Risks associated with obesity in pregnancy, for the mother and baby: A systematic review of reviews. *Obes Rev* 2015; 16:621–38.

31. Institute of Medicine and National Research Council. *Weight gain during pregnancy: Reexamining the guidelines.* Washington, DC: National Academies Press; 2009.

32. Weissgerber TL, Mudd LM. Preeclampsia and diabetes. *Curr Diab Rep* 2015; 15:579–4.

33. Blaak EE, Antoine JM, Benton D et al. Impact of postprandial glycaemia on health and prevention of disease. *Obes Rev* 2012; 13:923–84.

34. World Health Organization. *WHO recommendations on antenatal care for a positive pregnancy experience.* Geneva: World Health Organization; 2016.

35. Scholl T, Chen X, Stein TP. Vitamin D, secondary hyperparathyroidism, and pre-eclampsia. *Am J Clin Nutr* 2013; 98:787–93.

36. Palacios C, Gonzalez L. Is vitamin D deficiency a major global public health problem? *J Steroid Biochem Mol Biol* 2014; 144:138–45.

37. Harvey NC, Holroyd C, Ntani G et al. Vitamin D supplementation in pregnancy: A systematic review. *Health Technol Assess* 2014; 18:1–189.

38. Moon RJ, Harvey NC, Cooper C. Endocrininology in pregnancy: Influence of maternal vitamin D status on obstetric outcomes and the fetal skeleton. *Eur J Endocrinol* 2015; 173:R69–83.

39. Barrera D, Díaz L, Noyola-Martínez N et al. Vitamin D and inflammatory cytokines in healthy and preeclamptic pregnancies. *Nutrients* 2015; 7:6465–90.

40. Cohen JM, Kramer MS, Platt RW et al. The association between maternal antioxidant levels in midpregnancy and preeclampsia. *Am J Obstet Gynecol* 2015; 213:695e1–13.

41. Rayman MP, Bath SC, Westaway J et al. Selenium status in UK pregnant women and its relationship with hypertensive conditions of pregnancy. *Br J Nutr* 2015; 113:249–58.

42. Rayman MP, Searle E, Kelly L et al. Effect of selenium on markers of risk of pre-eclampsia in UK pregnant women: A randomised, controlled pilot trial. *Br J Nutr* 2014; 112:99–111.

43. Shim SM, Yun YU, Kim YS. Folic acid alone or multivitamin containing folic acid intake during pregnancy and the risk of gestational hypertension and preeclampsia through meta-analyses. *Obstet Gynecol Sci* 2016; 59:110–5.

44. Wen SW, Guo Y, Rodger M et al. Folic acid supplementation in pregnancy and the risk of pre-eclampsia: A cohort study. *PLoS One* 2016; 11:e0149818.

45. Wen SW, Champagne J, White RR et al. Effect of folic acid supplementation in pregnancy on preeclampsia: The folic acid clinical trial study. *J Pregnancy* 2013; 2013: 294312.
46. He FJ, Li J, Macgregor GA. Effect of longer term modest salt reduction on blood pressure: Cochrane systematic review and meta-analysis of randomised trials. *BMJ* 2013; 346:f1325.
47. Lopez-Jaramillo P, Lopez-Lopez J, Lopez-Lopez C. Sodium intake recommendations: A subject that needs to be reconsidered. *Curr Hypertens Rev* 2015; 11:8–13.
48. Ma Y, Shen X, Zhang D. 2015. The relationship between serum zinc level and preeclampsia: A meta-analysis. *Nutrients* 7:7806–20.
49. Okusanya B, Oladapo O, Long Q et al. Clinical pharmacokinetic properties of magnesium sulphate in women with pre-eclampsia and eclampsia: A systematic review. *BJOG* 2016; 123:356–66.
50. Gerrard J, Popeski D, Ebbeling L et al. Dietary omega 3 fatty acids and gestational hypertension in the Inuit. *Artic Med Res* 1991; Suppl:763–7.

12 Low Birth Weight and Small for Gestational Age in the Context of 1,000 Days

Amira M. Khan, Bianca Carducci,
and Zulfiqar A. Bhutta

CONTENTS

INTRODUCTION

The first 1,000 days, a period that spans from conception until 2 years of age, is a critical period of life. Optimal growth and development during this sensitive period is central to neonatal health and, ultimately, health throughout the lifespan. Furthermore, growth is dependent upon several elements, including maternal, fetal, placental, and environmental factors. However, in unfavorable prenatal conditions, it is hypothesized that permanent growth restriction and adult disease risk can occur [1].

This chapter will focus on the biology of the postnatal consequences of low birth weight (LBW) and small for gestational age (SGA). Related to these consequences is intrauterine growth restriction (IUGR), a prenatal observation of growth restriction proven either by a documented reduced fetal growth velocity and/or by the presence of specific causes such as fetal infection, compromised placental blood flow, or toxic effects. However, a child born SGA or LBW might have not suffered IUGR, and a baby born after a short span of IUGR might not be SGA or LBW.

BOX 12.1   KEY DEFINITIONS

Small for gestational age (SGA)—An infant who weighs <10th percentile of the birth weight for gestational age, gender-specific reference population.

Low birth weight (LBW)—An infant who weighs <2,500 grams regardless of gestational age.

Intrauterine growth restriction (IUGR)—A prenatal condition in which a fetus's estimated weight is below the recommended gender-specific weight for gestational age.

Preterm—Preterm is referred to as birth before 37 completed weeks of gestation or fewer than 259 days since the first day of a woman's last menstrual period.

LOW BIRTH WEIGHT

Historically, from the 1920s to the 1950s the terms *low birth weight* and *premature* were used synonymously in literature [2]. In 1950, the World Health Organization (WHO) Expert Group on Prematurity declared that an infant of birth weight under 5.5 pounds (~2,500 grams) should be termed *premature* [3]. However, epidemiological studies in the 1950s and 1960s made it evident that not all babies born small are premature, and not all premature babies are small. Recognizing this difference, in 1961 the WHO Expert Committee on Maternal and Child Health recommended that the two terms be used distinctly, and that the concept of prematurity in the definition should give way to that of LBW [4].

LBW is defined as a neonate who weighs <2,500 grams regardless of gestational age (until 1976, the WHO defined this as less than or equal to 2,500 grams) [5]. Birth weight

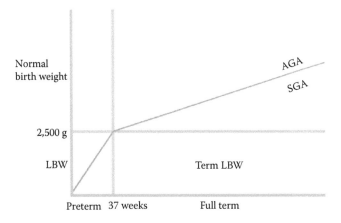

FIGURE 12.1 The relationship of birth weight and gestational age for classification of LBW and SGA. (Reproduced from Black RE, Victora CG, Walker SP et al., *Lancet* 2013, 382:427–51. With permission.)

is directly determined by the period of gestation and the rate of fetal growth [6]. Thus, LBW can be a consequence of being born too early (preterm) or too small (SGA). SGA is not always a consequence of a pathology and includes two groups: those who have achieved their growth potential but are constitutionally small, and those who experience growth restriction and do not achieve their genetic potential. Thus, the LBW group can include preterm neonates, at-term SGA neonates, and the preterm SGA neonates.

Figure 12.1 demonstrates that newborns who are at the expected weight (whether preterm or full term) are referred to as appropriate for gestational age (AGA), while those who are lower than expected weight for gestation (whether preterm or full term) are SGA. LBW can include both AGA and SGA, depending upon gestational age.

SMALL FOR GESTATIONAL AGE

There is no general consensus on the definition of SGA, thus several definitions exist. To measure SGA, gestational dating, precise measurements of weight and length at birth, and reference population data are key factors to consider. SGA children can be born either full term or premature. It is also important to note that being born SGA is neither indicative nor synonymous with IUGR, and vice versa, but an overlap can occur. The two most common definitions of SGA include [7]: (1) the birth weight for gestational age, sex specific, single/twin curve, where the 10th percentile of the curve should be used to classify SGA; and (2) the birth weight and/or length is less than two standard deviations below the mean for the neonate's gestational age, based on a reference population.

More recently, a multicountry project titled "INTERGROWTH–21st" aimed to develop international standards regarding sex-specific fetal (over 33 weeks gestational age), newborn, and postnatal growth (see Figure 12.2). Although highly robust and adopted by many countries, this new set of growth curves is not yet the standard [8].

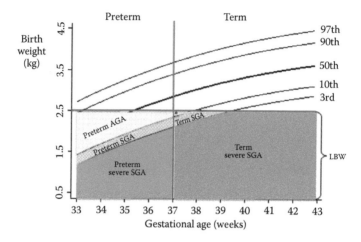

FIGURE 12.2 The INTERGROWTH–21st birth weight for gestational age standard. (Reproduced from Houk CP, Lee PA, *Int J Pediatr Endocrinol* 2012, 2012:11. With permission.)

THE BURDEN OF LOW BIRTH WEIGHT

According to the latest Global Nutrition Report, in 2014, globally 15% of infants were born LBW, or 20 million births per year [9]. Importantly, LBW is also an important population health indicator, as it occurs with greater prevalence in vulnerable populations. Estimates of LBW in low- and middle-income regions include 28% in South Asia, 13% in sub-Saharan Africa, and 9% in Latin America (see Table 12.1); however, intracountry variation exists. It is also worth noting that data on LBW remains limited or unreliable, as many deliveries occur in homes or small health clinics and are not reported in official figures, which may result in an underestimation of the prevalence of LBW [10].

TABLE 12.1
Percent of Infants Born Low Birth Weight

Region	Percent (%) of Infants Born Low Birth Weight	Percent (%) of Infants Not Weighed at Birth
Sub-Saharan Africa	13	54
Eastern and Southern Africa	11	46
West and Central Africa	14	60
Middle East and North Africa	–	–
South Asia	28	66
East Asia and Pacific	6	22
Latin America and Caribbean	9	10

Source: Adapted from UNICEF global databases, 2014, based on Multiple Indicator Cluster Surveys (MICS), Demographic and Health Surveys (DHS) and other nationally representative surveys, 2009–2013 (for % not weighed is 2008–2012), with the exception of India and Indonesia.

THE BURDEN OF SMALL FOR GESTATIONAL AGE

The prevalence and incidence of morbidity and mortality is often estimated to understand the impact of SGA and LBW worldwide. However, due to the lack of agreement on an SGA definition, there are no accurate estimates of SGA burden. In 2010, an estimated 32.4 million infants were born SGA in low- and middle-income countries (LMICs). Of these, 19 million were born at term, and 10.6 million were born both full term and LBW [11].

RISK FACTORS FOR LOW BIRTH WEIGHT (PRETERM AND/OR SMALL FOR GESTATIONAL AGE)

One of the earliest and most detailed reviews on the etiology of LBW, published in 1987, identified and classified 43 determinants of LBW [5]. The etiology and risk factors associated with LBW include those that contribute to preterm births and to impaired fetal growth (SGA). However, it is difficult to separate these risk factors as many remain common to both, and thus will be discussed together. Notably, nearly 40% of SGA do not have an identifiable cause. Of the 60% with known etiology, approximately 50% are a consequence of maternal factors, 5% a result of fetal factors, and 5% involve placental abnormalities [12].

The risk factors can be categorized into maternal, paternal, and fetal-related [1,13]:

- *Maternal*—Physiological and social factors (maternal height, ethnicity, low socioeconomic status, young or advanced maternal age, short interpregnancy intervals, poor maternal nutrition [low maternal body mass index, undernutrition, micronutrient deficiencies]), lifestyle factors (smoking, excess alcohol consumption, drug abuse, stress, excessive physical work), and inadequate antenatal care
  - Maternal mental health—Depression, domestic violence
  - Multiple pregnancy
  - Preeclampsia
  - Chronic medical illness—Diabetes, hypertension, anemia, asthma, thyroid disease, renal disease
  - Infections—Urinary tract infection, malaria, bacterial vaginosis, HIV, syphilis, Listeria monocytogenes, group B streptococcus, chorioamnionitis
  - Uterine factors—Bicornuate uterus, incompetent cervix
  - Uteroplacental factors—Insufficient uteroplacental perfusion, placenta previa, abruptio placentae
- *Paternal-related factors*—Paternal height or race
- *Fetal-related factors*—Gender (more common in males), congenital malformations, chromosomal abnormalities, inborn errors of metabolism, congenital infections

In addition to these factors, there are some circumstances that increase a predisposition to preterm births, including premature rupture of membranes, fetal distress, polyhydramnios, iatrogenic, trauma, and provider-initiated (induction of labor or elective caesarean for maternal or fetal indications).

THE ETIOLOGY OF INFANTS OF LOW BIRTH WEIGHT AND SMALL FOR GESTATIONAL AGE IN THE CONTEXT OF THE FIRST 1,000 DAYS

It is apparent from the risk factors mentioned that the etiology of LBW is multifactorial and complex, including both nutritional and nonnutritional elements. Moreover, the factors leading to LBW vary with different geographies and economies. Whereas nearly half of all LBW infants born in high-income countries are preterm, most LBW infants in LMIC are SGA and have been affected by IUGR.

Nutrition is the most influential environmental factor for fetal development. In fact, the prevalent underlying factors accountable for SGA in LMIC are nutrition-related, namely, poor weight gain in pregnancy, low preconception body mass index (BMI), and chronic maternal undernutrition (maternal short stature). It has been estimated that more than 50% of all LBW in LMIC can be attributed to poor maternal nutrition preconceptionally and early pregnancy [14]. Following recognition of the key role of maternal and fetal nutrition as causes of LBW, the World Health Assembly adopted the prevalence of LBW births as a nutrition indicator with a target of 30% reduction in prevalence by 2025 [15,16].

Thus, nutrition during the first 270 days of the 1,000 days is the key to healthy pregnancy outcomes. However, acknowledging the increasing importance of preconception nutrition, experts are now advocating a life-course approach for women's nutrition, rather than one focused solely on pregnancies.

MATERNAL WEIGHT GAIN DURING PREGNANCY

Of the nutrition-related factors, poor maternal weight gain during pregnancy has the strongest association with SGA. The WHO and FAO recommend that dietary intake during pregnancy should provide energy sufficient for "the full-term delivery of a healthy newborn baby of adequate size and appropriate body composition by a woman whose weight, body composition and physical activity level are consistent with long-term good health and well-being." Energy needs during pregnancy are increased to provide for growth of the fetus, placenta, and maternal tissues; the increased metabolic demands of pregnancy; maintaining maternal weight, body composition, and physical activity throughout pregnancy; and adequate energy stores for lactation.

BMI is an index for weight-for-height; the maternal prepregnancy BMI is used to assess maternal nutritional status. Recommended weight gain for each woman from preconception to delivery is based on her preconception BMI. For example, if a woman is underweight (BMI <18.5) she should gain between 12.5 and 18 kg during the pregnancy. Women who are overweight or obese preconception need not gain as much weight during pregnancy (Table 12.2).

TABLE 12.2

Recommended Weight Gain in Pregnancy Based on Prepregnancy Body Mass Index

Prepregnancy BMI (kg/m²)	Total Weight Gain (kg)
Underweight: BMI <18.5	12.5–18 kg
Normal weight: BMI 18.5–24.9	11.5–16 kg
Overweight: BMI 25–29.9	7–11.5 kg
Obese: BMI ≥30.0	5–9 kg

Source: Adapted from the Institute of Medicine, *Weight gain during pregnancy: Reexamining the guidelines*, Washington, DC: National Academies Press, 2009.

Note: BMI, body mass index.

Adequate weight gain is critical for optimal maternal (e.g., preventing maternal morbidity and mortality) and fetal outcomes (e.g., adequate growth and development). Maternal BMI, nutritional stores, diet, and placental blood flow determine the availability of nutrients to the fetus, which in turn influences fetal maturation and weight gain. Although a mother's body has improved utilization of nutrients during pregnancy, these changes might not be sufficient if her prenatal nutritional status or her diet during pregnancy is inadequate [17]. An inadequate supply of nutrients has a restrictive effect on fetal growth, leading to SGA and LBW [18].

Poor diet during pregnancy, as well as a lack of micronutrient supplementation, can lead to micronutrient deficiencies. Anemia (a low hemoglobin concentration) is prevalent among women in LMIC, where an estimated 40% prevalence among nonpregnant and 49% among pregnant women [19]. Anemia may be a cause of iron or other micronutrient deficiency, although it has many other genetic and nonnutritional causes. Anemia in pregnancy has been associated with an increased risk of maternal mortality and fetal growth restriction. It has been hypothesized that periconceptional anemia may influence hormone physiology, and that it may adversely affect fetal growth [20]. Maternal vitamin D deficiency, especially in early pregnancy, has also been shown to increase the risk of preterm birth and SGA [21]. Recent evidence suggests that preconceptional and periconceptional intake of multiple micronutrient supplements reduces the risk of LBW, SGA, and stillbirths to a greater extent than iron, with or without folic acid [22]. This indicates the importance of multiple micronutrients in the maternal diet for fetal growth and development [23].

MATERNAL BODY MASS INDEX

Maternal nutrition is generally a consequence of the interplay of many factors, including socioeconomic status, adequate nutrition, cultural norms, heavy labor, multiple pregnancies, and frequent infection. Preconception body mass index (BMI) is a strong predictor of pregnancy outcomes and neonatal birth weight. A low BMI (<18.5) is associated with a higher risk of preterm delivery; an even greater association is with neonatal weight outcomes and a higher risk of LBW and SGA.

The mechanism behind the association could be due to a direct consequence of poor nutrition and fewer available calories leading to fetal growth restriction [24]. However, other pathways, such as undernutrition leading to immune deficiency and infections, or physical labor leading to low BMI, could also explain poor fetal outcomes. Moreover, as discussed, maternal micronutrient status is critical for fetal growth and development. Preconception iron deficiency is associated with a fivefold increase in the risk of fetal growth restriction and a sevenfold increase in LBW [25].

MATERNAL SHORT STATURE

Maternal short stature, often a consequence of chronic undernutrition, has been shown to have a strong relationship with preterm birth, LBW, and SGA. In the mid-1990s, the WHO Collaborative Study of Maternal Anthropometry and Pregnancy Outcomes described a significant association of maternal short stature with LBW, SGA, and preterm birth [24]. More recently, in 2015, the Child Health Epidemiology Reference Group SGA/Preterm Birth Working Group published a meta-analysis in LMIC, in which they reported that nearly 6.5 million SGA and preterm births from these regions may be associated with low maternal height. According to the report, 5.5 million term SGA (18.6% of the global total), 550,800 preterm AGA (5.0% of the global total), and 458,000 preterm SGA (16.5% of the global total) births may be associated with maternal short stature. In turn, SGA infants are at a higher risk of stunted linear growth; 20% of stunting is attributable to fetal growth restriction [26].

Thus, many infants are born undernourished and underweight because their mothers are undernourished and underweight (Figure 12.3) [27]. This negative cycle underscores the importance of optimal nutrition for girls and women. Moreover, nearly 16 million women aged 15 to 19 years give birth annually [28], and these adolescent pregnancies are associated with increased maternal morbidity and mortality,

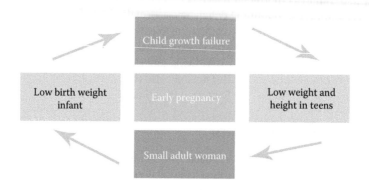

FIGURE 12.3   The intergenerational cycle of undernutrition. (Reproduced from World Food Programme, *WFP Bangladesh: Nutrition strategy 2012–2016*, Rome: World Food Programme, 2012. With permission.)

as well as preterm and LBW. In these young pregnancies, both mother and growing fetus are competing for nutrients, leading to a higher risk of poor outcomes.

MATERNAL CHRONIC DISEASES

Maternal hypertension and diabetes increase the risk of SGA infants. There is some evidence to suggest that prepregnancy hypertension and preeclampsia impair placental functioning, increasing the risk of poor nutrient and oxygen transfer to the fetus and, consequently, growth restriction (see Chapter 11 for more details). In terms of prepregnancy diabetes, as well as gestational diabetes, maternal hyperglycemia likely impacts the production of various placental proteins. These placental proteins affect placental metabolism, growth, and development [29].

MATERNAL COMMUNICABLE DISEASES

Maternal communicable diseases are another important risk factor, especially in LMIC, where infections can chronically affect maternal nutritional status and further expose fetuses to unfavorable growth environments [1]. Specifically, pregnant mothers infected with *Plasmodium falciparum* or *Plasmodium vivax* malaria are at increased risk of delivering an infant LBW, SGA, or both [30].

ENVIRONMENTAL FACTORS

Studies on urban air pollution, as well as tobacco smoke (firsthand and secondhand) have been well studied in relation to fetal outcomes, including growth and neurotoxic effects. Smoking during pregnancy has been adversely associated with reduced birth weight, with maternal smoking doubling the risk of having a LBW, SGA, and/or preterm infant. Secondhand tobacco smoke exposure has also been linked with these consequences, as well as with prenatal mortality.

In regards to air pollution or particulate matter, microscopic chemicals emitted from power plants, industries, and vehicles are easily inhaled by pregnant mothers, which can subsequently cause growth complications and other adverse maternal health consequences, such as deficiencies in lung function. In fact, as the third trimester is the most prolific period of growth, particulate matter has been positively associated with poor outcomes, such as growth restriction and LBW infants, most often when there is exposure within the third trimester window [31].

OUTCOMES OF LOW BIRTH WEIGHT AND SMALL FOR GESTATIONAL AGE

SHORT-TERM CONSEQUENCES

The association between LBW and SGA and increased risk of mortality is well documented. Several studies have demonstrated a greater relative risk for neonatal (1–28 days) and postneonatal (29–365 days) mortality associated with SGA, as compared to AGA [11]. The highest risk is for infants born preterm SGA relative to term

AGA. Moreover, over 50% of SGA babies, although not LBW, are still at double the risk of neonatal mortality than term AGA infants [32]. It is also estimated that 50% of unexplained stillbirths may be related to undiagnosed IUGR [33].

Many of the metabolic, hematological, and immunological disturbances associated with LBW and SGA could account for this high mortality risk, as indicated below:

- *Metabolic*—Hyperglycemia, hypoglycemia, hypocalcemia, hypothermia, late metabolic acidosis
- *Hematological*—Polycythemia (hyperviscosity)
- *Immunological*—Reduced immune response (reduced lymphocyte count, reduced immunoglobulin levels, increased risk of infection)

Moreover, preterm infants can suffer from many other complications in addition to the above disturbances. Preterm babies often have multisystem issues affecting their respiratory, gastrointestinal, cardiovascular, and central nervous systems. Respiratory distress syndrome, neonatal pneumonia, congenital heart defects, hyperbilirubinemia, intraventricular hemorrhage, and infections are some of the most common issues affecting preterm neonates.

INTERMEDIATE CONSEQUENCES

Linear Growth
Nearly all stunting occurs in the first 1,000 days after conception, and is most influenced by maternal and fetal nutrition. Maternal stunting puts infants at risk of being born SGA. In turn, SGA is one of the strongest predictors for childhood stunting. It has been estimated that nearly 20% of childhood stunting could be a consequence of being born SGA. Interestingly, being SGA carries a higher risk of stunting than being born preterm [11].

Neurodevelopmental
In the intermediate, there is consistent evidence for the association of SGA with impaired cognitive and motor development in early childhood (up to 36 months of age). Lower cognitive scores, poorer problem-solving abilities, and reduced activity have frequently been reported in infants and young children born SGA [34,35]. Further, several studies have identified a lasting adverse influence of LBW and SGA on IQ and academic achievement at school age and even during middle and late adolescence [1,36]. Behavioral problems, such as apathy and irritability, have been reported more frequently in SGA children, especially at preschool age. Substantial evidence also suggests an association of stunting with poor motor and cognitive development.

The underlying biology behind these neurodevelopmental consequences in infants born SGA is not fully understood. Cerebral cortex thinning and reduced cortical gray matter have been reported in very low birth weight and SGA births, suggesting that fetal growth restriction can affect brain development.

LONG-TERM CONSEQUENCES

The pathophysiological changes occurring in LBW and SGA babies puts them at risk of adverse health outcomes in the long term, which includes increased risk factors for chronic disease (e.g., obesity, hypertension, reduced glucose tolerance, increased insulin resistance, and dyslipidemia) and a higher incidence of noncommunicable diseases (e.g., type 2 diabetes, cardiovascular disease, and chronic lung and kidney disease).

THE DEVELOPMENTAL ORIGINS OF HEALTH AND DISEASE

The *developmental origins of health and disease* (DOHaD) hypothesis, originally described in 1989, proposed that an association existed between LBW and the risk of developing coronary heart disease later in life. Further studies suggested that prenatal and early postnatal undernutrition had permanent effects on the body's structure and function, predisposing it to chronic disease in adult life, and demonstrating the phenomenon of developmental plasticity and epigenetics.

The well-known Dutch famine studies demonstrated that individuals who had prenatal exposure to this 1944–1945 famine, especially those in the first trimester of pregnancy, were more likely to suffer from obesity, diabetes, and cardiovascular disease. Interestingly, in some cases, this was true even for infants not born LBW [37].

Evidence suggests that the fetus adapts to a limited supply of nutrients *in utero* by undergoing structural and physiological changes, such as the redistribution of blood flow to vital organs, alteration in the production of growth-influencing hormones, and changes in the glucose-insulin metabolism [33]. This forms the basis of the suggested "fetal programming" phenomenon. Some suggest that epigenetic changes inducing long-term alterations in gene expressions may be one of the mechanisms contributing to this "programming."

Although associated with short-term gains [38], rapid catch-up growth in LBW infants, leading to relatively quick weight gain in childhood, is linked to an increased risk of chronic disease in adults [33]. Though they possess reduced body fat mass, SGA children have been shown to have greater visceral adiposity and more centralized distribution of fat mass [39]. The fetal development process of adiposity is regarded as one of the factors influencing the increased insulin resistance associated with SGA infants in their adulthood [40].

However, although the pathophysiology for impaired renal and lung outcomes seen in SGA infants is not fully understood, it is postulated from the literature that fetal nutrient restriction influences the number of renal nephrons and the glomerular filtration rate, as well as the lung acinar and alveoli development [33].

STUNTING AND HUMAN CAPITAL

Height achieved in the first 1,000 days has been recognized as a good predictor of human capital [38]. Nearly 10% of children born SGA do not achieve catch-up growth after the first 1,000 days and remain stunted. Moreover, SGA children have presented with earlier starts to puberty and a short pubertal growth spurt, which further contributes to a shorter adult height. Stunting is associated with a loss of developmental potential, due to its adverse influence on cognitive function and economic yield at a population level.

INTERVENTIONS

In the presence of an SGA fetus during pregnancy, timely diagnosis of SGA, followed by optimal monitoring and surveillance, is important. Simple ultrasound as well as more modern techniques such as umbilical artery doppler, cardiotocography, and the assessment of amniotic fluid can be used for monitoring [41]. Diagnosis and monitoring is vital for the subsequent appropriate management and delivery of the SGA fetus. However, the prevention of SGA is of prime importance. Optimum nutrition, timely and effective antenatal care, improved management of high-risk pregnancies, and interventions targeting the underlying determinants of undernutrition are all pivotal for promoting optimal fetal development and preventing SGA.

A recent review of interventions across the continuum of care, and their effect on neonatal outcomes, identified key actions that have a significant effect in preventing SGA. These include multiple-micronutrient supplementation, balanced energy protein supplementation, intermittent preventive treatment for malaria in pregnancy, insecticide-treated bed nets, and antiplatelet medications for preeclampsia [42]. Furthermore, a previous review recognized antiplatelet agents and multiple-micronutrient supplementation as the most effective interventions for preventing SGA, and noted that they were appropriate to use by all pregnant women [43]. For high-risk pregnancies, the same review identified antiplatelets <16 weeks, cessation of smoking, and progesterone therapy as being most effective.

The Importance of Preconception Care and Interventions

Preconception care, as defined by the WHO, is the "provision of preventive, promotive or curative health and social interventions before conception occurs" [44]. It can be provided to both men and women; however, adolescent girls and women are especially vulnerable, and thus often the focus of these interventions. Preconception care for these groups is vital for the promotion of a healthy pregnancy, and the prevention of adverse fetal outcomes, including LBW. Research has recognized the association between maternal underweight and an increased risk of SGA [45], underscoring the significance of preconception interventions that address risk factors for maternal undernutrition [46]. Shorter interpregnancy intervals are associated with an increased risk of preterm and LBW [47]. A strong need exists for steps promoting optimum nutrition, family planning, and appropriate interpregnancy intervals. Moreover, the prevention of early and adolescent pregnancies is integral to the prevention of undernutrition in pregnant women [48]. On the other hand, childbearing past the age of 35 years is also associated with increased odds of LBW. More specifically, preconception folic acid supplementation has been shown to reduce the risk of SGA (<5th percentile) probably through epigenetic mechanisms [49].

MICRONUTRIENT SUPPLEMENTATION IN PREGNANCY

With the nutritional needs of both the mother and child being the greatest in the first 1,000 days, there has been significant research into the impact of supplementation. The 2016 WHO antenatal guidelines recommends routine iron and folic acid supplementation in pregnancy, where the dosage of iron is dependent upon the population-level prevalence of anemia [50]. The WHO is not currently recommending multiple-micronutrient supplementation during pregnancy [50]; however, over the past decade, a multiple-micronutrient supplement, UNIMMAP, has emerged as a means to cost-effectively combat other maternal micronutrient deficiencies during pregnancy and is being distributed in some countries. The UNIMMAP supplement contains vitamin A, vitamin B_1, vitamin B_2, niacin, vitamin B_6, vitamin B_{12}, folic acid, vitamin C, vitamin D, vitamin E, copper, selenium, and iodine, as well as 30 mg of iron and 15 mg of zinc.

According to a recent *Cochrane Review* [22], women who took UNIMMAP, compared to an iron supplement with or without folic acid, had a reduced risk of SGA infants (RR 0.90, 95% CI: 0.83, 0.97), LBW infants (RR 0.88, 95% CI: 0.85, 0.91), and stillbirths (RR 0.91, 95% CI: 0.85, 0.98). There were no differences found in preterm births, miscarriage, and anemia in the third trimester. These findings and previous reviews support the guidance to replace iron with folic acid supplements with multiple micronutrient supplements in pregnant mothers [51,52].

BALANCED ENERGY AND PROTEIN SUPPLEMENTATION

As mentioned, undernourishment in pregnancy increases the risk of adverse fetal outcomes. Gestational weight gain has been associated with increased birth weight and fewer SGA babies. To promote maternal weight gain, evidence indicates that balanced energy and protein supplementation increases mean birth weight and reduces the risk of stillbirths by 40% and SGA babies by 21%. However, findings regarding the reduction of preterm and LBW are inconclusive. Currently, there is no recommendation to increase energy and protein intake during pregnancy; however, guidance on appropriate weight gain has been well established [53,54].

CESSATION OF SMOKING AND ENVIRONMENTAL INTERVENTIONS

Tobacco smoking remains one of the few preventable factors associated with infants born preterm and/or LBW. Interventions regarding the promotion of smoking cessation include psychosocial interventions, pharmacological aids, and nicotine replacement therapy (NRT). There is only evidence to suggest that psychosocial interventions are associated with an increased proportion of women quitting smoking in late pregnancy, and subsequently reducing preterm and LBW infants [55,56]. Further studies are required to understand the impact of pharmacological and NRT interventions, as current reviews are inconclusive.

MALARIA PREVENTION

In moderate- to high-transmission settings, *P. falciparum* and *P. vivax* infection leading to placental parasitaemia and maternal anemia, can both contribute to LBW and SGA. The prevention and treatment of malaria is a priority intervention of the WHO in these areas. The use of insecticide-treated bed nets, intermittent preventive treatment of malaria with sulfadoxine-pyrimethamine (IPTp-SP), and appropriate case management are the components of this three-pronged approach. Three or more doses of IPTp-SP have been associated with higher mean birth weights and an estimated relative risk reduction for LBW by 20% [57]. Other reviews have estimated a 37% decrease in the risk of LBW infants in unprotected women (in their first or second pregnancy) as a result of IPTp-SP use [1]. Moreover, insecticide-treated bed nets have been associated with a 23% risk reduction for LBW.

CONCLUSION AND RECOMMENDATIONS

In conclusion, there has been significant progress in the delineation of risk factors, etiological pathways, and the consequences of SGA and LBW babies. In fact, the association between LBW, including SGA, and coronary heart disease and stroke in later life has been recognized. However, quite a few limitations exist in current data and research.

First and foremost, clarification and consistency around the use of SGA and IUGR definitions are necessary in order to observe accurate estimates of incidence and prevalence, as well as to understand the effectiveness and efficaciousness of interventions. Secondary to this is that, although maternal nutrition interventions that include general dietary advice are central in terms of reducing the risk of SGA, there is little hope of success unless the underlying determinants of undernutrition, including illiteracy, poverty, and the empowerment of women, are addressed.

Although several interventions that address fetal growth restriction and SGA have been identified, further research is needed to study the associated impact of these measures on maternal health and pregnancy outcomes [43]. Future research should particularly focus on the association between prepregnancy BMI and fetal growth restriction, as well as the relationship of maternal micronutrient deficiencies and poor fetal weight [58]. Finally, further studies should be conducted on the underlying mechanism of the long-term effects of IUGR and SGA on neurodevelopmental and metabolic outcomes [36].

REFERENCES

1. Kiess W, Chernausek SD, Hokken-Koelega ACS. *Small for gestational age causes and consequences.* Basel: Karger; 2009.
2. Wilcox AJ. On the importance and the unimportance of birthweight. *Int J Epi* 2001; 30(6):1233–41.
3. World Health Organization. Technical Report Series #27. Expert Group on Prematurity. Geneva: World Health Organization; 1950.
4. World Health Organization. *Public health aspects of low birth weight: Third report of the Expert Committee on Maternal and Child Health.* Geneva: World Health Organization; 1961.

5. Kramer MS. Determinants of low birth weight: Methodological assessment and meta-analysis. *Bull World Health Organ* 1987; 65:663–737.

6. Kramer MS. The epidemiology of low birthweight. *Nestle Nutr Inst Workshop Ser* 2013; 74:1–10.

7. Lee PA, Chernausek SD, Hokken-Koelega AC et al. International Small for Gestational Age Advisory Board Consensus Development Conference statement: Management of short children born small for gestational age, April 24–October 1, 2001. *Pediatrics* 2003; 111:1253–61.

8. Villar J, Cheikh Ismail L, Victora CG et al. International standards for newborn weight, length, and head circumference by gestational age and sex: The newborn cross-sectional study of the INTERGROWTH–21st project. *Lancet* 2014; 384:857–68.

9. International Food Policy Research Institute. *Global nutrition report—From promise to impact: Ending malnutrition by 2030.* Washington, DC: International Food Policy Research Institute; 2016.

10. Lee AC, Katz J, Blencowe H et al. National and regional estimates of term and preterm babies born small for gestational age in 138 low-income and middle-income countries in 2010. *Lancet Glob Health* 2013; 1:e26–36.

11. Black RE, Victora CG, Walker SP et al. Maternal and child undernutrition and over-weight in low-income and middle-income countries. *Lancet* 2013; 382:427–51.

12. Houk CP, Lee PA. Early diagnosis and treatment referral of children born small for gestational age without catch-up growth are critical for optimal growth outcomes. *Int J Pediatr Endocrinol* 2012; 2012:11.

13. Blencowe H, Cousens S, Chou D et al. Born too soon: The global epidemiology of 15 million preterm births. *Reprod Health* 2013; 10(Suppl 1):S2.

14. Ramakrishnan U. Nutrition and low birth weight: From research to practice. *Am J Clin Nutr* 2004; 79:17–21.

15. Black RE. Global prevalence of small for gestational age births. *Nestle Nutr Inst Workshop Ser* 2015; 81:1–7.

16. World Health Organization. *Global nutrition targets 2025: Low birth weight policy brief.* Geneva: World Health Organization; 2014.

17. Sharlin J, Edelstein S. *Essentials of life cycle nutrition.* Sudbury, MA: Jones and Bartlett Learning; 2011.

18. King JC. The risk of maternal nutritional depletion and poor outcomes increases in early or closely spaced pregnancies. *J Nutr* 2003; 133:S1732–6.

19. Save the Children. *Nutrition in the first 1,000 days: State of the world's mothers 2012.* London: Save the Children; 2012.

20. Allen LH. Biological mechanisms that might underlie iron's effects on fetal growth and preterm birth. *J Nutr* 2001; 131:S581–9.

21. Salam RA, Das JA Ali A et al. Maternal undernutrition and intrauterine growth restriction. *Expert Rev Obstet Gynecol* 2013; 8(6):559–67.

22. Haider BA, Bhutta ZA. Multiple-micronutrient supplementation for women during pregnancy. *Cochrane Database Syst Rev* 2015:CD004905.

23. Ramakrishnan U, Grant F, Goldenberg T et al. Effect of women's nutrition before and during early pregnancy on maternal and infant outcomes: A systematic review. *Paediatr Perinat Epidemiol* 2012; 26 Suppl 1:285–301.

24. Kozuki N, Lee AC, Black RE et al. Nutritional and reproductive risk factors for small for gestational age and preterm births. *Nestle Nutr Inst Workshop Ser* 2015; 81:17–28.

25. Dean SV, Imam AM, Lassi ZS et al. Importance of intervening in the preconception period to impact pregnancy outcomes. *Nestle Nutr Inst Workshop Ser* 2013; 74:63–73.

26. Kozuki N, Katz J, Lee AC et al. Short maternal stature increases risk of small-for-gestational-age and preterm births in low- and middle-income countries: Individual participant data meta-analysis and population attributable fraction. *J Nutr* 2015; 145:2542–50.

27. World Food Programme. *WFP Bangladesh: Nutrition strategy 2012–2016.* Rome: World Food Programme; 2012.

28. World Health Organization. Adolescent pregnancy fact sheet. Geneva: World Health Organization; 2014.

29. Vambergue A, Fajardy I. Consequences of gestational and pregestational diabetes on placental function and birth weight. *World J Diabetes* 2011; 2:196–203.

30. Rijken MJ, De Livera AM, Lee SJ et al. Quantifying low birth weight, preterm birth and small-for-gestational-age effects of malaria in pregnancy: A population cohort study. *PLoS One* 2014; 9:e100247.

31. Vinikoor-Imler LC, Davis JA, Meyer RE et al. Associations between prenatal exposure to air pollution, small for gestational age, and term low birthweight in a state-wide birth cohort. *Environ Res* 2014; 132:132–9.

32. Katz J, Lee AC, Kozuki N et al. Mortality risk among term and preterm small for gestational age infants. *Nestle Nutr Inst Workshop Ser* 2015; 81:29–35.

33. Salam RA, Das JK, Bhutta ZA. Impact of intrauterine growth restriction on long-term health. *Curr Opin Clin Nutr Metab Care* 2014; 17:249–54.

34. Tofail F, Hamadani JD, Ahmed AZ et al. The mental development and behavior of low-birth-weight Bangladeshi infants from an urban low-income community. *Eur J Clin Nutr* 2012; 66:237–43.

35. Walker SP, Wachs TD, Gardner JM et al. Child development: Risk factors for adverse outcomes in developing countries. *Lancet* 2007; 369:145–57.

36. Walker SP, Wachs TD, Grantham-McGregor S et al. Inequality in early childhood: Risk and protective factors for early child development. *Lancet* 2011; 378:1325–38.

37. Roseboom T, de Rooij S, Painter R. The Dutch famine and its long-term consequences for adult health. *Early Hum Dev* 2006; 82:485–91.

38. Victora CG, Adair L, Fall C et al. Maternal and child undernutrition: Consequences for adult health and human capital. *Lancet* 2008; 371:340–57.

39. Ibanez L, Lopez-Bermejo A, Suarez L et al. Visceral adiposity without overweight in children born small for gestational age. *J Clin Endocrinol Metab* 2008; 93:2079–83.

40. Jaquet D, Deghmoun S, Chevenne D et al. Dynamic change in adiposity from fetal to postnatal life is involved in the metabolic syndrome associated with reduced fetal growth. *Diabetologia* 2005; 48:849–55.

41. Royal College of Obstetricians & Gynaecologists. *The investigation and management of the small-for-gestational age fetus.* London: Royal College of Obstetricians and Gynaecologists; 2014.

42. Bhutta ZA, Das JK, Bahl R et al. Can available interventions and preventable deaths in mothers, newborn babies, and stillbirths, and at what cost? *Lancet* 2014; 384:347–70.

43. Morris RK, Oliver EA, Malin G et al. Effectiveness of interventions for the prevention of small-for-gestational age fetuses and perinatal mortality: A review of systematic reviews. *Acta Obstet Gynecol Scand* 2013; 92:143–51.

44. World Health Organization. *Meeting to develop a global consensus on preconception care to reduce maternal and childhood mortality and morbidity.* Geneva: World Health Organization; 2013.

45. Dean SV, Lassi ZS, Imam AM et al. Preconception care: Nutritional risks and interventions. *Reprod Health* 2014; 11(Suppl 3):S3.

46. Lassi ZS, Dean SV, Mallick D et al. Preconception care: Delivery strategies and packages for care. *Reprod Health* 2014; 11(Suppl 3):S7.

47. Lassi ZS, Kumar R, Mansoor T et al. Essential interventions: Implementation strategies and proposed packages of care. *Reprod Health* 2014; 11(Suppl 1):S5.

48. Lieberman E, Lang JM, Ryan KJ et al. The association of inter-pregnancy interval with small for gestational age births. *Obstet Gynecol* 1989; 74:1–5.

49. Hodgetts VA, Morris RK, Francis A et al. Effectiveness of folic acid supplementation in pregnancy on reducing the risk of small-for-gestational age neonates: A population study, systematic review and meta-analysis. *BJOG* 2015; 122:478–90.
50. World Health Organization. *WHO recommendations on antenatal care for a positive pregnancy experience*. Geneva: World Health Organization; 2016.
51. Ramakrishnan U, Grant FK, Imdad A et al. Effect of multiple micronutrient versus iron-folate supplementation during pregnancy on intrauterine growth. *Nestle Nutr Inst Workshop Ser* 2013; 74:53–62.
52. Bhutta ZA, Das JK, Rizvi A et al. Evidence-based interventions for improvement of maternal and child nutrition: What can be done and at what cost? *Lancet* 2013; 382:452–77.
53. Ota E, Hori H, Mori R et al. Antenatal dietary education and supplementation to increase energy and protein intake. *Cochrane Database Syst Rev* 2015; CD000032.
54. Imdad A, Bhutta ZA. Maternal nutrition and birth outcomes: Effect of balanced protein-energy supplementation. *Paediatr Perinat Epidemiol* 2012; 26(Suppl 1):178–90.
55. Chamberlain C, O'Mara-Eves A, Oliver S et al. Psychosocial interventions for supporting women to stop smoking in pregnancy. *Cochrane Database Syst Rev* 2013; CD001055.
56. Lassi ZS, Kumar R, Mansoor T et al. Essential interventions: Implementation strategies and proposed packages of care. *Reprod Health* 2014; 11(Suppl 1):S5.
57. World Health Organization. *Policy brief for the implementation of intermittent preventive treatment of malaria in pregnancy using sulfadoxine-pyrimethamine (IPTp-SP)*. Geneva: World Health Organization; 2013.
58. Neggers Y, Goldenberg RL. Some thoughts on body mass index, micronutrient intakes and pregnancy outcome. *J Nutr* 2003; 133:S1737–40.

Section VI

Pathophysiology and
Nutrition Requirements
in Child Malnutrition

13 Acute Malnutrition

André Briend

CONTENTS

INTRODUCTION

Children can be underweight because they have a low weight in relation to their height and/or because they are short in relation to their age. Waterlow introduced the term *wasted* for children with a low weight-for-height (WFH), and *stunted* for those who have a low height-for-age [1]. Children suffering from acute food shortage become wasted as they lose weight, but their height remains constant, and wasting is often considered as reflecting an episode of acute malnutrition (AM). In contrast, stunting is regarded as a more chronic form of malnutrition. Although wasting can also be chronic, and linear growth stops in the case of AM, AM is often defined by wasting, more specifically by a WFH <–2 z-score of the World Health Organization (WHO) growth standard [2]. Moderate acute malnutrition (MAM) is defined by a z-score between –2 and –3, and severe acute malnutrition (SAM) is defined by a WFH <-3 z-score or a mid-upper arm circumference (MUAC) <115 mm (in children aged 6 to 60 months) or the presence of nutritional edema [3]. The MUAC-based definition of SAM was introduced to include children with a low MUAC, because these children are at high risk of death [4]. In programs that aim to prevent malnutrition-associated mortality, MAM is also often defined by a MUAC between 115 and 125 mm [5]. The cutoffs for MAM and SAM are convenient for establishing programs, but there is no abrupt change of the pathophysiology or of the associated risk around these cutoffs, and there is a continuum between SAM and MAM.

Current estimates suggest that, globally, there are approximately 51 million wasted children defined by a WFH <–2. Of these, 19 million are severely wasted (WFH <–3), leading to approximately 540,000 deaths per year [6]. These estimates are based on prevalence data, whereas incidence would be needed to assess the number of deaths related to acute malnutrition. In addition, these estimates do not include children with a MUAC <115 mm with a WFH >–3, nor those with edema who have a high risk of death. The numbers of children affected every year by SAM and the associated mortality are, presumably, considerably higher.

PATHOPHYSIOLOGY

Children with AM have a history of insufficient energy and nutrient intakes compared to their needs. This can be due either to poor access to a nutrient-dense diet and/or inflammation (usually due to infections), which can induce anorexia and further increase nutritional needs. The organism adapts to this insufficient energy and nutrient intake by consuming its own tissues, mainly fat and muscle, which results in changes in body composition. The organism also adapts to the insufficient energy and nutrient by decreasing its energy and nutrient expenditure via a reduction in the activity of different organs, which impairs their function.

CHANGES IN BODY COMPOSITION

The weight deficit observed in malnourished affects tissues and organs in different ways [7]. Animal models show that the brain weight is relatively conserved compared to the thymus, heart, and kidneys, but that it is fat and muscle that are reduced the most in relation to body weight [8]. This decrease in fat and muscle can largely be explained by metabolic adjustments that take place when energy intake is insufficient [9]. In some children, these changes in body composition are associated with edema.

Decrease in Fat Mass

In well-nourished subjects, energy needs are largely provided by the metabolism of carbohydrate and fat, mainly carbohydrate following a meal, and then fat after a few hours of fasting. This switch between these two main fuels is due to a decrease of insulin secretion and to an increase of glucagon following the decrease of blood glucose concentration. This switch to fat as the main fuel is needed, as body carbohydrate reserves in the liver and muscle are limited and exhausted after a few hours of fasting [9]. In contrast, fat stores are usually large enough to maintain metabolism for several weeks. The brain, however, usually only consumes glucose as fuel, so a minimum must be synthesized by gluconeogenesis, either from glucogenic amino acids derived from muscle or from the glycerol derived from triglycerides. This process is set to a minimum after a few days, as the brain and other organs adapt and start using ketone bodies, produced during fat catabolism as fuel instead of carbohydrates. This switch to fat as the main fuel occurs much faster in children than in adults, presumably as the brain, the major glucose-consuming organ, is proportionally larger in children. Body fat seems to be the main factor limiting survival in malnourished children in the absence of infections [10]. Leptin, which is a marker of fat mass, is associated with survival in children treated for SAM [11]. In addition to its role as fuel, fat may also influence survival through its action on the immune system [12].

Decrease of Muscle Mass

Muscle is the main storage site of amino acids in the body and, when protein intake is insufficient to sustain metabolism, these amino acids are released, leading to a decrease in muscle mass. The release of amino acids is the result of the activation of ubiquitin-proteasome pathway, which are also activated in numerous conditions leading to muscle mass depletion, including trauma, diabetes, uremia, hyperadrenocorticolism, hyperthyroidism, immobilization, and sepsis [13]. In the event of infection, muscle breakdown can be accelerated by the need to synthesize acute phase proteins, which have a high content of aromatic amino acids. The synthesis of these acute phase proteins, which can account for up to 30% of total protein synthesis, requires a high level of protein catabolism, mainly from muscle [14].

Muscle, like all other lean tissues, contains a wide range of nutrients, which are essential for metabolism and are not stored in specific organs. Anorexia can occur during malnutrition, leading to lean tissue catabolism and the release of these nutrients, making them available for metabolism [15]. This may also contribute to muscle mass depletion.

Muscle mass has been shown to be linked with survival in a variety of clinical conditions in adults, including infections, cancer, and cirrhosis [16]. Epidemiological studies suggest that it is linked to survival in children as well [17], but this link may depend on the presence of infections associated with malnutrition as, in their absence, fat appears to be more important [18]. The critical role of muscle mass in making children more vulnerable to malnutrition is made likely by its low value in relation to body mass compared to adults [16].

The Presence of Edema

Edema is due to an expansion of the interstitial compartment, that is, of the extracellular water outside the vascular system. The association of edema with malnutrition has been known since biblical times, but only attracted the attention of clinicians after the publication of works by Cecily Williams, who introduced the term *kwashiorkor* to medical literature [19].

Edematous malnutrition is associated with a fatty liver, skin lesions, and apathy. To date, its pathophysiology is poorly understood, with none of the explanations proposed so far adequately explaining its clinical characteristics.

Possible Mechanisms of Edema

Protein deficiency was initially proposed because the protein intake of children with kwashiorkor was lower than what was then recommended for healthy children. The frequently decreased serum albumin levels in edematous malnutrition were attributed to insufficient protein intake, leading to decreased albumin synthesis, a decrease in plasma oncotic pressure, and edema. In this interpretation, a fatty liver was attributed to an insufficient synthesis of the lipoproteins needed to export fat from the liver [7].

However, all of the steps of this causal link were to be challenged. Protein requirements have been reestimated at lower levels, and the difference between intake and estimated requirements is not clear. A comparison of the diet of malnourished children with and without edema failed to show a lower protein intake in those who develop edema compared to healthy controls [20]. Albumin is an acute phase protein, which is influenced not only by protein intake but also by the presence of infection, and its relation with protein intake is not clear. The link between plasma albumin and edema remains debated [21,22].

Oxidative stress has been proposed as a possible cause of edema, based on the observation that circulating glutathione concentrations are low in edematous malnutrition [23]. As a result of infections or of the action of toxins, some malnourished children do not have the capacity to respond adequately to the associated oxidative stress. This may lead to membrane damage as a result of lipid oxidation or to changes in sulfated complex carbohydrates, which are present in the interstitial tissue, leading to edema [22]. This hypothesis is supported by consistent reports that edematous malnutrition is associated with increased oxidative stress [24,25]. Consistent with this hypothesis, supplementation of edematous children with glutathione or the antioxidant alpha-lipoic acid has a favorable effect on the clinical outcome of kwashiorkor [26]. It is not clear, however, whether oxidative stress is a cause or a consequence of edematous malnutrition. Several elements are not consistent with the primary role of oxidative stress as a cause of edema. Children infected with HIV, an infection associated with a high level of oxidative stress, more frequently have a nonedematous form of severe malnutrition [27]. There is no association between polymorphic variation in a series of redox enzyme genes associated with the response to an oxidative stress (catalase, superoxide dismutases, epoxide hydrolases, quinone oxidoreductase, and NADPH oxidase), and the presence of edema in children with SAM [28]. An attempt to prevent kwashiorkor via supplementation with an antioxidant including vitamins, minerals, and N-acetyl cysteine, which stimulates glutathione synthesis, did not show any effect [29].

One possible role of the gut microbiota, which plays an important role in the metabolism and immune response of the host, has been suggested by a study of twins in Malawi, only one of whom developed edematous malnutrition (discordant pairs). The twin with edematous malnutrition had an immature gut microbiota compared to that of the healthy twin, and this microbiota induced weight loss when transplanted to germ-free mice receiving a suboptimal diet [30]. However, another study did show that nonedematous SAM children have lower gut microbiota diversity compared to edematous SAM children, but no clear compositional differences were identified [31].

CHANGES IN BODY FUNCTIONS

In parallel with the change in body composition, the organism adapts to malnutrition by reducing the activity of virtually all of its body functions [7]. These changes have mainly been described in children suffering from SAM, but they also occur to a lesser extent in children with MAM.

Physical Activity

Malnourished children have reduced physical activity. This has been documented in children with SAM with accelerometers [32], but also with more moderate degrees of wasting, stunting, and anemia [33]. This reduction in physical activity can be regarded as an immediate adaptation to low energy intake, but the resulting decreased exploration of the environment may negatively affect cognitive development [34].

Sodium Pump Activity

The sodium pump maintains the concentrations of sodium and potassium in intra- and extracellular spaces. It expels sodium from the cells in exchange for potassium. This process is energy demanding and represents about 25% of the energy expenditure of the whole body at rest. The activity of the sodium pump is reduced in nonedematous malnutrition but is increased in edematous malnutrition, presumably to compensate for the increased cell membrane permeability to electrolyte [35]. The reduction of the sodium pump activity and membrane leakage are likely to explain the high sodium levels and low potassium levels observed within cells obtained from biopsies, and also the loss of total body potassium commonly observed in malnutrition, as potassium not entering the cells has to be excreted in the kidney [7].

Protein Metabolism

Early reports suggested that protein turnover slowed in children with SAM [36]. Later studies suggested, however, that a reduction of protein breakdown only occurred in children with edematous malnutrition and was maintained in a nonedematous form of malnutrition [37]. Protein breakdown, on the other hand, is increased in the presence of infection, which is also associated with an increased synthesis of acute phase proteins [14].

Renal Function

The kidney is one of the most energy-demanding organs, and its functions are reduced in malnutrition [38], demonstrated by a decreased capacity to concentrate or dilute urine, to excrete an acidic load, or to excrete excess sodium. This has implications for treatment, as these changes make children more vulnerable not only to dehydration but also to fluid overload in the event of aggressive rehydration.

Cardiac Function

There is a reduction of cardiac output in relation to body surface area, which is related to both a decreased heart rate and a decreased stroke volume. It is not clear whether this reduced function is due to a reduction of cardiac mass and volume, or to adaptation to a reduced demand [7]. Elevated levels of cardiac troponin in children with SAM suggest, however, that heart function is impaired by malnutrition, at least in its severe form [39]. This may explain the excess of deaths observed in SAM children receiving diets with a high sodium content [40].

NUTRITIONAL REQUIREMENTS OF CHILDREN WITH ACUTE MALNUTRITION

Children's nutritional requirements have two components: a maintenance component, which is needed to sustain metabolism, and a growth component, needed to synthesize new tissues. In malnourished children, the reestablishment of normal nutrient concentrations in different tissues for correcting preexisting deficiencies represents a third component.

In the first days of treatment, if the child has associated complications, especially infections, the priority is to treat these complications, and no attempt is made to promote weight gain. At this stage, requirements correspond to maintenance requirements and, for some nutrients, to the correction of associated deficiencies. During the recovery phase, growth (catch-up) can be 10 to 15 times faster than that of a well-nourished child of the same age. Nutritional requirements for growth are then increased in the same proportions. Nutrient requirements for maintenance and for growth vary in different proportions in relation to weight gain for energy, and for different nutrients. They also depend on the nature of the weight gain, mainly on the balance between lean and fat tissue.

ENERGY REQUIREMENTS

Requirements for Maintenance

Basal energy requirements in SAM children are influenced by two opposing factors. First, observed changes in body composition increase the proportion of organs, which have a high energy requirement per weight unit, such as the brain, heart, or kidneys. This tends to increase the energy requirements in relation to body weight.

On the other hand, most energy-demanding functions slow down as part of the adaptation to lower energy intakes, which tends to lower energy requirements. Studies describe a wide range of basal energy expenditure, ranging from 75 to 103 kcal/kg/day [7], slightly higher compared to about 75 kcal/kg in healthy children [41].

Requirements for Growth
The energy required for growth depends on the type of tissue being laid down [42]. Fat contains 9 kcal/g; about 10% extra is needed to incorporate each gram of fat into tissue, so that the total cost of 1 g of fat tissue deposited is about 10 kcal. Lean tissue consists of approximately 80% water and 20% protein. Protein contains 4 kcal/g, but the energy efficiency of protein deposition is only 50%, which means that for each gram of protein being deposited, about 8 kcal is needed. Hence, to deposit 1 g of lean tissue, which contains about 0.2 g of protein, 1.6 kcal is needed.

Total Energy Requirements
The total energy requirement (Etot) is the sum of energy needed for energy maintenance (Em) and for weight gain. This can be expressed as follows:

$$\text{Etot (kcal/kg/day)} = \text{Em (kcal/kg/day)} + (\text{Fg (g/kg/day)} \times 10\,\text{kcal})$$
$$+ (\text{Lg (g/kg/day)} \times 1.6\,\text{kcal})$$

where Fg is the fat gain and Lg is the lean tissue gain.

Total energy requirements increase rapidly with weight gain; during rapid catch-up, energy requirements approach twice the value of well-nourished children of the same age (Table 13.1).

TABLE 13.1
Energy Requirements in Relation to Weight Gain

Nature of Deposited Tissue	Total Energy Requirements (kcal/kg/day)		
	40% Lean, 60% Fat	50% Lean, 50% Fat	60% Lean, 40% Fat
Weight gain (g/kg/day)			
0	80	80	80
1	87	86	85
2	93	92	90
5	113	109	105
10	146	138	129
15	179	166	154

Note: The maintenance requirements were estimated at 80 kcal/kg/day.

PROTEIN REQUIREMENTS

Protein requirements also have two components: the basal requirement to maintain protein metabolism in the absence of weight gain, and a growth component, corresponding to the quantity of protein being deposited in new tissues.

Requirements for Maintenance

Maintenance requirements in protein are similar in children with SAM, as compared to well-nourished children and adults, at approximately 0.7 g/kg/day [7].

Requirements for Growth

Fat tissue contains only small amounts of proteins, and the effect of fat tissue deposition on protein requirements is negligible. Lean tissue contains about 0.2 g of protein per gram of tissue, but a higher amount is needed, as the amino acid profile of dietary protein is not identical to the proteins being laid down, and the efficiency of protein synthesis is also imperfect. An overall efficiency of 70% is a reasonable assumption [42].

Total Protein Requirements

The total protein requirement (Ptot) of a child recovering from SAM is the sum of requirements for maintenance and for growth:

$$\text{Ptot (g/kg/day)} = 0.70 \text{ (g/kg/day)} + (\text{Lg (g/kg/day)} \times 0.2)/0.70$$

where Lg is the lean tissue gain.

Protein requirements can be expressed as a percentage of energy in relation to weight gain and the nature of tissue deposition using the preceding equation. Figure 13.1 shows protein requirements for different proportions of lean tissue being

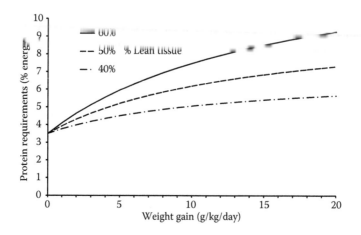

FIGURE 13.1 Protein requirements (% energy) in relation to weight gain and the percentage of lean tissue in weight gain. Energy maintenance estimated at 80 kcal/kg/day.

deposited during growth. It shows that protein represent about 3% of the energy requirement, in the absence of weight gain, and go up to nearly 9% when weight gain is very high. This percentage also increases with the proportion of lean tissue being laid down during growth.

These estimates are based on several approximations with regard to basal protein and energy requirements, and the efficiency of growth, but a change of parameters in the preceding equations shows that the proportion of energy that should be provided by protein is around 3% in children when weight gain is just maintained. During catch-up growth, protein requirements increase but never to more than 10% to 12% of energy. These estimates are to be compared with the diet currently recommended by World Health Organization at the beginning of treatment (F-75), of which protein provides 5% of total energy [43], and with the protein content of ready-to-use therapeutic foods (RUTF) used during the catch-up phase, which should be between 10% and 12% of total energy [44]. Of note, these protein requirements are calculated based on the quantity of proteins needed for synthesis of new tissues, and these estimates do not take into account the possible effects of protein quantity and quality on growth regulation and height gain [45]. The actual requirements for optimal height recovery may actually be higher.

Protein Quality

Children require essential amino acids to maintain their metabolism and synthesize new tissues. The proportion of essential amino acids needed for growth is higher than what is needed for maintenance (Table 13.2) [46].

Total amino acid requirement is the sum of amino acid needed for maintenance and for growth. Table 13.3 shows the effect of different rates of weight gain on the amino acid profile needed to cover both maintenance and growth requirements. As the proportion of essential amino acids needed for new tissue synthesis is higher than that for maintenance, the proportions of these essential amino acids increases with

TABLE 13.2

Amino Acid Pattern Needed for Maintenance of New Tissues Synthesized during Growth

	His	Ile	Leu	Lys	SAA	AAA	Thr	Trp	Val
Tissue amino acid pattern (mg/g protein)	27	35	75	73	35	73	52	12	49
Maintenance amino acid pattern (mg/g protein)	15	30	59	45	22	38	23	6	39

Source: Data from Food and Agriculture Organization of the United Nations (FAO), *Dietary protein quality evaluation in human nutrition: Report of an FAO expert consultation, 31 March–2 April, 2011, Auckland, New Zealand*, Rome: FAO, 2013.

TABLE 13.3

Amino Acid Scoring Pattern (mg/g Protein) for Well-Nourished Children Aged 6–36 Months, and for Recovering Malnourished Children with Different Rates of Weight Gain

	His	Ile	Leu	Lys	SAA	AAA	Thr	Trp	Val
Well-Nourished Children Aged 6–36 Months									
	20	32	66	57	27	52	31	8	43
Malnourished Children during Recovery with a Weight Gain of									
5 g/kg/day	22	33	68	61	29	58	34	9	45
10 g/kg/day	24	34	71	65	31	63	37	10	46
15 g/kg/day	25	34	72	67	32	66	38	11	47

Note: It was assumed that these children gained 50% of fat and 50% of lean tissue. The efficiency of protein synthesis was assumed to be 58%, as in the Food and Agriculture Organization of the United Nations (FAO), *Dietary protein quality evaluation in human nutrition: Report of an FAO expert consultation, 31 March–2 April, 2011, Auckland, New Zealand*, Rome: FAO, 2013.

weight gain. So, in children with catch-up growth, the quantity of essential amino acid needed for new synthesis is higher than that for "normal" children.

MINERAL AND VITAMINS

All forms of AM are associated with different degrees of mineral and vitamin deficiencies [7]. Mineral and vitamin requirements for children with AM were estimated based on the requirements of well-nourished children, adjusted to correct preexisting deficiencies and an increased rate of weight gain.

Potassium requirements are increased to correct a low tissue concentration, which can be due to low dietary intake, to the effect of repeated episodes of diarrhea, and to increased renal excretion. During the catch-up growth phase, the requirements associated with synthesis of lean tissues are also increased. Studies suggest that children with SAM require between 4 and 8 mEq/Kg/day of potassium. Magnesium deficiency is also very common, and its correction is needed to correct the associated potassium deficiency [7]. Zinc is needed for lean tissue synthesis and, if the amounts provided during catch-up growth are insufficient, excess fat is deposited during catch-up growth [47].

Iron requirements are mainly related to growth, and the requirement for maintenance is negligible in the absence of pathological blood loss. Children with SAM often have an excess of body iron at the beginning of their treatment, and the levels of these stores, assessed by ferritin plasma concentrations, are associated with an increased mortality [48]. Although this association between ferritin concentration and mortality can be partly explained by infections, iron is not given during the first few days of treatment in children with complicated SAM. During catch-up growth, however, iron requirements increase and may be higher than those for normal children.

PREVENTION AND TREATMENT OF ACUTE MALNUTRITION

PREVENTION

Prevention should be based on ensuring access to an adequate nutrient-dense diet, and preventing and treating infections, the latter being beyond the scope of this chapter. Ensuring an adequate diet should rely first on the promotion of exclusive breast-feeding for the first 6 months. Before this age, in poor settings, the introduction of other foods, beyond the risk of exposing the child to pathogens, always results in a decrease in the nutrient density of the diet. Beyond 6 months, it is necessary to introduce complementary feeding, following the WHO guiding principles [49]. The use of food supplements as a substantial part of protein and energy requirements is only needed to prevent AM in situations of food insecurity. Mineral and vitamin supplementation, or the use of foods fortified with problem nutrients (such as zinc, iron, and vitamin A), however, are often required to ensure the nutritional adequacy of the diet [49]. The nutrients most likely to be missing in the diet, and which should be provided by supplements or fortified foods, should be determined by analyzing locally available foods and local food habits [50].

TREATMENT

Severe Acute Malnutrition

Treatment of SAM is described in detail in national protocols that have been developed over the past 10 years and adapted to the local situation [51]. Uncomplicated forms, the most frequent, should be treated in the community, while only complicated forms should be referred to inpatient treatment.

Uncomplicated Forms

Children with uncomplicated forms of malnutrition, with good appetite and without clinical signs of infection, mainly need an energy- and nutrient-dense diet. The need for antibiotics is being debated and seems to depend on the context [52].

A diet high in energy and nutrients is difficult to achieve by combining local foods [53], and WHO and the United Nations Children's Fund (UNICEF) recommend the use of RUTF for these children [44]. RUTF provides 10% to 12% of its energy from protein, which, as discussed earlier, is sufficient to promote rapid weight gain. At least half of total protein should be derived from milk products [44]. Attempts to replace milk with plant proteins have demonstrated mixed results [54], possibly because a high proportion of essential amino acids is needed for rapid weight gain. Other factors, such as the presence of antinutrients associated with plant proteins (e.g., soy or beans), may also be involved [55].

High-fat diets are energy dense and facilitate the intake of the high quantities of energy needed to sustain rapid catch-up growth. RUTF provides 45% to 60% of its energy from fat. This should be compared with the energy content of the tissue being deposited: Assuming that the proportion is 50% fat and 50% lean tissue (with about 20% of proteins), the energy being deposited in 1 g of tissue is 4.5 kcal for fat and 0.4 kcal for proteins, corresponding, respectively, to 92% and 8% of the energy content of new tissues. These proportions of the fat and protein of new tissues and the corresponding energy requirements for growth are comparable to those observed during the first 3 months of life, when the child is breastfed, and receives about half of its energy from fat.

Initially, minerals and vitamins were given separately from the food and dosed in relation to weight, as was done for drugs. Minerals and vitamins were later incorporated into the F-100 rehabilitation diet in the 1999 WHO manual on management of SAM [43], following the common practice of feeding nonbreastfed children with infant formulas fortified with minerals and vitamins. An advantage of this approach is that children with the most rapid growth, with the highest requirements, have the highest intake of different vitamins and minerals. The formulation of RUTF was derived from the previously developed F-100 formula, with the addition of iron. The risk–benefit balance seems in favor of adding iron for children receiving RUTF who have rapid growth and very high iron requirements.

Compared to standard infant formulas, the nutrient density of RUTF differs mainly in terms of its higher potassium, magnesium, vitamin A, zinc, iron, and copper content (Table 13.4). These nutrients are specifically needed for children with SAM. Copper is also needed, as the high level of zinc provided to these children may result in copper deficiency.

TABLE 13.4

Comparison of the Content in Selected Nutrients of Ready-to-Use Therapeutic Food with That of the Codex Standard for Infant Formula

	RUTF[a]		Codex Standard for Infant Formula	
	Min	Max	Min	Max
Vitamin A (µg retinol equivalent/100 kcal)	145	205	60	180
Vitamin D (µg/100 kcal)	2.8	3.7	1	2.5
Vitamin E (mg/100 kcal)	3.7	NA	0.5	NA
Thiamin (mg/100 kcal)	NA	61	20	60
Potassium (mg/100 kcal)	203	440	60	180
Magnesium (mg/100 kcal)	15	26	5	NA
Iron (mg/100 kcal)	1.8	2.6	0.45	NA
Zinc (mg/100 kcal)	2	2.6	0.5	NA
Copper (µg/100 kcal)	260	340	35	NA
Selenium (µg/100 kcal)	3.7	7.5	1	NA
Linoleic acid (mg/100 kcal)	330	1111	300	NA
α-Linolenic acid (mg/100 kcal)	33	280	50	NA

Notes: Only nutrients that are markedly different between RUTF and the Codex standard for infant formula are shown in the table. Units were converted to facilitate the comparison with the Codex standard. NA, not available; RUTF, ready-to-use therapeutic food.

[a] RUTF nutrient composition is adapted from the World Health Organization, *Community-based management of severe acute malnutrition*, Geneva: World Health Organization, 2007.

Current RUTF specifications for essential fatty acids are very close to those of the Codex Alimentarius for infant formulas. Recent data suggest, however, that they may need revision, as children treated with current RUTF fail to maintain their status in long-chain polyunsaturated fatty acids, especially DHA; this is possibly as a result of too high levels of linoleic acid in RUTF with peanut butter as the major ingredient [56].

Complicated Forms

In complicated SAM, associated infections should be treated first. Asymptomatic infections are often present, particularly in children with anorexia, and antibiotics should be routinely given to these children. Until these infections are under control, children should receive just enough energy and nutrients to maintain their body weight and restore body functions. This is achieved by giving the low protein, low energy F-75 diet recommended by the WHO [43]. F-75 is fortified with all nutrients likely to be needed to correct deficiencies in concentrations similar to F-100. There is no added iron in F-75, as iron stores are often very high before treatment. Once their appetite is back to normal, these children should be treated as outpatients, as children with uncomplicated forms of SAM, and should receive RUTF.

Moderate Acute Malnutrition

The management of MAM relies on the same principle of giving a diet adequate to sustain rapid growth. Achieving a rapid weight gain, however, is less important than for children with SAM, and adequate weight gain allowing for catch-up growth can be achieved using local foods in situations where nutrient dense foods are available and accessible. Nutrient supplements can also be provided but, in this case, these should be given in addition to family foods, with the objective of providing only those nutrients missing in local foods. These supplements should provide less energy than those for SAM children, but should be more nutrient dense. Their desirable composition has been described in detail elsewhere [57].

CONCLUSION

AM affects millions of children globally and is associated with increased mortality. Prevention relies on providing a diet with nutrient-rich foods and in prevention of infections, and will remain difficult in poor populations with limited access to food and who are living in unhygienic conditions. Management of AM, however, especially in its severe form, has been greatly improved and simplified in recent years. SAM should be considered a public health priority, as it affects large numbers of affected children, is associated with a high risk of death, is easy to detect, and has an effective, affordable treatment.

REFERENCES

1. Waterlow JC. Note on the assessment and classification of protein-energy malnutrition in children. *Lancet* 1973; 2:87–9.
2. WHO Multicentre Growth Reference Study Group. WHO child growth standards based on length/height, weight and age. *Acta Paediatr Suppl* 2006; 450:76–85.

3. World Health Organization. *WHO Child Growth Standards and the identification of severe acute malnutrition in infants and children: Joint statement by the World Health Organization and the United Nations Children's Fund*. Geneva: World Health Organization; 2009.

4. Myatt M, Khara T, Collins S. A review of methods to detect cases of severely malnourished children in the community for their admission into community-based therapeutic care programs. *Food Nutr Bull* 2006; 27:S7–23.

5. United Nations High Commissioner for Refugees (UNHCR) and World Food Programme (WFP). *Guidelines for selective feeding: The management of malnutrition in emergencies*. Geneva: UNHCR, 2011.

6. Black RE, Victora CG, Walker SP et al. Maternal and child undernutrition and overweight in low-income and middle-income countries. *Lancet* 2013; 382:427–51.

7. Waterlow JC. *Protein-energy malnutrition*. 2nd ed. London: Edward Arnold; 1992.

8. Desai M, Crowther NJ, Lucas A et al. Organ-selective growth in the offspring of protein-restricted mothers. *Br J Nutr* 1996; 76:591–603.

9. Cahill GF Jr. Fuel metabolism in starvation. *Annu Rev Nutr* 2006; 26:1–22.

10. Kerr DS, Stevens MC, Robinson HM. Fasting metabolism in infants. I. Effect of severe undernutrition on energy and protein utilization. *Metabolism* 1978; 27:411–35.

11. Bartz S, Mody A, Hornik C et al. Severe acute malnutrition in childhood: Hormonal and metabolic status at presentation, response to treatment, and predictors of mortality. *J Clin Endocrinol Metab* 2014; 99:2128–37.

12. Procaccini C, Lourenco EV, Matarese G et al. Leptin signaling: A key pathway in immune responses. *Curr Signal Transduct Ther* 2009; 4:22–30.

13. Lecker SH, Solomon V, Mitch WE et al. Muscle protein breakdown and the critical role of the ubiquitin-proteasome pathway in normal and disease states. *J Nutr* 1999; 129:S227–37.

14. Reeds PJ, Fjeld CR, Jahoor F. Do the differences between the amino acid compositions of acute-phase and muscle proteins have a bearing on nitrogen loss in traumatic states? *J Nutr* 1994; 124:906–10.

15. Golden MHN. Severe malnutrition. In *Oxford textbook of medicine*, 3rd edition, Weatherall DJ, Ledington JGG, Warren DA, editors. Oxford, UK: Oxford University Press; 1996.

16. Briend A, Khara T, Dolan C. Wasting and stunting—Similarities and differences: Policy and programmatic implications. *Food Nutr Bull* 2015; 36:S15–23.

17. Briend A, Garenne M, Maire B, Fontaine O, Dieng K. Nutritional status, age and survival: The muscle mass hypothesis. *Eur J Clin Nutr* 1989; 43:715–26.

18. Van den Broeck J, Eeckels R, Hokken-Koelega A. Fatness and muscularity as risk indicators of child mortality in rural Congo. *Int J Epidemiol* 1998; 27:840–4.

19. Williams CD, Oxon BM, Lond H. Kwashiorkor: A nutritional disease of children associated with a maize diet. *Bull World Health Organ* 2003; 81:912–3.

20. Lin CA, Boslaugh S, Ciliberto HM et al. A prospective assessment of food and nutrient intake in a population of Malawian children at risk for kwashiorkor. *J Pediatr Gastroenterol Nutr* 2007; 44:487–93.

21. Coulthard MG. Oedema in kwashiorkor is caused by hypoalbuminaemia. *Paediatr Int Child Health* 2015; 35:83–9.

22. Golden MH. Nutritional and other types of oedema, albumin, complex carbohydrates and the interstitium—A response to Malcolm Coulthard's hypothesis: Oedema in kwashiorkor is caused by hypo-albuminaemia. *Paediatr Int Child Health* 2015; 35(2):90–109.

23. Golden MH, Ramdath D. Free radicals in the pathogenesis of kwashiorkor. *Proc Nutr Soc* 1987; 46:53–68.

24. Manary MJ, Leeuwenburgh C, Heinecke JW. Increased oxidative stress in kwashior-kor. *J Pediatr* 2000; 137:421–4.
25. Fechner A, Böhme C, Gromer S et al. Antioxidant status and nitric oxide in the malnu-trition syndrome kwashiorkor. *Pediatr Res* 2001; 49:237–43.
26. Becker K, Pons-Kühnemann J, Fechner A et al. Effects of antioxidants on glutathi-one levels and clinical recovery from the malnutrition syndrome kwashiorkor: A pilot study. *Redox Rep Commun Free Radic Res* 2005; 10:215–26.
27. Bachou H, Tylleskär T, Downing R et al. Severe malnutrition with and without HIV-1 infection in hospitalised children in Kampala, Uganda: Differences in clinical features, haematological findings and CD4+ cell counts. *Nutr J* 2006; 5:27.
28. Marshall KG, Swaby K, Hamilton K et al. A preliminary examination of the effects of genetic variants of redox enzymes on susceptibility to oedematous malnutrition and on percentage cytotoxicity in response to oxidative stress in vitro. *Ann Trop Paediatr* 2011; 31:27–36.
29. Ciliberto H, Ciliberto M, Briend A et al. Antioxidant supplementation for the preven-tion of kwashiorkor in Malawian children: Randomised, double blind, placebo con-trolled trial. *BMJ* 2005; 330:1109.
30. Smith MI, Yatsunenko T, Manary MJ et al. Gut microbiomes of Malawian twin pairs discordant for kwashiorkor. *Science* 2013; 339:548–54.
31. Kristensen KHS, Wiese M, Rytter MJH et al. Gut microbiota in children hospitalized with oedematous and non-oedematous severe acute malnutrition in Uganda. *PLoS Negl Trop Dis* 2016; 10:e0004369.
32. Faurholt-Jepsen D, Hansen KB, van Hees VT et al. Children treated for severe acute malnutrition experience a rapid increase in physical activity a few days after admission. *J Pediatr* 2014; 164(6):1421–4.
33. Aburto NJ, Ramirez-Zea M, Neufeld LM et al. Some indicators of nutritional status are associated with activity and exploration in infants at risk for vitamin and mineral deficiencies. *J Nutr* 2009; 139:1751–7.
34. Gardner JM, Grantham-McGregor SM, Himes J et al. Behaviour and development of stunted and nonstunted Jamaican children. *J Child Psychol Psychiatry* 1999; 40:819–27.
35. Patrick J, Golden M. Leukocyte electrolytes and sodium transport in protein energy malnutrition. *Am J Clin Nutr* 1977; 30:1478–81.
36. Golden MH, Waterlow JC, Picou D. Protein turnover, synthesis and breakdown before and after recovery from protein energy malnutrition. *Clin Sci Mol Med* 1977; 51:473–7.
37. Jahoor F, Budaloo A, Reid M et al. Protein metabolism in severe childhood malnutri-tion. *Ann Trop Paediatr* 2008; 28:87–101.
38. Klahr S, Alleyne GA. Effects of chronic protein-calorie malnutrition on the kidney. *Kidney Int* 1973; 3:129–41.
39. Faddan NHA, Sayh KIE, Shams H et al. Myocardial dysfunction in malnourished chil-dren. *Ann Pediatr Cardiol* 2010; 3:113–8.
40. Wharton BA, Howells GR, McCance RA. Cardiac failure in kwashiorkor. *Lancet* 1967; 2:384–7.
41. Butte NF, Wong WW, Hopkinson JM et al. Energy requirements derived from total energy expenditure and energy deposition during the first 2 years of life. *Am J Clin Nutr* 2000; 72:1558–69.
42. Ashworth A, Millward DJ. Catch-up growth in children. *Nutr Rev* 1986; 44:157–63.
43. World Health Organization. *Management of severe malnutrition: A manual for physi-cians and other senior health workers.* Geneva: World Health Organization; 1999.
44. World Health Organization. *Community-based management of severe acute mal-nutrition.* Geneva: World Health Organization; 2007.

45. Semba RD, Trehan I, Gonzalez-Freire M et al. Perspective: The potential role of essential amino acids and the mechanistic target of rapamycin complex 1 (mTORC1) pathway in the pathogenesis of child stunting. *Adv Nutr* 2016; 7:853–65.

46. Food and Agriculture Organization of the United Nations (FAO). *Dietary protein quality evaluation in human nutrition: Report of an FAO expert consultation, 31 March–2 April, 2011, Auckland, New Zealand.* Rome: FAO; 2013.

47. Golden MH, Golden BE. Effect of zinc supplementation on the dietary intake, rate of weight gain, and energy cost of tissue deposition in children recovering from severe malnutrition. *Am J Clin Nutr* 1981; 34:900–8.

48. Ramdath DD, Golden MH. Non-haematological aspects of iron nutrition. *Nutr Res Rev* 1989; 2:29–49.

49. Pan American Health Organization (PAHO) and the World Health Organization (WHO). *Guiding principles for complementary feeding of the breastfed child.* Geneva: WHO; 2003.

50. Ferguson EL, Darmon N, Fahmida U et al. Design of optimal food-based complementary feeding recommendations and identification of key "problem nutrients" using goal programming. *J Nutr* 2006; 136:2399–404.

51. The State of Severe Malnutrition. National guidelines and protocols for acute malnutrition [Internet]. Available from: http://www.severemalnutrition.org/en/home/Ge?type=environment&cat=00&date=2016. Accessed June 9 2017.

52. Isanaka S, Langendorf C, Berthé F et al. Routine amoxicillin for uncomplicated severe acute malnutrition in children. *N Engl J Med* 2016; 374:444–53.

53. Ferguson EL, Briend A, Darmon N. Can optimal combinations of local foods achieve the nutrient density of the F100 catch-up diet for severe malnutrition? *J Pediatr Gastroenterol Nutr* 2008; 46:447–52.

54. Oakley E, Reinking J, Sandige H et al. A ready-to-use therapeutic food containing 10% milk is less effective than one with 25% milk in the treatment of severely malnourished children. *J Nutr* 2010; 140:2248–52.

55. Briend A, Akomo P, Bahwere P et al. Developing food supplements for moderately malnourished children: Lessons learned from ready-to-use therapeutic foods. *Food Nutr Bull* 2015; 36:S53–58.

56. Brenna JT, Akomo P, Bahwere P et al. Balancing omega-6 and omega-3 fatty acids in ready-to-use therapeutic foods (RUTF). *BMC Med* 2015; 13:117.

57. Golden MH. Proposed recommended nutrient densities for moderately malnourished children. *Food Nutr Bull* 2009; 30:S267–342.

14 Management of Acute Malnutrition in Infants under 6 Months of Age

Marko Kerac and Marie McGrath

CONTENTS

INTRODUCTION

Both acute malnutrition and nutrition (breastfeeding) in infants under 6 months of age (infants <6 months) are important global health issues and have received much international attention over the years. However, it is only recently that the two in combination—the management of acute malnutrition in infants <6 months (MAMI)—have been examined [1]. This chapter outlines the background epidemiology, why acute malnutrition in this age group matters, key challenges around infant <6 months malnutrition, current assessment and treatment strategies, and, finally, directions for the future. Readers should look to other chapters of this book for added detail, as

MAMI has numerous links and synergies with other areas of malnutrition, with many opportunities to benefit both short- and long-term health.

EPIDEMIOLOGY OF A "FORGOTTEN PROBLEM"

For several decades, it was widely assumed that acute malnutrition in infants <6 months was a minor individual-level issue, rather than a significant public health problem. The logical fallacy went like this: Since breastfeeding is associated with good nutritional status, and since infants <6 months should be breastfed, poor nutrition among infants <6 months must therefore be rare, assuming it only occurs where infants are not breastfed or perhaps where there is early introduction of complementary foods. This was even expressed by authoritative sources, such as the World Health Organization (WHO) "Field Guide to Nutritional Assessment," which stated that "children under six months of age … are often still breast-fed and therefore satisfactorily nourished" [2]. Combined with the greater practical difficulties of conducting anthropometric measurements in young infants [3,4], this presupposition meant that infants <6 months were often omitted from nutrition surveys and surveillance activities [5,6]. As with any problem that is not being actively looked for, acute malnutrition in this age group was often simply overlooked. Specifically, the following factors were overlooked:

- Rates of breastfeeding are almost universally suboptimal [7].
- Despite being the cornerstone of good infant nutrition, breastfeeding is not 100% protective from nutrition-related problems.
- Nutritional status is dependent on many factors, not just good quality dietary intake [8]. Especially in young infants, there are a large number and variety of health problems that can adversely impact on nutrition. These can be challenging to diagnose and treat, even in high-income, well-resourced settings.

In 2010, in response to questions about infants <6 months by field-based practitioners, a report on MAMI [1] and a subsequent research paper [9] aimed to test previous assumptions and quantify the problem as an essential first step toward properly understanding this. An extrapolation of demographic and health survey data from 21 "high burden" low- and middle-income countries found an important burden of disease (Table 14.1). Other observations and issues arising from Table 14.1 include:

- Wasted infants <6 months constitute an important proportion of all wasted children aged <60 months. This is an argument for program planners and managers needing to take this group seriously and make provisions for their care.
- The 2006 WHO Child Growth Standards (WHO-GS; see Chapter 2 for more detail) really are the gold standard of good growth, setting the bar quite high. Using WHO-GS rather than the previous dominant National Center for Health Statistics (NCHS) growth standards thus results in more infants <6 months being recognized as "wasted."

TABLE 14.1
Global Epidemiology of Wasting in Infants under 6 Months of Age

	All Infants and Children (0 to 60 Months), $n = 556$ Million	Infants <6 Months (WHO Growth Standards), $n = 56$ Million	Infants <6 Months (NCHS Growth References), $n = 56$ Million
Total wasting (millions), weight-for-length z-score <–2	58	8.5	3.0
Moderate wasting (millions), WLZ ≥–2 to <–3	38	4.7	2.2
Severe wasting (millions), WLZ <–3	20	3.8	0.8

Source: Adapted from Kerac M, Blencowe H, Grijalva-Eternod C et al., *Arch Dis Child* 2011, 96(11):1008–13.

- This also challenges some prior assumptions that, since the WHO-GS were based on breastfed infants, they "will result in fewer breastfed babies diagnosed as growing poorly" [10,11].
- Figures for edematous malnutrition are not available. The table thus underestimates the total burden of disease of acute malnutrition and severe acute malnutrition (SAM; edematous malnutrition being part of that case definition) [12]. That said, anecdotal reports suggest that kwashiorkor is uncommon in this age group and that, if bilateral pitting edema is observed, another cause is more likely [13,14].

WHY MALNUTRITION MATTERS FOR INFANTS UNDER 6 MONTHS OF AGE

SHORT TERM

In the short term, mortality is the most serious risk faced by acutely malnourished infants <6 months. Acute malnutrition has a widely recognized, well-described high case fatality rate [15–17], but infants are at particular risk. Reasons include physiological and immunological immaturity, which make them more vulnerable in the first place and more likely to suffer severe adverse consequences. In one recent meta-analysis that compared infants <6 months with children 6–60 months in the same treatment programs, the infants' risk of death was significantly greater (risk ratio 1.30, 95% CI: 1.09, 1.56; P<0.01) [18]. Although biologically not unexpected, a key question is how much of this excess mortality can be avoided with improved or alternative treatment.

LONG TERM

The longer-term effect—and why infant <6 months malnutrition is a key topic in this book—is the increasing recognition that early-life nutritional exposures have clinically significant long-term "programming" effects on adult health and well being [19,20]. Although the best-known work focuses on exposures during prenatal life [21,22], the window of developmental plasticity (and hence the opportunity to make a positive difference) extends well beyond birth. Optimizing infant nutrition has a major role to play in reducing the current epidemic of noncommunicable disease [23,24]. Acute malnutrition represents an especially severe nutritional "insult" with a high likelihood of correspondingly severe long-term noncommunicable disease (NCD)-related risks [25]. There is a great need for interventions to help infants not only "survive" episodes of malnutrition but also to ultimately "thrive."

THE CHALLENGES OF ACUTE MALNUTRITION IN INFANTS UNDER 6 MONTHS OF AGE

MAMI currently lags behind great successes in treating older malnourished children [26]. This can be explained by the numerous challenges related to their needs and care.

A PERIOD OF RAPID MATURATION

Infants <6 months are not simply mini-children; the period represents a major transition from neonatal life, and the beginnings of independence from their mother's milk as the sole source of nutrition.

- Rapid physical and physiological maturation means that a 1-month-old, for example, is very different from a 4-month-old, even though only 3 months separate them in time. What is appropriate for some is not appropriate for all, for example, although exclusive breastfeeding is the target diet for all infants <6 months, some acute malnutrition treatment programs report a pragmatic decision to introduce early complementary feeds for those close to 6 months [1].
- There is also a spectrum of development that impacts on care, with some infants maturing faster or slower than most others

Staff who are skilled and experienced enough to successfully manage these subtleties of approach are often in short supply in settings where malnutrition is common. Any benefits of precisely age-tailored or developmentally tailored treatments thus need to be balanced against the added complexities that these impose on programs; guidelines that are too complex are likely to be poorly implemented in everyday practice. There is also a risk of mixed-messaging regarding feeding practices spilling over to the general population.

UNIQUE DIETARY NEEDS

Malnourished infants <6 months cannot be treated with simple top-up supplementary or therapeutic feeds, as can older malnourished children. Their target diet is exclusive breastfeeding. Even where the mother is around, establishing or reestablishing effective exclusive breastfeeding (the mainstay of treatment) is not always straight-forward [27–29]. It can require lots of time spent with well-trained, highly skilled support workers. Treatment is thus focused on software (skilled feeding support) rather than hardware (products such as ready-to-use therapeutic food). This often makes this less tangible and, thereby, often also less attractive as an investment for policy makers and program managers.

MANY AND COMPLEX UNDERLYING CAUSES

Rather than a diagnosis in its own right, acute malnutrition is often seen as a symptom of another problem. Only if the underlying cause is properly addressed can the malnutrition be definitively treated. Because the range of causes is greater in young infants (if left untreated, more severe cases can be fatal), assessment and treatment is more complex and, again, more resources are needed.

The underlying cause of infant malnutrition can include several contributing factors.

Reduced Nutrient Intake
- *An insufficient quantity and/or quality of food offered.* This can be due to a variety of social/cultural/economic factors. The child might be an orphan, with no wet nurse or a family unable to safely supply a sufficient quality/quantity of breast milk substitute, or the mother may be working, so is not available to breastfeed sufficiently (this can eventually result in secondary lactation failure).

  Although few conditions directly affect the maternal milk supply, many can reduce the frequency/duration of breastfeeding and thus cause secondary lactation failure. It is vital to properly understand the physiology of breastfeeding. More suckling will stimulate more breast milk production; if suckling is reduced for any reason, the breast milk supply will eventually be reduced.

  An anatomical abnormality or problem. For example, cleft palate, which can affect attachment (feeding suction) and allow milk to enter nasal passages, causing choking.
- *A functional problem.* For example, poor attachment to the breast due to poor technique; poor suckling or swallowing in prematurity, or cerebral palsy.
- *Feed aversion.* If an infant is in pain or discomfort when feeding (e.g., in severe reflux or where an anxious carer tries to force-feed), she or he can develop an aversion to feeds, and may try to spit out or otherwise refuse feeds.

Reduced Nutrient Absorption
- *Malabsorption.* This can happen in a variety of conditions (e.g., chronic diarrhea, environmental enteric dysfunction, celiac disease, and cystic fibrosis).

Increased Nutrient Loss
- Vomiting (e.g., severe gastroesophageal reflux or pyloric stenosis).

Increased or Impaired Nutrient Utilization

- Infections
- Acute (e.g., diarrhea, urinary tract infection)
- Chronic (e.g., HIV, TB)
- Congenital disease (e.g., congenital heart disease)
- Metabolic disorders

The frequency and severity of the underlying problem varies. Breastfeeding attachment problems are, for instance, relatively common; they may not always have major effects on nutritional status, but can be easily and rapidly rectified with skilled support. Congenital metabolic disorders are rare, but can be fatal and difficult to manage, even in resource-rich environments.

It is also vital to note the timing of the nutritional insult; some infants will be small because they were of low birth weight, but are now growing well and "catching up" postnatally. Others will be experiencing postnatal problems.

ASSESSING INFANTS UNDER 6 MONTHS OF AGE

Definitions of acute malnutrition in infants <6 months mirror those in older children, focusing on anthropometry (Table 14.2). This is combined with their clinical status (mainly the presence/absence of integrated management of childhood illness [IMCI] danger signs) to distinguish between complicated acute malnutrition (i.e., sick, clinically high-risk infants requiring more intensive, specialized care available in inpatient facilities) and uncomplicated malnutrition (i.e., clinically stable infants, suitable for outpatient care).

This current case definition has numerous limitations. Refining it is considered to be one of the most urgent future priorities to improve care [30], especially since mislabeling an infant with malnutrition carries serious risks, notably that a breastfeeding mother inappropriately introduces formula feeding, with the consequent risks of infection from unsafe water and contaminated bottles/feeding utensils. This

---

TABLE 14.2
Anthropometric Case Definitions of Acute Malnutrition in Infants under 6 Months of Age

Term	Defined By
Acute malnutrition (AM)	SAM + MAM
Moderate acute malnutrition (MAM)	Weight-for-length ≥−3 to <−2 z-scores
Severe acute malnutrition (SAM)	Weight-for-length <−3 z-scores and/or nutritional edema (kwashiorkor)

---

contrasts with the situation for older children, for whom a sensitive (rather than specific) case definition is perfectly acceptable, since the risks of a short period of top-up therapeutic or supplementary food are minimal, hence the benefit–risk balance of intervention is favorable [11].

Other difficulties for infants <6 months include:

- Weight-for-length is especially hard to measure in younger infants. Past studies and a recent review highlight poor reliability [3,4,31].
- Weight-for-length cannot be calculated at all for shorter infants (<45 cm), since the lookup tables do not extend below this level. For these, clinical assessment is carried out.
- There is currently no agreed mid-upper arm circumference (MUAC)-based case definition, as there is for older children. This is a major gap, since MUAC enables quick, effective case finding and is key to the achievement of high coverage by treatment programs [32]. Promising MUAC-based studies do exist, but need replicating in additional settings in order for global policy to change [33].
- Infant <6 months malnutrition is special (and thus especially challenging to manage), in that it is critically dependent on the mother. Her prenatal and postnatal nutrition and physical/mental health affect the nutritional status and growth of her infant [34]. Assessing the mother–infant dyad, rather than the infant alone, should be common practice.
- A single anthropometric measurement is difficult to interpret, but sequential growth monitoring-type measures are often not available. Some infants with low weight-for-length will catch up and become "normal" without intervention. Conversely, other infants can be nutritionally vulnerable and not gaining/losing weight, even if their current weight-for-length measurement is still in the "normal" range.

The recent 2013 WHO Guidelines on Managing Acute Malnutrition also note that "infants who have been identified to have poor weight gain and who have not responded to nutrition counselling and support should be admitted for further investigation and treatment" [35].

Recognizing the more complex nature of infant <6 months malnutrition, and thus the need for more comprehensive assessment, an IMCI-style assessment checklist has recently been developed (Figure 14.1). This is an attempt to translate the WHO 2013 technical guidelines into an operational handbook for use by field-level health care workers [35]. It also aims to harmonize the WHO malnutrition [35] and IMCI guidelines [36]. Using the IMCI "traffic light" approach (red = urgent problem; yellow = problem but suitable for community-level care; green = no acute problem), it leads the user through key areas, including anthropometry, feeding practice, clinical condition, and maternal well-being. This includes the mother's physical, mental, and nutritional health.

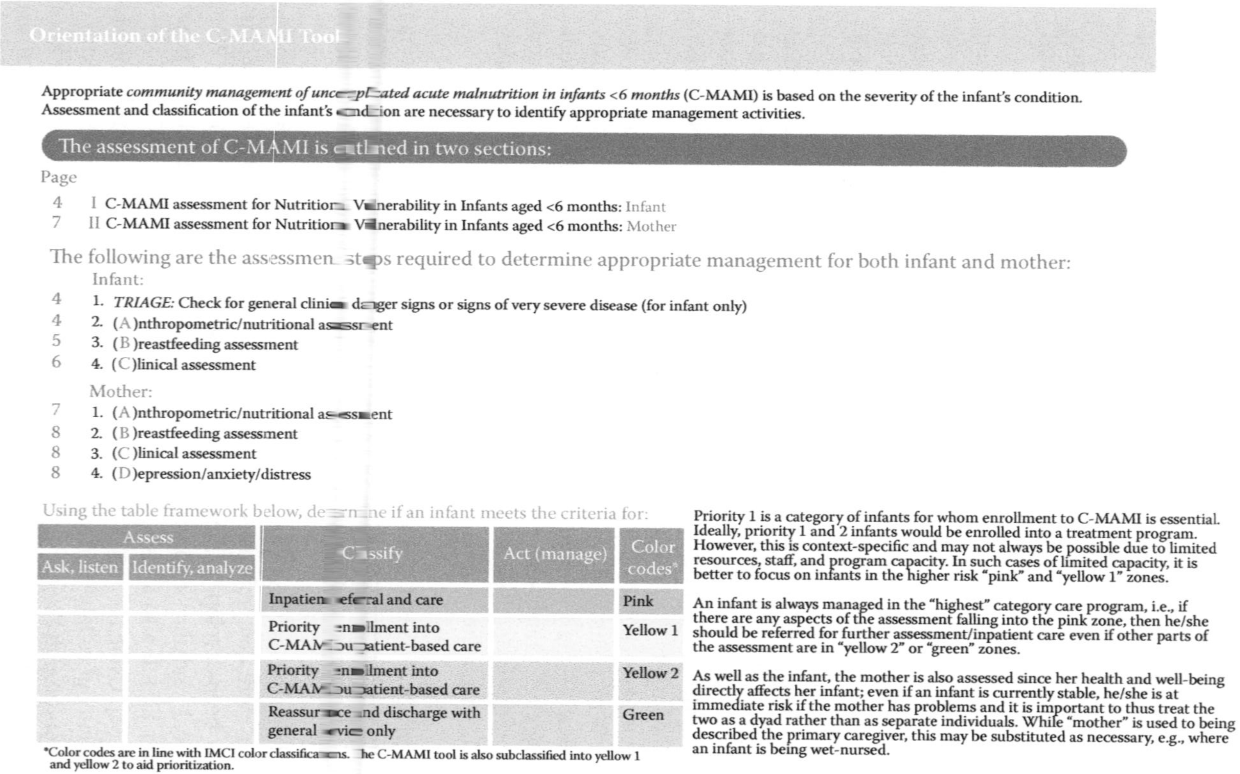

Orientation of the C-MAMI Tool

Appropriate *community management of uncomplicated acute malnutrition in infants <6 months* (C-MAMI) is based on the severity of the infant's condition. Assessment and classification of the infant's condition are necessary to identify appropriate management activities.

The assessment of C-MAMI is obtained in two sections:

Page

4   I C-MAMI assessment for Nutrition Vulnerability in Infants aged <6 months: Infant

7   II C-MAMI assessment for Nutrition Vulnerability in Infants aged <6 months: Mother

The following are the assessment steps required to determine appropriate management for both infant and mother:

Infant:

1. *TRIAGE:* Check for general clinical danger signs or signs of very severe disease (for infant only)
2. (A)nthropometric/nutritional assessment
3. (B)reastfeeding assessment
4. (C)linical assessment

Mother:

1. (A)nthropometric/nutritional assessment
2. (B)reastfeeding assessment
3. (C)linical assessment
4. (D)epression/anxiety/distress

Using the table framework below, determine if an infant meets the criteria for:

Assess		Classify	Act (manage)	Color codes*
Ask, listen	Identify, analyze			
			Inpatient referral and care	Pink
			Priority enrollment into C-MAMI outpatient-based care	Yellow 1
			Priority enrollment into C-MAMI outpatient-based care	Yellow 2
			Reassurance and discharge with general advice only	Green

*Color codes are in line with IMCI color classifications. The C-MAMI tool is also subclassified into yellow 1 and yellow 2 to aid prioritization.

Priority 1 is a category of infants for whom enrollment to C-MAMI is essential. Ideally, priority 1 and 2 infants would be enrolled into a treatment program. However, this is context-specific and may not always be possible due to limited resources, staff, and program capacity. In such cases of limited capacity, it is better to focus on infants in the higher risk "pink" and "yellow 1" zones.

An infant is always managed in the "highest" category care program, i.e., if there are any aspects of the assessment falling into the pink zone, then he/she should be referred for further assessment/inpatient care even if other parts of the assessment are in "yellow 2" or "green" zones.

As well as the infant, the mother is also assessed since her health and well-being directly affects her infant; even if an infant is currently stable, he/she is at immediate risk if the mother has problems and it is important to thus treat the two as a dyad rather than as separate individuals. While "mother" is used to being described the primary caregiver; this may be substituted as necessary, e.g., where an infant is being wet-nursed.

FIGURE 14.1 C-MAMI tool. Community management of uncomplicated acute malnutrition in infants <6 months (C-MAMI). Version 1.0, November 2015. Full version free to download from http://www.ennonline.net/c-mami.

MANAGING ACUTELY MALNOURISHED INFANTS UNDER 6 MONTHS OF AGE

For many years, normative guidance recommended inpatient care for all acutely malnourished infants <6 months [1]. This is still the case in almost all national acute malnutrition guidelines [1]. A major international-level development in the latest 2013 WHO Guidelines on Severe Acute Malnutrition [35] was the introduction of community-based care for infants <6 months with "uncomplicated" SAM. This brought their classification and management in line with that of older children [26].

Key principles that guide management include:

- Early identification of malnourished and nutritionally vulnerable infants in the community, and at any health-related contact point (e.g., when attending for immunizations, or when seeking clinic or hospital treatment for an acute or chronic illness).
- Early treatment in the community wherever possible (uncomplicated cases).
- Early referral and treatment of sick infants (complicated cases).
- Breastfeeding support to restore effective exclusive breastfeeding wherever possible.
- Careful use of breast milk substitutes for nonbreastfed infants, if wet-nursing or banked breast milk is not available.
- General medical care for any underlying problems.
  - All severely malnourished infants, including those with uncomplicated diseases, should receive a course of broad-spectrum antibiotics (oral if possible, parenteral if needed).
  - Other conditions (e.g., HIV, sepsis, surgical problems, or disability) should be dealt with as needed.
- Follow-up on infants postdischarge in the community.
- Targeted support to the mother that considers her nutritional, physical, and mental condition.
- Considering and engaging with the broader household and community and societal influences on the mother–infant dyad that may affect feeding and/ or care practices.

INPATIENT MANAGEMENT [37]

The cornerstone of inpatient care is supplementary suckling. This aims to restore exclusive breastfeeding, and involves taping a small nasogastric tube onto the breast, with the tip adjacent to the nipple. Supplementary milk (infant formula, F-75 therapeutic milk, or diluted F-100 therapeutic milk—there is no hard evidence as to which of these is best) [36] is given via the tube. Crucially, however, the infant is placed on the breast, and attempts to suckle at the same time, thereby stimulating breast milk production. As the quantity of breast milk supply begins to increase, the quantity of supplementary milk can be decreased and, eventually, stopped entirely. A galactagogue (e.g., metoclopramide) can be given to aid this process.

Although supplementary sucking can be a great success in centers with experience, success rates vary [1] and depend on many factors, including motivation, breast-feeding views, practicalities, understanding, and perceptions of hospital-based medicine [27].

The management of nonbreastfed infants is not well catered for in terms of guidance, and creates huge practical challenges in resource-poor settings. It begins by feeding infants with the same supplementary milks (infant formula, F-75, or diluted F-100) to get them back to their normal weight. The biggest difficulty then comes in terms of what to do longer term. Identifying a wet nurse is a priority when culturally acceptable and available. Where wet nursing is not viable, infant formula is necessary for infants <6 months; nutritional indication for infant formula in children 6 months to 2 years depend on the complementary foods available, in particular animal source foods [38]. Where infant formula is not a viable option, a 5-month-old nonbreastfed infant recovering from malnutrition might start early complementary feeds based on skilled assessment as part of a pragmatic care approach, but a 2- or 3-month-old needs a longer term appropriate breast milk substitute, which, in many settings, may not be affordable or available, and requires a package of support services, including water, sanitation, and hygiene (WASH), and access to nutrition and health monitoring. The risks of relapse and of chronic undernutrition in these infants are high. More evidence is needed as to what works for this especially vulnerable group of infants and their carers.

OUTPATIENT MANAGEMENT

The WHO 2013 guidance recommends the following for the outpatient care of infant <6 months with severe acute malnutrition [35]:

- "Counselling and support for optimal infant and young child feeding, based on general recommendations for feeding infants and young children, including for low birth weight infants,
- Weekly weight monitoring to observe changes,
- Referral into inpatient care if the infant does not gain weight, or loses weight, and
- Assessment of the physical and mental health status of mothers or caregivers with appropriate treatment and support provided."

Since these are all low-risk interventions, it is appropriate to extend the same approach to infants with moderate acute malnutrition and even those who are nutritionally vulnerable but do not (yet) have weight-for-length <–2 z-scores. Again, the C-MAMI tool brings together and reformats guidance that has been successful in other settings for other infant populations. Figure 14.2 is an example from the C-MAMI tool of how to identify and address poor attachment, which is a common problem.

1 Firsttime Breastfeeding Counseling and Support Actions

Image	Symptoms of good practice	Good practice: Yes or No	Counseling and support actions

1.1 Good attachment

Observe breastfeeding:

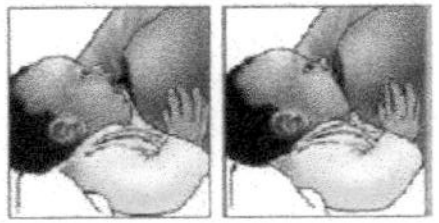

1. Infant's mouth wide open
2. Lower lip turned outward
3. Chin touches breast
4. More areola visible above than below mouth

Note: A good position is comfortable for the mother

Show mother how to position the infant:

☐ Infant's body should be in a straight line (head, neck, trunk, and legs).
- *Illustrate how vital this is by asking the mother/caregiver to drink or eat something herself with her head and neck flexed forward or turned to the side: swallowing will be difficult. Try the same with head/neck/body straight or tipped very slightly back—swallowing is much easier. Infants will similarly swallow much easier when held straight.*

☐ *Infant's body should be facing the breast.*
☐ *Infant should be held close to mother.*
☐ *Position infant nose to nipple as infant is brought onto breast.*
☐ *Mother should support the infant's whole body, not just neck and shoulders.*
  See video: http://globalhealthmedia.org/portfolio-items/breastfeedingpositions/?portfolioID=10861

Show mother how to help the infant attach deeply:

☐ Touch her infant's lips with her nipple.
☐ Wait until her infant's mouth is opening wide.
☐ Move her infant quickly onto her breast, aiming the infant's lower lip well below the nipple, so that the nipple goes to the top of the infant's mouth and infant's chin will touch her breast.
  See video: http://globalhealthmedia.org/portfolio-items/breastfeeding-attachment/?portfolioID=10861

☐ If infant not alert/doesn't open mouth, hand express drops of milk and apply on infant's lips to stimulate mouth opening.
☐ If latch is inadequate, try alternative position.
☐ If not able to attach well immediately, demonstrate breast milk expression and feeding by a cup.

FIGURE 14.2 Guidance on how to identify and address poor attachment in the "C-MAMI" tool. C-MAMI tool. Version 1.0, November 2015. Full version free to download from http://www.ennonline.net/c-mami.

FUTURE DIRECTIONS

Most WHO recommendations for managing infants <6 months are "strong" recommendations [35]. It is vital, however, to acknowledge that the underlying evidence is "low quality" [39]. This represents a problem but also an opportunity to rapidly move things forward for this vulnerable group.

In a recent research prioritization exercise [30], the three top-ranked research questions were

1. *How should infant <6 months SAM be defined?* As discussed, there are serious problems with the current anthropometry weight-for-length focused case definition of acute malnutrition in infants <6 months. The need for a refined and improved case definition offers a great opportunity to embrace more holistic terms of reference, assessing infants and their mothers and their social/home environments. Only thus can their problems be understood, and truly effective care packages planned and put in place for each individual.

2. *What are/is the key opportunities/timing when infant SAM management can be incorporated with other health care programs?* MAMI offers an opportunity to connect numerous emergency and development programs and stakeholders across many sectors. Rather than being a narrow vertical program, it has to ensure links with other health and social programs, so that every occasion where there is contact with an infant provides an opportunity for identifying those who are nutritionally vulnerable and supporting/treating them.

3. *What are the priority components of a package of care for outpatient treatment of infant <6 months SAM?* MAMI requires interventions at many levels, from individual case management to societal level interventions, such as social protection/maternity pay.

CONCLUSION

MAMI is a new and rapidly evolving area. Although much progress has been made, much still needs to take place. The 2013 WHO guidelines on infant <6 months are a major step forward, but still need to be translated into country-level guidelines and, most important, to front-line clinical settings. Despite the many challenges and difficulties, addressing MAMI offers an opportunity to escape from siloed nutrition thinking and programming. Strengthening the early identification and management of nutritionally vulnerable infants has to go past nutrition and involve other sectors, especially reproductive health, neonatal health, and WASH; has to include the nutrition and health of mothers and, within that, adolescents; and has to innovate and look to development programming for management strategies.

REFERENCES

1. Emergency Nutrition Network/University College London/Action Contre la Faim. Management of Acute Malnutrition in Infants (MAMI) project. Available at: http://www.ennonline.net/mamitechnicalreview (accessed February 5, 2017).

2. World Health Organization. *Field guide on rapid nutritional assessment in emergencies.* Geneva: World Health Organization; 1995.

3. Mwangome MK, Berkley JA. The reliability of weight-for-length/height z scores in children. *Matern Child Nutr* 2014; 10(4):474–80.

4. Mwangome M, Berkley J. Measuring infants aged below 6 months: Experience from the field. *ENN Field Exchange* 2014; 47:34.

5. Prudohn, C. Including infants in nutrition surveys. *ENN Field Exchange* 2000; 9:15.

6. Lopriore C, Dop MC, Solal-Celigny A et al. Excluding infants under 6 months of age from surveys: Impact on prevalence of pre-school undernutrition. *Public Health Nutr* 2007; 10(1):79–87.

7. Victora CG, Bahl R, Barros AJ et al. Breastfeeding in the 21st century: Epidemiology, mechanisms, and lifelong effect. *Lancet* 2016; 387(10017):475–90.

8. Goh LH, How CH, Ng KH. Failure to thrive in babies and toddlers. *Singapore Med J* 2016; 57(6):287–91.

9. Kerac M, Blencowe H, Grijalva-Eternod C et al. Prevalence of wasting among under 6-month-old infants in developing countries and implications of new case definitions using WHO growth standards: A secondary data analysis. *Arch Dis Child* 2011; 96(11):1008–13.

10. Waterlow JC. *Protein-energy malnutrition.* 2nd ed. London: Edward Arnold; 1992.

11. Ahmad UN, Yiwombe M, Chisepo P et al. Interpretation of World Health Organization growth charts for assessing infant malnutrition: A randomised controlled trial. *J Paediatr Child Health* 2014; 50(1):32–39.

12. Frison S, Checchi F, Kerac M. Omitting edema measurement: How much acute malnutrition are we missing? *Am J Clin Nutr* 2015; 102(5):1176–81.

13. Karunaratne R. Management of acute malnutrition in Infants aged <6 months (MAMI) in Malawi: Prevalence and risk factors in an observational study. Royal College of Paediatrics & Child Health Annual Meeting 2015. Birmingham, UK.

14. Emergency Nutrition Network. Edema in infants <6 months and PLW [Internet]. Accessible at: http://www.en-net.org/question/1659.aspx (accessed February 5, 2017).

15. Black RE, Victora CG, Walker SP et al. Maternal and child undernutrition and overweight in low-income and middle-income countries. *Lancet* 2013; 382(9890):427–51.

16. Schofield C, Ashworth A. Why have mortality rates for severe malnutrition remained so high? *Bull World Health Organ* 1996; 74(2):223–9.

17. Heikens GT. How can we improve the care of severely malnourished children in Africa? *Proc Nutr* 1007; 4(1):e1J.

18. Grijalva-Eternod CS, Kerac M, McGrath M et al. Admission profile and discharge outcomes for infants aged less than 6 months admitted to inpatient therapeutic care in 10 countries. A secondary data analysis. *Matern Child Nutr* 2016 (in press) doi: 10.1111/mcn.12345.

19. Ruemmele FM. Early programming effects of nutrition: Life-long consequences? *Ann Nutr Metab* 2011; 58(Suppl 2):5–6.

20. Tarry-Adkins JL, Ozanne SE. Mechanisms of early life programming: Current knowledge and future directions. *Am J Clin Nutr* 2011; 94(6):S1765–71.

21. Fernandez-Twinn DS, Ozanne, SE. Early life nutrition and metabolic programming. *Ann N Y Acad Sci* 2010; 1212:78–96.

22. Eriksson JG. Early programming of later health and disease: Factors acting during prenatal life might have lifelong consequences. *Diabetes* 2010; 59(10):2349–50.

23. Lanigan J, Singhal A. Early nutrition and long-term health: A practical approach. *Proc Nutr Soc* 2009; 68(4):422–9.

24. Singhal A. The global epidemic of noncommunicable disease: The role of early-life factors. *Nestle Nutr Inst Workshop Ser* 2014; 78:123–32.

25. Lelijveld N, Seal A, Wells JC et al. Chronic disease outcomes after severe acute mal-nutrition in Malawian children (ChroSAM): A cohort study. *Lancet Glob Health* 2016; 4(9):e654–62.

26. Trehan I, Manary MJ. Management of severe acute malnutrition in low-income and middle-income countries. *Arch Dis Child* 2015; 100(3):283–7.

27. Lelijveld N, Mahebere-Chirambo C, Kerac M. Carer and staff perspectives on supple-mentary suckling for treating infant malnutrition: Qualitative findings from Malawi. *Matern Child Nutr* 2014; 10(4):593–603.

28. Lutter CK, Morrow AL. Protection, promotion, and support and global trends in breast-feeding. *Adv Nutr* 2013; 4(2):213–9.

29. Sinha B, Chowdhury R, Sankar MJ et al. Interventions to improve breastfeeding out-comes: A systematic review and meta-analysis. *Acta Paediatr* 2015; 104(467):114–34.

30. Angood C, McGrath M, Mehta S et al. Research priorities to improve the management of acute malnutrition in infants aged less than six months (MAMI). *PLoS Med* 2015; 12(4):e1001812.

31. Mwangome MK, Fegan G, Mbunya R et al. Reliability and accuracy of anthropometry performed by community health workers among infants under 6 months in rural Kenya. *Trop Med Int Health* 2012; 17(5):622–9.

32. Myatt M, Khara T, Collins S. A review of methods to detect cases of severely malnour-ished children in the community for their admission into community-based therapeutic care programs. *Food Nutr Bull* 2006; 27(3 Suppl):S7–23.

33. Mwangome MK, Fegan G, Fulford T et al. Mid-upper arm circumference at age of routine infant vaccination to identify infants at elevated risk of death: A retrospective cohort study in the Gambia. *Bull World Health Organ* 2012; 90(12):887–94.

34. Stewart RC. Maternal depression and infant growth: A review of recent evidence. *Matern Child Nutr* 2007; 3(2):94–107.

35. World Health Organization. *Updates on the management of severe acute malnutrition in infants and children*. Geneva: World Health Organization; 2013.

36. World Health Organization. *Integrated management of childhood illness (IMCI) chart booklet*. Geneva: World Health Organization; 2014.

37. Kerac M, Tehran I, Lelijveld N et al. *Inpatient treatment of severe acute malnutrition in infants aged <6 months*. Geneva: World Health Organization; 2012.

38. World Health Organization. *Guiding principles for feeding non-breastfed children 6–24 months of age*. Geneva: World Health Organization; 2003.

39. Guyatt GH, Oxman AD, Vist GE et al. GRADE: An emerging consensus on rating quality of evidence and strength of recommendations. *BMJ* 2008; 336(7650):924–6.

15 Chronic Malnutrition

Saskia de Pee

CONTENTS

INTRODUCTION

Chronic malnutrition refers to a process of suboptimal nutrition, which is the result of nutrient needs not being met and/or suffering from disease or infection, which continues over a prolonged period of time. Undernutrition has several short- and longer-term consequences due to the effects of accumulated nutritional deficits on growth, health, and development, including higher morbidity and mortality, increased risk of noncommunicable diseases, poorer school performance, and lower income earning potential due to suboptimal cognitive and physical development early in life.

The main indicator of chronic malnutrition at the population level is the prevalence of stunting, or shortness for age, among children under 5 years old, which is assessed by comparing length (0–23 months) or height (24 months and older) to that of a reference population of the same age and sex [1].

However, although the prevalence of low length/height-for-age is the key indicator of the extent to which young children in a population are affected by chronic malnutrition, shortness is just one consequence of prolonged undernutrition. Stunting is related to an increased risk of morbidity and mortality, poorer school performance, lower income earning potential as an adult, and an increased risk of noncommunicable diseases later in life, because it indicates inadequate nutrition during a critical phase of development, that is, the first 1,000 days (from conception until the second birthday) [2–6]. In addition to linear growth, inadequate nutrition during this period also affects epigenetic programming during fetal life [7], brain growth and development, the performance of the immune system, and factors that increase the risk of noncommunicable disease later in life [2–6].

A child whose nutrient intake does not meet his or her needs over a prolonged period of time will become stunted and also suffer from micronutrient deficiencies. This underlies the relationship between stunting and the increased risk of morbidity and mortality, as well as suboptimal brain development. Thus, stunting is an indicator of vulnerability and suboptimal development, especially at the population level.

Measures for the prevention of chronic malnutrition should target all of those who are at risk. This may be the entire population in the window of the first 1,000 days in areas where stunting prevalence is high (e.g., above 40%), or it may be a subgroup of the population amongst whom prevalence is high (e.g., in a specific geographic area or in households with a low socioeconomic status). The prevention of chronic malnutrition requires the improvement of both diets and health of groups implicated throughout the 1,000 days window, pregnant and lactating women, and children under 24 months of age. It should also be ensured that women of reproductive age, including adolescent girls, have a good nutritional status, do not start their reproductive life until they are themselves fully grown and developed (i.e., 19 years of age or older), and that their pregnancies are well spaced (i.e., do not occur too closely together).

Improvements of both diets and health can occur during the economic development of a country. However, the extent to which this occurs, and who benefits, varies, and specific targeted efforts may be required to ensure that those who most need the improvements are not left behind. The same applies to populations of countries that experience little economic development or face a difficult period (e.g., an economic crisis, or an emergency), during which those individuals passing through this critical 1,000 days window should be protected against undernutrition. In order to sustain improvements to diet and health, the underlying and basic causes of malnutrition such as access to recourses and economic and political climate should also be improved, requiring action by sectors that have not traditionally focused on nutrition.

This chapter reviews the causes of and approaches to prevent chronic malnutrition.

PREVALENCE OF STUNTING, AND TARGETS AND COMMITMENTS FOR ITS REDUCTION

In 2016, approximately 23% of the world's children under 5 years old were stunted, coming to around 155 million children [8]. This percentage was considerably higher in the past, that is, approximately 40% in 1990, 33% in 2000, and 27% in 2010 [1]. Thus, the proportion of the world's population that today suffers the consequences of chronic malnutrition during early childhood is around 35%, or more than 2.5 billion people. This is comparable to the number of people who are estimated to suffer from micronutrient deficiencies and is also close to the number suffering from overweight or obesity. It is important to note that these forms of malnutrition can occur simultaneously in one individual, that is, most often stunting and micronutrient deficiencies during childhood, and overweight or obesity accompanied by micronutrient deficiencies during adulthood, but sometimes these three can even occur simultaneously during childhood [9].

In 2012, the World Health Assembly (WHA) set a target of 40% for the reduction of stunting between 2010 and 2025, which is equivalent to an annual, year-on-year, reduction of around 3%, and applies to all countries [10]. Thus, a country with 30% stunting in 2010 should reduce prevalence to 18% or less by 2025, and a country with 45% in 2010 should reduce prevalence to 27% or less by 2025.

Preventing malnutrition is also a key component of the Sustainable Development Goals (SDGs) [11]. SDG2, Zero Hunger, aims to end hunger, achieve food security and improved nutrition, and promote sustainable agriculture. It includes the following nutrition targets for 2030: "End hunger and ensure access by all people in particular the poor and people in vulnerable situations, including infants, to safe, nutritious, and sufficient food all year round; end all forms of malnutrition," including achieving by 2025, the WHA targets on stunting and wasting in children under 5 years of age, and addressing the nutritional needs of adolescent girls, pregnant and lactating women, and older persons [11].

The Scaling Up Nutrition (SUN) Movement [12] was formed in 2010 to work together across constituents and sectors to improve nutrition. Governments of countries that have joined the SUN Movement are in the lead, and organizations at national as well as global levels are organized in networks for civil society, the United Nations (UN), donors, businesses, and researchers to work in a collective manner to improve nutrition. The UN established the Zero Hunger Challenge in 2012 [13], prior to the establishment of the SDGs, to galvanize action focused on improving food and nutrition.

CAUSES OF CHRONIC MALNUTRITION

Direct causes of undernutrition can include an inadequate intake of required nutrients, illness, and infection. Nutrient requirements are diverse and high during the first 1,000 days, relative to energy requirements, and are therefore challenging to meet (see Chapters 3 to 5 for further details). Illness increases nutrient needs, and it can cause anorexia, which decreases food intake, and also increases losses of certain nutrients, such as vitamin A [14]. Furthermore, inflammation and subclinical infection, such as environmental enteropathic dysfunction (EED), also interfere with nutrient absorption and utilization and, hence, also contribute to malnutrition [15]. The underlying causes of undernutrition lead to the direct causes of undernutrition, and include food insecurity, poor hygiene, water and sanitation, limited curative and preventive health services as well as inadequate caring practices. These underlying causes are in turn affected by a country's economic situation, governance, climate, and gender roles.

In addition, intrauterine growth retardation (IUGR), which results in the small for gestational age births, which underlie 20% of stunting in low- and middle-income countries [16], is caused by poor maternal nutrition and health. This cannot completely be addressed by good nutrition and health during the start of the 1,000 days window, that is, during pregnancy. Factors that should be modified prior to pregnancy include the age of the mother (i.e., child bearing should start after adolescence) and pregnancies need to be well spaced (at least 24 months between delivery and the start of the next pregnancy). Girls should be well nourished in order to not have short stature and/or poor nutritional status as they become pregnant themselves. In order to ensure adequate nutrition at the time of conception, women of reproductive age need to be well nourished in general.

RELATIONSHIP BETWEEN CHRONIC AND ACUTE MALNUTRITION

The increased vulnerability to morbidity among children who suffer from chronic malnutrition, whether they are yet classified as stunted or not, also increases the risk of acute malnutrition, which is measured by low weight-for-height or small mid-upper arm circumference, and is generally caused by weight loss [17]. Morbidity leads to weight loss, reduced appetite and, hence, reduced food consumption, lower absorption, and increased utilization and also loss of nutrients. The reverse, that is, morbidity and acute malnutrition leading to chronic malnutrition, is not necessarily true, but frequent and/or longer lasting periods may add up and, as such, contribute to chronic malnutrition, especially when the in-between periods of recovery are too short and do not allow for accumulated nutrient deficits to be caught up on.

STUNTING SHOULD PROMPT PREVENTIVE ACTION AT POPULATION LEVEL

As mentioned earlier, stunting is a population-level indicator of vulnerability and suboptimal development. Acute malnutrition, on the other hand, is an indicator of an increased risk of (further) morbidity and mortality at an individual level, based upon which the individual can be treated to recover from acute malnutrition (i.e., gaining weight by rebuilding lean tissue and fat storage) and any concurrent illness.

As a population-level indicator, stunting should prompt actions to ensure that nutrient needs are met, and disease and poor birth outcomes are prevented, especially among individuals in the period of the 1,000 days (i.e., pregnant and lactating women and children under 2 years of age).

Furthermore, because of the increased risk of morbidity and mortality, a high level of stunting among a population that is exposed to an adverse situation, such as an increased risk of illness in the rainy season (e.g., malaria or diarrhea), crowded conditions with higher infection pressure due to an emergency or refugee situation, or increased food insecurity and, hence, reduced dietary quality, should also prompt specific actions to prevent illness. Such actions should focus both on health (e.g., immunization, adequate water and sanitation facilities, and hygiene practices) and ensuring the consumption of a nutritious diet, including an adequate micronutrient intake to support the immune system. This can be in the form of supplements (e.g., vitamin A and zinc, in particular), home fortificants, or fortified foods for the general population and special fortified foods for nutritionally vulnerable groups.

There are three reasons why stunting is not diagnosed and treated at an individual level, but needs to be prevented at population level. First, a child is diagnosed as stunted when his or her length is lower than that of 2.25% of the reference population (two standard deviations below the median, or −2 z-scores), and it takes time for that growth deficit to accumulate. Some children may already be below that cutoff at birth due to intrauterine growth restriction, which results in a small for gestational age birth (a birth weight that is below that of 10% of the reference population for that gestational age) [18]. For other children, growth is slower than that of the reference population, due to which they gradually deviate more and more from the reference curve. In some populations, stunting prevalence is already 20% to 25% at the age of 6 months, whereas in others it mainly develops between the ages of 6 and 18 months. In all countries, the prevalence of stunting increases until the age of 24 months, after which it stabilizes. Thus, the process of growth faltering until the point that the child's length (or height) has fallen below −2 z-scores of that of the reference population and the child is classified as stunted is gradual. While it is ongoing, however, the child is already at a disadvantage due to unmet nutrient needs and, hence, at an increased risk of morbidity and mortality, and of not achieving his or her full potential in terms of growth and development.

Second, while catching up on height after the age of 2 years is biologically possible, many of the other consequences of undernutrition during the first 1,000 days, such as suboptimal brain growth and development, and the consequences of intra-uterine growth retardation in terms of increased risk of noncommunicable disease cannot be undone or "repaired." The increased risk of morbidity and mortality is also current, and it should be prevented at the moment it occurs while the child may be in the process of becoming stunted. In addition, it is very important to note that moving from below the −2 z-score to above that cutoff may not necessarily indicate that the child is catching up on linear growth, as the accumulated deficit compared to the median height of the reference population may still be increasing [19].

The third reason is that it is not only stunted children (i.e., those who fall below the −2 z-score cutoff) who are not developing to their full potential. Where stunting prevalence is high, the distribution of the whole population is shifted toward the left (see Figure 15.1). Thus, children who have a z-score of, for example, −1 may also be

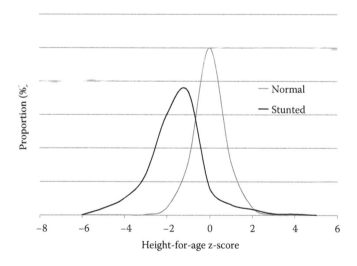

FIGURE 15.1 The distribution of height-for-age z-scores of a stunted and a normal population.

well below their own potential and would thus benefit from population-level interventions for the prevention of chronic malnutrition. These three reasons basically indicate that

- Stunting is just one indicator of chronic malnutrition, at a population level.
- Ongoing exposure to undernutrition, which is known as "chronic malnutrition," has both immediate and longer-term consequences.
- It is important to prevent undernutrition throughout the period when the risk is highest (i.e., the 1,000 days period), because of the immediate as well as the irreversible consequences.
- Indicators of success in preventing undernutrition are not limited to the reduction of stunting, but also include reductions of morbidity and micronutrient deficiencies, and intermediate indicators such as improved dietary diversity, a lower incidence of adolescent pregnancies, and some of which can occur without a simultaneous measurable change of stunting.

NUTRIENT REQUIREMENTS DURING THE FIRST 1,000 DAYS

Nutrient requirements need to be met in order to prevent undernutrition. During the first 1,000 days, these needs are high in comparison to energy needs due to the rapid growth of the fetus, infant, and young child. For example, a child aged 6 to 8 months needs 9 times as much iron and 4 times as much zinc per 100 kcal of complementary food, as that required by an adult man per 100 kcal of his diet [20].

Recommended Nutrient Intakes (RNI) [21] or Recommended Dietary Allowances (RDA) [22] have been set for individual nutrients and for specific age, sex, and physiological status (e.g., pregnant or breastfeeding) groups. These recommended intakes are for normal, healthy individuals, and are set at a level that would meet or exceed the needs of 97.5% of the specific group. The Estimated Average Requirement (EAR) is the level at which the needs of 50% of the population would be met or exceeded, which for most nutrients is at a level of approximately 70% of the RNI. For nutrients and age groups for which an RNI or EAR has not been established, which is particularly the case for young children, an Adequate Intake (AI) has been set, which is based on the actual intakes of apparently healthy individuals [22].

In order to meet the recommended intakes of some 40 nutrients, a diverse diet needs to be consumed, because no single food contains all required nutrients. Such a diet must include plant-source foods (i.e., staple foods, vegetables, and fruits), animal source foods (i.e., eggs, fish, dairy, and meat), as well as fortified foods (e.g., iodized salt and flour fortified with folic acid) [23–25]. Consuming a diet that meets not only energy and protein needs, which can be fulfilled with quite a basic diet, but also the needs of vitamins, minerals, essential fatty acids, essential amino acids, and other nutrients is more challenging and costs more, as it needs to be more diverse [26,27].

Certain subgroups of individuals have higher needs than the rest of their population group, including individuals who suffer from undernutrition (e.g., moderate acute malnutrition), infants born small for gestational age, people with micronutrient deficiencies, and people with chronic infections. Precise nutrient requirements under these circumstances are not known, but it is likely, for example, that children who suffer frequent episodes of illness have needs that are somewhere between those of normal, healthy children and children with moderate acute malnutrition [28]. In addition, it is likely that nutrient needs vary between days and between periods with less or more bodily growth [29].

At an individual level, these varying nutrient needs are difficult to take into account, except during nutritional treatment, such as for severe acute malnutrition. At a population level, they are important to take into account when setting targets for nutrient intake that aim to prevent deficiencies, where it is important to ensure that the recommended intake is met, including by individuals with the lowest intakes from their current diet. This means that the population distribution of intake needs to be shifted to be above the EAR, while not exceeding the tolerable upper limit of the intake of individual nutrients by far, and for a prolonged period of time [22,30–32].

PREVENTING CHRONIC MALNUTRITION: APPROACHING IT FROM A CONCEPTUAL AND EVIDENCE-BASED POINT OF VIEW

Any intervention to prevent chronic malnutrition should aim to achieve an adequate nutrient intake, prevent disease, and ensure healthy, well-spaced, nonadolescent pregnancies. Conceptually, this may sound straightforward. However, realizing this at a population level, in a specific context, in a way that complements existing diets, health services, and hygiene measures; is accepted and well-practiced by the intended target group and its caretakers; and is cost-effective (i.e., achieves the most at the lowest cost) is very challenging.

As far as achieving optimal nutrient intake is concerned, improvements to food intake can be achieved in two ways that can complement each other: (1) people can be encouraged to make better food choices, including ensuring optimal breastfeeding practices and consuming a diverse diet that includes a range of foods from different food groups; and (2) people can be provided with specific nutritious food complements to add to their existing diet. Systematic reviews have assessed the impact of both approaches, and it has been concluded that nutrition education about good complementary feeding can reduce stunting and improve height-for-age z-scores (HAZ) in food-secure contexts and that nutrition education in combination with food supplements reduces stunting and improves HAZ in food-insecure contexts [33,34].

This makes sense. Where people can afford to make better choices and more nutritious foods are available to choose from, their diet can improve. This also underlies the relationship seen across countries between the level of a country's gross domestic product (GDP) and stunting prevalence. For every 10% higher the GDP is, the stunting prevalence is, on average, approximately 6% lower [35]. A higher GDP not only indicates greater purchasing power but also a more developed food system that results in greater access to a diverse diet, including animal source foods and processed foods for specific target groups, as well as a range of other factors that can improve nutrition and health. Where and for whom purchasing power remains relatively low, specific assistance may be required to enable improvements of dietary intake toward greater diversity and inclusion of more animal source and fortified foods in order to meet nutrient requirements.

Where specific assistance is required to improve nutrient intake, the aim is to do so most efficiently (i.e., by making the most difference at the lowest cost), including with the smallest possible modification of the daily diet. This often means that a special nutritious, fortified food is selected to complement the diet of children aged 6 to 23 months, because these are more cost-effective in terms of meeting nutrient needs, as compared to a combination of locally available fresh foods. A young child's need for iron, zinc, and often also calcium are particularly difficult to meet without fortification [20,25,36].

Income support, in the form of cash or vouchers for purchasing food, is another way to improve access to food, but the extent to which it improves nutrient intake of specific individuals in the household depends on what foods can be purchased, how much purchasing power is increased (i.e., how much cash is provided), and what choices are made by the consumer. The main difference is that cash is targeted at poor households and is likely to benefit most of its members in several ways, whereas a specific nutritious food for a specific individual is very targeted, can be well tailored to specific nutrient needs, and costs less. In fact, to make social safety nets more nutrition sensitive, the inclusion of specific nutritious foods for certain target groups can be recommended [37]. This approach is similar to that of the Women, Infants, and Children (WIC) program in the United States [38].

The impact that the distribution of a special nutritious food will have on the desired outcome (i.e., reduction of undernutrition) depends on several factors, including [39]:

- Its specific composition, that is, which nutrients it contains and in what amounts
- Its actual intake, that is, what quantity is consumed and how frequently, which depends on actual distribution and communication/marketing messages, as well as acceptance and understanding by caretakers
- The difference that it makes to total nutrient intake, that is, where the nutrient intake gap is greater and when it is consumed in addition to (rather than as a replacement for) the existing diet, the impact is likely to be larger

For these reasons, the magnitude of impact that a particular food type, of a well-defined composition, can have varies by context, as it depends on the difference that it can make in terms of closing the nutrient intake gap, whether consumption is as intended and whether it adds, rather than replaces, existing nutrient intake. Therefore, systematic reviews that assess the magnitude of the impact of particular types of commodities for reducing undernutrition based on studies conducted in different contexts and with commodities of somewhat different composition are of limited value, as these studies have very low external validity (i.e., their findings may not apply to other settings).

When the aim is to improve nutrient intake substantially among the targeted population, rather than selecting a commodity based on evidence of impact from other contexts, the choice should be guided by knowledge of the local context with regard to the likely magnitude and characteristics of the nutrient intake gap, and local food preferences and practices, as well as manufacturing possibilities, costs, and possible distribution mechanisms, including whether certain consumers can be reached through market channels and others through public sector distribution (e.g., health, social protection, and community outreach). Box 15.1 lists the types of considerations

---

BOX 15.1   CONSIDERATIONS WHEN SELECTING FOOD-BASED INTERVENTIONS TO IMPROVE NUTRIENT INTAKE

When selecting a food-based intervention to improve nutrient intake from the diet in a specific context, which can range from providing specific nutritious foods to certain target groups or increasing their availability in markets to recommending dietary changes possibly accompanied by homestead food production or other food security interventions, it is important to consider

- Quantitatively the magnitude of the difference that needs to be made in order to achieve recommended nutrient intakes.
- Biologically the nutrients that will need to be released from the foods during digestion and become available for absorption (i.e., the nutrients need to be bioavailable and the consumer in good health).
- Demand, that is, consumers need to obtain (purchase or receive for free), like, and consume the food(s) and at the recommended frequency and in the recommended amount.
- Supply, that is, how can it be ensured that the foods that are produced or manufactured become available for the intended consumers, are purchased or distributed to the beneficiaries, and so forth.

---

that are important for selecting suitable food(s) for improving the nutrient intake of specific target groups, in a context where specific nutritional assistance is required.

When an approach has been chosen and is gradually being rolled out, it is important to monitor the acceptability and actual intake of the commodities, as well as concurrent changes to the existing diet. Nutrition surveillance is a good way to monitor programs in context [40].

PREVENTING CHRONIC MALNUTRITION: APPROACHING IT FROM A PRACTICAL, CONTEXT-SPECIFIC, POINT OF VIEW—"FILLING THE NUTRIENT GAP"

Considering that meeting nutrient requirements is a prerequisite for preventing undernutrition, that many factors affect whether nutrient intake targets are being met, that we mostly have proxy indicators for the extent to which nutrient needs are being met, and that approaches for improving nutrient intake need to be specific to their target group and context, the design of context-specific approaches for prevention of undernutrition should start with a thorough situation analysis.

Such a situation analysis, in combination with an approach for multisectoral decision making, has been developed in the form of the Fill the Nutrient Gap (FNG) tool by the World Food Programme (WFP), with input from the University of California–Davis, Harvard University, Epicentre, the International Food Policy Research Institute (IFPRI), Mahidol University, and UNICEF [41]. The FNG was piloted in three countries in 2016, and will be consolidated and expanded in 2017 to approximately 12 countries, and more in 2018.

The premise of the FNG approach is that recommended nutrient intakes have been defined and that many people, especially where stunting and micronutrient deficiencies are highly prevalent, do not meet their nutrient needs. Foods are the main source of nutrients, and actual food consumption is determined by context-specific factors such as food production, manufacturing, import and export, transport and distribution, prices relative to income, culture, knowledge, and so on. In order to improve food consumption in such a way that nutrient needs can be met, including among target groups with higher needs, the specific context needs to be better understood, and different sectors need to work together to improve the situation.

The FNG consists of an analytical and decision-making component. The analytical component has two parts: (1) a situation analysis using available secondary data that focuses on the type and scale of nutrient intake deficits and enablers and constraining factors; and (2) a linear programming component, using the Cost of Diet tool developed by Save the Children UK, which assesses the costs to a household of meeting the nutrient intake recommendations of its members at the lowest possible cost, using locally available foods, and comparing that to their level of food expenditure [26,27].

The analytical components provide rich context-specific insights into the likelihood that nutrient needs are met, and the main bottlenecks and main opportunities to improve the situation, considering both nutrition-specific and sensitive intervention opportunities as well as potential delivery channels for reaching specific target groups.

The process through which the analyses are undertaken is the other essential component of the approach. The process starts when national-level stakeholders decide that they would like to have the FNG analyses conducted. This is typically when policy revisions (e.g., of nutrition or social protection) are due. In an inception meeting, the stakeholders jointly decide on the desired focus of the analysis (e.g., adolescent girls, pregnant and lactating women, and children under 2) and the specific, distinct geographical areas for which secondary data are likely to be available (e.g., urban versus rural and different regions). The inception meeting is attended by stakeholders from different sectors, including health, agriculture, social protection, education, trade and industry, and different constituencies including national government, UN agencies, academia, private sector, and civil society. The meeting aims to achieve a joint understanding of the approach, and a commitment to contribute to the analysis by sharing relevant sources of secondary data and reviewing the findings that are being put together by a technical team. Based on the results, stakeholders then jointly formulate recommendations for policies, as well as the programs that different sectors can undertake.

So far, the experiences have been very positive. Focusing discussions, which are informed by existing secondary data, on improving nutrient intake in the country's context creates a common understanding and allows for the formulation of concrete recommendations for next steps among stakeholders. This multisectoral approach also aligns very well with that of the SUN movement.

CAN STUNTING BE AN INDICATOR OF BOTH THE NEED FOR NUTRITION INTERVENTIONS AND THEIR IMPACT?

Stunting is an indicator of chronic malnutrition used to monitor the undernutrition situation within a country over time, and across countries. It is the preferred indicator because it reflects the cumulative impact of nutrition insults (i.e., inadequate nutrient intake and disease across the first 1,000 days), and because it also reflects the impact of underlying and basic causes of malnutrition. Furthermore, it reflects other short- and longer-term consequences of undernutrition as well, including an increased risk of morbidity and mortality, suboptimal cognitive development and, hence, poorer schooling performance and income-earning later in life, and an increased risk of noncommunicable diseases. Stunting is a good indicator for monitoring improvements to a population's nutritional status over a longer period of time, at least a few years, as related to the cumulative exposure to nutrition-specific and sensitive policies and strategies, including proper safety nets and changes to the economic situation.

However, for two reasons, stunting may not be the best primary outcome indicator of the impact of a specific intervention. First, stunting accumulates over a substantial period of time, covering prepregnancy, which affects nutritional status at conception, pregnancy, early lactation, and the complementary feeding period, and is affected, both positively and negatively, by many different factors. Therefore, while a specific intervention may make a contribution to improving nutrition, it may not cover a long enough period of the first 1,000 days or be insufficient on its own to lead to a substantial change of stunting without concurrent improvement of other conditions as well. Second, as stunting is defined as inadequate linear growth (being too short for one's age), it is only when specific nutrients that enable the adequate growth of bones and muscles are also consumed in required amounts that stunting will reduce [42]. These nutrients are provided by a diverse diet that includes plant-source foods, animal source foods, and fortified foods. Further, specific nutrition interventions may be a better source of certain nutrients (e.g., micronutrients) than of specifically those nutrients that are required to achieve linear growth. In that case, the intervention may have a limited impact on stunting but still have a positive impact on other forms and consequences of chronic malnutrition, such as micronutrient deficiencies, cognitive development, and morbidity.

Therefore, it is very important that interventions to reduce undernutrition in a population are judged not only by their impact on stunting, but also by their impact on other forms and on the consequences of undernutrition, including micronutrient status and morbidity, and that targets are set and changes monitored for dietary intake, disease and infection, and maternal nutrition, as well as their underlying factors.

CONCLUSION

Chronic malnutrition refers to a process of suboptimal nutrition, due to nutrient needs not being met and suffering from disease and infection, which continues over a prolonged period of time and to a state of undernutrition that has several short- and longer-term consequences. The consequences are due to the effects of the accumulated nutritional deficits on growth, health, and development, including higher morbidity and mortality, poorer school performance, and lower income-earning potential due to suboptimal cognitive and physical development early in life, and an increased risk of noncommunicable diseases.

Stunting, or shortness for age, is the main population-level indicator of prolonged or "chronic" malnutrition. However, in order for it to be interpreted, it is important to understand and use it correctly. Ongoing exposure to undernutrition has both immediate and longer-term consequences, which are revealed in diverse ways, not only as a height deficit that also takes time to accumulate. Stunting among the under-fives indicates the risk of individuals currently in the 1,000 days window, among whom undernutrition should be prevented because of immediate and irreversible consequences, and the vulnerability of those 5 years and under to adverse conditions, such as an increased risk of malaria and diarrhea during the rainy season. Indicators of success of preventing undernutrition are not limited to a reduction of stunting, but also include other effects of improved nutritional status, such as reductions of morbidity, micronutrient deficiencies, and low birth weights.

Meeting nutrient requirements is a prerequisite for the prevention of undernutrition and requires the consumption of a diverse diet, which should also include animal source foods and fortified foods. The limited availability and affordability of a diverse diet are major constraints to meeting nutrient requirements. In order to substantially improve nutrient intake among the targeted population, rather than select a commodity based on evidence of its impact from other contexts, the choice should be guided by knowledge of the local context. Context-specific information is required with regard to the likely magnitude and characteristics of the nutrient intake gap, food preferences and practices, manufacturing possibilities and costs of specific nutritious foods, and possible distribution mechanisms, including whether certain consumers can be reached through market channels or public sector distribution. Such context-specific solutions should be designed and implemented in a coordinated manner by governments, the UN, donors, the private sector, and civil society stakeholders across health, food, and social safety-net systems.

Interventions for reducing undernutrition in a population should be judged not only by their impact on stunting, but also by their impact on other forms and consequences of undernutrition, including micronutrient status and morbidity. A country's progress in reducing chronic malnutrition needs to be monitored over time, and should use a range of indicators of chronic malnutrition and health as well as of dietary intake (e.g., diversity and special nutritious foods), factors affecting the consumption of different foods (e.g., price, affordability, availability, knowledge), factors related to maternal nutrition (e.g., age at first pregnancy, or the school enrollment of adolescents), coverage of preventive health services, and hygiene and sanitation practices.

REFERENCES

1. De Onis M, Blössner M, Borghi E. Prevalence and trends of stunting among pre-school children, 1990–2020. *Public Health Nutr* 2012; 15(1):142–8.
2. Frongillo EA, Jr. Symposium: Causes and etiology of stunting. *J Nutr* 1999; 129:S529–30.
3. Victora CG, Adair L, Fall C et al. Maternal and child undernutrition: Consequences for adult health and human capital. *Lancet* 2008; 371:340–57.
4. Dewey KG, Begum K. Why stunting matters. Alive and Thrive technical brief, issue 2. 2010. Available at: http://aliveandthrive.org/wp-content/uploads/2014/11/Brief-2-Why -stunting-matters_English.pdf (accessed February 5, 2017).
5. Black RE, Victora CG, Walker SP et al. Maternal and child undernutrition and over-weight in low-income and middle-income countries. *Lancet* 2013; 382:427–51.
6. Adair LS, Fall CHD, Osmond C et al. Associations of linear growth and relative weight gain during early life with adult health and human capital in countries of low and middle income: Findings from five birth cohort studies. *Lancet* 2013; 382:525–34.
7. Moore SE. Early life nutritional programming of health and disease in the Gambia. *J Dev Orig Health Disease* 2016; 7:123–31.
8. UNICEF/WHO/World Bank Group. Joint child malnutrition estimates. Levels and trends in child malnutrition. Geneva: World Health Organization; 2016. Accessible at: http://www.who.int/nutgrowthdb/jme_brochoure2017.pdf?ua=1 (accessed June 13, 2017).
9. Tzioumis E, Adair LS. Childhood dual burden of under- and over-nutrition in low- and middle-income countries: A critical review. *Food Nutr Bull* 2014; 35:230–43.
10. World Health Organization. Global nutrition targets 2025: Stunting policy brief (WHO/ NMH/NHD/14.3). Geneva: World Health Organization; 2014.
11. Sustainable Development Goals (SDGs). [Internet]. Available at: http://www.un.org /sustainabledevelopment/hunger/ (accessed February 5, 2017).
12. Scaling Up Nutrition movement. [Internet]. Available at: http://scalingupnutrition.org /sun-countries/about-sun-countries/ (accessed February 5, 2017).
13. Zero Hunger Challenge (ZHC). [Internet]. Available at: https://www.un.org/zerohunger /content/challenge-hunger-can-be-eliminated-our-lifetimes (accessed February 5, 2017).
14. Stephensen CB, Alvarez JO, Kohatsu J et al. Vitamin A is excreted in the urine during infection. *Am J Clin Nutr* 1994; 60:388–92.
15. Owino V, Ahmed T, Freemark M et al. Environmental enteric dysfunction and growth failure/stunting in global child health. *Pediatrics* 2016; 138(6): e20160641.
16. Christian P, Lee SE, Angel MD et al. Risk of childhood undernutrition related to small for gestational age and preterm birth in low- and middle-income countries. *Int J Epidemiol* 2013; 42:1340–55.
17. De Onis M. Child growth and development. In de Pee S, Taren D, Bloem MW, editors, *Nutrition and health in a developing world*, 3rd ed. Totowa, NJ: Humana Press; 2017.
18. Lee ACC, Katz J, Blencowe H et al. National and regional estimates of term and pre-term babies born small for gestational age in 138 low-income and middle-income countries in 2010. *Lancet Global Health* 2013; 1:e26–36.
19. Leroy JL, Ruel M, Habicht J-P, Frongillo EA. Using height-for-age differences (HAD) instead of height-for-age z-socres (HAZ) for the meaningful measurement of population-level catch-up in linear growth in children less than 5 years of age. *BMC Pediatr* 2015; 15:145
20. Dewey, KG. The challenge of meeting nutrient needs of infants and young children during the period of complementary feeding: An evolutionary perspective. *J Nutr* 2013; 143:2050–4.
21. World Health Organization/Food and Agriculture Organization. *Vitamin and mineral requirements in human nutrition*, 2nd ed. Geneva: World Health Organization; 2004.

22. Institute of Medicine (IOM). *Dietary reference intakes: The essential reference for dietary planning and assessment.* Washington, DC: IOM; 2006.

23. De Pee S, Bloem MW. Current and potential role of specially formulated foods and food supplements for preventing malnutrition among 6–23 month-old children and for treating moderate malnutrition among 6–59 month-old children. *Food Nutr Bull* 2009; 30:S434–63.

24. Dror DK, Allen LH. The importance of milk and other animal source foods for children in low-income countries. *Food Nutr Bull* 2011; 32:227–43.

25. Osendarp S, Broersen B, van Liere MJ et al. Complementary feeding diets made of local foods can be optimized, but additional interventions will be needed to meet iron and zinc requirements in 6–23 month old children in low and middle income countries. *Food Nutr Bull* 2016; 37:544–70.

26. Baldi G, Martini E, Catharina M, de Pee S. Cost of the diet (CoD) tool: First results from Indonesia and applications for policy discussion on food and nutrition security. *Food Nutr Bull* 2013; 34:S35–42.

27. Geniez P, Mathiassen A, de Pee S et al. Integrating food poverty and minimum cost diet methods into a single framework: A case study using a Nepalese household expenditure survey. *Food Nutr Bull* 2014; 35:151–9.

28. Golden, MH. Proposed recommended nutrient densities for moderately malnourished children. *Food Nutr Bull* 2009; 30:S267–342.

29. Garza C. Commentary: Please sir, I want some more (and something else). *Int J Epidemiol* 2015; 44(6):1876–8.

30. De Pee S, Timmer A, Martini E et al. Programmatic guidance brief on use of micronutrient powders (MNP) for home fortification. Technical Advisory Group (HF-TAG). Geneva: World Food Programme; 2011.

31. Bruins MJ, Mugambi G, Verkaik-Kloosterman J et al. Addressing the risk of inadequate and excessive micronutrient intakes: Traditional versus new approaches to setting adequate and safe micronutrient levels in foods. *Food Nutr Res* 2015; 59:26020.

32. De Pee S. Nutrient needs and approaches to meeting them. In De Pee S, Taren D, Bloem MW, editors, *Nutrition and health in a developing world*, 3rd ed. Totowa, NJ: Humana Press; 2017.

33. Dewey KG, Adu-Afarwuah S. Systematic review of the efficacy and effectiveness of complementary feeding interventions in developing countries. *Matern Child Nutr* 2008; 24–85.

34. Bhutta ZA, Das JK, Rizvi A et al. Evidence-based interventions for improvement of maternal and child nutrition: What can be done and at what cost? *Lancet* 2013; 382(9890):452–77.

35. Ruel MT, Alderman H, Maternal and Child Nutrition Study Group. Nutrition sensitive interventions and programmes: How can they help to accelerate progress in improving maternal and child nutrition? *Lancet* 2013; 382(9891):536–51.

36. Skau JKH, Bunthang T, Chamnan C et al. The use of linear programming to determine whether a formulated complementary food product can ensure adequate nutrients for 6- to 11-month-old Cambodian infants. *Am J Clin Nutr* 2014; 99:130–8.

37. Langendorf C, Roederer T, de Pee S et al. Preventing acute malnutrition among young children in crises: A prospective intervention study in Niger. *Plos Med* 2014; 11:e1001714.

38. Schultz DJ, Byker Shanks C et al. The impact of the 2009 special supplemental nutrition program for women, infants, and children food package revisions on participants: A systematic review. *J Acad Nutr Diet* 2015; 115:1832–46.

39. De Pee S. Special nutritious solutions to enhance complementary feeding. Editorial. *Matern Child Nutr* 2015; 11:i–viii.

40. Bloem MW, de Pee S, Semba RD. How much do data influence programs for health and nutrition? Experience from health and nutrition surveillance systems. In *Nutrition and health in developing countries*, 2nd ed., Semba RD, Bloem MW, editors. Totowa, NJ: Humana Press; 2008.

41. World Food Programme. Fill the nutrient gap tool. Situation analysis, consensus, decision making. [Internet]. Available at: http://documents.wfp.org/stellent/groups/public /documents/communications/wfp288102.pdf (accessed February 5, 2017).

42. Golden MH. Specific deficiencies versus growth failure: Type I and type II nutrients. *SCN News* 1995; 12:15–21.

16 Nutritional Regulation of the Growth Plate

Julian C. Lui

CONTENTS

INTRODUCTION

During embryonic development, bone formation begins with the condensation of mesenchymal stem cells. In a number of places in the body, such as the flat bones of the skull, bone formation is driven by a process called intramembranous ossification, where mesenchymal stem cells differentiate directly into bone-forming osteoblasts. In most other places, however, bones are formed by a different process known as endochondral ossification (Figure 16.1; for review, see Kronenberg [1]). In this process, mesenchymal stem cells first differentiate into chondrocytes. These chondrocytes secrete cartilage matrix composed mainly of type II collagen. Proliferation of chondrocytes leads to an overall expansion of this cartilage tissue and, in the center of the cartilage, cells stop dividing and start enlarging to become type X collagen-producing hypertrophic chondrocytes. These hypertrophic chondrocytes drive cartilage matrix mineralization, and eventually undergo apoptosis, leaving a cartilage matrix scaffold for the invasion of blood vessels and osteoblasts to lay down bone matrix in the center of the cartilage, known as the primary ossification center. As chondrocytes continue

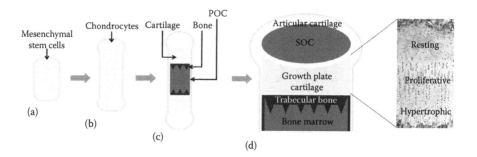

FIGURE 16.1 Endochondral ossification and histology of the growth plate. (a) Endochondral bone formation starts with a mesenchymal condensation. (b) Mesenchymal stem cells differentiate into chondrocytes, and chondrocytes at the center of the cartilage mold start to hypertrophy. (c) Hypertrophic chondrocytes undergo apoptosis and draw the invasion of blood vessels and osteoblasts, building calcified bone in the primary ossification center (poc). (d) Later in bone development, the epiphysis also undergoes calcification, forming the secondary ossification center (soc), and the growth plate cartilage remains near the ends of long bone continues to drive bone elongation.

to proliferate, undergo hypertrophy, and are then invaded by osteoblasts, the long bones continue to lengthen and ossify in the center, resulting in longitudinal bone growth and an overall increase in body size and height. Gradually, as bones continue to grow in length, endochondral ossification becomes increasingly restricted to the cartilaginous structure found near the two opposite ends of long bones known as the growth plate. Within each growth plate, chondrocytes are arranged into three histologically distinct zones called resting, proliferative, and hypertrophic zones. Closest to the epiphysis, round and slowly dividing resting chondrocytes serve as precursor cells capable of self-renewing and giving rise to new clones of proliferative chondrocytes. In the proliferative zones, cells are arranged in columns parallel to the long axis of the bone and, within each column, chondrocytes undergo rapid proliferation, pushing themselves gradually toward the center of the bone, where they undergo the same process of hypertrophy and apoptosis. In late adolescence, as adult height is gradually attained, the growth plate continues to narrow, until growth potential is depleted, the epiphyseal fuses, and growth in height stops.

ENDOCRINE REGULATION OF BONE GROWTH

The integrated processes of chondrocyte proliferation and hypertrophy are regulated and coordinated by a complex network of endocrine signaling molecules, including growth hormone (GH), insulin-like growth factor I (IGF-I), glucocorticoid, thyroid hormone, estrogen, and androgen (Figure 16.2).

GROWTH HORMONE INSULIN-LIKE GROWTH FACTOR-I AXIS

Both GH and IGF-I are potent stimulators of longitudinal bone growth (for review, see Lui and Baron [2]). GH is produced by the anterior pituitary and secreted into

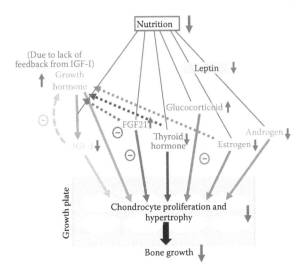

FIGURE 16.2 A schematic model for nutritional regulation of endocrine signals important for bone growth. The effects of malnutrition are annotated by red lines and red arrows. The negative sign represents inhibitory effects.

circulation. When GH travels to the liver, it stimulates hepatic production of IGF-I, which in turn acts as an endocrine factor to stimulate chondrocyte proliferation and hypertrophy at the growth plate. GH can also stimulate the local production of IGF-I at the growth plate to stimulate growth in a paracrine/autocrine fashion. In addition, GH may also directly stimulate growth at the growth plate, independent of its effect on circulating IGF-I, mainly through recruiting resting zone chondrocytes into the proliferative columns. GH excess due to pituitary adenomas in childhood therefore results in gigantism and, conversely, GH deficiency in children results in decreased IGF-I and poor linear growth. Similarly, mutations in the GH receptor cause GH insensitivity syndrome, which can cause low IGF-I levels and poor linear growth. Normally, IGF-I provides a negative feedback to the pituitary to limit GH secretion; children with GH insensitivity therefore have elevated GH levels. Surprisingly, GH appears to be dispensable for fetal growth, as suggested by the relatively normal prenatal growth in individuals with GH deficiency or insensitivity (with modestly reduced IGF-I). In contrast, IGF-I appears to be important for both pre- and post-natal growth, as suggested by the much more significant intrauterine growth retardation and postnatal growth failure found in individuals with mutation of the *IGF1* gene or its receptor. Insulin-like growth factor II (IGF-II), which bears structural similarities with IGF-I, and also acts through the IGF-I receptor, is a paracrine/autocrine factor long known as an important stimulator of prenatal growth [3]. However, some recent clinical findings suggest that IGF-II may also be important for postnatal growth, as individuals with nonsense mutations of *IGF2* demonstrated severe intrauterine growth retardation as well as postnatal growth restriction [4].

Normal linear growth requires an adequate intake of calories, protein, vitamins, and certain other micronutrients, such as iodine, phosphate, and zinc. One of the major effects of malnutrition, or negative energy balance, on limiting body growth is due to its effect on the GH–IGF-I axis. Importantly, inadequate calorie intake slows linear growth by downregulating GH receptor expression in the liver [5], thereby inducing a state of GH insensitivity that is similar to that observed in individuals with GH receptor mutations. Undernourished children therefore show decreased IGF-I levels and increased GH levels due to the lack of negative feedback.

Recent evidence suggests that malnutrition-induced GH insensitivity may, in part, be mediated by fibroblast growth factor 21 (FGF21). Unlike many other FGFs, which only act locally as paracrine/autocrine factors, FGF21 lacks the FGF heparin-binding domain and therefore can still act as a paracrine factor but also diffuse from the tissue of synthesis like a hormone. FGF21 can activate FGFR1 and FGFR3 [6], both of which transmit a growth-inhibitory effect in chondrocytes. Thus, for example, the activation of a mutation of FGFR3 causes achondroplasia, which causes severe short stature in children. Consistently, transgenic mice overexpressing Fgf21 exhibit reduced bone growth, partly by increasing FGF signaling at the growth plate and partly by inducing GH insensitivity at the liver [7]. Recent studies suggest that FGF21 plays an important role in fasting-induced GH insensitivity and growth inhibition. Prolonged fasting increases circulating FGF21 in both mice [8] and humans [9] and, when placed under food restriction, Fgf21 knockout mice showed significantly increased linear growth and growth plate thickness as compared to wild-type mice, strongly suggesting that growth inhibition by fasting is driven by Fgf21 [10]. Most importantly, decreased GH receptor expression and GH insensitivity were also corrected by ablation of Fgf21 [10], therefore suggesting that, at least in mice, decreased GH action during fasting is driven by fasting-induced expression of Fgf21.

In additional to inadequate calorie intake leading to GH insensitivity, protein deficiency alone may also impair bone growth by suppressing the GH–IGF-I axis. Previous studies have shown that protein deficiency, even when compensated with normal calorie intake, still led to impaired childhood growth and poor bone quality [11], in part due to decreased levels of GH, and thereby IGF-I production [12].

GLUCOCORTICOID EXCESS

Apart from the GH-IGF-I axis, other hormonal systems may also directly regulate growth at the growth plate. Glucocorticoids, which are a subclass of steroid hormones, are quite commonly used as anti-inflammatory and immunosuppressive drugs in children. Treatment with glucocorticoids at a high dose for an extended period of time often leads to decreased linear growth in children, which is in part due to the suppression of chondrocyte proliferation and increased apoptosis by glucocorticoid [13]. Normally, the glucocorticoid produced in our body is called cortisol, which comes from the adrenal gland, and its production is stimulated by the adrenocorticotrophic hormone (ACTH) from the anterior pituitary. Individuals with a rare genetic disorder called Cushing's disease, usually caused by a pituitary adenoma producing an excessive amount of ACTH, therefore also exhibit poor linear growth. In addition to glucocorticoid's direct growth-inhibitory effect on the growth plate, it could also inhibit growth indirectly through other endocrine signals. Prolonged glucocorticoid excess negatively regulates GH production [14] and also increases the release of free fatty acid [15], which in turn negatively regulates GH actions.

Endogenous glucocorticoid production and sensitivity increases under stressful conditions, such as illness and prolonged food restriction [16], which directly suppresses growth at the growth plate, and also indirectly suppresses growth, by suppressing the GH–IGF-I axis.

THYROID HORMONE

Thyroid hormones produced by the thyroid gland, namely, 3,5,3'-triiodothyronine (T3) and thyroxine (T4), are important regulators of metabolism. T4 serves as a prohormone, which can be converted to the bioactive T3 by the enzyme type II iodothyronine deiodinase (DIO2) expressed in the thyroid and most other tissues, including the growth plate. Thyroid hormones are an important stimulator of normal linear growth. Hypothyroidism delays longitudinal bone growth and endochondral ossification, while thyrotoxicosis, which results in excessive production of thyroid hormones, accelerates both processes. Nonetheless, both hypothyroidism and thyrotoxicosis eventually lead to short stature. Thyroid hormones support bone growth by promoting the recruitment of resting chondrocyte into a proliferative zone, as well as stimulating chondrocyte hypertrophy [17]. Thyroid hormones are also essential for the normal deposition of the extracellular matrix by stimulating the production of type II and type X collagens and the synthesis of alkaline phosphatase [18]. Thyroid hormones also indirectly stimulate growth by modulating the production of GH and IGF-I. In humans with hypothyroidism and in thyroidectomized mice, both GH and IGF-I levels are decreased, and replacement of GH in hypothyroid rats or thyroid hormone receptor knockout mice partially rescues bone growth [19]. Among the four different isoforms of thyroid hormone receptors (TR-α_1, -α_2, -β_1, -β_2, encoded by two genes, *THRA* and *THRB*) found in humans, TR-α_1 and -α_2 are more highly expressed in the growth plate, and therefore appear to elicit much of the effect of thyroid hormone on longitudinal growth. Consistently, the homozygous deletion of *THRB* showed some delay in skeletal maturation, but exhibited normal growth [20], while homozygous deletion of *THRA* has not yet been described. So far, most cases of thyroid hormone resistance found in humans that showed postnatal growth retardation is caused by a dominant negative mutation of either TR-α or TR-β receptors [21]. Studies in humans showed that malnutrition decreases both T3 and T4 levels, and thereby slows linear growth [22]. Similarly, iodine deficiency, which is very common in some developing countries, leads to thyroid hormone deficiency and decreased linear growth. In several animal models, however, it was shown that food restriction mostly reduces T3 levels rather than T4 [23], suggesting that malnutrition may affect the conversion of T4 to T3 rather than thyroid hormone production.

ESTROGEN, ANDROGEN, AND LEPTIN

Another important hormonal system that regulates linear growth is the sex steroids. During puberty, estrogens and androgens are produced by the gonads and are associated with the pubertal growth spurt. This growth acceleration is primarily induced by estrogen rather than androgen, because a near-normal growth spurt is observed in individuals with androgen insensitivity [24], but little or no growth spurt occurs in individuals with deficiency of aromatase [25], an adrenal enzyme that is responsible for the conversion of androgens into estrogens. This growth-stimulatory effect of estrogen during puberty is mediated in part through increasing GH production, as well as its direct action on growth plate chondrocytes [26]. Although estrogen stimulates longitudinal growth, it also accelerates skeletal maturation, in part through accelerating the depletion of resting chondrocytes in the growth plate [27]. In children with precocious puberty, premature estrogen exposure therefore leads to premature epiphyseal fusion and decreased final height and, conversely, in individuals with hypogonadism, a lack of estrogen leads to delayed epiphyseal fusion and tall stature. Androgens (like testosterone) per se, without conversion to estrogen, may also help stimulate bone growth. In boys, dihydrotestosterone, a nonaromatizable androgen, can accelerate linear growth [28]. Apparently, unlike estrogen, this effect is not associated with an increase in circulating GH or IGF-I.

Malnutrition also strongly attenuates the reproductive axis, therefore decreasing the levels of both estrogen and androgen. Some recent studies have shown that this is driven, in part, by decreased levels of a hormone from the adipose cells called leptin (for review, see Sanchez-Garrido and Tena-Sempere [29]). Leptin was originally described as a circulating hormone involved in feeding behavior and energy homeostasis, but was later also found to be important for controlling the timing of puberty, and thus indirectly affecting bone growth. Along with kisspeptin, leptin is a major permissive signal for the onset of puberty. In malnourished children, decreased body fat mass leads to decreased leptin, and insufficient leptin in the body thereby delays the onset of puberty and the pubertal growth spurt. Whether leptin has an effect on bone growth, independent of its regulation of puberty, remains controversial. Mice with a loss-of-function mutation of the leptin gene (ob/ob mice) have shorter femur length but increased overall body length [30], and leptin administration in ob/ob mice increases bone length, suggesting a growth-promoting effect [31]. Nevertheless, leptin administration has many indirect effects, such as altering GH, IGF-I, and glucocorticoid levels, therefore it is unclear whether leptin has a direct effect on the growth plate. More importantly, clinical cases of leptin or leptin receptor mutations in humans are either associated with tall stature or have no effect on linear growth [32]. Therefore, it remains unclear whether leptin stimulates or suppresses bone growth in humans.

CALCIUM, PHOSPHATE, AND VITAMIN D

One particular micronutrient that is important for bone homeostasis and bone growth is vitamin D. Vitamin D can be obtained from some foods as well as dietary supplements, but most people are reliant on endogenous synthesis through exposure to sunlight. Sunlight allows the photoconversion of 7-dehydrocholesterol to cholecalciferol (vitamin D_3) in the skin, which is then stored in adipose tissues. Vitamin D deficiency can therefore result from inadequate exposure to sunlight or prolonged food restriction (less common).

Cholecalciferol itself is not biologically active. It is metabolized in the liver by vitamin D 25-hydroxylase to produce calcifediol (also known as 25-hydroxyvitamin D).

Calcifediol is then converted in the kidney by 1-α-hydroxylase into calcitriol, or 1,25-dihydroxyvitamin D (1,25-OHD), the biologically active hormonal form.

Vitamin D is important in calcium homeostasis as it stimulates the absorption of calcium in the intestine. Because decreased calcium levels in blood increase parathyroid hormone levels, and consequently increase bone resorption, vitamin D is important for proper bone mineralization, and vitamin D (or calcium) deficiency results in undermineralization of the bones. In adults, it could lead to osteoporosis; in childhood, it causes a skeletal disorder known as rickets, which is one of the most common noncommunicable childhood diseases in developing countries [33]. In children with rickets, the hypertrophic zone is expanded due to decreased mineralization and defective apoptosis of the terminal hypertrophic chondrocytes. These cellular changes thus help explain the bone deformity and widening metaphysis often seen in rachitic bones.

Genetic studies in mice have provided important insights into the molecular mechanisms by which vitamin D deficiency impairs childhood bone formation and growth. Targeted ablation of vitamin D receptor (Vdr) in mice leads to hypocalcemia, hyperparathyroidism, hypophosphatemia (due to hyperparathyroidism), and rickets [34]. However, prevention of abnormal mineral ion homeostasis in Vdr-null mice using a high-calcium and high-phosphorus diet has prevented the development of rickets, suggesting that impaired mineral ion homeostasis, rather than Vdr-mediated signaling itself, is the primary cause of impaired bone growth [35]. Importantly, mutations in PHEX in mice, which causes hypophosphatemia (but only slightly lower calcium), and mice fed with a low-phosphorus high-calcium diet, both developed rachitic bones [36]. Similarly, children with X-linked hypophosphatemia, which results from increased urinary phosphate excretion, have decreased phosphorus but normal levels of calcium and vitamin D in blood, but also developed rickets and short stature. Taken together, evidence from mouse models and human disease collectively suggest that impaired growth in rickets is primarily caused by phosphate deficiency, rather than the lack of vitamin D or calcium per se.

INTRACELLULAR REGULATION OF BONE GROWTH

Clearly, chondrogenesis and normal chondrocyte functions are regulated by multiple different intracellular signaling pathways, genes, and epigenetic mechanisms. The purpose of this section is not, therefore, to exhaustively describe all these intracellular regulatory mechanisms. Rather, it focuses on a few that have recently been found to be relevant to the nutritional regulation of bone growth.

mTOR SIGNALING PATHWAY

The mammalian target of rapamycin (mTOR) signaling pathway is perhaps the most relevant signaling pathway connecting nutritional status and cell growth (for review, see Laplante and Sabatini [37]). mTOR is a member of the phosphatidylinositol 3-kinase (PI3K)-related kinase protein family and was initially discovered in yeast to be the mediator of toxic effects of a macrolide called rapamycin. Later, two mTOR-containing protein complexes were identified in mammalian cells, and were named mTORC1 and mTORC2. Both mTORC1 and mTORC2 are highly sensitive to rapamycin but, in addition, mTORC1 also responds to other intracellular and extracellular signals, such as growth factors, stress, amino acids, and energy levels, and is known to mediate multiple downstream signaling pathways to regulate protein and lipid synthesis, cell cycle progression, and autophagy. In contrast, mTORC2 is mainly involved in the regulation of cytoskeletal organization in response to growth factors. The mTORC1-mediated signaling pathway therefore appears to be an important intracellular mechanism, allowing mammalian cells to assess nutritional status and make cellular decisions on cell survival, growth, and proliferation.

One of the key immediate upstream regulators of mTORC1 is a heterodimer of tuberous sclerosis 1 and 2 (TSC1 and TSC2) (Figure 16.3). TSC1/2 acts as a GTP-ase-activating protein (GAP) for a Ras family protein called Rheb (or Ras homolog enriched in the brain). When bound to GTP, Rheb interacts with mTORC1 to stimulate its kinase activity. Like most GAPs, TSC1/2 negatively regulates the activity of Rheb (or any Ras proteins) by catalyzing the conversion of the active GTP-Rheb back to the inactive GDP-Rheb. Consequently, TSC1/2 serves as a negative regulator of mTORC1 activity. One of the many growth factors that stimulate the mTOR pathway through TSC1/2 is IGF-I. IGF-I stimulates both Ras and PI3K signaling pathways, which subsequently lead to increased phosphorylation of Erk (or extracellular signal-regulated kinases) and Akt, respectively. Because both phospho-Erk and phosphor-Akt inhibit TSC1/2, IGF-I signaling stimulates mTORC1 activity. As described in the previous section, IGF-I levels decrease during malnutrition.

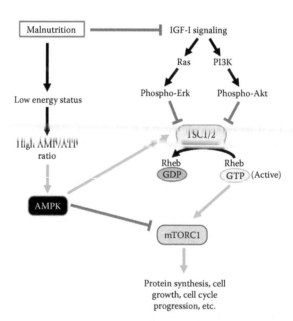

FIGURE 16.3 A schematic model for nutritional regulation of mTORC1 signaling. Green arrows denote activation, black arrows denote effects, and red blunted arrows denote inhibition.

The decreased chondrocyte proliferation and hypertrophy during food restriction may therefore be in part mediated intracellularly through decreased IGF-I and inhibition of mTOR signaling in chondrocytes.

Another important upstream regulator of mTORC1 is adenosine monophosphate-activated protein kinase (AMPK). In a low energy state, the ratio of AMP to ATP increases in the cell, which in turn leads to an increase in AMPK activity. AMPK directly phosphorylates and activates TSC2, which leads to the inhibition of Rheb. In addition, AMPK also phosphorylates and allosterically inhibits one of the key components in mTORC1, which is called regulatory-associated protein of mammalian target of rapamycin (raptor). The end results of both events ensure the suppression of the mTOR pathway in response to a decreased energy level in cells. Similar to decreased IGF-I levels, this alternative mechanism mediated by AMPK likely contributes in part to the inhibition of mTOR signaling during malnutrition.

The role of mTOR signaling specifically in chondrocytes and in longitudinal bone growth has been elucidated by the infusion of the mTOR inhibitor, rapamycin, directly at the tibia. Rapamycin treatment inhibits bone growth by strongly suppressing hypertrophy differentiation and type X collagen production [38]. These findings were consistent with the cellular effects of mTOR signaling described earlier and support its potential role in terms of growth inhibition during malnutrition.

EPIGENETICS AND SIRTUINS

Epigenetics broadly refers to changes in gene functions without involving changes in the genomic DNA sequence. More specifically, it is often used to describe changes in DNA modifications and histone modifications, such as acetylation, methylation, and ubiquitination. The involvement of epigenetics in childhood growth is not a new concept, in general, but has gathered much attention due to some recent clinical findings of mutations in histone- and DNA-modifying genes causing overgrowth syndromes [39].

Several independent studies collectively suggested the involvement of histone deacetylases (HDAC), which are enzymes that remove acetylation from histones, in the regulation of bone growth of individuals in a malnourished state. First, HDAC in general suppresses growth in chondrocytes, because knockout mice of HDAC3 showed decreased bone growth [40], while the treatment of HDAC inhibitors in chondrocytes stimulated expression of type II collagen as well as Sox9, which is an important transcription factor for chondrogenesis [41]. In contrast, HDAC overexpression downregulates the expression of aggrecan and type II collagen in chondrocytes [42]. Importantly, some members of a specific class of HDAC called sirtuins, which are NAD-dependent acetylases, were shown to be stimulated by food restriction. It was found that the expression of SIRT1 and 6 was upregulated by food restriction, both *in vitro* and *in vivo*. The enzymatic activity of SIRT1 was also positively regulated by NAD+ (or nicotinamide adenine dinucleotide), which reflects energy status and increases during fasting (for review, see Guarente [43]). There are many downstream effectors of SIRT1 [43], but that which linked sirtuins back to the mTOR signaling is AMP kinase [44]. SIRT1 deacetylates and activates STK11 (also known as LBK1), which is a kinase that activates AMPK [45]. SIRT1 upregulation during fasting may

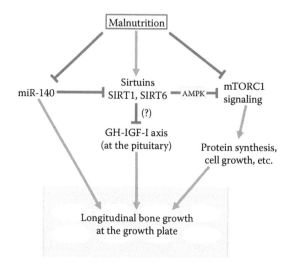

FIGURE 16.4 A proposed model for nutritional regulation of intracellular mechanisms important for bone growth. Green arrows denote activation and red blunted arrows denote inhibition. The effect of sirtuins on the GH–IGF-I axis remains unclear, as it has been supported by experimental evidence but has been contradicted by clinical findings.

thus suppress growth in part by inhibiting mTOR signaling though AMPK, as discussed earlier (Figure 16.4).

Interestingly, some evidence has suggested that sirtuins may also be involved in the intracellular mechanisms that downregulate the GH–IGF-I axis during fasting. Resveratrol, which is a SIRT1 activator, significantly decreases GH secretion in the pituitary [46]. Similarly, neural-specific SIRT6 knockout mice have reduced GH and IGF-I levels, which have contributed to postnatal growth retardation in these mice [47]. However, it is important to point out that these findings were not consistent with what is seen in malnourished children, where GH insensitivity (and thus increased GH levels), rather than decreased GH levels, is usually observed.

MicroRNAs

MicroRNAs (miRNAs) are small (18–24 nucleotides long) noncoding RNAs involved in the regulation of gene expression. Longitudinal bone growth requires the expression of certain miRNA at the growth plate, as a cartilage-specific knockout of Dicer, which encodes for the endoribonuclease needed to process miRNAs, has led to severe growth retardation. Genetic studies in mice showed that the two major miRNAs involved in bone growth are miR-140 and the let-7 family (for review, see Lui [48]). Importantly, the expressions of several miRNAs at the growth plate, including miR-140, were downregulated in response to general food restriction in mice [49]. It is yet unclear which and how many downstream target genes are important for mediating the effect of miR-140 on the growth plate; however, one predicted target of miR-140 is SIRT1 [49]. Luciferase assay showed that SIRT1 expression can indeed be suppressed by miR-140 *in vitro*, which provided an interesting possible intracellular mechanism for growth suppression during fasting. Food restriction suppresses the expression of miR-140, which in turn leads to derepression of SIRT1 (along with other molecular targets). Consequently, the upregulation of SIRT1 leads to acetylation and inhibition of TSC2, thus suppressing the mTOR signaling pathway important for cell growth (Figure 16.4).

GROWTH PLATE SENESCENCE AND CATCH-UP GROWTH

GROWTH DECELERATION AND GROWTH PLATE SENESCENCE

In humans, longitudinal bone growth is rapid in early life, but decelerates with age and gradually ceases during adolescence. Although linear growth is regulated by various endocrine factors, as mentioned earlier, this age-associated decline in growth rate does not appear to be caused by changes in hormonal levels; rather, it depends on a mechanism intrinsic to the growth plate itself, called growth plate senescence [50]. Clinically, the degree of growth plate senescence is perhaps defined as a child's bone age, which is assessed by taking a single x-ray of the child's left hand and wrist, and the degree of skeletal ossification is evaluated by pediatric endocrinologists. Growth plate senescence is characterized by both structural and functional changes, including the gradual depletion of resting zone chondrocytes, a decreasing proliferation rate in the proliferative zone, and the diminishing size of the terminal hypertrophic cells with age. These changes occur with increasing age but do not appear to be driven by time or age per se, and instead depend on growth itself [51]. This observation implies a negative feedback mechanism by which linear growth is regulated: growth leads to the depletion of growth potential and advances growth plate senescence, therefore driving growth to gradually slow and cease, setting a fundamental limit on linear bone growth.

GROWTH PLATE SENESCENCE MODEL FOR CATCH-UP GROWTH

Growth plate senescence may also help explain the clinical phenomenon of catch-up growth, which is defined as a greater-than-normal linear growth rate for chronological age following a period of growth inhibition [52]. Clinically, catch-up growth has been observed after prior growth inhibiting conditions such as malnutrition, hypothyroidism, glucocorticoid treatment, and celiac disease, is resolved. Previously, catch-up growth had been ascribed to a homeostatic mechanism driven by the central nervous system. Subsequent studies have shown, instead, that catch-up growth results from delayed growth plate senescence [51]. Because of the negative feedback described earlier, normal growth gradually uses up growth potential and drives growth plate senescence. Therefore, growth-inhibiting conditions may instead preserve growth potential and delay growth plate senescence. When the condition resolves, the growth plate grows more rapidly than is normal for chronological age, resulting in catch-up growth (Figure 16.5). In malnourished children, growth plate senescence could be delayed by two independent mechanisms. First, the general inhibition of linear growth by malnutrition could help preserve growth

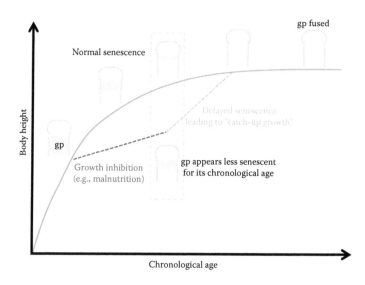

FIGURE 16.5   The growth plate senescence model for catch-up growth. gp, growth plate.

potential in the growth plate. Second, because estrogen accelerates growth plate senescence and epiphyseal fusion by depleting the number of resting chondrocytes [27], malnutrition also delays senescence by shutting down the reproductive axis, and thereby estrogen production, through leptin deficiency [29]. Based on the concept of growth plate senescence, catch-up growth should completely correct the growth deficiency over time, once the underlying problem is resolved. However, children with prior growth deficiency often end up shorter than is expected of their genetic potential, even after catch-up growth. Future research into growth plate senescence may help better understand and overcome this permanent loss in growth potential.

CONCLUSION

Longitudinal bone growth is driven by chondrogenesis, and the proliferation and hypertrophic differentiation of chondrocytes at the growth plate cartilage. The growth plate is regulated by complex interactions between different endocrine systems, epigenetic mechanisms, and intracellular signaling pathways. Nutrition plays an important role in modulating these multiple regulatory mechanisms of the growth plate, thereby only permitting rapid linear growth when the organism is able to take in plentiful nutrients. As children grow and continue to increase in height, the growth plate gradually ages (becomes senescent), leading to functional and structural declines of the growth plate and a decrease in growth velocity, until growth stops and adult height is attained. The temporary inhibition of bone growth during childhood may also slow down aging of the growth plate, which may help explain the spike in growth rate when growth inhibition is resolved, a clinical phenomenon also known as catch-up growth.

ACKNOWLEDGMENTS

This work was supported by the Intramural Research Program of the Eunice Kennedy Shriver National Institute of Child Health and Human Development (NICHD), U.S. National Institutes of Health (NIH).

REFERENCES

1. Kronenberg HM. Developmental regulation of the growth plate. *Nature* 2003; 423(6937):332–6.
2. Lui JC, Baron J. Mechanisms limiting body growth in mammals. *Endocr Rev* 2011; 32(3):422–40.
3. DeChiara TM, Efstratiadis A, Robertson EJ. A growth-deficiency phenotype in heterozygous mice carrying an insulin-like growth factor II gene disrupted by targeting. *Nature* 1990; 345(6270):78–80.
4. Begemann M, Zirn B, Santen G et al. Paternally inherited IGF2 mutation and growth restriction. *N Engl J Med* 2015; 373(4):349–56.
5. Straus DS, Takemoto CD. Effect of fasting on insulin-like growth factor-I (IGF-I) and growth hormone receptor mRNA levels and IGF-I gene transcription in rat liver. *Mol Endocrinol* 1990; 4(1):91–100.
6. Suzuki M, Uehara Y, Motomura-Matsuzaka K et al. betaKlotho is required for fibroblast growth factor (FGF) 21 signaling through FGF receptor (FGFR) 1c and FGFR3c. *Mol Endocrinol* 2008; 22(4):1006–14.
7. Inagaki T, Lin VY, Goetz R et al. Inhibition of growth hormone signaling by the fasting-induced hormone FGF21. *Cell Metab* 2008; 8(1):77–83.
8. Galman C, Lundasen T, Kharitonenkov A et al. The circulating metabolic regulator FGF21 is induced by prolonged fasting and PPARalpha activation in man. *Cell Metab* 2008; 8(2):169–74.
9. Fazeli PK, Lun M, Kim SM et al. FGF21 and the late adaptive response to starvation in humans. *J Clin Invest* 2015; 125(12):4601–11.
10. Kubicky RA, Wu S, Kharitonenkov A et al. Role of fibroblast growth factor 21 (FGF21) in undernutrition-related attenuation of growth in mice. *Endocrinology* 2012; 153(5):2287–95.
11. Orwoll E, Ware M, Stribrska L et al. Effects of dietary protein deficiency on mineral metabolism and bone mineral density. *Am J Clin Nutr* 1992; 56(2):314–9.
12. Glick PL, Rowe DJ. Effects of chronic protein deficiency on skeletal development of young rats. *Calcif Tissue Int* 1981; 33(3):223–31.
13. Chrysis D, Ritzen EM, Savendahl L. Growth retardation induced by dexamethasone is associated with increased apoptosis of the growth plate chondrocytes. *J Endocrinol* 2003; 176(3):331–7.
14. Mazziotti G, Giustina A. Glucocorticoids and the regulation of growth hormone secretion. *Nat Rev Endocrinol* 2013; 9(5):265–76.
15. Macfarlane DP, Forbes S, Walker BR. Glucocorticoids and fatty acid metabolism in humans: Fuelling fat redistribution in the metabolic syndrome. *J Endocrinol* 2008; 197(2):189–204.
16. Manary MJ, Muglia LJ, Vogt SK et al. Cortisol and its action on the glucocorticoid receptor in malnutrition and acute infection. *Metabolism* 2006; 55(4):550–4.
17. Miura M, Tanaka K, Komatsu Y et al. Thyroid hormones promote chondrocyte differentiation in mouse ATDC5 cells and stimulate endochondral ossification in fetal mouse tibias through iodothyronine deiodinases in the growth plate. *J Bone Miner Res* 2002; 17(3):443–54.

18. Bassett JH, Swinhoe R, Chassande O et al. Thyroid hormone regulates heparan sulfate proteoglycan expression in the growth plate. *Endocrinology* 2006; 147(1):295–305.
19. Kindblom JM, Gothe S, Forrest D et al. GH substitution reverses the growth phenotype but not the defective ossification in thyroid hormone receptor alpha 1-/-beta-/- mice. *J Endocrinol* 2001; 171(1):15–22.
20. Takeda K, Sakurai A, DeGroot LJ et al. Recessive inheritance of thyroid hormone resistance caused by complete deletion of the protein-coding region of the thyroid hormone receptor-beta gene. *J Clin Endocrinol Metab* 1992; 74(1):49–55.
21. Bochukova E, Schoenmakers N, Agostini M et al. A mutation in the thyroid hormone receptor alpha gene. *N Engl J Med* 2012; 366(3):243–9.
22. Turkay S, Kus S, Gokalp A et al. Effects of protein energy malnutrition on circulating thyroid hormones. *Indian Pediatr* 1995; 32(2):193–7.
23. Herlihy JT, Stacy C, Bertrand HA. Long-term food restriction depresses serum thyroid hormone concentrations in the rat. *Mech Ageing Dev* 1990; 53(1):9–16.
24. Zachmann M, Prader A, Sobel EH et al. Pubertal growth in patients with androgen insensitivity: Indirect evidence for the importance of estrogens in pubertal growth of girls. *J Pediatr* 1986; 108(5 Pt 1):694–7.
25. Grumbach MM. Estrogen, bone, growth and sex: A sea change in conventional wisdom. *J Pediatr Endocrinol Metab* 2000;13(Suppl 6):1439–55.
26. Blanchard O, Tsagris L, Rappaport R et al. Age-dependent responsiveness of rabbit and human cartilage cells to sex steroids in vitro. *J Steroid Biochem Mol Biol* 1991; 40(4–6):711–6.
27. Nilsson O, Weise M, Landman EB et al. Evidence that estrogen hastens epiphyseal fusion and cessation of longitudinal bone growth by irreversibly depleting the number of resting zone progenitor cells in female rabbits. *Endocrinology* 2014; 155(8):2892–9.
28. Keenan BS, Richards GE, Ponder SW et al. Androgen-stimulated pubertal growth: the effects of testosterone and dihydrotestosterone on growth hormone and insulin-like growth factor-I in the treatment of short stature and delayed puberty. *J Clin Endocrinol Metab* 1993; 76(4):996–1001.
29. Sanchez-Garrido MA, Tena-Sempere M. Metabolic control of puberty: Roles of leptin and kisspeptins. *Horm Behav* 2013; 64(2):187–94.
30. Hamrick MW, Pennington C, Newton D et al. Leptin deficiency produces contrasting phenotypes in bones of the limb and spine. *Bone* 2004; 34(3):376–83.
31. Steppan CM, Crawford DT, Chidsey-Frink KL et al. Leptin is a potent stimulator of bone growth in ob/ob mice. *Regul Pept* 2000; 92(1-3):73–78.
32. Farooqi IS, Wangensteen T, Collins S et al. Clinical and molecular genetic spectrum of congenital deficiency of the leptin receptor. *N Engl J Med* 2007; 356(3):237–47.
33. Elder CJ, Bishop NJ. Rickets. *Lancet* 2014; 383(9929):1665–76.
34. Li YC, Pirro AE, Amling M et al. Targeted ablation of the vitamin D receptor: An animal model of vitamin D-dependent rickets type II with alopecia. *Proc Natl Acad Sci USA* 1997; 94(18):9831–5.
35. Li YC, Amling M, Pirro AE et al. Normalization of mineral ion homeostasis by dietary means prevents hyperparathyroidism, rickets, and osteomalacia, but not alopecia in vitamin D receptor-ablated mice. *Endocrinology* 1998; 139(10):4391–6.
36. Sabbagh Y, Carpenter TO, Demay MB. Hypophosphatemia leads to rickets by impairing caspase-mediated apoptosis of hypertrophic chondrocytes. *Proc Natl Acad Sci USA* 2005; 102(27):9637-42.
37. Laplante M, Sabatini DM. mTOR signaling in growth control and disease. *Cell* 2012; 149(2):274–93.
38. Phornphutkul C, Wu KY, Auyeung V et al. mTOR signaling contributes to chondrocyte differentiation. *Dev Dyn* 2008; 237(3):702–12.

39. Ko JM. Genetic syndromes associated with overgrowth in childhood. *Ann Pediatr Endocrinol Metab* 2013; 18(3):101–5.

40. Bradley EW, Carpio LR, Westendorf JJ. Histone deacetylase 3 suppression increases PH domain and leucine-rich repeat phosphatase (Phlpp)1 expression in chondrocytes to suppress Akt signaling and matrix secretion. *J Biol Chem* 2013; 288(14):9572–82.

41. Furumatsu T, Tsuda M, Yoshida K et al. Sox9 and p300 cooperatively regulate chromatin-mediated transcription. *J Biol Chem* 2005; 280(42):35203–8.

42. Hong S, Derfoul A, Pereira-Mouries L et al. A novel domain in histone deacetylase 1 and 2 mediates repression of cartilage-specific genes in human chondrocytes. *FASEB J* 2009; 23(10):3539–52.

43. Guarente L. Calorie restriction and sirtuins revisited. *Genes Dev* 2013; 27(19):2072–85.

44. Fulco M, Cen Y, Zhao P et al. Glucose restriction inhibits skeletal myoblast differentiation by activating SIRT1 through AMPK-mediated regulation of Nampt. *Dev Cell* 2008; 14(5):661–73.

45. Lan F, Cacicedo JM, Ruderman N et al. SIRT1 modulation of the acetylation status, cytosolic localization, and activity of LKB1: Possible role in AMP-activated protein kinase activation. *J Biol Chem* 2008; 283(41):27628–35.

46. Monteserin-Garcia J, Al-Massadi O, Seoane LM et al. Sirt1 inhibits the transcription factor CREB to regulate pituitary growth hormone synthesis. *FASEB J* 2013; 27(4):1561–71.

47. Schwer B, Schumacher B, Lombard DB et al. Neural sirtuin 6 (Sirt6) ablation attenuates somatic growth and causes obesity. *Proc Natl Acad Sci USA* 2010; 107(50):21790–4.

48. Lui JC. Regulation of body growth by microRNAs. *Mol Cell Endocrinol* 2017; pii: S0303-7207(16)30435-X. doi:10.1016/j.mce.2016.10.024. [Epub ahead of print]

49. Pando R, Even-Zohar N, Shtaif B et al. MicroRNAs in the growth plate are responsive to nutritional cues: Association between miR-140 and SIRT1. *J Nutr Biochem* 2012; 23(11):1474–81.

50. Lui JC, Nilsson O, Baron J. Growth plate senescence and catch-up growth. *Endocr Dev* 2011; 21:23–29.

51. Forcinito P, Andrade AC, Finkielstain GP et al. Growth-inhibiting conditions slow growth plate senescence. *J Endocrinol* 2011; 208(1):59–67.

52. Boersma B, Wit JM. Catch-up growth. *Endocr Rev* 1997; 18(5):646–61.

Section VII

Body Composition

17 Gestational Weight Gain and Body Composition in Pregnant and Postpartum Women

Elisabet Forsum

CONTENTS

GESTATIONAL WEIGHT GAIN AND ITS COMPONENTS

There has been considerable controversy over the years as to what is desirable weight gain during pregnancy, and guidelines have varied considerably over time. For example, the authors of an American textbook in obstetrics in 1966 stated that "excessive weight gain in pregnancy is highly undesirable for several reasons; it is essential to curtail the increment in gain to 28 lb (12.5 kg) at most and preferably 15 lb (6.8 kg)" [1]. Later, however, this policy of severe weight restriction of 6.8 kg was challenged when it was recognized that it was associated with low infant birth weight and with health problems in the offspring [1].

Based on a review of evidence available at the time, Hytten and Leitch in 1971 [2] defined optimal weight gain during pregnancy. These authors described the physiological norms for gestational weight gain (GWG), and the weight gain associated with optimal reproductive performance with respect to infant birth weight, infant survival, and incidence of preeclampsia. This weight gain was found to be 650 g at 10 weeks, 4,000 g at 20 weeks, 8,500 g at 30 weeks, and 12,500 g at term [2]. Hytten and Leitch [2] also described the components of this gain in terms of organs and tissues, as well as in terms of molecules (i.e., water, fat, and protein). Figure 17.1 shows how the product of conception (i.e., fetus, placenta, and amniotic fluid) and maternal tissues (i.e., uterus, breasts, blood, tissue fluid, and maternal stores) contribute to GWG. According to Prentice et al. [3], the weight gained at term consists of a little more than 8 kg water, about 990 g protein, and slightly more than 3 kg body fat. As indicated in Figure 17.1, pregnancy is associated with the retention of water or tissue fluid. It is important to point out that Hytten and Leitch [2] emphasize that a successful outcome of pregnancy, for women as well as for infants, may well occur despite large variations in GWG.

In its 1990 report, the Institute of Medicine (IOM) [4] recommended that lean and underweight women should gain more weight during pregnancy than normal-weight women, who, in turn, were recommended to gain more weight than overweight women. The impetus for this is that a low GWG is associated with adverse effects for the infant, such as low birth weight and, possibly, preterm birth, while a high GWG is associated with adverse outcomes for the mother, such as preeclampsia, gestational diabetes, and caesarean section, as well as large infants. Another important reason was that optimal GWG is a function of the women's prepregnant nutritional status, which was described by a woman's category of body mass index (BMI). By acknowledging this important effect of the prepregnancy BMI, this report [4] introduced a novel way to recommend GWG. The IOM guidelines were revised in 2009 [5] and Table 17.1 shows GWG recommendations for American women in accordance with the two reports [4,5]. In 1990, no recommendation was given to obese women, reflecting a lack of sufficient scientific data [4]. An important reason behind the revision of the recommendations in 2009 was that the number of overweight and obese women had increased. Furthermore, more women were becoming pregnant at an older age [5] and consequently entered pregnancy with conditions such as

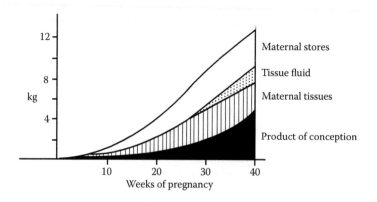

FIGURE 17.1 The components of weight gained during a normal pregnancy. (Reproduced from Hytten FE, Leitch I, *The physiology of human pregnancy*, London: Blackwell Scientific Publications, 1971. With permission.)

TABLE 17.1

Recommendations by the Institute of Medicine for Gestational Weight Gain of Women in Accordance with Their Body Mass Index (BMI) before Conception

IOM 1990		IOM 2009	
BMI Category	GWG Recommendation (kg)[a]	BMI Category[b]	GWG Recommendation (kg)[a]
Low (BMI <19.8 kg/m^2)	12.5–18	Underweight (BMI <18.5 kg/m^2)	12.5–18
Normal (BMI = 19.8–26.0 kg/m^2)	11.5–16	Normal weight (BMI = 18.5–24.9 kg/m^2)	11.5–16
High (BMI >26–29.0 kg/m^2)	7–11.5	Overweight (BMI = 25.0–29.9 kg/m^2)	7–11.5
		Obesity (BMI ≥30 kg/m^2)	5–9

Source: Institute of Medicine, *Nutrition during pregnancy*, Washington, DC: The National Academy Press, 1990; and Institute of Medicine and National Research Council, *Weight gain during pregnancy: Reexamining the guidelines,* Rasmussen KR, Yaktine AL, editors, Washington, DC: The National Academies Press; 2009.

[a] Refers to weight (kilograms) gained during the complete pregnancy.

[b] Represents BMI cutoff as defined by the World Health Organization, Obesity: Preventing and managing the global epidemic, Geneva: Report of a WHO Consultation, 2000, p. 9.

hypertension or diabetes, which increased their risk of adverse pregnancy outcomes. Recommendations for the GWG of obese women are therefore an important addition to the 2009 IOM report [5]. In spite of the extensive scientific evidence underpinning the IOM guidelines, they are not universally accepted. Many countries have no GWG recommendations, and those who have country specific recommendations may not use the IOM 2009 guidelines [5,7]. Recommended GWG for maternal BMI subgroups also varies across countries [7]. This may, to some extent, be explained by the fact that women from different populations, and of different ethnic backgrounds, may need to gain different amounts of weight. However, the paper by Alavi et al. [7] gives the impression that these discrepancies in GWG guidelines stem in part from a lack of appropriate evidence and the failure to recognize the importance of such guidelines.

Individual women may gain very different amounts of weight during pregnancy, and the size of the fat component of GWG, estimated by Prentice et al. [3] to be slightly more than 3 kg for healthy women, is also highly variable. In a study of pregnant women from New York, Lederman et al. [8] identified a significant positive relationship between weight gain and fat retention ($r = 0.81$, $p < 0.0001$). Based on data from 63 American women, Butte et al. [9] also reported such a relationship ($r = 0.76$, $p = 0.001$). Expressing GWG (x) and fat mass gain during pregnancy (y) in kilograms, this relationship was described by the following equation: $y = 0.675x - 3.89$.

ASSESSMENT OF BODY COMPOSITION OF WOMEN DURING REPRODUCTION

Information regarding composition of the human body is required to understand its biology and function, relevant in relation to human health, and fundamental when assessing nutritional requirements. Pregnancy affects the female body in several ways, which has important consequences when methodology commonly used in human body composition research is applied to women during reproduction. Many methods will assess the composition of the mother and fetus as a complete unit, which may or may not be relevant for a particular research question. Furthermore, many commonly applied methods may be associated with health hazards (e.g., radiation), which limits their use during pregnancy. Based on knowledge of the composition of the human body, Wang et al. [10] published a five-level model, providing fundamental information on body composition and organizational rules for research in this area. The model is intended to improve terminology and to describe how commonly used body composition concepts are associated with each other. As described next, pregnancy affects conditions at all of these five levels.

LEVEL I (ATOMIC)

The human body is composed of about 50 elements, but oxygen, carbon, hydrogen, nitrogen, calcium, and phosphorus account for >98% of body weight [10]. The whole body content of most major elements can be measured *in vivo* (e.g., potassium by whole body counting or nitrogen by prompt-γ neutron activation) [10]. Such techniques have been used to study women during reproduction. For example, using whole body estimates of potassium and nitrogen, Butte et al. [9] assessed protein gain and loss during pregnancy and postpartum for women with different BMI before pregnancy.

LEVEL II (MOLECULAR)

The major components of the molecular level are water, fat or lipid, protein, mineral, and glycogen. Mineral may be divided into osseous and extraosseous or nonosseous. Glycogen represents a minor component and is often ignored. Only water and osseous mineral can be directly estimated, while measurement of components included in one of the other four levels is required to assess the remaining molecular components [10]. Sowers et al. [11] studied bone minerals in women before and after pregnancy to investigate how the fetal demand for calcium (approximately 30 g during a complete pregnancy) affects the femoral bone density of the mother. Body water is commonly assessed by means of dilution methods [12], often using stable isotopes of water (^{2}H or ^{18}O) as tracers. The body water content of pregnant women has also been successfully measured by means of bioelectrical impedance [13]. However, recent reports indicate that this methodology is associated with problems such as lack of appropriate validation studies and poor precision during gestation [14,15]. Based on a two-component model (a model where the body is considered to consist of two components, i.e., fat and fat-free mass) [12], estimates of total body water can be used to calculate total body fat by assuming that the water content of the fat-free mass is constant (usually 73% in healthy nonpregnant adults). Thus, total body fat is equal to body weight minus total body water divided by 0.73. However, during pregnancy, the water content of the fat-free mass in the body increases. In 1988, van Raaij et al. [16] published figures describing how this content increases throughout gestation. The IOM recommends using these figures when assessing body composition of pregnant women based on total body water and a two-component model [5].

LEVEL III (CELLULAR)

The human body is composed of the following three main components at the cellular level: cells, extracellular fluid, and extracellular solids [10]. Although conceptually important, little research has been directed at this level, possibly because some of these components are difficult to measure [10]. All components at this level are altered in pregnancy when the number of cells in the body increases. The amount of extracellular fluid and, consequently, extracellular solids also increases. Considerable attention has been devoted to measuring extracellular water during gestation due to its relationship with edema, which often is present in pregnant women and is associated with hypertensive complications. The amount of extracellular water can be assessed by means of the tracer sodium bromide [12], and intracellular water can be calculated as total body water minus extracellular water.

LEVEL IV (TISSUE SYSTEM)

Cells are organized in tissues, such as muscle, adipose, and connective tissue. Many tissues can be further divided into subclasses. For example, adipose tissue can be divided into subcutaneous and nonsubcutaneous or visceral. During pregnancy, several maternal tissues increase in size, and new tissues are formed as the fetus and its adnexa develop. Details of how pregnancy affects the size and function of various tissues, organs, and physiological systems have been published [17]. Magnetic resonance imaging has been used to assess the volume of adipose tissue in different parts of the maternal body throughout a reproductive cycle [18].

LEVEL V (WHOLE BODY)

Important variables at the whole-body level are body weight and height, which may be combined to calculate BMI, an index intended to describe the level of body fatness. For obvious reasons, BMI is less useful in pregnant than in nonpregnant individuals. The assessment of skinfold thickness is another commonly used method to study body fat during pregnancy, which has the advantages of being simple and can describe the distribution of subcutaneous fat or adipose tissue (Figure 17.2). Marshall et al. [15] provided evidence suggesting that this methodology may be acceptable in larger studies. Equations to convert skinfolds to total body fat are available for pregnant women [14]. However, during pregnancy, several problems are associated with such methodology [14].

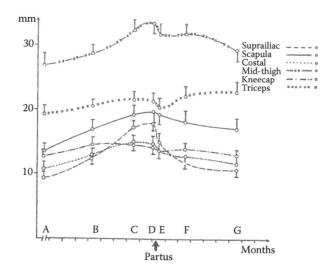

FIGURE 17.2 Skinfold thickness of various body regions at various stages of reproduction. Thickness (mm) of six skinfolds at different stages of reproduction. A, before pregnancy; B, gestational week 16–18; C, gestational week 36 30; D, gestational week; E, 5–10 days postpartum; F, 2 months postpartum; G, 6 months postpartum. Values are averages with SDs and obtained from 24 healthy Swedish women. (Data from Forsum E, Sadurskis A, Wager J, Estimation of body fat in healthy Swedish women during pregnancy and lactation, *Am J Clin Nutr* 1989, 50:465–73.)

The assessment of body volume and body density, that is, body weight per body volume, is very important in body composition methodology. Estimates of body density are often used to assess body fat based on a two-component model [19], if the densities of body fat and fat-free mass are known. For fat, this figure is considered to be 0.9007, with little variation, while the figure for fat-free mass decreases during pregnancy as a consequence of the body water retention associated with this condition. Previously, body volume and density were estimated using underwater weighing, but air displacement plethysmography is presently more commonly applied [19]. A specific device, the Bod Pod™, can be used for this purpose and is commercially available. Air displacement plethysmography is applicable in pregnant women if performed in an appropriate way [20]. For example, this technique requires information about the thoracic gas volume of the subject at the time of testing. The procedure to estimate this volume is, however, quite demanding, and many subjects are unable to produce an acceptable value. Henriksson et al. [21] demonstrated that equations to predict thoracic gas volume from age and height and developed for nonpregnant women resulted in only a minor error in the estimate of body fat when applied in women in gestational week 32. Marshall et al. [15] have also presented data demonstrating the usefulness of air displacement plethysmography when assessing body composition during pregnancy.

An important factor to consider when body fat is assessed using the two-component model based on body density is that the density of fat-free mass is variable and estimates in individual subjects may therefore deviate from reference estimates of body fat. During pregnancy, the biological variation in fat-free mass density appears to increase initially and then be followed by a decrease [22]. Similar observations were made for variations in the fat-free mass water content during pregnancy [23]. This is not surprising, since water is the main component of fat-free mass [19] and has a major influence on its density. These observations indicate that two-component models, based on the assessment of body water or body density, are as valid in late pregnancy as they are in nonpregnant individuals, if appropriate figures for fat-free mass hydration or density are used. However, more studies are needed to confirm and extend this statement.

A common goal in body composition studies during reproduction is to assess total body fat. Such estimates have limitations, as indicated earlier, and tend to require assumptions that may be affected by pregnancy. To avoid some frequently made assumptions, it is often recommended that estimates of women's body fat during reproduction should be obtained using methods based on three- or four-component models. A commonly used three-component model has been published by Going [19], and has also been used during reproduction [22–25]. Three- or four-component models used in women during reproduction have been published by Fuller et al. [26], Pipe et al. [27], Kopp-Hoolihan et al. [25], and Lederman et al. [8]. Unfortunately, studies based on three- or four-component models tend to be costly and demanding for subjects and investigators. This limits their application in population studies.

As obvious from the preceding text, the influence of pregnancy on body composition has been extensively studied at the various levels described by Wang et al. [10]. When conducting body composition studies in postpartum women, it is often assumed that the basic conditions are the same as for other nonpregnant women. This is reasonable, since there is limited information indicating that such differences exist at all, and further, there is no specific reason to suspect that important differences exist in the nonpregnant versus postpartum state. However, studies of fat-free mass hydration of women 2 weeks after delivery [23,24] showed that this value was still slightly higher than the value generally considered to be appropriate for nonpregnant women. These observations probably reflect that the transition back to the nonpregnant state requires a certain amount of time.

FUNCTIONAL ASPECTS ON BODY FAT DURING REPRODUCTION

BODY FAT AND ENERGY METABOLISM DURING PREGNANCY IN RELATION TO BIRTH WEIGHT

After conception, several metabolic and physiological changes occur in the pregnant body. These are mediated by hormones secreted by the corpus luteum, the embryo, and later the placenta. The increase in resting energy metabolism, as well as alterations in glucose and lipid metabolism, represent important metabolic changes. The circulatory system adapts in response to the increased demand for transportation, which, together with the retention of water and fat in the body, represent important physiological changes. These changes create a demand for energy, and pregnancy is associated with an increased requirement for dietary energy [28]. The main part of the increased energy needs is due to the rise in resting energy metabolism and the retention of body fat, while the energy content of the fetus is relatively small. According to recent estimates the energy requirements for an entire pregnancy are 321 MJ, the "energy cost of pregnancy" [28]. About 50% is required for the increase in energy metabolism and approximately 45% for body fat retention [28]. However, studies on pregnant women from different countries have shown that these figures vary in response to the nutritional situation in which women live. Table 17.2 shows the energy costs of fat retention and increases in the resting energy metabolism for pregnant women from different countries, ranging from well-nourished Swedish women to undernourished Gambian women. The data presented in this table suggest a relationship between the increase in resting energy metabolism and the energy equivalent of the amount of body fat retained during pregnancy. Of special interest is that undernourished Gambian women showed a decreased resting energy metabolism, loss of body fat during pregnancy, and delivery of babies with low birth weight. However, when supplemented with extra dietary energy during gestation, these Gambian women maintained their prepregnant resting energy metabolism and even retained some body fat. The birth weight of their babies was also slightly higher than the corresponding values for unsupplemented Gambian women. The data in Table 17.2 have been interpreted to represent a functional plasticity in maternal metabolism of benefit for fetal survival under harsh dietary conditions [29].

Table 17.2 also indicates that women from affluent countries tend to retain more fat during pregnancy than their counterparts from poor countries. In addition, there is some evidence to suggest that, when average data from different populations are used, the body fat content before conception is positively related to the total energy

---

TABLE 17.2

Increase in Maintenance Energy and Energy Equivalent of Retained Body Fat during a Complete Pregnancy

Country	Number of Women	Maintenance Energy (MJ)	Retained Body Fat (MJ)	Birth Weight of Infants (Mean ± SD; kg)
Sweden	11	210	239	3.56 ± 0.44
England	12	158	155	3.77 ± 0.58
The Netherlands	57	144	89	3.46 ± 0.53
Scotland	88	126	103	3.37 ± 0.40
Thailand	44	98	57	2.98 ± 0.35
Philippines	40	89	57	2.89 ± 0.40
The Gambia, unsupplemented	10	−43	−20	2.94 ± 0.14
The Gambia, supplemented[a]	12	6	92	3.13 ± 0.10

Source: Data from Prentice AM, Goldberg GR, Energy adaptations in human pregnancy: Limits and long-term consequences, *Am J Clin Nutr* 2000, 71(5 Suppl):1226S–32S; and from references therein.

[a] During pregnancy women were supplemented with approximately 670 kJ per day for 6 days a week.

cost of pregnancy [29]. The data in Table 17.2 indicate that there may be a relationship between fat retention during pregnancy and the birth weight of infants when population averages are used. However, this relationship is much less evident for women within a population. For example, Butte et al. [9] found no such relationship when 63 healthy American women were studied using appropriate body composition methodology. Similar observations were made by Kopp-Hoolihan et al. [25].

Available evidence suggests that pregnancy is associated with a strong tendency to retain body fat. It may therefore appear surprising that, as indicated earlier, the correlation between the amount of fat retained and infant birth weight is weak or nonexistent. However, it may be relevant to note that, although the amount of fat retained by a healthy pregnant woman (i.e., slightly more than 3 kg) may appear considerable, it is only a fraction of the total amount of fat in her body. Therefore, when the impact of maternal body fat on fetal growth is studied, the total body fat content, rather than the amount of fat retained during pregnancy, may be the relevant variable to consider. This suggestion is supported by observations showing that the risk of having a large for gestational age infant increased with increasing maternal BMI [30]. However, BMI is not a perfect estimate of body fatness, and corresponding studies using appropriate body composition methodology are needed. A study based on air displacement plethysmography of women in gestational week 32, as well as of their newborn infants, identified a relationship between maternal and infant body fatness for girls but not for boys [20]. It may be argued that estimates of body fatness in gestational week 32 include the fat content of the fetus. In this study, however, this amount represented only about 0.4% of the total body fat of these women [20]. Also, a study of women with gestational diabetes mellitus (GDM) indicated that maternal fatness may influence fatness of male and female infants differently [31].

MATERNAL BODY FAT VERSUS INSULIN RESISTANCE AND LEPTIN

It is generally considered that the resting metabolic rate correlates significantly with fat-free mass in humans, while the contribution of fat mass to the variation in resting energy expenditure is much smaller [32]. However, there is some evidence to indicate that the latter relationship is enhanced by pregnancy. For example, in 22 healthy women, Löf et al. [33] found a significant correlation between total body fat and resting energy expenditure in both gestational weeks 14 and 32, but not before pregnancy. Furthermore, Bronstein et al. [34] found that fat mass, but not fat-free mass, was a highly significant predictor of the resting energy expenditure in pregnant women, whereas the opposite was found in nonpregnant women. These observations indicate that the metabolic activity of adipose tissue increases during pregnancy. It is therefore of interest to consider the metabolic variables associated with the body fat content of women during reproduction.

In healthy pregnancy, there is a slight decrease in fasting blood glucose levels and an increased insulin response to a glucose load, which is due to an increased insulin resistance [35]. The latter increase helps to direct glucose in the maternal blood to the placenta and, ultimately, to the fetus. In women with GDM, blood glucose levels and insulin resistance are increased, and such women tend to deliver large infants [35]. Overweight and obesity are well-known risk factors for GDM, and it is therefore not surprising that, for healthy women in gestational week 32, variables describing body fatness were significantly associated with glucose homeostasis variables in the circulation, such as glucose, insulin, homeostasis model assessment–insulin resistance (HOMA-IR), hemoglobin A_{1c}, and insulin-like growth factor binding protein-1 [36]. These observations can be reconciled with the fact that an effect of obesity, a condition characterized by a high body fat content, is to increase insulin resistance. In a study on 17 healthy women, the relationship between body fatness and insulin resistance (measured as HOMA-IR) was steeper in gestational week 32 than it was before pregnancy, which suggests that pregnancy enhances the relationship between body fatness and insulin resistance [37].

The so-called adipokines represent bioactive compounds produced by human adipose tissue and acting locally and distantly through autocrine, paracrine, and endocrine mechanisms [38]. They impact on multiple functions such as immunity, insulin sensitivity, blood pressure, and lipid metabolism, as well as on appetite regulation and energy balance [38]. It is therefore not surprising that their relationship with body fat has been studied during reproduction. Of special interest is the adipokine leptin, which is considered to have effects on food intake and energy expenditure [38]. Serum leptin concentrations have been found to correlate with body fatness before and during pregnancy, and changes in body fat during reproduction also correlated with corresponding changes in serum leptin [37]. The significance of these observations is, however, unclear and the area deserves further study.

Body Fat throughout a Reproductive Cycle

After delivery, women tend to lose weight, and the mobilization of body fat is often a part of this weight loss. In fact, it is often considered that the fat gained during pregnancy represents a physiological energy store to be used during lactation, when demands for dietary energy are higher than during pregnancy. However, available data do not support the statement that fat loss is a biologically programmed part of human lactation [3]. Prentice et al. [3] suggest that about 500 g body fat may be mobilized per month to contribute to the energy cost of lactation, and also that this cost may well be covered in other ways, such as an increased energy intake or a decrease in physical activity. Available studies indicate that, on average, lactating women mobilize less than 500 g body fat per month [39,40].

It has long been recognized that adipose tissue in different body locations differs in metabolic activity [41]. Such differences have also been demonstrated during reproduction [41,42]. It is therefore of interest to investigate where the fat retained during pregnancy is located and from which depots it is mobilized postpartum. In a group of healthy Swedish women, adipose tissue volume was investigated by means of magnetic resonance imaging before conception, as well as at 5 to 10 days and 2, 6, and 12 months after delivery [18]. In this way it was possible to demonstrate, as shown in Figure 17.3, that these women placed most of the fat retained during pregnancy in adipose tissue on the trunk and a smaller amount on the thighs. Furthermore, for the thigh compartment, changes in size postpartum were correlated with a score based on a combination of lactation length and intensity [18]. For the other compartments shown in Figure 17.3, no such correlations were found [18].

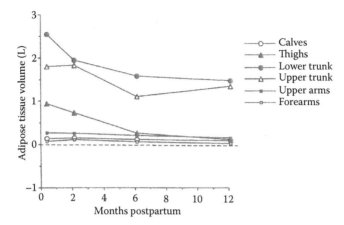

FIGURE 17.3 Adipose tissue volume in different body compartments of healthy Swedish women 5–10 days and 2 months postpartum ($n = 15$) as well as 6 ($n = 13$) and 12 months ($n = 10$) postpartum [18]. Figures are averages and represent differences versus the corresponding prepregnant figures. (Reproduced from Sohlström A, Body fat during reproduction in a nutritional perspective: Studies in women and rats, thesis, Karolinska Institute, Stockholm, 1993. With permission.)

These observations support the statement that the thighs have a specific biological function to store fat during pregnancy for use during lactation, a statement that can be reconciled with observations regarding changes during reproduction in enzymatic activities of adipocytes from the thigh region [41]. Furthermore, the observation that women placed a significant amount of fat on the trunk can be reconciled with recent results showing that, during pregnancy, normal-weight women accumulated fat in adipocytes from subcutaneous abdominal adipose tissue [42].

That fat retained during pregnancy is found in specific depots was already demonstrated by Taggart et al. in 1967 [44]. These authors found that thigh and trunk skinfolds increase during pregnancy, while triceps skinfold tend to decrease around parturition, and then increase. Later, Forsum et al. [45] made similar observations in Swedish women and confirmed that the triceps skinfold, which decreased at the end of pregnancy, showed an increase during the postpartum period (Figure 17.2). Collectively these observations suggest that the amount of fat in different body depots varies during the reproductive cycle, a statement that can be reconciled with observations suggesting an increased metabolic activity of adipose tissue during pregnancy and lactation.

REPRODUCTION IN OVERWEIGHT AND OBESE WOMEN

The currently used definitions of overweight and obesity are based on BMI and are given in Table 17.1. Recent data [46] show that the prevalence of these conditions has been increasing worldwide over the last decades in all population groups. Consequently, the number of women of reproductive age who are overweight or obese has also increased. According to the World Health Organization (WHO), this is a concern, since "growing rates of maternal overweight are leading to higher risks of pregnancy complications, and heavier birth weight and obesity in children" [47]. In 2009, the IOM stated, "Evidence from the scientific literature is remarkably clear that pre-pregnancy BMI is an independent predictor of many adverse outcomes of pregnancy" [5]. Maternal overweight and obesity are associated with complications in the mother during pregnancy, as well as at and after delivery. Excessive GWG adds to these problems. For the mother, there are increased risks of preeclampsia, GDM, and complications during delivery. For the infant, there is a risk of being large-for-gestational-age at birth, possibly associated with an increased risk of overweight and obesity later in life [5]. Therefore, overweight and obesity in women of reproductive age is clearly becoming more of a public health issue. An additional concern is that there is a negative association between maternal obesity and the initiation as well as the continuation of lactation [48]. A large study among Danish women demonstrated that the risk of early termination of full or any breastfeeding rose with the increasing prepregnant BMI of the women [49]. In conclusion, overweight and obesity represent serious threats to the health of women and their children. However, the mechanisms involved are not completely known. Body composition studies of women during the reproductive cycle may help to identify and describe such mechanisms.

REFERENCES

1. Abrams B, Altman SL, Pickett KE. Pregnancy weight gain: Still controversial. *Am J Clin Nutr* 2000; 71(5 Suppl):1233S–41S.
2. Hytten FE, Leitch I. *The physiology of human pregnancy.* London: Blackwell Scientific Publications; 1971.
3. Prentice AM, Spaaij CJK, Goldberg GR et al. Energy requirements of pregnant and lactating women. *Eur J Clin Nutr* 1996; 50(Suppl 1):S82–S111.
4. Institute of Medicine. *Nutrition during pregnancy.* Washington, DC: The National Academy Press; 1990.
5. Institute of Medicine and National Research Council. *Weight gain during pregnancy: Reexamining the guidelines.* Rasmussen KR, Yaktine AL, editors. Washington, DC: The National Academies Press; 2009.
6. World Health Organization. Obesity: Preventing and managing the global epidemic. Geneva: Report of a WHO Consultation; 2000, p. 9.
7. Alavi N, Haley S, Chow K et al. Comparison of national gestational weight gain guidelines and energy intake recommendations. *Obes Rev* 2013; 14:68–85.
8. Lederman SA, Paxton A, Heymsfield SB et al. Body fat and water changes during pregnancy in women with different body weight and weight gain. *Obstet Gynecol* 1997; 90:483–8.
9. Butte NF, Ellis KJ, Wong WW et al. Composition of gestational weight gain impacts maternal fat retention and infant birth weight. *Am J Obstet Gynecol* 2003; 189:1423–32.
10. Wang, Z-M, Pierson RN, Heymsfield SB. The five-level model: A new approach to organizing body composition research. *Am J Clin Nutr* 1992; 56:19–28.
11. Sowers M, Crutchfield M, Jannausch M et al. A prospective evaluation of bone mineral change in pregnancy. *Obstet Gynecol* 1991; 77:841–5.
12. Schoeller DA. Hydrometry. In *Human body composition*, Heymsfield SB, Lohman TG, Wang Z et al., editors. Champaign, IL: Human Kinetics; 2005, pp. 35–49.
13. Lukaski HC, Siders WA, Nielsen EJ et al. Total body water in pregnancy: Assessment by using bioelectrical impedance. *Amn J Clin Nutr* 1994; 59:578–85.

14. Widen EM, Gallagher D. Body composition changes in pregnancy: Measurements, predictors and outcomes. *Eur J Clin Nutr* 2014; 68:643–52.

15. Marshall NE, Murphy EJ, King JC et al. Comparison of multiple methods to measure maternal fat mass in late gestation. *Am J Clin Nutr* 2016; 103:1055–63.

16. van Raaij JM, Peek ME, Vermaat-Miedema SH et al. New equations for estimating body fat mass in pregnancy from body density or total body water. *Am J Clin Nutr* 1988; 47:942–7.

17. Chamberlain G, Pipkin FB. *Clinical physiology in obstetrics.* London: Blackwell Science Ltd.; 1998.

18. Sohlström A, Forsum E. Changes in adipose tissue volume and distribution during reproduction in Swedish women as assessed by magnetic resonance imaging. *Am J Clin Nutr* 1995; 61:287–95.

19. Going SB. Hydrodensitometry and air displacement plethysmography. In *Human body composition*, Heymsfield SB, Lohman TG, Wang Z et al., editors. Champaign, IL: Human Kinetics; 2005, pp. 17–33.

20. Henriksson P, Löf M, Forsum E. Parental fat-free mass is related to the fat-free mass of infants and maternal fat mass is related to the fat mass of infant girls. *Acta Paediatr* 2015; 104:491–7.

21. Henriksson P, Löf M, Forsum E. Assessment and prediction of thoracic gas volume in pregnant women: An evaluation in relation to body composition assessment using air displacement plethysmography. *Br J Nutr* 2013; 109:111–7.

22. Forsum E, Henriksson P, Löf M. The two-component model for calculating total body fat from body density: An evaluation in healthy women before, during and after pregnancy. *Nutrients* 2014; 6:5888–99.

23. Löf M, Forsum E. Hydration of fat-free mass in healthy women with special reference to the effect of pregnancy. *Am J Clin Nutr* 2004; 80:960–5.

24. Hopkinson JM, Butte NF, Ellis KJ et al. Body fat estimation in late pregnancy and early postpartum: Comparison of two-, three-, and four-compartment models. *Am J Clin Nutr* 1997; 65:432–8.

25. Kopp-Hoolihan LE, van Loan MD, Wong WW et al. Fat mass deposition during pregnancy using a four-component model. *J Appl Physiol* 1999; 87:196–202.

26. Fuller NJ, Jebb SA, Laskey MA et al. Four-component model for the assessment of body composition in humans: Comparison with alternative methods, and evaluation of the density and hydration of fat-free mass. *Clin Sci (Lond)* 1992; 82:687–93.

27. Pipe NGJ, Smith T, Halliday D et al. Changes in fat, fat free mass and body water in human normal pregnancy. *Br J Obstet Gynaecol* 1979; 86:929–40.

28. Butte NF, King J. Energy requirements during pregnancy and lactation. *Public Health Nutr* 2005; 8(7A):1010–27.

29. Prentice AM, Goldberg GR. Energy adaptations in human pregnancy: Limits and long-term consequences. *Am J Clin Nutr* 2000; 71(5 Suppl):1226S–32S.

30. Cnattingius S, Villamor E, Lagerros YT et al. High birth weight and obesity: A vicious circle across generations. *Int J Obes (Lond)* 2012; 36:1320–4.

31. Lingwood BE, Henry AM, d'Emden MC et al. Determinants of body fat in infants of women with gestational diabetes mellitus differ with fetal sex. *Diabetes Care* 2011; 34:2581–5.

32. Nelson KM, Weinsier RL, Long CL et al. Prediction of resting energy expenditure from fat-free mass and fat mass. *Am J Clin Nutr* 1992; 56:848–56.

33. Löf M, Olausson H, Boström K et al. Changes in basal metabolic rate during pregnancy in relation to changes in body weight and composition, cardiac output, insulin-like growth factor I, and thyroid hormones and in relation to fetal growth. *Am J Clin Nutr* 2005; 81:678–85.

34. Bronstein M, Mak R, King J. Unexpected relationship between fat mass and basal metabolic rate in pregnant women. *Br J Nutr* 1996; 75:659–68.

35. O'Hare JP. Carbohydrate metabolism. In *Clinical physiology in obstetrics*, Chamberlain G, Pipkin FB, editors. London: Blackwell Science Ltd.; 1998, pp. 192–211.

36. Henriksson P, Löf M, Forsum E. Glucose homeostasis variables in pregnancy versus maternal and infant body composition. *Nutrients* 2015; 7:5615–27.

37. Eriksson B, Löf M, Olausson H et al. Body fat, insulin resistance, energy expenditure and serum concentrations of leptin, adiponectin and resistin before, during and after pregnancy in healthy Swedish women. *Br J Nutr* 2010; 103:50–57.

38. Ronti T, Lupattelli G, Mannarino E. The endocrine function of adipose tissue: An update. *Clin Endocrinol (Oxf)* 2006; 64:355–65.

39. van Raaij JMA, Schonk CM, Vermaat-Miedema SH et al. Energy cost of lactation, and energy balance of well-nourished Dutch lactating women: Reappraisal of the extra energy requirements of lactation. *Am J Clin Nutr* 1991; 53:612–9.

40. Goldberg GR, Prentice AM, Coward WA et al. Longitudinal assessment of the components of energy balance in well-nourished lactating women. *Am J Clin Nutr* 1991; 54:788–98.

41. Rebuffe-Scrive M, Enk L, Crona N et al. Fat cell metabolism in different regions in women. Effect of menstrual cycle, pregnancy and lactation. *J Clin Invest* 1985; 75:1973–6.

42. Svensson H, Wetterling L, Bosaeus M et al. Body fat mass and the proportion of very large adipocytes in pregnant women are associated with gestational insulin resistance. *Int J Obes (Lond)* 2016; 40(4):646–53.

43. Sohlström A. Body fat during reproduction in a nutritional perspective: Studies in women and rats. Thesis, Karolinska Institute, Stockholm; 1993.

44. Taggart NR, Holliday RM, Billewicz WZ et al. Changes in skinfolds during pregnancy. *Br J Nutr* 1967; 21:439–51.

45. Forsum E, Sadurskis A, Wager J. Estimation of body fat in healthy Swedish women during pregnancy and lactation. *Am J Clin Nutr* 1989; 50:465–73.

46. Ng M, Fleming T, Robinson M et al. Global, regional, and national prevalence of overweight and obesity in children and adults during 1980–2013: A systematic analysis for the Global Burden of Disease Study 2013. *Lancet* 2014; 384(9945):766–81.

47. World Health Organization. Nutrition: Challenges. 2016. http://www.who.int/nutrition/challenges/en/ (accessed March 15, 2016).

48. Rasmussen KM. Association of maternal obesity before conception with poor lactation performance. *Annu Rev Nutr* 2007; 27:103–21.

49. Baker JL, Michaelsen KF, Sorensen TIA et al. High prepregnant body mass index is associated with early termination of full and any breastfeeding in Danish women. *Am J Clin Nutr* 2007; 86:404–11.

18 Body Composition Changes from Infancy to Young Childhood

Daniel J. Hoffman and Pamela L. Barrios

CONTENTS

INTRODUCTION

Human growth is a dynamic process that involves the complex regulation of tissue development from the moment of conception through adulthood. Fundamental to this process is the need for sufficient substrates required for cellular growth and energy to facilitate growth and development. Thus, when nutrients are sufficiently available and readily absorbed, growth and development occur in a healthy manner. However, when there are interruptions to either nutrient availability, and/or absorption or existing pathologies that influence the endocrine system, growth is disrupted, delayed, or ceases. Central to growth, from conception through early childhood, is the development of tissues that form organs and support normal homeostatic processes, promoting health and decreasing the risk of various diseases. For the majority of children, body tissues are most often grouped with the two major depots of body composition: fat mass (FM) and lean body mass (LBM). Yet these compartments are not constant in terms of their proportion as a child grows, and changes in the relative proportion of one or the other can create a condition in which "healthy" growth is altered and the risk of disease increases.

The study of human body composition is thought to date back to work by Adolphe Quetelet, the French mathematician who, in 1835, used body weight (kilograms) divided by the squared term for height (meters) as a means of normalizing differences in body size between people [1]. This simple calculation came to be known as the body mass index (BMI) and was the first attempt to compare the body weight of adults; children were compared using a similar index, with the exception that height was calculated as the cubic term of length (meters), or the ponderal index.

It is worth noting that neither BMI nor the ponderal index are meant to describe body fatness, but they are both very highly correlated with FM in large samples of adults. However, as science advanced and the various body compartments became more clearly identified in terms of their physiological role, the need to characterize body weight, rather than simply comparing it between persons or over time, became more important. Thus, the use of various anthropometric measures, including waist circumference and skinfold measures at specific landmarks of the body, complemented BMI as a means of describing the body weights of children and adults. Recently, advanced methods to assess body compartments, namely, FM, LBM, and total body water (TBW), have allowed for more precise methods to fully characterize body composition in children and adults.

Body composition may also play a role in the programming in early life of a variety of health outcomes, such as type 2 diabetes, obesity, and cardiovascular disease [2]. While it is unclear as to the mechanism by which poor growth *in utero* or during early childhood influences the risk of these diseases, there is some suggestion that the development of body composition may play a role. In short, it is hypothesized that alterations in specific tissue development during key windows, such as the first trimester of gestation or the first 1,000 days may "program" tissue function, such as altered pancreatic function or dysfunctional response to stress by the hypothalamic-pituitary axis, increasing the risk of chronic diseases later in life. Therefore, understanding how different nutritional environments in early life affect both body composition and weight gain is of great importance. The following sections describe the changes in body composition in infancy and early childhood, growth patterns and body composition, and the relationship between poor growth and body composition as a risk factor for adult diseases.

ELEMENTS OF HUMAN BODY COMPOSITION

Early knowledge of human body composition was based on whole body carcass analyses of human fetuses, infants, and adults from the 1900s [3–6]. However, no similar data exists between infancy and adulthood, except for one 4.5-year-old male child who died of tuberculosis meningitis [7]. Fortunately, several methods have been developed to measure body composition indirectly, increasing our current knowledge of this missing age period [8,9]. Still, when it comes to infants, these methods have limited application as they are fairly invasive, may need active participant cooperation, and may involve radiation exposure. Nonetheless, the development of precise methods that are not invasive has expanded greatly since the early 1990s, including the use of stable isotopes and air displacement, allowing for newborn infants and young children to be measured over the course of months or years. These methods have advanced the field of neonatal body composition greatly, as discussed below.

Whole body composition is often assessed using a variety of models. The earliest, simplest, and most widely used model to assess body composition is the two-compartment model (2C) model, where body weight is divided into FM and the remaining tissues categorized as LBM [10]. Due to the challenges of directly measuring FM and LBM, the 2C model calculates LBM by extrapolating one or more of its subcompartments, TBW or total body potassium (TBK), with the assumption that their relative concentrations are constant at a given age. After determining LBM, FM is then calculated as the difference between body weight and LBM. Aiming to reduce some of the limitations from the 2C model, the three-compartment (3C) was developed, in which body weight is divided into FM, TBW, and the remaining solids, such as skeletal mass. The four-compartment (4C) model further divides remaining solids into proteins and minerals [11,12]. Finally, the five-compartment (5C) model is the most comprehensive means of describing body composition, using the following five levels: atomic, molecular, cellular, tissue system, and whole body [12]. Each additional level adds a layer of complexity to the measurement of body composition, often requiring a new technique (e.g., dual-energy x-ray absorptiometry [DXA], isotope dilution), yet the 5C model is a robust means of accounting for the interindividual variability of FFM. Nonetheless, the 4C model is the most robust practical method for measuring body composition [11], and has been used in children as young as 5 years of age [13]. Among many advantages of using the 4C model, the most important for children is that hydration, bone mineral content (BMC), and body density are measured directly and not assumed.

BODY COMPOSITION DURING INFANCY

Body composition from birth through early childhood often undergoes rapid changes associated with normal growth processes [14]. Understanding these changes and the relative contributions of LBM and FM to weight during infancy can enhance our knowledge of the nutritional needs of both healthy and sick infants, and provide important physiological insight into anthropometric measures. Over the past few decades, there has been a great interest in the relationship between early nutrition, growth, and future health outcomes, due to the large volume of evidence associating a number of diseases in adulthood with experiences in early life [2,15,16].

Whereas body composition references exist for adults, there is very limited reference data for the neonatal and early childhood periods [17]. One of the earliest analyses documented a rapid decrease of LBM hydration during the first few weeks of postnatal life [6], and cadaver studies showed that newborn infants have a higher total body water mass (TBW) relative to weight [18], consistent with data indicating that hydration decreases as one ages [19]. In 1982, Fomon and colleagues published age- and gender-specific body composition data from birth to 10 years of age in a model with FM, LBM, and TBW, with the assumption that boys weighed 0.15 kg more than girls at birth [20].

In particular, for the male reference infant, there was a distinct increase in FM as a percentage of body weight from birth to one year, from approximately 14% to 23%. At the same time, the percentage of body weight as protein remained constant at 13%, while the hydration of LBM decreased from almost 70% at birth to 61% at one year. When reported as a percentage of FFM, there was an increase in protein from 15% to almost 17%, and a negligible decrease in TBW from 81% to 79%, from birth to one year. Similar changes in the same components of body composition were reported for girls. Although this study provided extremely important body composition data, it was a preliminary and crude representation of body composition during infancy and later years, as it extrapolated information for all ages based on three time periods (birth, 6 months, and 9 and 10 years for boys and girls, respectively). However, the foundations were laid for future studies to test the findings reported by Fomon to better understand changes in body composition during the first 1,000 days. With the advent of advanced imaging techniques and the wider use of stable isotopes, more data are now available to build on the work of Fomon and colleagues.

In 2000, Butte et al. published body composition reference data for infants in a longitudinal study of 76 children, studied for the first 2 years of life [21]. Body composition was assessed using a multicompartment model, with TBW measured using deuterium dilution, bone mineral content (BMC) using dual energy x-ray absorptiometry, and total body potassium using whole body counting [21]. Age-related changes in LBM and FM were characterized and the rapid "chemical maturation" during the developmental period was modeled as previously described by Fomon [20]. Briefly, similar changes in the percentage of FM and the percentage of body weight as protein and TBW for both boy and girl reference infants, as well as the percentage of LBM as TBW, were consistent with those reported by Fomon [20]. More important are the data from the longitudinal aspect of this study, as the incremental changes in body composition were also reported. In terms of body weight, both boy and girl infants gained weight at a lower rate from birth through the first 2 years of postnatal life, from approximately 32 g/d to 6 g/d. When considering the specific components of body composition, the accretion of FM and LBM occurred at a slower rate from birth through 2 years, 20 g/d to 2 g/d and 13 g/d to 3 g/d, respectively. The very rapid growth rate *in utero* and during the first few months of postnatal life is consistent with the decreased rate of protein accretion as reported from 2.6 g/d to 1.1 g/day from birth to 24 months. In human infants, there is a rapid increase of adipose tissue during the third trimester of pregnancy, which reaches its peak around 6 months postpartum [8,17]; at this time there is a decline in %FM as fat deposition begins to slow compared to LBM (Figure 18.1) [8,21,22,23,24].

Changes in body composition for healthy infants are mostly attributed to loss of TBW, given that most infants lose between 5% and 10% of their birth weight in the first postnatal days [19]. In fact, some weight loss even occurs during the first 24 hours after birth, mostly due to water loss from skin and/or changes in renal function. In one study, the mean weights at days 1 and 2 were significantly less than the mean weight at day 0 [25]. Verma reported that a strong determinant of maximum weight loss in babies is maturation, and that gestational age is inversely related to transepidermal water loss in the first month of life [26]. Adaptation to life outside the uterus for infants includes the keratinization of their skin after a week [27], while the release of natriuretic hormones called atrial natriuretic peptides prompts the contraction of the extracellular compartment [28]. As the child grows, the amount and distribution of TBW changes with a decrease in extracellular volume and an increase in intracellular volume [19]. Reference values for FM and LBM of boys and girls from published data are summarized in Tables 18.1 and 18.2 [6,20–25,23,29,24–34], respectively, and clearly show the consistency of values reported from a number of different studies.

With regard to body fat distribution in infants, little has been reported in the literature given the challenges of accurately measuring body fat distribution. However, one study was conducted in which body composition and adipose tissue distribution

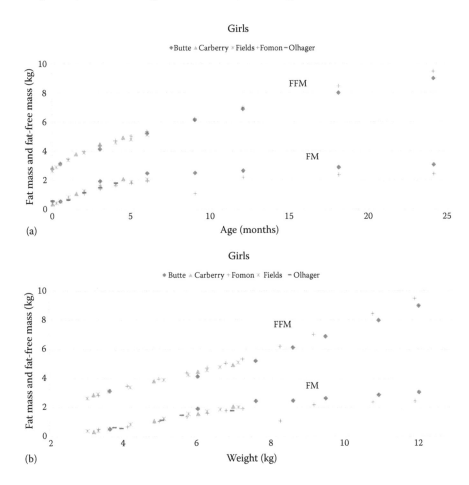

FIGURE 18.1 Fat mass and fat-free mass at different body weights according to four different cohort studies.   (Continued)

was measured using MRI in newborns, and during the first 4 months of life [23]. The advantage of using MRI over other approaches is the fact that FM is generally derived from calculations that use a number of assumptions regarding LBM. Using MRI, such assumptions are no longer needed, and a more direct assessment of FM is possible. Using MRI, Harrington et al. showed that babies born appropriate-for-gestational-age have very little internal-abdominal adipose tissue (20 ± 11 g), with the majority of the fat being subcutaneous (693 ± 205 g) [28]. These findings were confirmed by Modi et al. [30] in a study with full-term Asian Indian and white European infants, where the predominant adipose tissue at birth was superficial subcutaneous nonabdominal FM (445 g and 504 g, respectively), and the intraabdominal adipose tissue was very low (40 g and 24 g, respectively). The same study highlighted ethnic differences in adipose tissue distribution that may help to detect a propensity for abdominal obesity at birth [30]. Thus, it is important that future studies of

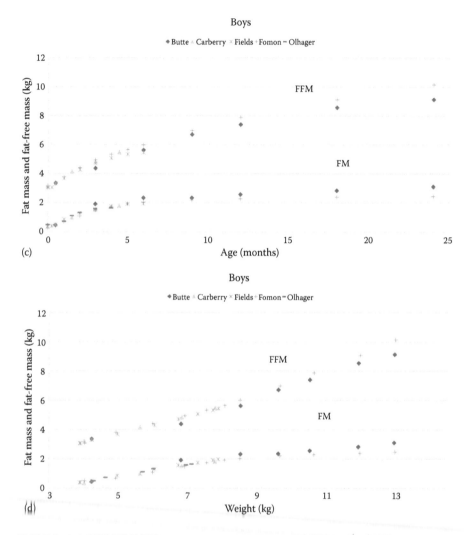

FIGURE 18.1 (CONTINUED) Fat mass and fat-free mass at different body weights according ing to four different cohort studies. (Data from Fomon SJ, Haschke F, Ziegler EE et al., *Am J Clin Nutr* 1982, 35(5 Suppl):1169–75; Butte NF, Hopkinson JM, Wong WW et al., *Pediatr Res* 2000, 47(5):578–85; Fields DA, Gilchrist JM, Catalano PM et al., *Obesity* 2011, 19(9):1887–91; Olhager E, Flinke E, Hannerstad U et al., *Pediatr Res* 2003, 54(6):906–12; Carberry AE, Colditz PB, Lingwood BE, *Pediatr Res* 2010, 68(1):84–88.)

changes in body composition during growth make every attempt to include a sample of boys and girls from diverse ethnic groups. In fact, one of the major criticisms of reference models to date is that most infant cohorts are too homogenous in terms of race and ethnicity, but the logistics and expense of measuring a large number of diverse infants often limits the heterogeneity and size of such studies.

Finally, more recent studies using a number of different techniques than those in previous work show that there is variability in %FM reported in boys and girls

TABLE 18.1

Body Composition Changes during the First 24 Months in Boys

Author	Year	Age (months)	Method	Ethnicity	N	Weight (kg)	FM (kg)	% FM	T. ATV (mL)	TBW (%)	TBW (% of FFM)	FFM (kg)	% Protein
Moulton	1923	Birth	Chemical Analyses							81.90			
Fomon	1982	Birth	Multicompartment			3.55	0.49	13.70		69.60	80.60	3.06	12.90
Fomon	1982	1	Multicompartment			4.45	0.67	15.10		68.40	80.50	3.78	12.90
Fomon	1982	2	Multicompartment			5.51	1.09	19.90		64.30	80.30	4.41	12.30
Fomon	1982	3	Multicompartment			6.44	1.49	23.20		61.40	80.00	4.94	12.00
Fomon	1982	4	Multicompartment			7.06	1.74	24.70		60.10	79.90	5.32	11.90
Fomon	1982	5	Multicompartment			7.58	1.91	25.30		59.60	79.70	5.66	11.90
Fomon	1982	6	Multicompartment			8.03	2.03	25.40		59.40	79.60	5.99	12.00
Fomon	1982	9	Multicompartment			9.18	2.19	24.00		60.30	79.30	6.98	12.40
Fomon	1982	12	Multicompartment			10.15	2.28	22.50		61.20	79.00	7.86	12.90
Fomon	1982	18	Multicompartment			11.47	2.38	20.80		62.20	78.50	9.08	13.50
Fomon	1982	24	Multicompartment			12.49	2.46	19.50		62.90	78.10	10.13	14.00
Butte	2000	0.5	Multicompartment		33	3.76	0.44	11.40		73.90	82.70	3.35	12.50
Butte	2000	3	Multicompartment		33	6.33	1.91	30.20		56.50	81.00	4.37	10.60
Butte	2000	6	Multicompartment		32	8.04	2.32	29.10		57.20	80.70	5.63	10.90
Butte	2000	9	Multicompartment		33	9.13	2.34	25.70		59.20	19.70	6.71	12.00
Butte	2000	12	Multicompartment		32	10.03	2.56	25.60		59.00	79.30	7.40	12.30
Butte	2000	18	Multicompartment		30	11.43	2.83	24.50		59.10	78.30	8.55	12.90
Butte	2000	24	Multicompartment		29	12.46	3.10	25.40		58.10	77.00	9.13	13.50
Harrington	2003	Birth	MRI	Multiethnic		3.30	0.76	23.40					
Olhager	2003	Birth	MRI/Doubly labeled water	Swedish	4	3.82	0.49	12.80	961				

(Continued)

TABLE 18.1 (CONTINUED)
Body Composition Changes during the First 24 Months in Boys

Author	Year	Age (months)	Method	Ethnicity	N	Weight (kg)	FM (kg)	% FM	T. ATV (mL)	TBW (%)	TBW (% of FFM)	FFM (kg)	% Protein
Olhager	2003	1	MRI/Doubly labeled water	Swedish	3	4.16	0.72	17.00	1114				
Olhager	2003	1.5	MRI/Doubly labeled water	Swedish	3	5.23	1.12	21.50	1675				
Olhager	2003	2	MRI/Doubly labeled water	Swedish	5	5.54	1.32	23.80	1971				
Olhager	2003	3	MRI/Doubly labeled water	Swedish	6	6.47	1.59	24.30	2370				
Olhager	2003	4	MRI/Doubly labeled water	Swedish	4	6.63	1.63	24.10	2434				
Carberry	2010	Birth	Pea Pod	Australian	25	3.54	0.34	9.42				3.20	
Carberry	2010	1.5	Pea Pod	Australian	25	5.15	1.00	19.10				4.15	
Carberry	2010	3	Pea Pod	Australian	25	6.35	1.53	23.76				4.81	
Carberry	2010	4.5	Pea Pod	Australian	25	7.30	1.80	22.50				5.48	
Fields	2011	Birth	Pea Pod	Multiethnic	20	3.41	0.37	10.66				3.05	
Fields	2011	0.25	Pea Pod	Multiethnic	54	3.44	0.39	11.14				3.05	
Fields	2011	0.5	Pea Pod	Multiethnic	71	3.76	0.49	12.67				3.28	
Fields	2011	1	Pea Pod	Multiethnic	72	4.50	0.83	18.27				3.67	
Fields	2011	2	Pea Pod	Multiethnic	68	5.56	1.28	22.70				4.28	
Fields	2011	3	Pea Pod	Multiethnic	60	6.26	1.56	24.65				4.70	
Fields	2011	4	Pea Pod	Multiethnic	54	6.81	1.74	25.29				5.07	
Fields	2011	5	Pea Pod	Multiethnic	44	7.24	1.91	26.15				5.33	
Fields	2011	6	Pea Pod	Multiethnic	30	7.40	1.95	25.94				5.45	

(Continued)

TABLE 18.1 (CONTINUED)
Body Composition Changes during the First 24 Months in Boys

Author	Year	Age (months)	Method	N	Ethnicity	Weight (kg)	FM (kg)	% FM	T. ATV (mL)	TBW (%)	TBW (% of FFM)	FFM (kg)	% Protein
Hull	2011	Birth	Pea Pod		Multiethnic	3.25	0.41	12.30				2.84	
Fields	2012	6	Pea Pod	37		7.11	1.92	26.70				5.19	
Fields	2012	6	DEXA	37		7.11	2.28	31.10				5.02	
Andersen	2013	Birth	Pea Pod	170	Ethiopian	3.11		7.30					
Andersen	2013	1.5	Pea Pod	155	Ethiopian	4.78		21.60					
Andersen	2013	2.5	Pea Pod	165	Ethiopian	5.57		25.00					
Andersen	2013	3.5	Pea Pod	166	Ethiopian	6.30		26.70					
Andersen	2013	4.5	Pea Pod	157	Ethiopian	6.87		27.50					
Andersen	2013	6	Pea Pod	182	Ethiopian	7.52		26.00					
Au	2013	Birth	Pea Pod		Multiethnic	3.42		9.20				2.95	
Paley	2015	Birth	Pea Pod	21	African American	3.05	0.36	11.61					
Paley	2015	Birth	Pea Pod	14	Asian	3.46	0.43	12.24					
Paley	2015	Birth	Pea Pod	88	Caucasian	3.36	0.43	12.66					

Source: Moulton CR, *J Biol Chem* 1923, 57(1):79–97; Fomon SJ, Haschke F, Ziegler EE et al., *Am J Clin Nutr* 1982, 35(5 Suppl):1169–75; Butte NF, Hopkinson JM, Wong WW et al., *Pediatr Res* 2000, 47(5):578–85; Fields DA, Gilchrist JM, Catalano PM et al., *Obesity* 2011, 19(9):1887–91; Hull HR, Thornton JC, Ji Y et al., *Am J Obstet Gynecol* 2011, 205(3):211 e1–7; Olhager E, Flinke E, Hannerstad U et al., *Pediatr Res* 2003, 54(6):906–12; Harrington TA, Thomas EL, Frost G et al., *Pediatr Res* 2004, 55(3):437–41; Carberry AE, Colditz PB, Lingwood BE, *Pediatr Res* 2010, 68(1):84–88; Au CP, Raynes-Greenow CH, Turner RM et al., *Early Hum Dev* 2013, 89(10):839–43; Andersen GS, Girma T, Wells JC et al., *Am J Clin Nutr* 2013, 98(4):885–94; Paley C, Hull H, Ji Y et al., *Pediatr Obes* 2015, 11(5):361–8; and Fields DA, Demerath EW, Reobelli A et al., *Obesity* 2012, 20(11):2302–6.

Note: DXA, dual-energy x-ray absorptiometry; FFM, fat-free mass; FM, fat mass; MRI, magnetic resonance imaging; TATV, total adipose tissue volume; TBW, total body water.

TABLE 18.2

Body Composition Changes during the First 24 Months in Girls

Author	Year	Age (months)	Method	Ethnicity	N	Weight (kg)	FM (kg)	% FM	T.ATV (mL)	TBW (%)	TBW (% of FFM)	FFM (kg)	% Protein
Moulton	1923	Birth	Chemical Analyses							81.90			
Fomon	1982	Birth	Multicompartment			3.33	0.49	14.90		68.60	80.60	2.83	12.80
Fomon	1982	1	Multicompartment			4.13	0.67	16.20		67.50	80.50	3.46	12.70
Fomon	1982	2	Multicompartment			4.99	1.05	21.10		63.20	80.20	3.93	12.20
Fomon	1982	3	Multicompartment			5.74	1.37	23.80		60.90	79.90	4.38	12.00
Fomon	1982	4	Multicompartment			6.30	1.59	25.20		59.60	79.70	4.72	11.90
Fomon	1982	5	Multicompartment			6.80	1.77	26.00		58.80	79.50	5.03	11.90
Fomon	1982	6	Multicompartment			7.25	1.92	26.40		58.40	79.40	5.34	12.00
Fomon	1982	9	Multicompartment			8.27	2.07	25.00		59.30	79.00	6.20	12.50
Fomon	1982	12	Multicompartment			9.18	2.18	23.70		60.10	78.80	7.01	12.90
Fomon	1982	18	Multicompartment			10.78	2.35	21.80		61.30	78.40	8.43	13.50
Fomon	1982	24	Multicompartment			11.91	2.43	20.40		62.20	78.20	9.48	13.90
Butte	2000	0.5	Multicompartment		43	3.64	0.52	14.20		73.20	83.10	3.12	12.20
Butte	2000	3	Multicompartment		43	6.03	1.90	31.50		55.60	81.10	4.11	10.20
Butte	2000	6	Multicompartment		43	7.60	2.44	32.00		54.90	80.70	5.21	10.40
Butte	2000	9	Multicompartment		41	8.62	2.47	28.80		56.90	79.80	6.12	11.40
Butte	2000	12	Multicompartment		42	9.50	2.62	27.60		56.90	78.80	6.88	12.20
Butte	2000	18	Multicompartment		41	10.94	2.87	26.30		57.80	78.20	7.99	12.70
Butte	2000	24	Multicompartment		43	12.02	3.05	25.40		57.70	78.00	8.99	13.10
Harrington	2003	Birth	MRI	Multiethnic		3.30	0.76	23.40					
Olhager	2003	Birth	MRI/Doubly labeled water	Swedish	6	3.96	0.58	14.20	1101.20				

(Continued)

TABLE 18.2 (CONTINUED)
Body Composition Changes during the First 24 Months in Girls

Author	Year	Age (months)	Method	Ethnicity	N	Weight (kg)	FM (kg)	% FM	T.ATV (mL)	TBW (%)	TBW (% of FFM)	FFM (kg)	% Protein
Olhager	2003	1	MRI/Doubly labeled water	Swedish	3	3.76	0.63	17.90	1046.70				
Olhager	2003	2	MRI/Doubly labeled water	Swedish	6	5.07	1.13	22.20	1689.80				
Olhager	2003	3	MRI/Doubly labeled water	Swedish	4	5.60	1.47	26.20	2192.30				
Olhager	2003	4	MRI/Doubly labeled water	Swedish	2	6.96	1.79	25.80	2681.00				
Carberry	2010	Birth	Pea Pod	Australian	20	3.20	0.33	10.09				2.87	
Carberry	2010	1.5	Pea Pod	Australian	20	4.86	1.07	21.84				3.80	
Carberry	2010	3	Pea Pod	Australian	20	6.04	1.58	26.07				4.46	
Carberry	2010	4.5	Pea Pod	Australian	20	6.99	2.06	29.33				4.93	
Fields	2011	Birth	Pea Pod	Multiethnic	15	3.03	0.40	13.19				2.63	
Fields	2011	0.25	Pea Pod	Multiethnic	58	3.32	0.42	12.47				2.90	
Fields	2011	0.5	Pea Pod	Multiethnic	66	3.61	0.52	14.07				3.09	
Fields	2011	1	Pea Pod	Multiethnic	63	4.21	0.83	19.38				3.38	
Fields	2011	2	Pea Pod	Multiethnic	60	5.12	1.24	24.03				3.88	
Fields	2011	3	Pea Pod	Multiethnic	55	5.78	1.55	26.56				4.23	
Fields	2011	4	Pea Pod	Multiethnic	49	6.29	1.73	26.99				4.57	
Fields	2011	5	Pea Pod	Multiethnic	40	6.64	1.86	27.72				4.78	
Fields	2011	6	Pea Pod	Multiethnic	35	7.13	2.03	28.28				5.10	
Hull	2011	Birth	Pea Pod	Multiethnic		3.25	0.41	12.30				2.84	
Fields	2012	6	Pea Pod	Multiethnic	47	7.11	1.92	26.70				5.19	

(Continued)

TABLE 18.2 (CONTINUED)
Body Composition Changes during the First 24 Months in Girls

Author	Year	Age (months)	Method	Ethnicity	N	Weight (kg)	FM (kg)	% FM	T.ATV (mL)	TBW (%)	TBW (% of FFM)	FFM (kg)	% Protein
Fields	2012	6	DEXA		47	7.11	2.28	31.10					5.02
Andersen	2013	Birth	Pea Pod	Ethiopian	178	3.00		7.80					
Andersen	2013	1.5	Pea Pod	Ethiopian	171	4.54		21.10					
Andersen	2013	2.5	Pea Pod	Ethiopian	175	5.38		25.80					
Andersen	2013	3.5	Pea Pod	Ethiopian	179	6.05		28.70					
Andersen	2013	4.5	Pea Pod	Ethiopian	169	6.57		28.50					
Andersen	2013	6	Pea Pod	Ethiopian	196	7.29		27.70					
Au	2013	Birth	Pea Pod	Multiethnic		3.42		9.20					
Paley	2015	Birth	Pea Pod	African-American	23	3.09	0.48	15.40				2.95	
Paley	2015	Birth	Pea Pod	Asian	16	3.06	0.37	11.72					
Paley	2015	Birth	Pea Pod	Caucasian	98	3.21	0.47	14.28					
Paley	2015	Birth	Pea Pod	Hispanic	32	3.09	0.44	13.99					

Source: Moulton CR. *J Biol Chem* 1923, 57(1):7–9; Fomon SJ, Haschke F, Ziegler EE et al., *Am J Clin Nutr* 1982, 35(5 Suppl):1169–75; Butte NF, Hopkinson JM, Wong WW et al., *Pediatr Res* 2000, 47(5):578–85; Fields DA, Gilchrist JM, Catalano PM et al., *Obesity* 2011, 19(9):1887–91; Hull HR, Thornton JC, Ji Y et al., *Am J Obstet Gynecol* 2011, 205(3):211 e1–7; Olhager E, Flinke E, Hannerstad U et al., *Pediatr Res* 2003, 54(6):906–12; Harrington TA, Thomas EL, Frost G et al., *Pediatr Res* 2004, 55(3):437–41; Carberry AE, Colditz PB, Lingwood BE, *Pediatr Res* 2010, 68(1):84–88; Au CP, Raynes-Greenow CH, Turner RM et al., *Early Hum Dev* 2013, 89(10):839–43; Andersen GS, Girma T, Wells JC et al., *Am J Clin Nutr* 2013, 98(4):885–94; Paley C, Hull H, Ji Y et al., *Pediatr Obes* 2015, 11(5):361–8; and Fields DA, Demerath EW, Pietrobelli A et al., *Obesity* 2012, 20(11):2302–6.

Note: DXA, dual-energy x-ray absorptiometry; FFM, fat-free mass; FM, fat mass; MRI, magnetic resonance imaging; TATV, total adipose tissue volume; TBW, total body water.

at birth, most likely due to the ethnic differences in the populations from which the study samples were selected (Tables 18.1 and 18.2 [6,20–25,23,29,24–34]). As shown in Figure 18.1, there is an overall agreement between the various approaches used to measure body composition in infants. While there are clear divergences in body composition data based on the technique used and when analyzed by age and body weight, such differences are most likely due to inherent differences in populations used in the study, and suggest that gender is a major determinant of whole body composition during the first year of life.

POOR GROWTH AND BODY COMPOSITION

Aside from changes in body composition during "healthy" growth, the importance of the first 1,000 days is most relevant when one considers not only the acute impact of poor nutrition and infectious diseases on the health of a child, but also the relationship with health and disease in adulthood. For almost two decades, research has revealed clear associations between poor growth *in utero* or during early childhood and an increased risk of chronic diseases in adulthood [35–38]. At the same time, new research is providing convincing evidence that supports a number of mechanisms to explain earlier associations, giving support to the notion that the first 1,000 days are important for a lifetime of health.

In terms of body composition, poor growth during gestation and in the first 2 years of life is not only associated with metabolic adaptions that favor fat deposition, but also with unhealthy fat accumulation and central adiposity [39,40]. However, the precise mechanisms, be they intrauterine nutrition or a postnatal, environmental factor (e.g., postnatal diet, breastfeeding, or physical activity) remain unclear. Nonetheless, a large number of studies have reported that small size at birth is associated with a high BMI or FM, compared to adults or peers of normal birth weight [41,42]. A study of birth weight and body composition reported that birth weight was negatively correlated with central adiposity in children from the United States [43]. It was also found that birth weight is associated with higher lean tissue mass, but not higher adipose tissue mass [44]. These results may be interpreted to suggest that poor growth in utero is related to the poor development of lean body mass, a factor that may explain greater fat deposition during growth following *in utero* nutrient restriction.

Considering the components of body composition, children who were born small for gestational age are reported to have greater central FM, independent of total FM, when assessed using a precise imaging technique for body composition (dual energy x-ray absorptiometry, DEXA) [45]. A similar study found that low birth weight was associated with higher central adiposity [46], even after controlling for physical activity and socioeconomic status, postnatal variables that are known to influence body composition. A limitation to both studies was that DEXA does not discriminate visceral fat, which is reported to have a high turnover of fatty acids, promoting chronic risk of disease, and subcutaneous fat. Regardless, it is clear that poor growth *in utero* is a risk factor for unhealthy body fat distribution even among children who are not obese.

While the association between birth weight and FM is of great importance, one needs to determine whether birth weight is associated with LBM. For example, a cohort study of infants who were followed through adolescence reported that each standard deviation increase in birth weight was associated with a 1.2 kg increase in LBM, but not FM [47]. Based on this study, relative to long-term health, poor growth *in utero* may be implicated in the poor accumulation of metabolically active tissue, and may result in abnormal substrate metabolism (e.g. insulin resistance) or reduced energy expenditure for the same body weight. Such findings are consistent with research on children who were stunted, defined as a height-for-age z-score more than 2 standard deviations (SD) below the mean of a reference population [48]. Studies of stunted children in Brazil and South Korea have consistently reported that children who have experienced growth deficits have lower rates of fat metabolism compared to normal height children [49,50]. In the Brazilian cohort, stunted children who had their body composition measured over a 4-year period deposited more fat in their central region compared to normal height children, independent of total body FM [51]. Thus, it is clear that nutritional insults, either a lack of proper nutrients or infectious diseases that interrupt nutrient absorption, early in life have profound metabolic and body composition implications later in life.

POOR GROWTH AND ADULT CHRONIC DISEASE

Of great importance is the determination of when and how poor nutrition during "critical periods" of development, such as the first 1,000 days, influence disease risk and adult health. Perhaps the most well-known of such studies is that of the Dutch famine, in which a cohort of men born before, during, and after the famine in Holland of 1944–1945 were studied for body weight in relation to exposure to famine during gestation [52]. Men who were exposed to the famine during late gestation and early infancy were less likely to be obese as adults than men who were exposed to the famine during the first two trimesters of gestation. These results clearly suggest that early, prenatal exposure to undernutrition is related to weight gain later in adulthood.

While deficits in weight can be recovered, linear growth deficits are less likely to be recovered, even with adequate nutriture, especially if the deficit occurs between birth and 2 years of age [53]. A study of 228 Brazilian children who attended a nutrition rehabilitation center reported that age at the beginning of treatment affected the likelihood of recovery [54]. Briefly, children aged 12 to 23 months and 24 months and older were 30% and 51% less likely to recover from undernutrition, respectively, than children who were treated up to 12 months. In this study, children 24 months and older showed a slower rate of recovery and had to be treated longer to fully recover from undernutrition [55]. Based on these studies, it is clear that proper nutrition during the first 1,000 days is essential not only for maintaining normal changes in body composition during rapid growth, but also for preventing permanent growth retardation (i.e., stunting). Nonetheless, stunting is a challenge for health professionals, as the ability to improve height in countries with extreme social and economic hardships is complicated, and often met with little success, despite efforts to improve overall nutrition. Thus, as previously discussed, the timing of nutrition on health is critical in terms of later disease risk, as well as the ability to recover from nutritional insults.

With regard to body fat distribution later in childhood, a number of studies have investigated body composition in children who were growth retarded in early life. For example, in Senegal, adolescent girls who were stunted before the age of 2 were more likely to accumulate subcutaneous fat on the trunk and arms than nonstunted girls, despite having the same BMI [56]. A longitudinal study in Guatemala was among one of the first to report that stunting was associated with central fat distribution [57], such that men and women who had been severely stunted as children had significantly greater abdominal fatness as adults, independent of total FM and other confounding factors. These studies support the hypothesis that stunting increases not only the risk of obesity, but also of anthropomorphic phenotypes associated with chronic diseases.

In terms of how poor nutrition in the first 1,000 days has lifelong effects on body composition and health, it is well documented that nutritional deficits during the first two years of childhood that are severe enough to result in moderate to severe growth retardation are often associated with an increased risk of chronic diseases, including obesity in later life [58,59]. Moreover, broad economic changes in transitional countries have seen changes in food systems that favor excess energy intake; thus, many countries are hosting a "double burden" of both under- and overnutrition. Furthermore, the recurrence of infectious diseases is also related to stunting and, in some world regions, there has been little progress in preventing such diseases. It is worth noting that considerable effort is now being made to understand how infections, as well as the microbiome, may mediate changes in growth relative to diet. Regardless, much more needs to be done to improve our understanding of how body composition changes during early childhood, and the role of such changes on health outcomes and later risk for chronic diseases.

CONCLUSION

From the moment of conception through the first two years of postnatal life, the dynamic process of growth and development occurs at a rapid pace, and the body composition of a newborn child undergoes a series of changes as the process of "healthy" growth proceeds. In general, early models of body composition changes from birth to age 2 years relied on a number of assumptions to allow for the precise assessment of LBM and FM in neonates and infants. As new methods were developed, such assumptions were no longer needed, and more accurate estimates of body composition were available. Although such differences in methods could have resulted in marked divergences between early and later cohorts of children, no such observations were reported. Instead, there is a clear consistency in the data among a large number of cohorts from across the world, although some differences in body fat distribution have been reported for different ethnic/racial groups or children who are growth retarded. Briefly, it can be seen that infants increase their %FM during the first year, coinciding with a decreased rate of accretion of FM and LBM. Such changes are accompanied by a decrease in the hydration of LBM, reflecting tissue changes during the first year of postnatal life.

Perhaps more important than these broad changes in body composition in the first 1,000 days is the body of work that supports the concept that disruptions to "healthy" growth may interrupt the growth process in such a way that body composition and metabolic changes increase the risk of chronic diseases later in life. What is still unclear is how interactions between diet and the gut influence growth, and the degree to which micronutrient deficiencies alter bone growth as a moderating factor in lower LBM accretion during periods of undernutrition. These questions have great potential to advance the understanding of how normal or abnormal growth influence body composition throughout life.

REFERENCES

1. Quetelet A. *Sur l'homme et le développement de ses facultés ou essai de physique sociale*. Paris: Bachelier; 1835.
2. Wells JC. The programming effects of early growth. *Early Hum Dev* 2007; 83(12):743–8.
3. Fee BA, Weil WB Jr. Body composition of infants of diabetic mothers by direct analysis. *Ann NY Acad Sci* 1963; 110(2):869–97.
4. Iob V, Swanson WW. Mineral growth of the human fetus. *Am J Dis Child* 1934; 47(2):302–6.
5. Givens MH, Macy IG. The chemical composition of the human fetus. *J Biol Chem* 1933; 102:7–17.
6. Moulton CR. Age and chemical development in mammals. *J Biol Chem* 1923; 57(1):79–97.
7. Widdowson EM, Spray CM. Chemical development in utero. *Arch Dis Child* 1951; 26(127):205–14.
8. Fomon SJ, Nelson SE. Body composition of the male and female reference infants. *Annu Rev Nutr* 2002; 22:1–17.
9. Rigo J, de Curtis M, Pieltain C. Nutritional assessment in preterm infants with special reference to body composition. *Semin Neonatol* 2001; 6(5):383–91.
10. Behnke A, Feen B, Welham W. The specific gravity of healthy men: Body weight ÷ volume as an index of obesity. *JAMA* 1942; 118(7):495–8.
11. Heymsfield SB, Lichtman S, Baumgartner RN et al. Body composition of humans: Comparison of two improved four-compartment models that differ in expense, technical complexity, and radiation exposure. *Am J Clin Nutr* 1990; 52(1):52–8.
12. Wang ZM, Pierson RN Jr, Heymsfield SB. The five-level model: A new approach to organizing body-composition research. *Am J Clin Nutr* 1992; 56(1):19–28.
13. Ellis KJ. Reference infant and child models of body composition. *Nutr Today* 2009; 44(2):54–61.
14. Demerath EW, Fields DA. Body composition assessment in the infant. *Am J Hum Biol* 2014; 26(3):291–304.
15. Barker DJ. Maternal nutrition, fetal nutrition, and disease in later life. *Nutrition* 1997; 13(9):807–13.
16. Yu ZB, Han SP, Zhu GZ et al. Birth weight and subsequent risk of obesity: A systematic review and meta-analysis. *Obes Rev* 2011; 12(7):525–42.
17. Ziegler EE, O'Donnell AM, Nelson SE et al. Body composition of the reference fetus. *Growth* 1976; 40(4):329–41.
18. Toro-Ramos T, Paley C, Pi-Sunyer FX et al. Body composition during fetal development and infancy through the age of 5 years. *Eur J Clin Nutr* 2015; 69(12):1279–89.
19. Méio MDBB, Moreira MEL. Total body water in newborns. In *Handbook of anthropometry: Physical measures of human form in health and disease*, Preedy RV, editor. New York: Springer New York; 2012, pp. 1121–35.
20. Fomon SJ, Haschke F, Ziegler EE et al. Body composition of reference children from birth to age 10 years. *Am J Clin Nutr* 1982; 35(5 Suppl):1169–75.

21. Butte NF, Hopkinson JM, Wong WW et al. Body composition during the first 2 years of life: An updated reference. *Pediatr Res* 2000; 47(5):578–85.
22. Fields DA, Gilchrist JM, Catalano PM et al. Longitudinal body composition data in exclusively breast-fed infants: A multicenter study. *Obesity* 2011; 19(9):1887–91.
23. Hull HR, Thornton JC, Ji Y et al. Higher infant body fat with excessive gestational weight gain in overweight women. *Am J Obstet Gynecol* 2011; 205(3):211 e1–7.
24. Verma RP, Shibli S, Fang H et al. Clinical determinants and utility of early postnatal maximum weight loss in fluid management of extremely low birth weight infants. *Early Hum Dev* 2009; 85(1):59–64.
25. Sridhar S, Baumgart S. Water and electrolyte balance in newborn infants. In *Neonatal nutrition and metabolism*, 2nd ed., Thureen PJ, Hay WW, editors. Cambridge: Cambridge University Press; 2006, pp. 104–14.
26. Modi N. Clinical implications of postnatal alterations in body water distribution. *Semin Neonatol* 2003; 8(4):301–6.
27. Olhager E, Flinke E, Hannerstad U et al. Studies on human body composition during the first 4 months of life using magnetic resonance imaging and isotope dilution. *Pediatr Res* 2003; 54(6):906–12.
28. Harrington TA, Thomas EL, Frost G et al. Distribution of adipose tissue in the newborn. *Pediatr Res* 2004; 55(3):437–41.
29. Modi N, Thomas EL, Uthaya SN et al. Whole body magnetic resonance imaging of healthy newborn infants demonstrates increased central adiposity in Asian Indians. *Pediatr Res* 2009; 65(5):584–7.
30. Byrne CD, Phillips DI. Fetal origins of adult disease: Epidemiology and mechanisms. *J Clin Pathol* 2000; 53(11):822–8.
31. Curhan GC, Willett WC, Rimm EB et al. Birth weight and adult hypertension, diabetes mellitus, and obesity in U.S. men. *Circulation* 1996; 94(12):3246–50.
32. Godfrey KM, Barker DJ. Fetal nutrition and adult disease. *Am J Clin Nutr* 2000; 71(5 Suppl):1344S–52S.
33. Hales CN. Fetal and infant origins of adult disease. *J Clin Pathol* 1997; 50(5):359.
34. Dietz WH. Critical periods in childhood for the development of obesity. *Am J Clin Nutr* 1994; 59(5):955–9.
35. Parsons TJ, Power C, Manor O. Fetal and early life growth and body mass index from birth to early adulthood in 1958 British cohort: Longitudinal study. *BMJ* 2001; 323(7325):1331–5.
36. Sichieri R, Siqueira KS, Moura AS. Obesity and abdominal fatness is associated with undernutrition early in life in a survey in Rio de Janeiro. *Int J Obes Relat Metab Disord* 2000; 24(5):614.
37. Gerstner GE. Haloperidol affects guinea pig chewing burst durations. *Brain Behav Evol* 1996; 48(2):94–102.
38. Okosun IS, Liao Y, Rotimi CN et al. Impact of birth weight on ethnic variations in subcutaneous and central adiposity in American children aged 5–11 years. A study from the Third National Health and Nutrition Examination Survey. *Int J Obes Relat Metab Disord* 2000; 24(4):479–84.
39. Kahn HS, Narayan KM, Williamson DF et al. Relation of birth weight to lean and fat thigh tissue in young men. *Int J Obes Relat Metab Disord* 2000; 24(6):667–72.
40. Dolan MS, Sorkin JD, Hoffman DJ. Birth weight is inversely associated with central adipose tissue in healthy children and adolescents. *Obesity* 2007; 15(6):1600–8.
41. Labayen I, Moreno LA, Blay MG et al. Early programming of body composition and fat distribution in adolescents. *J Nutr* 2006; 136(1):147–52.
42. Singhal A, Wells J, Cole TJ et al. Programming of lean body mass: A link between birth weight, obesity, and cardiovascular disease? *Am J Clin Nutr* 2003; 77(3):726–30.

43. World Health Organization. *WHO Child Growth Standards: Length/height-for-age, weight-for-age, weight-for-length, weight-for-height and body mass index-for-age: Methods and development.* Geneva: WHO, 2006.

44. Hoffman DJ, Sawaya AL, Verreschi I et al. Why are nutritionally stunted children at increased risk of obesity? Studies of metabolic rate and fat oxidation in shantytown children from Sao Paulo, Brazil. *Am J Clin Nutr* 2000; 72(3):702–7.

45. Lee SK, Nam SY, Hoffman DJ. Growth retardation at early life and metabolic adaptation among North Korean children. *J Dev Orig Health Dis* 2015; 6(4):291–8.

46. Hoffman DJ, Martins PA, Roberts SB et al. Body fat distribution in stunted compared with normal-height children from the shantytowns of Sao Paulo, Brazil. *Nutrition* 2007; 23(9):640–6.

47. Ravelli GP, Stein ZA, Susser MW. Obesity in young men after famine exposure in utero and early infancy. *N Engl J Med* 1976; 295(7):349–53.

48. UNICEF. *Tracking progress on child and maternal nutrition: A survival and development priority.* New York: UNICEF, 2009.

49. Lindsay RS, Dabelea D, Roumain J et al. Type 2 diabetes and low birth weight: The role of paternal inheritance in the association of low birth weight and diabetes. *Diabetes* 2000; 49(3):445–9.

50. Fernandes MB, Lopez RV, de Albuquerque MP et al. A 15-year study on the treatment of undernourished children at a nutrition rehabilitation centre (CREN), Brazil. *Public Health Nutr* 2012; 15(6):1108–16.

51. Benefice E, Garnier D, Simondon KB et al. Relationship between stunting in infancy and growth and fat distribution during adolescence in Senegalese girls. *Eur Journal Clin Nutr* 2001; 55(1):50–8.

52. Schroeder DG, Martorell R. Fatness and body mass index from birth to young adulthood in a rural Guatemalan population. *Am J Clin Nutr* 1999; 70(1):137s–44s.

53. Lucas A, Fewtrell MS, Cole TJ. Fetal origins of adult disease–the hypothesis revisited. *BMJ* 1999; 319(7204):245–9.

54. Stein AD, Zybert PA, van de Bor M et al. Intrauterine famine exposure and body proportions at birth: The Dutch Hunger Winter. *Int J Epidemiol* 2004; 33(4):831–6.

55. Carberry AE, Colditz PB, Lingwood BE. Body composition from birth to 4.5 months in infants born to non-obese women. *Pediatr Res* 2010; 68(1):84–8.

56. Au CP, Raynes-Greenow CH, Turner RM et al. Fetal and maternal factors associated with neonatal adiposity as measured by air displacement plethysmography: A large cross-sectional study. *Early Hum Dev* 2013; 89(10):839–43.

57. Andersen GS, Girma T, Wells JC et al. Body composition from birth to 6 months of age in Ethiopian infants: Reference data obtained by air displacement plethysmography. *Am J Clin Nutr* 2013; 98(4):885–94.

58. Paley C, Hull H, Ji Y et al. Body fat differences by self-reported race/ethnicity in healthy term newborns. *Pediatr Obes* 2015; 11(5):361–8.

59. Fields DA, Demerath EW, Pietrobelli A et al. Body composition at 6 months of life: Comparison of air displacement plethysmography and dual-energy x-ray absorptiometry. *Obesity* 2012; 20(11):2302–6.

Section VIII

The Gut Microbiome

19 Early Life Gut Microbiome

Christopher J. Stewart and Stephen P. Cummings

CONTENTS

INTRODUCTION

The human body has trillions of microbes inhabiting every surface exposed to the environment, both inside and out, termed the *microbiome* (*micro* = small, *biom* = community in a distinct environment). The largest community of microbes is found within the gastrointestinal tract where the gut microbiome in healthy adults is estimated to harbor 10^{14} bacterial cells, which roughly equates to 100 to 300 times the number of genes in the human genome. This collection of microbes assembles largely through chance, but the assembly is not without design. Microbes have existed on earth for 3.5 billion years, based on the oldest discovered fossilized evidence; humanity has therefore coevolved with its microbial flora, and they play a pivotal role in the neonate, facilitating maturation of the immune system, protection from potential pathogens, and digestion of gut contents for the synthesis of essential vitamins and nutrients.

Improvements in high-throughput methodologies, coupled to reducing costs, have profoundly improved our ability to survey the microbiome. To date, the vast majority of human microbiome studies have focused on the bacterial community in the gut. Typically, a variable region of the universal 16S rRNA gene is amplified, giving 250bp fragments that allow identification of bacterial genera. Large-scale studies of human populations, such as the Human Microbiome Project (HMP), have revealed that the human microbiome is unique to an individual and remains highly stable through adulthood in healthy individuals. Research is now focused on how perturbations in normal microbiome development occur and the subsequent risk factors associated with a range of diseases.

The initial development of the microbiome has important consequences for immediate and long-term health. The infant gut microbiome is thought to reach adultlike profiles at around 3 years of age. Thus, the first 1,000 days of life are critical, and this chapter addresses the factors influencing gut microbiome development through this time.

FACTORS AFFECTING THE NEONATAL GUT MICROBIOME: DAY 0 TO 30

Under normal circumstances neonates are generally thought to be sterile in the intrauterine environment, with the first exposure to microbes occurring during birth. However, intriguing emerging evidence also suggests that bacterial DNA might be present in the placenta. Although the question remains elusive of when exactly the first exposure of the neonate to microbes takes place, the diversity of the human microbiome immediately after birth is low and increases from this point on. The first microbes to interact with the neonate have significant roles in educating the immune system, programming healthy immune development and metabolic programming. Primary colonizers of the gut microbiome include mixture of cutaneous and enteric facultative anaerobes, dominated by lactic acid bacteria including *Lactobacillus*, *Enterococcus*, *Streptococcus*, and coagulase-negative staphylococci. In a matter of days, these primary colonizing bacteria deplete the oxygen supplies within the gut, resulting in the growth of strictly anaerobic bacteria including *Bifidobacterium*, *Clostridium*, and *Bacteroides* [1].

During the first month of life the development of the microbiome is highly chaotic and largely stochastic. The main factors contributing to initial microbial exposure are related to the maternal microbiome, with mode of delivery and gestation also playing crucial roles. This section explores the prenatal and postnatal phases, discussing the importance of these influences on the developing neonatal microbiome.

MATERNAL MICROBIOME

There are many factors that influence the maternal microbiome at various body sites, which in turn influence the neonatal microbiome (Figure 19.1). The main factors shaping the maternal microbiome include diet, environment (e.g., socioeconomic status), and antibiotic exposure [2]. For instance, diet is known to have significant effects on the gut microbiome, due to altering favorable growth conditions from undigested dietary components, and likely has some influence on the oral community. Given the placental microbiome seems linked to the oral community, such effects may have direct implications to the developing fetus [3]. It should be noted that evidence of a placental microbiome is an active area of study with particular focus on whether these organisms are viable (or only detected due to presence of DNA) and to what extent the findings are biased by contaminants.

The maternal gut, vaginal, and skin microbiomes all contribute to the initial exposure of viable organisms to the neonate, which is discussed in detail in the next section. Again, the environment of the mother will influence the bacterial communities in each of these body sites, with variation between different geographies, socioeconomic classes, and cultural traditions having the most pronounced effects. Independent of birth mode, maternal skin microbiome is transferred to the neonate through the initial

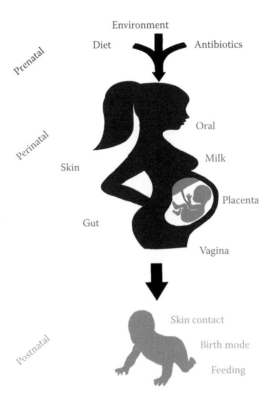

FIGURE 19.1 The maternal influence of the neonatal microbiome. Schematic shows the major influences to the mother's microbiome in the prenatal phase, which are then changed in the perinatal phase, and ultimately affect the neonate in the postnatal phase.

stages of life from holding, cuddling, kissing, and breastfeeding. Importantly, the skin microbiome varies drastically based on the specific site, with sebaceous sites dominated by *Propionibacterium* and moist areas dominated by *Staphylococcus* and *Corynebacterium* [4]. The role of feeding in modulating the developing offspring microbiome is discussed later in the chapter, but noteworthy for this section is that the maternal milk microbiome is influenced by several factors, including maternal weight, diet, and antibiotics. Thus, even after birth, the maternal diet continues to be a significant driver of the early infant microbiome, with reduced diversity in obese and high-fat-consuming mothers [5]. It is also well understood that the milk microbiome is not stable and changes over the course of lactation [6], with evidence suggesting a correlation between the early milk microbiome with the oral microbiome [5]. In light of such evidence and associations of the placental microbiome with oral bacterial communities, it is important to determine the true extent of oral communities in seeding other sites, and how therapeutic manipulation can be tailored to maximize oral health in the prenatal and postnatal stages.

Despite the numerous ways in which the maternal microbiome is altered, it is well established that full-term, vaginally delivered, breastfed neonates represent the "gold standard" for short- and long-term health.

DELIVERY MODE

In the first 24 hours of life, neonatal skin and gut microbiomes resemble the bacterial community from the mode of birth [7]. Vaginally delivered neonates are initially colonized by bacterial communities that resemble the vaginal microbiome, resulting in dominant bacteria genera including *Lactobacillus*, *Prevotella*, and *Atopobium* [7]. Specifically, the vaginal microbiome is usually dominated by one of the following species of *Lactobacillus*: *L. crispatus*, *L. iners*, *L. jensenii*, or *L. gasseri* [8]. Intriguing evidence suggests that a lack of dominance by *Lactobacillus* spp., accompanied by increased *Gardnerella* and *Ureaplasma* abundance, is associated with an increased risk of preterm birth [8]. In contrast, cesarean delivery results in the offspring microbiome being more comparable to the skin microbiome [7]. Rates of caesarean delivery are increasing worldwide and are more common in preterm infants. This has been shown to result in delayed colonization, with a gut initially dominated by environmental bacteria, specifically a high prevalence of *Clostridium*, *Escherichia*, *Streptococcus*, and *Staphylococcus* [9]. Following neonates longitudinally has revealed higher levels of Bifidobacteria and *Bacteroides*, with less *Clostridium* (importantly *Clostridium difficile*), in vaginally delivered infants compared to cesarean section [10].

It is noteworthy that the infant gut microbiota more resembles its own mother's vaginal microbiota than that of nonrelated mothers, but there is a lack of distinct similarity with the respective maternal skin microbiota [7]. What this means in terms of the long-term health status of the infant remains unknown, but the consensus is that initial acquisition of the maternal vaginal microbiome is beneficial to the infant and may protect against some diseases and conditions, including allergy, obesity, and asthma. This hypothesis was tested in a recent pilot study aimed to seed microbiome of neonates born by caesarian section with the maternal vaginal community [11]. This involved incubating a gauze in the mothers vagina for one hour prior to cesarean delivery; immediately following birth the gauze was rubbed in the neonate's mouth, face, and body. Although the cohort was small (*n* = 4), in all cases where the cesarean delivery babies were exposed to vaginal fluids, the microbiome more closely represented the vaginal community and vaginally delivered infants, compared to other cesarean delivered infants [11]. Providing the mother has been screened for group B *Streptococcus*, such seeding techniques are unlikely to have negative consequences for neonates, but further work is needed to confirm the benefit of such procedures. Indeed, recent evidence suggests the effect of birth mode is lost in the first 6 weeks of life [12], but it is possible that the immune development in the first weeks of have long-term consequences [13].

The profound effects of birth mode on the neonatal gut microbiome are generally lost over time, typically within the first 30 days of life, with other factors overriding this initial microbiome composition.

GESTATIONAL AGE

The gestational age refers to the number of weeks from the last menstrual cycle of the mother to the birth of the infant. A normal pregnancy, where an infant is delivered full term, will last 38 to 42 weeks. Babies born at less than 37 weeks gestation are considered preterm. However, most research exploring preterm infants is focused on very low birth weight (VLBW; <1500 g) infants born at less than 32 weeks gestation. The management and care of preterm infants has resulted in a relatively high survival rate compared to that two decades ago. Preterm neonates generally have a unique and unnatural beginning to life, where they are immediately housed in incubators with limited environmental exposure. In comparison, term neonates do not generally have restricted environmental contact and will be exposed to a much greater number of microbes in the days immediately following birth. Given the importance of the developing neonatal microbiome, a number of studies have been performed to characterize the preterm gut microbiome and determine how this differs from that of full-term infants.

The overall acquisition of a stable and diverse gut microbiome is delayed in preterm infants [14], with preterm infants expected to have lower diversity in the first month of life compared to term infants. In general, a low diversity is perceived as a negative to an individual's overall health status. Perhaps more alarming for preterm infants are the specific bacterial genera that are delayed or missing from the gut microbiome. One or a combination of *Staphylococcus*, *Klebsiella*, *Enterococcus*, and *Escherichia* typically dominate the preterm gut microbiome [15]. This is in contrast to term infants, who have a much more even community and abundant genera, such as *Bifidobacterium*, *Clostridium*, *Lactobacillus*, and *Bacteroides*. Importantly, "beneficial" bacteria, which are thought to confer positive influences on the host's well-being, are more abundant in term infants. Specifically, preterm infants show a delayed and reduced development of bifidobacteria and lactobacilli [16].

Interestingly, a recent study by La Rosa et al. [17] showed that the preterm gut microbiome underwent a patterned development per increasing postconceptional age, once the cohort was stratified by gestational age, with the slowest microbial assembly in the most preterm infants. The study suggested that antibiotics, feeding, and single- versus open-room housing did not significantly affect overall gut microbiome profiles. Although this study has yet to be replicated or validated, these findings suggest that assembly of the preterm gut microbiome is nonrandom, and the pace of assembly is dependent on the degree of prematurity (gestational age).

Coupled to the increased prevalence of antibiotics in the preterm population, it is important to consider that limited environmental exposure is likely to be the primary reason for much of the contrast between preterm and term neonates. For preterm neonates, discharge from the neonatal intensive care unit (NICU) represents a significant change in microbial exposure, comparable in some respects to that of term infants following birth. It was recently shown that following discharge from the NICU, preterm infants are able to establish a gut microbiome comparable to healthy term infants in both overall diversity and the individual bacterial genera that reside in the gut [18]. However, the point at which the preterm gut microbiome converges with that of term infants remains elusive (Figure 19.2). The same study also showed that despite increasing bacterial variability between infants over time, the functional metabolic profiles of each infant converged. Such findings suggest that, in the absence of a specific pathogen, the exact identities of the bacteria are of less importance than the capacity of the gut microbiome to have the necessary overall function.

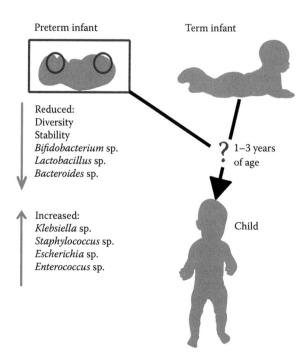

FIGURE 19.2 The development of the gut microbiome in preterm compared to term infants. Factors on the left describe what is reduced or increased in preterm populations. The question mark represents the unknown time at which the preterm microbiome converges to that of term infants.

FACTORS AFFECTING THE INFANT GUT MICROBIOME: DAY 31 TO 364

The initial seeding of the microbiome in the prenatal stages and through birth represents an important period in the colonization of the newborn infant. However, at this stage the deliberate manipulation of the gut microbiome is generally not feasible, with little scope for clinical intervention. In other words, while a range of variables will affect the initial exposure to microorganisms, such variables are often difficult to control (e.g., birth mode and gestation). While the long-term influence of such variables is important to understand, clinical trials are typically focused on variables that can be controlled. From a clinical perspective, it is of paramount importance to comprehensively understand how a single or combination of interventions will influence the developing gut microbiome in the neonate and infant stages of life.

Following birth and over the first months of life, the gut microbiome is chaotic and highly dynamic, with large shifts in the overall bacterial community potentially occurring from day to day [15,19]. This early stage of life represents a key window for the manipulation and long-term establishment of a stable and diverse community, which is typically regarded as stable from childhood and throughout adulthood. The gold standard for establishing a healthy gut microbiome is thought to result from full term, vaginally delivered, and breastfed (FTVDBF) infants. During the infant stage, FTVDBF infants tend to have relatively high abundances of *Bifidobacterium*, *Bacteroides*, *Clostridium*, and *Atopobium*. The percentage of facultative to strict anaerobes in the FTVDBF infant gut microbiome at day 90 is 60% to 40%, respectively [20].

This section portrays this key phase of life, focusing on variables that can generally be controlled, such as antibiotics and supplementation. This represents an exciting area of research, and clinical trials aimed to more comprehensively understand the mechanisms of these interventions offer important avenues for understanding the role of the gut microbiome in programming long-term host health.

FEEDING

Given the importance of seeding the microbiome with beneficial bacteria, it is not surprising that breast milk contains a range of beneficial microorganisms, including lactobacilli and bifidibacteria [21]. Historically assumed to be sterile, more than 700 different bacterial species have been detected in human milk, commonly including species from the genera *Weissella, Leuconostoc, Staphylococcus, Streptococcus,* and *Lactococcus* [5]. The importance of a feeding regime for the early development of the gut microbiome became apparent when researchers realized that the bacteria in stool did not resemble the mode of delivery after the initial week of life. As discussed, lactobacillus represents a major genus of the vaginal microbiome, but this genus is relatively low in abundance in the gut microbiome. By analyzing bacterial taxa at the strain level, it has been shown that specific bacterial strains transferred from maternal breast milk successfully colonize the neonatal gut [22].

Although relatively conserved compared to other sample types (e.g., stool), the microbial community in breast milk is different between different mothers and can change over time within the same individual [23]. Hunt et al. [23] found the most abundant genera of bacteria in human milk from American mothers were *Streptococcus, Staphylococcus, Serratia,* and *Corynebacteria,* and reported notably lower levels of lactobacilli and bifidiobacteria compared to European cohorts. Even within the same geographical location, differences in the number and diversity of bacteria in the breast milk microbiome arising from dietary influences have been reported in both human and animals. In macaques, a high-fat diet in the maternal pre- and postnatal stages influenced the developing offspring gut microbiome, most notably in the abundance of *Campylobacter* [24]. Importantly, this was not related to obesity per se, rather it was a direct effect of the high-fat diet. However, in a human population Cabrera-Rubio et al. [5] found that the weight of the mother negatively correlated with the diversity of bacteria in her offspring. The long-term influences on the gut microbiome from the epigenetic signatures of previous generations' dietary habits will be important to understand, but such data is currently elusive.

As well as the direct passage of viable microbes, human milk also confers protection against pathogens through the transmission of maternal immunoglobulin A (IgA) [25]. IgA is the first source of antibody-mediated immune protection in the neonatal gut and has been shown to prevent the translocation of bacteria as well as promote long-term gut homeostasis by regulating the gut microbiome [26]. Human milk oligosaccharides (HMOs), which have no direct nutritional value for the infant, constitute a major proportion of breast milk. The role of HMOs is to serve as a prebiotic for the successful colonization and growth of beneficial bacteria typically derived directly from the breast milk. Furthermore, breast milk contains lactoferrin, which is an iron-binding protein that reduces the availability of iron in the gut, preventing the overgrowth of potentially pathogenic iron sequestering bacteria [27]. Thus, breast milk serves as a symbiotic by providing both prebiotic and probiotic properties to the infant enteric ecosystem.

Infant formula is a mix of lactose, fat, HMOs, and protein, and is manufactured to complement or replace human milk when exclusive breastfeeding is not possible or opted against. The effects of maternal breast milk, compared to infant formula, on the neonatal microbiome are summarized in Table 19.1. The influence of breast milk or formula on the diversity remains inconsistent, with both reported to increase bacterial diversity in the gut microbiome of the neonate [28,29].

Interestingly, after the introduction of complementary foods, the gut microbiota of breastfed infants changes to reflect that of formula-fed infants. This occurs as the result of a significant increase in the abundance of the genera *Enterococcus* and *Enterobacter*, and the appearance of facultative and obligate, particularly anaerobes *Bacteroides*, *Clostridium*, and other anaerobic Streptococci [30]. Furthermore, transition to family (table) foods has been shown to be a key event that drives infant gut microbiome composition, eclipsing previously important influences such as birth mode, gestational age, antibiotic use, and maternal weight [31].

Receipt of formula has been associated with infectious diseases, inflammatory mediated conditions, and long-term allergy risk. For example, feeding with an exclusive human milk-based diet significantly reduced the incidence of necrotizing enterocolitis (NEC) and late onset sepsis (LOS), resulting in significantly lower mortality compared to infants fed a bovine-based diet and formula [32]. This might reflect the lower numbers of *Bifidobacterium* common to formula-fed infants [33]. Thus, an active area of research is currently focused on determining the best alternative to breast milk, with promise shown by donor milk, allowing an exclusive human breast milk diet.

TABLE 19.1

Comparison of Breast Milk and Formula on the Neonatal Microbiome

	Breast Milk	Infant Formula
Abundant bacterial genera in the neonatal gut microbiome following feed	*Lactobacillus*	*Bacteroides*
	Bifidobacterium	*Clostridium*
	Weissella	*Streptococcus*
	Leuconostoc	*Enterobacteria*
	Streptococcus	*Veillonella*
Notable components which influence the neonatal gut microbiome	HMOs	Lactose
	Immunoglobulins	Fat
	Lactoferrin	
Risk of neonatal disease	Reduced	Increased

DISEASE AND ANTIBIOTICS

The number of infant deaths in the United States is around 6 infants for every 1,000 births. A major cause of infant death is prematurity, and the increasing number of infants surviving preterm birth has presented new problems. Despite increased and novel research efforts [34], little improvement has been made in the rate of mortality due to preterm disease such as NEC and sepsis, which account for >20% of all preterm mortality [35]. Significant morbidity associated with prematurity, such as delayed growth and cognitive and neurological functioning, are also common to infants delivered preterm. To reduce the risk of infections, vulnerable neonates may undergo a short course of antibiotics. Concerns regarding antibiotic treatment related to the gut microbiota include the spread of antibiotic resistance among pathogens, and that alteration of the microbiota will interfere with human–microbe interactions that are fundamental to human development.

Antibiotics have been demonstrated to play a significant and long-term role in altering the bacterial composition within the gut microbiota. Antibiotics are commonly prescribed to neonates, particularly preterm infants, where standard practice is to administer antibiotics for 48 hours following birth, and longer if infection is suspected. This is likely to result in the delayed development of a diverse gut microbiota in preterm infants. Understanding the direct impact of antibiotic administration on the developing gut microbiome is extremely challenging. The type of antibiotics, the combination used, their dosage, and length of time of administration vary hugely between individual infants, reflecting the patient's needs and the preferences and experiences of the clinicians treating the individual. These variables and issues with robust sample collection mean that monitoring the exact effects on the gut microbiota *in vivo* is extremely difficult. Nonetheless, when the role of ceftriaxone was studied in term breastfed infants, a decreased count of total bacteria was observed, particularly *Enterobacteriaceae*, enterococci, and lactobacilli [36]. Ceftriaxone was also shown to cause a disappearance of *Bifidobacterium* spp. with a preservation of potentially pathogenic *Streptococcus* spp. and *Staphylococcus* spp. The total bacterial load was also significantly reduced following antibiotic administration for NEC in preterm infants, however, by 2 weeks postdiagnosis the load was comparable to that of controls [37].

SUPPLEMENTATION

As appreciation for the significance of the gut microbiome in health and disease has risen in recent years, so too has the interest in manipulating its development with probiotics and prebiotics. Probiotics are live microorganisms, which can colonize the host and provide benefits to health, both directly (e.g., producing a substrate for the host) and indirectly (e.g., preventing overgrowth of potential pathogens). Prebiotics are substrates that promote the growth of beneficial bacteria. Symbiotics are a combination of prebiotics and probiotics.

The neonatal and infant stages represent the most important windows for establishing beneficial microbes in the gastrointestinal tract. This is currently a highly active area of research, with substantial pharmaceutical investment. Although further work is needed, it is not unreasonable to suggest that in the next decade the use of probiotics in early life will be routine practice in the most vulnerable populations. The use of probiotics and prebiotics in preterm infants seems particularly important for establishing long-term health, although recent large trials have produced opposing results. An Australian study of 1,099 preterm infants showed a significant reduction in NEC (but not sepsis) resulting from a probiotic containing *Bifidobacterium infantis*, *Bifidobacterium lactis*, and *Streptococcus thermophiles* [38]. Conversely, a UK study of 1,310 preterm infants using a probiotic containing a single strain of *Bifidobacterium infantis* found no significant effects on the incidence of neonatal disease [39]. While these two studies represent large observational findings, more focused mechanistic-based studies are ongoing to systematically determine the optimal combination of microbes, as well as the dosage and prebiotic substrates. For instance, recent work has shown that administration of *L. reuteri* competitively excludes enteropathogenic *E. coli*, a major cause of diarrheal infant death, from mucus and epithelium intestinal models; however, the efficacy of *L. reuteri* was strain specific [40].

While results are still emerging, it has been shown that probiotics can profoundly shift the gut microbiome in neonates, unsurprisingly due to much larger relative abundances of the probiotic species in the probiotic recipients [16,41–43]. The evenness of the community also tends to increase, indicating increased ecological stability. A recent novel finding explored preterm infants who received probiotics matched to controls through the NICU and postdischarge, showing that the *Bifidobacterium* strain used in the probiotic colonized long-term at three times the expected level, but the *Lactobacillus* strain used was only detected while probiotics were being administered (suggesting unsuccessful colonization) [16]. Such studies highlight important issues around the selection of the most appropriate probiotic to use for successful long-term colonization. It should also be noted that probiotics have been trialed for the prevention of allergic disease in mothers and infants, with an apparent reduced incidence of eczema, but not asthma, allergy, or other allergic conditions [44]. Overall, probiotics are rarely shown to have negative consequences and, at worst, tend to have no overall effect.

Although less researched than probiotics, the potential for prebiotics to modulate a health gut microbiome is huge. As discussed, much of the beneficial nature of breast milk is thought to result from its prebiotic properties. However, supplementation of HMOs to the feed of 12 preterm infants did not result in the expected increase in bifidobacteria [45]. For early infancy, one of the most prominent probiotics is lactoferrin, which is known to help protect against infection and promote nutritional status [27]. Lactoferrin has been shown to specifically inhibit pathogenic bacteria in the gut microbiome by scavenging free iron to such an extent that remaining concentrations are too low to enable the growth pathogenic *E. coli* [46]. The effects of prebiotics on the overall gut microbiome structure will emerge in the coming years.

TRANSITION FROM INFANT TO CHILD: DAY 365 TO 1,000 AND BEYOND

Neonatal and, to a lesser degree, infant gut microbiome development is chaotic, dynamic, and highly individual. As discussed in the preceding sections, the temporal development of the bacterial community during this time can be influenced significantly by environmental, host-related, and clinical variables. By 1 to 3 years of age there is generally less clinical intervention, and the gut microbiome reaches a "climax community" of microbes that will generally remain stable through adulthood [47] (Figure 19.3). Although there is no specific gut microbiome in adults, compared to neonates and infants there is a notable increase in the Bacteroidetes at the expense of Proteobacteria, while Firmicutes remains relatively dominant from birth. The more adultlike gut microbiome is predominantly characterized by increased abundances of the genera *Clostridium*, *Faecalibacterium*, *Blautia*, *Ruminococcus*, *Lactobacillus* (Firmicutes), *Bacteroides*, and *Prevotella* (Bacteroidetes), as well as *Bifidobacterium* [1].

The exact age at which the adultlike community establishes is not yet known and likely varies between individuals, especially driven by nutritional status and geography, with Western populations reaching maturation by 1 year of life [48]. In a study comparing the gut microbiome of children aged 0 to 3 years between three extreme groups (American Indians from the Amazon, Rural Malawains, and urban Americans) the bacterial profiles were distinct between the groups, but the Western American children were the greatest outliers [47]. Notably, bifidobacteria still dominated the communities from all locations. Although the gut microbiome is still developing, there are many factors contributing to the microbes that colonize it. Aside from the geographical and dietary extremes discussed earlier, more subtle influences within a given population may have profound long-term effects. Having siblings results in a higher proportion of *Bifidobacterium* compared to single infants [10].

Perhaps unsurprisingly, the gut microbiomes of twins are more comparable to each other than those of other nonrelated infants, but it is unclear if this is driven by comparable environmental exposure or genetic influence [49]. Exposure to animals, such as household pets, will also shape the developing gut microbiome [50]. A recent study showed that the gut microbiome profiles of preterm infants at 1 to 3 years of age was comparable to that of term infants in terms of both diversity and the abundant bacterial genera [10]. The same study also found no profound lasting effects on the gut microbiome from a previous diagnosis of disease, number of days of antibiotics, gestational age, birth mode, and mode of delivery. These findings are further supported by a recent study, which found that the same variables do not influence the

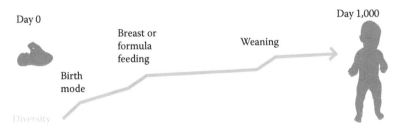

FIGURE 19.3 The schematic shows that diversity increases from day 0 to day 1,000 of life. Birth mode, initial feeds, and weaning are major events in which the diversity increases more rapidly.

gut microbiome long term, with the additional finding that maternal weight also has no lasting effect [31].

There is currently a progressive increase in metabolic and immune-mediated diseases in infants and children for which the developing gut microbiome might have important consequences. Studies exploring the gut microbiome in allergic disease and obesity have risen in recent years, but the direct effect of the gut microbiome on these conditions is not well determined [51]. Still, it is reasonable to postulate that the gut microbiome will have important influences. For example, a reduced abundance of Bacteroides, which is common to the Western diet associated with low dietary fiber intake, was shown to predict atopic eczema [52]. Likewise for obesity, a reduced abundance of *Bifidobacterium* and an increased prevalence of *Staphylococcus* was found at 1 year of age in infants who went on to become obese at 7 years of age [53].

The neonatal and infants phases seem the most applicable for modulation of the gut microbiome. However, in certain circumstances, such as following antibiotic administration, there may be a clear rationale for seeding a beneficial gut microbiome though the use of probiotics and prebiotics. Although microbiome research has offered unprecedented information on the role of microbes in health and disease, the field as a whole is still emerging. Further work is needed to ascertain the extent to which microbiome findings are causative and the subsequent mechanisms, proving causality over existing correlations. The next decade offers huge promise for microbiome research, no more so than for the developing infant, where the first 1,000 days offers a key window to establish a beneficial population contributable to long-term health.

REFERENCES

1. Matamoros S, Gras-Leguen C, Le Vacon F et al. Development of intestinal microbiota in infants and its impact on health. *Trends Microbiol* 2013; 21:167–73.
2. Munyaka PM, Khafipour E, Ghia J-E. External influence of early childhood establishment of gut microbiota and subsequent health implications. *Front Pediatr* 2014; 2:109.
3. Aagaard K, Ma J, Antony KM et al. The placenta harbors a unique microbiome. *Sci Transl Med* 2014; 6:237ra65.
4. Grice, EA, Segre JA. The skin microbiome. *Nat Rev Microbiol* 2013; 9:244–53.
5. Cabrera-Rubio R, Collado MC, Laitinen K et al. The human milk microbiome changes over lactation and is shaped by maternal weight and mode of delivery. *Am J Clin Nutr* 2012; 96:544–51.
6. Stewart CJ, Marrs ECL, Nelson A et al. Development of the preterm gut microbiome in twins at risk of necrotising enterocolitis and sepsis. *PLoS One* 2013; 8:e73465.
7. Dominguez-Bello MG, Costello EK, Contreras M et al. Delivery mode shapes the acquisition and structure of the initial microbiota across multiple body habitats in newborns. *Proc Natl Acad Sci USA* 2010; 107:11971–5.
8. DiGiulio DB, Callahan BJ, McMurdie PJ et al. Temporal and spatial variation of the human microbiota during pregnancy. *Proc Natl Acad Sci* 2015; 112:11060–5.
9. Thompson-Chagoyán OC, Maldonado J, Gil A. Colonization and impact of disease and other factors on intestinal microbiota. *Dig Dis Sci* 2007; 52:2069–77.
10. Penders J, Thijs C, Vink C et al. Factors influencing the composition of the intestinal microbiota in early infancy. *Pediatrics* 2006; 118:511–21.

11. Dominguez-Bello MG, De Jesus-Laboy KM, Shen N et al. Partial restoration of the microbiota of cesarean-born infants via vaginal microbial transfer. *Nat Med* 2016; 1–4.

12. Chu DM, Ma J, Prince AL et al. Maturation of the infant microbiome community structure and function across multiple body sites and in relation to mode of delivery. *Nat Med* 2017; 23:314–26.

13. Aagaard K, Stewart CJ, Chu D. Una destinatio, viae diversae: Does exposure to the vaginal microbiota confer health benefits to the infant, and does lack of exposure confer disease risk? *EMBO Rep* 2016; 17:1679–1684.

14. Stewart CJ, Marrs ECL, Magorrian S et al. The preterm gut microbiota: Changes associated with necrotizing enterocolitis and infection. *Acta Paediatr* 2012; 101(11):1121–7.

15. Stewart CJ, Embleton ND, Marrs ECL et al. Temporal bacterial and metabolic development of the preterm gut reveals specific signatures in health and disease. *Microbiome* 2016; 4:67.

16. Abdulkadir B, Nelson A, Skeath T et al. Routine use of probiotics in preterm infants: Longitudinal impact on the microbiome and metabolome. *Neonatology* 2016; 109:239–47.

17. La Rosa PS, Warner BB, Zhou Y et al. Patterned progression of bacterial populations in the premature infant gut. *Proc Natl Acad Sci USA* 2014; 111(34):12522–7.

18. Stewart CJ, Skeath T, Nelson A et al. Preterm gut microbiota and metabolome following discharge from intensive care. *Sci Rep* 2015; 5:17141.

19. Stewart CJ, Nelson A, Scribbins D et al. Bacterial and fungal viability in the preterm gut: NEC and sepsis. *Arch Dis Child Fetal Neonatal Ed* 2013; 98:F298–303.

20. Arboleya S, Solís G, Fernández N et al. Facultative to strict anaerobes ratio in the preterm infant microbiota: A target for intervention? *Gut Microbes* 2012; 3(6):583–8.

21. Collado MC, Delgado S, Maldonado A et al. Assessment of the bacterial diversity of breast milk of healthy women by quantitative real-time PCR. *Lett Appl Microbiol* 2009; 48:523–8.

22. Martín V, Maldonado-Barragán A, Moles L et al. Sharing of bacterial strains between breast milk and infant feces. *J Hum Lact* 2012; 28(1):36–44.

23. Hunt KM, Foster JA, Forney LJ et al. Characterization of the diversity and temporal stability of bacterial communities in human milk. *PLoS One* 2011; 6:1–8.

24. Ma J, Prince AL, Bader D et al. High-fat maternal diet during pregnancy persistently alters the offspring microbiome in a primate model. *Nat Commun* 2014; 5:3889.

25. Donovan SM, Wang M, Li M et al. Host-microbe interactions in the neonatal intestine: Role of human milk oligosaccharides. *Adv Nutr* 2012, 3(3):450S–5S.

26. Rogier EW, Frantz AL, Bruno MEC et al. Secretory antibodies in breast milk promote long-term intestinal homeostasis by regulating the gut microbiota and host gene expression. *Proc Natl Acad Sci USA* 2014; 111:3074–9.

27. Embleton ND, Berrington JE, McGuire W et al. Lactoferrin: Antimicrobial activity and therapeutic potential. *Semin Fetal Neonatal Med* 2013; pii: S1744-165X(13)00009-7.

28. Praveen P, Jordan F, Priami C et al. The role of breast-feeding in infant immune system: A systems perspective on the intestinal microbiome. *Microbiome* 2015; 3:41.

29. Schwartz S, Friedberg I, Ivanov IV et al. A metagenomic study of diet-dependent interaction between gut microbiota and host in infants reveals differences in immune response. *Genome Biol* 2012; 13:r32.

30. Adlerberth I, Wold AE. Establishment of the gut microbiota in Western infants. *Acta Paediatr* 2009; 98(2):229–38.

31. Laursen MF, Andersen LBB, Michaelsen KF et al. *Infant gut microbiota development is driven by transition to family foods independent* of maternal obesity. *mSphere* 2016; 1(1):e00069-15.

32. Hair AB, Peluso AM, Hawthorne KM et al. Beyond necrotizing enterocolitis prevention: Improving outcomes with an exclusive human milk-based diet. *Breastfeed Med* 11(2):70–4.
33. Roger LC, Costabile A, Holland DT et al. Examination of faecal Bifidobacterium populations in breast- and formula-fed infants during the first 18 months of life. *Microbiology* 2010; 156:3329–41.
34. Stewart CJ, Nelson A, Treumann A et al. Metabolomic and proteomic analysis of serum from preterm infants with necrotising enterocolitis and late onset sepsis. *Pediatr Res* 2015; 79:425–31.
35. Berrington J, Stewart CJ, Embleton N et al. Gut microbiota in preterm infants: Assessment and relevance to health and disease. *Arch Dis Child Fetal Neonatal Ed* 2013; 98:F286–90.
36. Savino F, Roana J, Mandras N et al. Faecal microbiota in breast-fed infants after antibiotic therapy. *Acta Paediatr* 2011; 100:75–8.
37. Abdulkadir B, Nelson A, Skeath T et al. Stool bacterial load in preterm infants with necrotising enterocolitis. *Early Hum Dev* 2016; 95:1–2.
38. Jacobs SE, Tobin JM, Opie GF et al. Probiotic effects on late-onset sepsis in very preterm infants: A randomized controlled trial. *Pediatrics* 2013; 132:1055–62.
39. Costeloe K, Hardy P, Juszczak E et al. Bifidobacterium breve BBG-001 in very preterm infants: A randomised controlled phase 3 trial. *Lancet* 2016; 387:649–60.
40. Walsham ADS, MacKenzie DA, Cook V et al. Lactobacillus reuteri inhibition of enteropathogenic *Escherichia coli* adherence to human intestinal epithelium. *Front Microbiol* 2016; 7:1–10.
41. Cox MJ, Huang YJ, Fujimura KE et al. Lactobacillus casei abundance is associated with profound shifts in the infant gut microbiome. *PLoS One* 2010; 5:e8745.
42. Eisenhauer N, Scheu S, Jousset A. Bacterial diversity stabilizes community productivity. *PLoS One* 2012; 7:e34517.
43. Preidis GA, Saulnier DM, Blutt SE et al. Probiotics stimulate enterocyte migration and microbial diversity in the neonatal mouse intestine. *FASEB J* 2012; 26:1960–9.
44. Bridgman SL, Kozyrskyj AL, Scott JA et al. Gut microbiota and allergic disease in children. *Ann Allergy Asthma Immunol* 2016; 116(2):99–105.
45. Stenger MR, Reber KM, Giannone PJ et al. Probiotics and prebiotics for the prevention of necrotizing enterocolitis. *Curr Infect Dis Rep* 2011; 13:13–20.
46. Bullen JJ, Rogers HJ, Leigh L. Iron-binding proteins in milk and resistance to *Escherichia coli* infection in infants. *Br Med J* 1972; 1:69–75.
47. Yatsunenko T, Rey FE, Manary MJ et al. Human gut microbiome viewed across age and geography. *Nature* 2012; 486:222–7.
48. Bäckhed F, Roswall J, Peng Y et al. Dynamics and stabilization of the human gut microbiome during the first year of life. *Cell Host Microbe* 2015; 17:690–703.
49. Turnbaugh PJ, Hamady M, Yatsunenko T et al. A core gut microbiome in obese and lean twins. *Nature* 2009; 457:480–4.
50. Song SJ, Lauber C, Costello EK et al. Cohabiting family members share microbiota with one another and with their dogs. *Elife* 2013; 2:e00458.
51. Rodríguez JM, Murphy K, Stanton C et al. The composition of the gut microbiota throughout life, with an emphasis on early life. *Microb Ecol Health Dis* 2015; 26:26050.
52. Abrahamsson TR, Jakobsson HE, Andersson AF et al. Low diversity of the gut microbiota in infants with atopic eczema. *J Allergy Clin Immunol* 2012; 129(2):434–40, 440. e1-2.
53. Kalliomäki M, Collado M. Early differences in fecal microbiota composition in children may predict overweight. *Am J Clin Nutr* 2008; 87(3):534–8.

20 Impact of Different Exposures, Including Environmental Enteropathies, on Gut Flora and Integrity

Fayrouz A. Sakr Ashour

CONTENTS

INTRODUCTION

The gastrointestinal tract is one of the primary lines of defense for the body against pathogens. In a state of health, the gut is able to respond to a pathogen or allergen through various mechanisms. The gut microbiota enhance these defense mechanisms, improve intestinal immunity, aid in nutrient and drug metabolism, and synthesize some nutrients [1]. A disturbance in any of the finely tuned mechanical and immunological components of the intestinal tract, or the microbiota, is likely to disrupt these functions [2]. Prominent causes of gut disruption include inflammation, infection, chemotherapy, radiation, parental nutrition, pharmaceuticals (e.g., antibiotic administration), various diseases, and environmental exposures. Variations can result from diet, gastric acidity, intestinal motility, overall immune status, and general hygiene [3]. Infants and young children are particularly vulnerable to these disruptions because of their developing immune system, and rapid growth and development phase [4]. This chapter will discuss some of the important factors that affect the gut flora, their possible impact on health outcomes, and the remaining gaps in our knowledge. Environmental enteropathy (EE), as one of the most prominent conditions affecting the gut flora and function, will be addressed in some detail. This chapter will describe the epidemiology and pathophysiology of EE, its health implications, diagnostic methods, and trials for treatment. Promising avenues for new diagnostic tools and interventions will be highlighted.

EPIDEMIOLOGY

Tropical enteropathy is a subclinical disorder seen in the gastrointestinal tract that affects millions of individuals, particularly those living in low-income countries [5]. Tropical enteropathy was first described in adults, both "natives" and "Westerners," living in the tropics in Africa, Asia, and South America, from which the disorder obtained its name [6–9]. It was absent in residents of temperate countries, and resolved for those who moved or relocated to Western countries [10,11]. Residents of semitropical regions, in affluent communities with good sanitation, had a similar intestinal morphology to those of temperate regions. Ethnicity was not an important determinant of the differences seen in gut function, with a greater resemblance found between those residing in the same region, as opposed to those from a similar country of origin. People with a higher socioeconomic status were found to have a better intestinal absorptive capacity [12]. This suggested that the quality of the environment played a more important role than the climate in the pathogenesis, and so the name was expanded from *tropical enteropathy* to *environmental enteropathy*, with *environmental* to more widely mean "acquired" [13]. A more recent term, *environmental enteropathic dysfunction* (EED), is being used to encompass the functional effects of enteropathy.

Environmental enteric dysfunction occurs after birth [14], and many of the histopathological findings accompany cases of malnutrition [15]. Unlike tropical sprue, which is characterized by diarrhea, steatorrhea, and fatigue, EED is largely asymptomatic. Disruption of the intestinal functions seen in EED could be the result of bacterial overgrowth, repeated infection from various pathogens (bacterial, viral, or protozoal), water contamination, environmental toxins, food allergens, and/or malnutrition, yet the exact etiology remains unknown [15]. It is likely that a combination of these rather than a single factor are responsible EED. For example, while bacterial infection is endemic in many of these environments, it has only been associated with enteropathy in a few cases [16]. Food and water contamination, and poor sanitation have been proposed as important underlying factors [17].

DEFINITION

Environmental enteropathy is defined as a subclinical disorder resulting in structural and functional changes to the small intestine. The morphologic changes are proposed to occur through a T-cell mediated process characterized by a disruption in the balance between the proinflammatory (e.g., tumor necrosis factor alpha and interferon gamma) and regulatory cytokine-producing cells (e.g., transforming growth factor beta) [18]. The alterations include villous shortening and atrophy, crypt hyperplasia, and inflammatory cell infiltrate (e.g., lymphocytes and plasma cells) in the epithelium and lamina propria [16] (Figure 20.1). Gut function is consequently impaired. The damage to epithelial cells results in a loss of essential enzymes, and the decrease in the functional surface area of the intestinal epithelium causes maldigestion and malabsorption. An increase in gut permeability occurs due to the epithelial disruption leading to microbial translocation. This translocation into the systemic circulation consequently causes a state of chronic inflammation [16,19]. and altered intestinal resistance. The resulting in lowered immunity, increased susceptibility to infection, and loss of appetite, propagates the inflammation and malnutrition cycle (Figure 20.2) [20,21]. Continuous exposure to the enteric pathogens results in a state of perpetual hyperstimulation of the mucosal immune system. This is likely

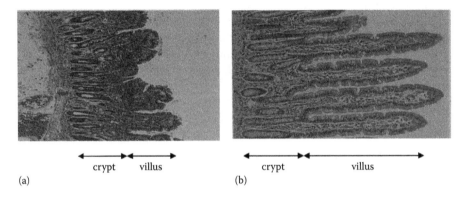

FIGURE 20.1   Histological changes in (a) mucosa of children with EED show villous atrophy, crypt hyperplasia (crypt to villous ratio < 1), and increased inflammatory cells in lamina propria compared to (b) normal intestinal mucosa. (Reproduced from McKay S, Gaudier E, Campbell DI et al., *Int Health* 2010, 2:172–80. With permission.)

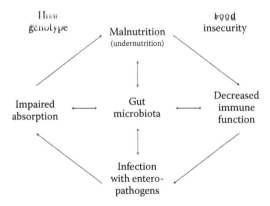

FIGURE 20.2   The relationship between undernutrition, the gut microbiota, the immune system, and environmental enteropathy. (Reproduced from Kau AL, Ahern PP, Griffin NW et al., *Nature* 2011, 474:327–36. With permission.)

why oral vaccines, such as polio and rotavirus vaccines, are less effective in children with EED [22].

The well-established role of zinc in the immune response has prompted interest in examining its role in EED [23]. It is likely that the relationship between zinc deficiency and EED is bidirectional. As a result of the decreased functional surface area of the intestine in EED, zinc homeostasis is impaired. There is a decrease in dietary zinc absorption and a failure to conserve endogenous zinc [24]. Additionally, zinc deficiency contributes to EED by disrupting the intestinal barrier function, inducing intestinal inflammation, increasing intestinal permeability, and therefore susceptibility to infections, including EED [24,25]. Zinc deficiency in children with EED also increases the burden of illness by decreasing absorption of other nutrients, impairing adaptive immune function, and aggravating the state of chronic systemic inflammation [2].

The emerging science of nutritional metabolomics carries the promise of better defining the pathways involved in enteropathy. In a recent targeted metabolomics study, researchers found alterations in serum metabolites among enteropathic children, including lower serum phosphatidylcholines, sphingomyelins, tryptophan, and higher serum glutamate, histidine, serine, taurine, serotonin, and acetylcarnitine, compared to healthy children [26]. These metabolites are involved in key roles in differentiation, growth and development, gut health, and energy metabolism. Future trials are necessary to better elucidate the role these metabolites have in the pathological pathway of EED.

HEALTH IMPLICATIONS

If EED is asymptomatic, why does it matter? Although initially, EED is asymptomatic, the bacterial proliferation, inflammatory infiltration, and changes in absorptive functions indicate that there is an underlying state of ill health [27]. Combined with recurrent diarrhea and other infections, and poor living conditions that threaten adequate nutrient intake and sanitary conditions, the negative impacts on child growth and development seem inevitable. Superadded diarrhea may result from bacterial overgrowth or lactose intolerance due to the loss of lactase from intestinal brush border damage, with subsequent dehydration and acidosis if not promptly treated [28].

Many of the risk factors associated with EED, such as poverty, water contamination, and unsanitary conditions have also been proposed to impact early childhood development [17]. The state of chronic inflammation, and altered microbiota and immune response makes infants particularly prone to repeated infections [20].

Stunting, a simple proxy for healthy child growth, is an outcome variable of EED [13]. The exact mechanisms through which EED contributes to the pathogenesis of growth faltering are not fully understood. Most evidence stems from observational studies, but the postulated mechanisms include intestinal dysfunction and disturbance of the gut environment, a concurrent state of malnutrition, and intestinal and systemic inflammatory response [18,29]. Bacterial translocation, which occurs with EED, was also found to be positively associated with growth failure [18]. Future studies are needed to understand the exact relationship between EED and growth failure. Studies should be initiated early in life (at birth or during the postnatal period) to determine whether EED is a cause of, a result of, or is aggravated by, malnutrition [30].

DIAGNOSIS

A definitive diagnosis of EED-altered intestinal architecture requires an intestinal biopsy, which was first developed by Crosby–Kugler [31]. However, in infants and young children who suffer no clear clinical symptoms, this would obviously be a measure that is neither feasible nor acceptable, especially in resource poor settings [30]. Diagnosing EED using an intestinal biopsy would also bring about other challenges, such as practicality, safety, cost, uncertainties regarding the number of samples required, and risk of sampling error [32]. Several other measures have been proposed to serve as a proxy [33,34], yet no single validated and reproducible diagnostic biomarker exists [30].

The dual absorption test is the most commonly used measure in the field. It allows the examination of both the absorptive capacity and epithelial integrity/gut barrier function [32], with lactulose to mannitol (or rhaminose) ratio most commonly used. Lactulose is a disaccharide which, in the case of a normal intestinal lumen, is nonabsorbable, and mannitol is a monosaccharide that is readily absorbed through a normal mucosa via a passive transcellular route not requiring a transporter, but not metabolized in the body. Both sugars are filtered through glomeruli and not reabsorbed, and hence are measurable in urine. Measuring the concentration of the monosaccharide alone is used to assess the absorptive capacity of the intestine; disaccharide alone is used to assess intercellular tight junctions. Therefore, in enteropathy there will be a decrease in mannitol and an increase in lactulose absorption, as a result of mucosal cell injury and the disruption of tight epithelial junctions, respectively [35]. There is no consensus on the diagnostic cutoff for lactulose:mannitol (L:M) ratios, which may be the result of the variations in many test parameters, as discussed later, but 0.12 is most frequently utilized in the literature [32]. In addition to varying test parameters, it may be unclear how to identify "normal" in conditions where adaptive and pathological conditions may be very close. A 5-hour urine collection period is commonly the standard but is often not practical in field situations [32]. The doses of the sugars used in the test should be adjusted according to body weight; otherwise, side effects such as diarrhea and vomiting can occur. As gastric emptying could affect absorption kinetics, standard fasting should be followed. A reverse solvent drag may occur as a result of this hyperosmolar solution, which could be remedied by administering the sugars with a liquid meal [36].

A great deal of heterogeneity exists with the use of this biomarker, which is the result of many factors. Controllable factors include differences in inclusion/exclusion criteria, pretest requirements, sugar dosage, posttesting collection methods and times, detection techniques/methods for assessing analyst concentrations, and displaying and interpreting results [30,37] (Box 20.1). These factors may be improved and standardized to allow for better reproducibility of the L:M ratio diagnostic test. Bacteriuria may also alter test results by lowering sugar concentrations in urine [38]. The sugar loads can also disrupt intestinal permeability, giving false results [29].

What defines normal test values have been developed from "healthy" children in Western countries; these values should be validated against ones obtained from healthy children living in the same regions as those suffering from EED. Other non-controllable influences may be the result of different etiological factors, geographic settings, and variations in individual response [39]. Although little can be done to change the latter, defining the parameters for possible causes of heterogeneity accurately over time may provide some useful patterns. Noteworthy, a reduced lactulose and mannitol excretion can occur concurrently giving an unchanged ratio; hence, there is value in reporting the sugar absorption test results in addition to the ratio [40]. There is certainly some merit in using the L:M test as a measure of gut dysfunction, including its safety, and longstanding use in the literature [32].

BOX 20.1   SOURCES OF HETEROGENEITY IN LACTULOSE:MANNITOL (L:M) RATIO TEST

- Inclusion/exclusion criteria of participants (e.g., breastfeeding, HIV positive, severe malnutrition, suffering other illnesses)
- Pretest requirements (e.g., fasting versus nonfasting, fasting duration)
- Sugar load (i.e., not all studies dose sugar load to body size)
- Posttesting collection methods and times
- Detection techniques/methods for assessing analyst concentrations (e.g., high-performance liquid chromatography–mass spectrometry)
- Displaying results (e.g., arithmetic versus geometric mean)
- Interpreting results (i.e., lack of cutoff standardization)

Citrulline, a nonprotein amino acid produced by enterocytes, has been proposed as a potential biomarker for EED but has not been formally examined [13]. Glutamine and derivative amino acids are the main precursors of citrulline, which is almost exclusively produced from bowel epithelial cells in the upper and middle portions of the intestinal villi [41]. Loss of the small bowel epithelial cell mass has shown decreased circulating levels of citrulline in celiac and Crohn's disease, and intestinal dysfunction [42,43]. Dietary citrulline is only significantly found in watermelon and is thus not affected by exogenous sources, since plasma citrulline levels are reflective of *de novo* gut epithelial cell synthesis [44]. Low citrulline concentrations are also correlated with hypoalbuminemia and anemia [45]. More research is needed to explore the potential of plasma citrulline as a biomarker for the functional absorptive capacity of the small intestine in children.

Fecal biomarkers reg1B, alpha-1-antitrypsin, myeloperoxidase, calprotectin, and neopterin are biomarkers of enteric inflammation and have been more commonly used in combination to assess EED. Their low cost and feasibility, especially in children, has been appealing, although they are nonspecific and have not been systematically evaluated. However, they do hold promise in terms of their use in combination with other biomarkers. It has been suggested that a composite score should be set up, providing a set of diagnostic criteria consisting of a number of parameters for the assessment of inflammation, immune activation, mucosal permeability, and gut absorptive function. This "enteropathic index" should be developed in consideration of the different purposes it may need to serve such as population screening, individual diagnosis, and clinical management. It should be noted that these parameters should also be evaluated for their potential to serve as predictors of subsequent stunting, risk of diarrhea, and vaccine responsiveness [13,43,46].

TREATMENT

Informed trials for the treatment of EED have been increasing over the past decade, using a number of interventions, either separate or in combination, to strengthen immune defenses, combat the different proposed etiologies, and/or remedy/reverse the negative health outcomes (Table 20.1). Yet, to date, none of these have proven successful in the treatment of EED. The structural and functional changes of EED— disrupted mucosal integrity, impaired immune function, and compromised digestion and absorption—are also thought to impact the success of malnutrition interventions and vaccine programs [22,47].

Many of the intervention trials have used surrogate outcome measures such as stunting and child development, and have generally failed to improve these outcomes. Although this is the ultimate goal of EED control, other factors contribute to these outcomes, making the failure or success of treatment not EED-specific, and again emphasizes the lack of knowledge with respect to the exact pathophysiology of enteropathy. This could explain why a handwashing trial in Pakistan was able to reduce diarrhea, while there was no effect on stunting [48]. Pilot interventions being evaluated to reduce enteropathy include the administration of antibiotics, single or multiple micronutrient supplementation, and the improvement of water and sanitation. However, if children globally received interventions of vitamin A and zinc supplementation, balanced energy protein supplementation, complementary

TABLE 20.1
Treatment Targets for Environmental Enteropathic Dysfunction

	Treatment Targets	
Combat Proposed Etiologies	**Control Inflammation/ Strengthen Immune System**	**Minimize Health Implications**
Safe and clean water supply	Probiotics	Breastfeeding promotion
Maternal/caregiver education (e.g., promote handwashing)	Polyunsaturated fatty acids Vitamins A, D, and C supplementation	Multiple micronutrient supplementation Protein-energy supplementation
Antibiotics	Zinc supplementation	Vitamin A supplementation
Better hygiene practices	Glutamine supplementation	Zinc supplementation
Multiple micronutrient supplementation		
Protein-energy supplementation		

feeding, breastfeeding promotion, and prenatal micronutrient supplementation for their mothers, with 90% coverage, stunting would only decrease by less than a quarter [49]. More experimental evidence is clearly needed to delineate the different causal pathways to stunting, and outline the relative contribution of enteropathy to its pathogenesis.

Water, sanitation, and hygiene (WASH) efforts have not been able to completely protect children in the critical first 3 years of life. This may, arguably, be due to the lack of specificity of these interventions for this particular age group, or the fact that many of these interventions were focused on improved sanitation, water treatment, and maternal/caregiver handwashing [50]. Ngure et al. argue that WASH needs to encompass important vectors of soil, animal feces, and infant foods to be more effective [50].

Nutrition has been a significant player in many of the interventions geared toward the prevention, treatment, and control of EED. Effective nutrition interventions can (1) promote healing of the mucosal barrier and strengthen immunity; (2) replenish the body and reallocate key nutrients lost through malabsorption and inflammation; (3) allow for catch-up growth; (4) enhance appetite; and (5) restore the microecology of the gut [51]. Although malnourished children who took part in a 4-week nutrition feeding program showed an improvement in growth indicators and some histologic morphology, their intestinal function failed to recover. Furthermore, the mucosal architecture and growth faltering returned 1 year after follow-up, indicating that the persistence of environmental insults make short-term solutions ineffective [52]. Multiple micronutrient supplements given to children aged 1 to 3 years in Malawi showed a transient modest improvement, yet there was no change in linear growth [53]. A focus on diet-based interventions is under way using, for example, common beans and cowpeas to ameliorate the inflammatory response associated with EED, as well as to provide a more nutritious replacement for low-quality complementary foods, and one that is a good source of protein and micronutrients [54]. Improved nutrition could reduce the impact of infections on growth and development by multiple mechanisms that include (1) strengthening the immune system; (2) compensating for malabsorption and replenishing key nutrients in the body; (3) allowing for catch-up growth; (4) improving appetite; and (5) restoring healthy gut flora [51].

The benefits of short-chain fatty acids (SCFA) in the control of the inflammatory response and aiding mucosal healing provides another promising avenue for treatment. Probiotics are another avenue for treatment to correct underlying dysbiosis and to restore the ecological balance of the microbiota from the use of antibiotics [55]. Probiotics have showed some promise in restoring normal gut flora and inhibiting pathogenic overgrowth [56]. However, human trials failed to show any evidence of benefit from the administration of omega-3 fatty acids and probiotic supplements in the treatment of EED [33,55].

Human trials have shown that zinc supplementation decreased intestinal permeability and partially alleviated EED, albeit transiently [57,58]. Amino acids also play an important role in maintaining gut health, and supporting gut barrier integrity and function [59,60]. In addition, 9% to 12% of body protein is synthesized in the gut [61]. Mainly based on animal models, suggestions for supplementation with specific amino acids (e.g., arginine, glutamine, glutamate, sulfur containing amino acids and their metabolites, glycine, lycine, threonine, methionine, serine, proline, and the amino acid-derived compounds, polyamines) have been made to aid in the process of mucosal healing, and resume normal gut physiology and metabolism [62]. Evidence also supports the important role that polyamines have in restoring enterocyte function, and mucosal healing [63]. Glutamine has also been associated with improved mucosal barrier function in some animal and human studies; however, findings have not been consistent, and it has not been successful in improving growth indicators in children with EED [64–66]. It should be noted that in many low-income countries the human immunodeficiency virus (HIV) remains a huge burden, and 30% of pregnant women with HIV in low-income countries do not have access to antiretroviral therapy [67]. Mother-to-child transmission rates increase from 2%–5% to 15%– 45% in the absence of any intervention. Although the number of cases of children with HIV is declining, 134,000 new cases were registered in 2013 [68]. It is possible that EED with HIV may be a distinct pathophysiological entity, and one that requires a different approach to treatment, as seen in adults [69].

Certain pathogens have been associated with EED, such as *Citrobacter rodentium* or hookworm [70], which require the use of antihelminths and antibiotics as other routes for treatment. The administration of a broad-spectrum antibiotic to children in Malawi failed to improve gut barrier function and, in some instances, disrupted the gut microbiome balance. It could be that bacterial overgrowth is not an essential etiological factor in the pathogenesis of EED or that repeated exposures prevent a single antibiotic course from eliminating EED [52,71]. In India, the administration of azithromycin to infants showed an improvement in calprotectin and myeloperoxidase, yet the immunogenicity of the oral polio vaccine showed no improvement, suggesting that EED may not always be an important player in the immunogenicity of oral vaccines. Conversely, a single dose of albendazole offset the progression of EED in a randomized control trial among rural Malawian children [57].

In impoverished societies, EED is entrenched in the downward spiral of malnutrition and inflammation/infections. Although many well-informed efforts have been made to remedy EED, they are still speculative to a great extent, as seen by their modest, inconsistent, or short-term results. This is likely the consequence of the intertwined relationship between child health growth, development, and nutrition, with EED impacting all of these domains. Some outstanding research gaps are highlighted in Box 20.2. A better understanding of the biological mechanisms underlying EED are needed to be able to develop and evaluate intervention packages that target these links.

**BOX 20.2   HIGHLIGHTS OF THE MAIN KNOWLEDGE
GAPS IN THE UNDERSTANDING OF GUT
MICROBIOME, EED, AND OTHER EXPOSURES**

- Identify range of sequelae of EED for childhood and adulthood
- Ideal macronutrient (e.g., essential amino acids and fatty acids) and micronutrient (e.g., zinc and vitamin A) profiles for EED treatment

- Carrying out clinical studies to test/demonstrate the efficacy of the nutrient combinations for EED treatment
- Relative contribution of the proposed etiological factors to the pathogenesis of EED
- Delineate the relationship between EED and malnutrition (cause, consequence, or symbiotic)
- Establishing testing protocols and standardization of L:M testing to improve reproducibility and allow for comparability across studies (e.g., urine collection period, saccharide dosage)
- Establishing consensus on diagnostic criteria that include carefully chosen parameters of function and injury
- Discovering field-friendly, cheap, and validated biomarkers of function for screening, surveillance, and interventions
- Exploring possible early exposures of the microbiome affecting food intolerance and allergies later in childhood
- Better understanding of how the metabolic changes seen in EED impact nutrient requirements and homeostasis, including altered intake, bioavailability, utilization, and losses
- Monitoring effect of antibiotic exposure on the developing intestinal immune system
- Measuring duration of the microbial disruptions after the various exposures
- Mapping colonization pattern of infant microflora exposed to a mixture of breast milk and formula milk

OTHER EXPOSURES

A newborn slowly develops a complex and diverse microbial ecosystem that closely resembles the adult microbiota by approximately 2 years of age [72]. Recent evidence suggests that microbial colonization of the previously presumed sterile infant gut occurs *in utero*. Mothers, therefore, share much of their microbiome with the fetus *in utero*, during delivery and lactation, and possibly later, by being exposed to similar environments. The infant's microbiome is influenced by the mother's diet, health status, mode of delivery, gestational weight gain, type of feeding, genetics, and the use of antibiotics [73,74]. In vaginal delivery, the newborn likely acquires the bacteria through the mother's fecal material, which passes during birth and comes in contact via the perineal and anal area. Infants born by cesarean section have different flora than infants born vaginally, and their microbiota is somewhat similar to that of the mother's skin, likely because of contact through the abdominal incision [75]. The different bacterial colonization acquired through these delivery methods at many times persists beyond the neonatal period, and may contribute to the differential susceptibility to allergies and other diseases in later childhood [76].

Gestational age is another factor that impacts gut flora. Premature infants show a delay in the establishment of normal microbiota, which is further impacted by the use of perinatal antibiotics [77]. Given that many premature infants are in neonatal intensive care units postdelivery, hospitalization is likely a confounder in the relationship between microbiota and prematurity through the acquisition of nosocomial bacteria. Interestingly, it is not only premature babies that have a different microbiota pattern; newborns delivered at 37 weeks have also shown a different microbiota picture than those born later [78]. More research is needed to map the gut microbiome of preterm infants through early childhood to examine the potential impacts on growth, development, and overall health.

Exposure to antibiotics is another important factor that can disrupt the intestinal microbiota balance. This is especially true in early life, since the body's immune system is just beginning to develop and establish immunological tolerance [4]. Broad-spectrum antibiotics can affect 30% of the gut bacterial load, resulting in a substantial decline in taxonomic richness, diversity, and composition [79]. This may result in the overgrowth of nonsensitive organisms, the colonization of new organisms, or the development of resistance across existing strains. The magnitude, duration, nature of the disruption, and the time taken to restore gut flora will likely depend on many factors, such as the antimicrobial spectrum, and the route and duration of the administration. Infants' gastrointestinal dysbiosis can also be caused by antibiotic administration to mothers during the perinatal and postnatal periods. During pregnancy, mothers take antibiotics against infection or the premature rupture of membranes, resulting in the different microbiota composition of infants. Interestingly, the microbiota of breastfed infants showed a faster recovery, as compared to those who were formula fed [80]. The gastrointestinal tract dysbiosis also appears to alter the antiviral immune response to systemic viral infections, even under strict standards of hygiene [81].

Another source of antibiotic exposure is through the consumption of farm crops and animals that have been treated with antibiotics. This overexposure to antibiotics can disrupt many physiologic equilibria and the intestinal milieu's overall inflammatory tone [82], which has body-wide and long-term repercussions, such as obesity, metabolic syndrome, insulin resistance, altered immunity, increased susceptibility to infection, and inflammatory bowel disease [83–85].

A flora with a diet dependent pattern can be seen as early as 4 days of age, as seen by the infant microbiota's similarities to the mother's colostrum within the first few days of life [86]. A distinct bacterial intestinal colonization pattern is seen in breastfed neonates, as compared to formula-fed neonates [87]. Generally, "beneficial" microbiota were seen in term infants born vaginally, at home, and who were exclusively breastfed [87].

The mother's health and dietary practices also impact the infant's gut flora. Poor diet, obesity, and diabetes interfere with the development of a stable, healthy intestinal microbiome in infants, affecting cell programming, and the metabolic and biochemical pathways that impact a child's weight. This dysbiosis has also been shown to have other long-term health consequences [74]. Maternal probiotic supplementation also reported differences in the colonization of the newborn gut that persisted up to 2 years of age [88].

Diet is an important influencer of the gut microflora, and its interactions with gut flora have been increasingly studied in recent years [89]. As the child transitions

TABLE 20.2
Important Factors That Affect the Gut Flora Prenatally, at Birth, and Postnatally

	Prenatally	At Birth	Postnatally
Factors impacting gut flora composition	Maternal weight gain Maternal diet Antibiotics Maternal chronic diseases (e.g., diabetes, obesity) Probiotics	Mode of delivery Location of delivery (i.e., hospitalization) Gestational age Sanitation	Breastfeeding Formula feeding Type of complementary food Antibiotics Sanitation Macro- and micronutrient deficiencies Probiotics

to table foods, there is an upsurge in the abundance of Bacteroidetes, which are positively correlated with SCFA levels (acetate, propionate, and butyrate) and other changes consistent with a more stable adult microbiome profile. SCFA levels and bacterial load are generally higher after the introduction of solid foods [81].

Infant feeding practices continue to influence immunity for up to 3 to 5 years, and sometimes well into adulthood, mediated by the gut microflora [90]. Some studies have also shown differences in gut environment composition in those exposed to vegetarian foods and fermented vegetables showing more diverse *lactobacillus* strains [91]. Animal models have also shown that protein energy malnutrition changes the intestinal flora by causing bacterial overgrowth in various parts of the gut. Similarly, substituting a low-fat, plant polysaccharide-rich diet with a Western high-fat, high-sugar diet altered the structure of the microbiota in mice [92]. Iron fortification has had adverse effects on Kenyan infants' gut microbiome, increasing pathogenic bacteria and inducing inflammation [93]. The alterations in nutrient availability result in an adaptation mechanism of nutrient absorption, and many of these alterations have been proposed in the pathogenesis of obesity, such as changes involving increased extraction of additional calories from ingested foods [84]. Similarly, gut microbiota play a key role in the metabolism of dietary choline and phosphatidylcholine through the enzyme trimethylamine oxidase, producing trimethylamine N-oxide, which has been linked to atherosclerosis and cardiovascular disease events [90] (Table 20.2).

CONCLUSION

The gut flora plays an important role in many body functions. Environmental enteropathy remains one of the most important, and challenging, conditions to impact the gut flora in low-income countries, proving to be a particular challenge in infants and children with developing immunity and increased susceptibility to infection. This is an added burden in countries struggling with malnutrition, repeated infection, HIV, and poor environmental conditions. Although we have described various factors that impact gut flora composition, and possibly function, the exact pathophysiology of these pathways remain largely unknown. Without an understanding of the metabolic and pathologic pathways, and the presence of some consensus on ideal biomarker or a combination of biomarkers to assess the different aspects of gut function (e.g., inflammation, absorption, permeability) and developmental outcomes, designing successful interventions seems like an unachievable task.

REFERENCES

1. Canani RB, Costanzo MD, Leone L et al. Potential beneficial effects of butyrate in intestinal and extraintestinal diseases. *World J Gastroenterol* 2011; 17:1519–28.
2. Lindenmayer GW, Stoltzfus RJ, Prendergast AJ. Interactions between zinc deficiency and environmental enteropathy in developing countries. *Adv Nutr* 2014; 5:1–6.
3. Salminen S, Isolauri E, Onnela T. Gut flora in normal and disordered states. *Chemotherapy* 1995; 41(Suppl 1):5–15.
4. Francino MP. Early development of the gut microbiota and immune health. *Pathogens* 2014; 3:769–90.
5. Lindenbaum J, Harmon JW, Gerson CD. Subclinical malabsorption in developing countries. *Am J Clin Nutr* 1972; 25:1056–61.
6. Lindenbaum J, Kent TH, Sprinz H. Malabsorption and jejunitis in American Peace Corps volunteers in Pakistan. *Ann Intern Med* 1966; 65:1201–9.
7. Sheehy TW, Legters LJ, Wallace DK. Tropical jejunitis in Americans serving in Vietnam. *Am J Clin Nutr* 1968; 21:1013–22.
8. Brunser O, Eidelman S, Klipstein FA. Intestinal morphology of rural Haitians. A comparison between overt tropical sprue and asymptomatic subjects. *Gastroenterology* 1970; 58:655–68.
9. Falaiye JM. Present status of subclinical intestinal malabsorption in the tropics. *Br Med J* 1971; 4:454–8.
10. Klipstein FA, Falaiye JM. Tropical sprue in expatriates from the tropics living in the continental United States. *Medicine (Baltimore)* 1969; 48:475–91.
11. Gerson CD, Kent TH, Saha JR et al. Recovery of small-intestinal structure and function after residence in the tropics. II. Studies in Indians and Pakistanis living in New York City. *Ann Intern Med* 1971; 75:41–48.
12. Menzies IS, Zuckerman MJ, Nukajam WS et al. Geography of intestinal permeability and absorption. *Gut* 1999; 44:483–9.
13. Keusch GT, Denno DM, Black RE et al. Environmental enteric dysfunction: Pathogenesis, diagnosis, and clinical consequences. *Clin Infect Dis* 2014; 59(Suppl 4):S207–12.
14. Chacko CJ, Paulson KA, Mathan VI et al. The villus architecture of the small intestine in the tropics: A necropsy study. *J Pathol* 1969; 98:146–51.
15. Burman D. The jejunal mucosa in kwashiorkor. *Arch Dis Child* 1965; 40:526–31.
16. Sullivan PB, Marsh MN, Mirakian R et al. Chronic diarrhea and malnutrition-histology of the small intestinal lesion. *J Pediatr Gastroenterol Nutr* 1991; 12:195–203.
17. Humphrey JH. Child undernutrition, tropical enteropathy, toilets, and handwashing. *Lancet* 2009; 374:1032–5.
18. Campbell DI, Murch SH, Elia M et al. Chronic T cell-mediated enteropathy in rural west African children: Relationship with nutritional status and small bowel function. *Pediatr Res* 2003; 54:306–11.
19. Ramakrishna BS, Venkataraman S, Mukhopadhya A. Tropical malabsorption. *Postgrad Med J* 2006; 82:779–87.

20. Prendergast A, Kelly P. Enteropathies in the developing world: Neglected effects on global health. *Am J Trop Med Hyg* 2012; 86:756–63.
21. Kau AL, Ahern PP, Griffin NW et al. Human nutrition, the gut microbiome and the immune system. *Nature* 2011; 474:327–36.
22. Korpe PS, Petri WA. Environmental enteropathy: Critical implications of a poorly understood condition. *Trends Mol Med* 2012; 18:328–36.
23. Raiten DJ, Sakr Ashour FA, Ross AC et al. Inflammation and Nutritional Science for Programs/Policies and Interpretation of Research Evidence (INSPIRE). *J Nutr* 2015; 145:1039S–108S.
24. Manary MJ, Abrams SA, Griffin IJ et al. Perturbed zinc homeostasis in rural 3-5-y-old Malawian children is associated with abnormalities in intestinal permeability attributed to tropical enteropathy. *Pediatr Res* 2010; 67:671–5.
25. Prasad AS. Zinc in human health: Effect of zinc on immune cells. *Mol Med* 2008; 14:353–7.
26. Semba RD, Shardell M, Sakr Ashour FA et al. Child stunting is associated with low circulating essential amino acids. *EBioMedicine* 2016; 6:246–52.
27. Fagundes Neto U, Martins MC, Lima FL et al. Asymptomatic environmental enteropathy among slum-dwelling infants. *J Am Coll Nutr* 1994; 13:51–6.
28. Kukuruzovic RH, Haase A, Dunn K et al. Intestinal permeability and diarrhoeal disease in Aboriginal Australians. *Arch Dis Child* 1999; 81:304–8.
29. Goto R, Mascie-Taylor CG, Lunn PG. Impact of intestinal permeability, inflammation status and parasitic infections on infant growth faltering in rural Bangladesh. *Br J Nutr* 2009; 101:1509–16.
30. Keusch GT, Rosenberg IH, Denno DM et al. Implications of acquired environmental enteric dysfunction for growth and stunting in infants and children living in low- and middle-income countries. *Food Nutr Bull* 2013; 34:357–64.
31. Crosby WH, Kugler HW. Intraluminal biopsy of the small intestine; the intestinal biopsy capsule. *Am J Dig Dis* 1957; 2:236–41.
32. Denno DM, VanBuskirk K, Nelson ZC et al. Use of the lactulose to mannitol ratio to evaluate childhood environmental enteric dysfunction: A systematic review. *Clin Infect Dis* 2014; 59(Suppl 4):S213–9.
33. van der Merwe LF, Moore SE, Fulford AJ et al. Long-chain PUFA supplementation in rural African infants: A randomized controlled trial of effects on gut integrity, growth, and cognitive development. Am *J Clin Nutr* 2013; 97:45–57.
34. Kosek M, Haque R, Lima A et al. Fecal markers of intestinal inflammation and permeability associated with the subsequent acquisition of linear growth deficits in infants. *Am J Trop Med Hyg* 2013; 88:390–6.
35. Victora CG, Adair L, Fall C et al. Maternal and child undernutrition. Consequences for adult health and human capital. *Lancet* 2008; 371:340–57.
36. Peeters M, Hiele M, Ghoos Y et al. Test conditions greatly influence permeation of water soluble molecules through the intestinal mucosa: Need for standardisation. *Gut* 1994; 35:1404–8.
37. Lee GO, Kosek P, Lima AA et al. Lactulose: Mannitol diagnostic test by HPLC and LC-MSMS platforms: Considerations for field studies of intestinal barrier function and environmental enteropathy. *J Pediatr Gastroenterol Nutr* 2014; 59:544–50.
38. Ukabam SO, Homeida MM, Cooper BT. Small intestinal permeability in normal Sudanese subjects: Evidence of tropical enteropathy. *Trans R Soc Trop Med Hyg* 1986; 80:204–7.
39. Lostia AM, Lionetto L, Principessa L et al. A liquid chromatography/mass spectrometry method for the evaluation of intestinal permeability. *Clin Biochem* 2008; 41:887–92.
40. Lima AA, Soares AM, Lima NL et al. Effects of vitamin A supplementation on intestinal barrier function, growth, total parasitic, and specific *Giardia* spp. infections in Brazilian children: A prospective randomized, double-blind, placebo-controlled trial. *J Pediatr Gastroenterol Nutr* 2010; 50:309–15.

41. Wu G, Knabe DA, Flynn NE. Synthesis of citrulline from glutamine in pig enterocytes. *Biochem J* 1994; 299(Pt 1):115–21.
42. Lee EH, Ko JS, Seo JK. Correlations of plasma citrulline levels with clinical and endoscopic score and blood markers according to small bowel involvement in pediatric Crohn disease. *J Pediatr Gastroenterol Nutr* 2013; 57:570–5.
43. Basso MS, Capriati T, Goffredo BM et al. Citrulline as marker of atrophy in celiac disease. *Intern Emerg Med* 2014; 9:705–7.
44. Collins JK, Wu G, Perkins-Veazie P et al. Watermelon consumption increases plasma arginine concentrations in adults. *Nutrition* 2007; 23:261–6.
45. Crenn P, Vahedi K, Lavergne-Slove A et al. Plasma citrulline: A marker of enterocyte mass in villous atrophy-associated small bowel disease. *Gastroenterology* 2003; 124:1210–9.
46. Kosek M, Guerrant RL, Kang G et al. Assessment of environmental enteropathy in the MAL-ED cohort study: Theoretical and analytic framework. *Clin Infect Dis* 2014; 59(Suppl 4):S239–47.
47. McKay S, Gaudier E, Campbell DI et al. Environmental enteropathy: New targets for nutritional interventions. *Int Health* 2010; 2:172–80.
48. Luby SP, Agboatwalla M, Painter J et al. Combining drinking water treatment and hand washing for diarrhoea prevention, a cluster randomised controlled trial. *Trop Med Int Health* 2006; 11:479–89.
49. Black RE, Victora CG, Walker SP et al. Maternal and child undernutrition and overweight in low-income and middle-income countries. *Lancet* 2013; 382:427–51.
50. Ngure FM, Reid BM, Humphrey JH et al. Water, sanitation, and hygiene (WASH), environmental enteropathy, nutrition, and early child development: Making the links. *Ann N Y Acad Sci* 2014; 1308:118–28.
51. Dewey KG, Mayers DR. Early child growth: How do nutrition and infection interact? *Matern Child Nutr* 2011; 7(Suppl 3):129–42.
52. Sullivan PB. Studies of the small intestine in persistent diarrhea and malnutrition: The Gambian experience. *J Pediatr Gastroenterol Nutr* 2002; 34(Suppl 1):S11–3.
53. Smith HE, Ryan KN, Stephenson KB et al. Multiple micronutrient supplementation transiently ameliorates environmental enteropathy in Malawian children aged 12–35 months in a randomized controlled clinical trial. *J Nutr* 2014; 144:2059–65.
54. Trehan I, Benzoni NS, Wang AZ et al. Common beans and cowpeas as complementary foods to reduce environmental enteric dysfunction and stunting in Malawian children: Study protocol for two randomized controlled trials. *Trials* 2015; 16:520.
55. Doron SI, Hibberd PL, Gorbach SL. Probiotics for prevention of antibiotic-associated diarrhea. *J Clin Gastroenterol* 2008; 42(Suppl 2):S58–63.
56. Walker AW, Lawley TD. Therapeutic modulation of intestinal dysbiosis. *Pharmacol Res* 2013; 69:75–86.
57. Ryan KN, Stephenson KB, Trehan I et al. Zinc or albendazole attenuates the progression of environmental enteropathy: A randomized controlled trial. *Clin Gastroenterol Hepatol* 2014; 12:1507–13.e1.
58. Wessells KR, Hess SY, Rouamba N et al. Associations between intestinal mucosal function and changes in plasma zinc concentration following zinc supplementation. *J Pediatr Gastroenterol Nutr* 2013; 57:348–55.
59. Amin HJ, Zamora SA, McMillan DD et al. Arginine supplementation prevents necrotizing enterocolitis in the premature infant. *J Pediatr* 2002; 140:425–31.
60. Wang WW, Qiao SY, Li DF. Amino acids and gut function. *Amino Acids* 2009; 37:105–10.
61. Reeds PJ, Burrin DG, Stoll B et al. Enteral glutamate is the preferential source for mucosal glutathione synthesis in fed piglets. *Am J Physiol* 1997; 273:E408–15.
62. dos Santos R, Viana ML, Generoso SV et al. Glutamine supplementation decreases intestinal permeability and preserves gut mucosa integrity in an experimental mouse model. *J Parenter Enteral Nutr* 2010; 34:408–13.

63. Wang JY, Johnson LR. Polyamines and ornithine decarboxylase during repair of duo-denal mucosa after stress in rats. *Gastroenterology* 1991; 100:333–43.

64. Wang B, Wu Z, Ji Y et al. L-glutamine enhances tight junction integrity by activating CaMK kinase 2-AMP-activated protein kinase signaling in intestinal porcine epithelial cells. *J Nutr* 2016; 146:501–8.

65. Lima AA, Brito LF, Ribeiro HB et al. Intestinal barrier function and weight gain in malnourished children taking glutamine supplemented enteral formula. *J Pediatr Gastroenterol Nutr* 2005; 40:28–35.

66. Williams EA, Elia M, Lunn PG. A double-blind, placebo-controlled, glutamine-supplementation trial in growth-faltering Gambian infants. *Am J Clin Nutr* 2007; 86:421–7.

67. The Gap Report; 2014.

68. Murray CJ, Ortblad KF, Guinovart C et al. Global, regional, and national incidence and mortality for HIV, tuberculosis, and malaria during 1990–2013: A systematic anal-ysis for the Global Burden of Disease Study 2013. *Lancet* 2014; 384:1005–70.

69. Louis-Auguste J, Greenwald S, Simuyandi M et al. High dose multiple micronutrient supplementation improves villous morphology in environmental enteropathy without HIV enteropathy: Results from a double-blind randomised placebo controlled trial in Zambian adults. *BMC Gastroenterol* 2014; 14:15.

70. Kelly P, Menzies I, Crane R et al. Responses of small intestinal architecture and func-tion over time to environmental factors in a tropical population. *Am J Trop Med Hyg* 2004; 70:412–9.

71. Trehan I, Shulman RJ, Ou CN et al. A randomized, double-blind, placebo-controlled trial of rifaximin, a nonabsorbable antibiotic, in the treatment of tropical enteropathy. *Am J Gastroenterol* 2009; 104:2326–33.

72. Koenig JE, Spor A, Scalfone N et al. Succession of microbial consortia in the develop-ing infant gut microbiome. *Proc Natl Acad Sci USA* 2011; 108(Suppl 1):4578–85.

73. Mueller NT, Shin H, Pizoni A et al. Birth mode-dependent association between pre-pregnancy maternal weight status and the neonatal intestinal microbiome. *Sci Rep* 2016; 6:23133.

74. Soderborg TK, Borengasser SJ, Barbour LA et al. Microbial transmission from moth-ers with obesity or diabetes to infants: An innovative opportunity to interrupt a vicious cycle. *Diabetologia* 2016; 59:895–906.

75. Dominguez-Bello MG, Costello EK, Contreras M et al. Delivery mode shapes the acquisition and structure of the initial microbiota across multiple body habitats in new-borns. *Proc Natl Acad Sci USA* 2010; 107:11971–5.

76. Kuitunen M, Kukkonen K, Juntunen-Backman K et al. Probiotics prevent IgE associated allergy until age 5 years in cesarean-delivered children but not in the total cohort. *J Allergy Clin Immunol* 2009; 123:335–41.

77. Arboleya S, Sánchez B, Solís G et al. Impact of prematurity and perinatal antibiotics on the developing intestinal microbiota: A functional inference study. *Int J Mol Sci* 2016; 17(5):649.

78. Dogra S, Sakwinska O, Soh SE et al. Dynamics of infant gut microbiota are influenced by delivery mode and gestational duration and are associated with subsequent adipos-ity. *MBio* 2015; 6(1):e02419-14.

79. Dethlefsen L, Huse S, Sogin ML et al. The pervasive effects of an antibiotic on the human gut microbiota, as revealed by deep 16S rRNA sequencing. *PLoS Biol* 2008; 6:e280.

80. Azad MB, Konya T, Persaud RR et al. Impact of maternal intrapartum antibiotics, method of birth and breastfeeding on gut microbiota during the first year of life: A pro-spective cohort study. *BJOG* 2016; 123:983–93.

81. Gonzalez-Perez G, Hicks AL, Tekieli TM et al. Maternal antibiotic treatment impacts development of the neonatal intestinal microbiome and antiviral immunity. *J Immunol* 2016; 196:3768–79.

82. Francino MP. Antibiotics and the human gut microbiome: Dysbioses and accumulation of resistances. *Front Microbiol* 2015; 6:1543.

83. Hildebrand H, Malmborg P, Askling J et al. Early-life exposures associated with antibiotic use and risk of subsequent Crohn's disease. *Scand J Gastroenterol* 2008; 43:961–6.

84. Turnbaugh PJ, Ley RE, Mahowald MA et al. An obesity-associated gut microbiome with increased capacity for energy harvest. *Nature* 2006; 444:1027–31.

85. Turnbaugh PJ, Gordon JI. The core gut microbiome, energy balance and obesity. *J Physiol* 2009; 587:4153–8.

86. Collado MC, Rautava S, Aakko J et al. Human gut colonisation may be initiated in utero by distinct microbial communities in the placenta and amniotic fluid. *Sci Rep* 2016; 6:23129.

87. Penders J, Thijs C, Vink C et al. Factors influencing the composition of the intestinal microbiota in early infancy. *Pediatrics* 2006; 118:511–21.

88. Gueimonde M, Sakata S, Kalliomäki M et al. Effect of maternal consumption of lactobacillus GG on transfer and establishment of fecal bifidobacterial microbiota in neonates. *J Pediatr Gastroenterol Nutr* 2006; 42:166–70.

89. De Filippo C, Cavalieri D, Di Paola M et al. Impact of diet in shaping gut microbiota revealed by a comparative study in children from Europe and rural Africa. *Proc Natl Acad Sci USA* 2010; 107:14691–6.

90. Stock J. Gut microbiota: An environmental risk factor for cardiovascular disease. *Atherosclerosis* 2013; 229:440–2.

91. Alm JS, Swartz J, Björkstén B et al. An anthroposophic lifestyle and intestinal microflora in infancy. *Pediatr Allergy Immunol* 2002; 13:402–11.

92. Turnbaugh PJ, Ridaura VK, Faith JJ et al. The effect of diet on the human gut microbiome: A metagenomic analysis in humanized gnotobiotic mice. *Sci Transl Med* 2009; 1:6ra14.

93. Jaeggi T, Kortman GA, Moretti D et al. Iron fortification adversely affects the gut microbiome, increases pathogen abundance and induces intestinal inflammation in Kenyan infants. *Gut* 2015; 64:731–42.

Section IX

Effects of Early Life Exposures and Nutrition

21 Effects of Early Diet on Childhood Allergy

Merryn Netting and Maria Makrides

CONTENTS

INTRODUCTION

Food allergies are classified into two categories: those that are mediated by immuno globulin E (IgE) antibodies and those mediated by immune cells (non-IgE mediated) [1]. Although it is possible to develop an IgE-mediated allergy to any food, most individuals with allergies react to one, or a combination, of nine common foods: cow's milk, soy, egg, wheat, peanut, tree nuts, sesame, fish, and shellfish [2]. The most common symptoms associated with food allergy in children include urticaria (hives), angioedema, eczema, enterocolitis, enteropathy, irritability, vomiting, diarrhea, and anaphylaxis [1].

Allergic disease, and in particular food allergies, significantly impact general health perception, parental emotional distress, and family activities [3]. Young children are particularly at risk of developing food allergy, and it is estimated that up to 10% of toddlers have food allergy, compared with 1% to 2% of adults [1,4]. Why young children are becoming increasingly sensitized to food allergens is the focus of ongoing research. The development of tolerance to multiple foods during early life is essential to survival, and the immune mechanisms that enable the development of tolerance are highly developed and regulated. Food allergy in an infant or child represents a failure to develop tolerance to a food protein and is associated with aberrant T helper cell 2 (Th2) balance [5]. Environmental factors such as maternal dietary patterns and micronutrient supplements, as well as postnatal factors such as the mode of birth, feeding type, early infections, timing, and type of solid foods, and exposure to allergens all influence the developing immune system [6]. However, it has become clear that these environmental factors may exert differing effects, depending on genetic predisposition, and are modulated by epigenetic effects on gene expression.

Fewer allergies are observed in low-income countries; however, rates appear to be increasing as these countries adopt more Westernized lifestyle patterns [7,8]. There also appear to be important migration effects; the HealthNut cohort in Melbourne had a high rate of food allergy and eczema among children born to Asian immigrants [9]. This difference in allergy rate was significantly higher than that in children born in Asia who migrated to Australia with their parents [9]. The reasons for this increased incidence is unclear, and is likely to be due to multiple mechanisms. In this chapter, following an overview of the hypotheses of allergy development, we examine the evidence relating to dietary influences on development of food allergy, and discuss the most recent infant feeding guidelines for allergy prevention.

HYPOTHESES OF ALLERGY DEVELOPMENT

The reasons why some individuals develop allergies are complex and multifactorial. The main hypotheses about the etiology of allergy include the hygiene and dual allergen exposure hypotheses.

HYGIENE HYPOTHESIS

The hygiene hypothesis is based on the observation that people living in European farming communities have lower rates of food allergy and asthma compared to those in the city [10]. It is hypothesized that the farming environment and lifestyle leads to greater exposure during infancy to bacterial markers called endotoxins that influence the microbiome, which in turn promotes tolerogenic immune pathways.

In a Swedish study, washing dishes by hand was associated with a lower incidence of allergic disease as compared with using automated dishwashers [11]. The protective effect of dishwashing by hand was stronger in children eating fermented foods and foods purchased from the farm door [11]. This effect may be related to exposure to beneficial microbial exposures not destroyed by high-temperature machine dishwashing. In Westernized countries, there has been an increased consumption of virtually sterile processed foods. The influence of the consumption of highly processed foods on the gut microbiome may be one of the mechanisms by which Western-style diets are associated with increases in atopy [12]. The hygiene hypothesis has been further supported by observations that birth by cesarean section is associated with an increased risk of food allergy or atopy [13], with the mechanism postulated to be related to the lack of inoculation with maternal gut flora that occurs during a normal vaginal delivery.

DUAL ALLERGEN EXPOSURE HYPOTHESIS

The dual exposure to allergen hypothesis proposes that tolerance to antigens occurs in the neonate through high-dose oral exposure, and that allergic sensitization occurs through low-dose cutaneous exposure [14]. It is hypothesized that there is a balance between these processes and is an issue particularly for babies with eczema who have filament aggregating protein (filaggrin) mutations, resulting in altered skin barrier function and abnormal immune reactivity [14]. It is proposed that food proteins (from the household environment) pass through the disrupted skin barrier, leading to Th2 responses, IgE production, and sensitization to the allergen [15]. This may be particularly problematic for infants at risk of developing allergy (including babies with eczema), where allergenic foods are consumed by household members but avoided in the infant's diet as the baby is exposed to the allergen via their skin, but not the potentially protective oral route.

DIETARY INFLUENCES ON DEVELOPMENT OF ALLERGY

Our diet has the potential to alter the tendency toward allergy by immunomodulatory effects of dietary components and exposures to allergens. Infants may be exposed to nutritional influences via the maternal diet while they are *in utero*, during breast-feeding, or as they start consuming solid foods [16].

Maternal dietary patterns associated with decreased sensitization in their children include Mediterranean diets, and the consumption of fruits, vegetables, and fish [12]. These nutrient-rich foods and dietary patterns contain immunomodulatory factors, including polyunsaturated fatty acids, antioxidants, vitamin D, prebiotics, and probiotics. These factors may foster a healthy immune system by modifying the immune response or the functioning of the immune system. It is also possible that women who consume these types of diets have other associated lifestyle factors that are associated with a reduced risk of allergy development, such as the avoidance of tobacco smoke, and less exposure to environmental pollutants [17].

LONG-CHAIN POLYUNSATURATED FATTY ACIDS

Omega-3 long-chain polyunsaturated fatty acids (LCPUFA), found predominantly in marine oils, modulate the immune system by the inhibition of inflammatory pathways and may be the reason diets containing fish are associated a lower risk of atopic disorders [12]. Diets rich in omega-3 LCPUFA may alter the immune system by affecting Th cell balance, specifically inhibiting Th2 cell differentiation, and thereby reducing IgE-mediated allergy. The effect of omega-3 LCPUFA in the maternal diet during pregnancy and lactation on development of childhood allergy has been tested in many randomized controlled trials (RCTs), with inconsistent results. The most recent *Cochrane Review* of maternal supplementation with omega-3 LCPUFA concluded that, although there was little effect of omega-3 supplementation during pregnancy and/or breastfeeding in the reduction of allergic disease in children, there were some reductions in atopic disease outcomes, including food allergy and eczema in the first year in children born to women at high risk of allergy [18]. Since this review, the results of the Copenhagen Prospective Studies on Asthma in Childhood[2010] (COPSAC[2010]) have been published [19]. In the COPSAC[2010], pregnant women were randomized to large doses (2.4 g) of omega-3 LCPUFA or a placebo (olive oil) per day during the last trimester of pregnancy, and their offspring were followed to 5 years of age. The study reported that, for children born to the intervention group, the absolute risk of a persistent wheeze or asthma was reduced by one-third, with further analysis suggesting that this effect was strongest in the children of women whose blood eicosapentaenoic acid and docosahexaenoic acid levels were in the lowest third of the trial population. Supplementation with omega-3 LCPUFA may be not recommended for all women, but might be useful for subgroups, such as those with a family history of food allergy or with poor omega-3 LCPUFA status. LCPUFA have many effects, and supplementation may be not without risk, as it has been associated with longer gestational length [20].

The effect of direct supplementation of infants' diets with omega-3 LCPUFA has also been investigated. The Childhood Asthma Prevention Study (CAPs) was an RCT conducted in children with a family history of asthma, which supplemented diets with omega-3 LCPUFA and restricted the dietary intake of dietary omega-6 fatty acids from 6 months of age to 5 years of age [21]. The study found no effect on asthma, eczema, or atopy at 5 years of age. Schindler et al. [22] reviewed nine more studies, including a total of 2,704 infants, that assessed the effect of higher versus lower intake of LCPUFA on allergic outcomes in infants for the Cochrane database. The *Cochrane Review* concluded that there is no evidence that PUFA supplementation in infancy has an effect on infant or childhood allergy, asthma, dermatitis/eczema, or food allergy; however, the authors noted that the studies were of variable quality and heterogeneous in nature.

ANTIOXIDANTS

There is an association between maternal fruit and vegetable intake and less atopy in their babies [12], which may be due to the antioxidant content of diets rich in fruits and vegetables. The antioxidant vitamins (C, A, and E) and minerals (zinc and selenium) present in fruit and vegetables may protect against atopic disease of the airway by protection against oxidant damage and inflammation of the airways [23]. However, there have been no antioxidant supplementation trials to test the link between antioxidants in the maternal diet and atopy development.

VITAMIN D

Vitamin D is a fat-soluble vitamin and hormone with many roles, including immunomodulation, a process in which the immune response is altered. Vitamin D is obtained through the action of sunlight on skin and, in smaller amounts, through diet. Interest in poor vitamin D status and increased allergy risk began with the observation that people living at lower latitudes with greater sun exposure have fewer allergies; this has been shown in the United States and Australia [24]. However, this association has not been demonstrated in cohort studies or RCTs with vitamin D supplementation. The Barwon infant study [25] found no association between low vitamin D levels at 6 months of age and egg allergy at 1 year of age. Higher cord blood vitamin D was associated with less eczema, but not with food sensitization or food allergy at 1 year of age in a subset of infants enrolled in an RCT comparing the effect of fish oil supplementation during infancy on allergic outcomes [26].

ALLERGEN EXPOSURE: TIMING AND TYPE OF ALLERGENS

MATERNAL ALLERGEN AVOIDANCE DURING PREGNANCY

Avoidance of common allergens in the maternal diet during pregnancy is not recommended as a strategy for allergy prevention in Australia and other countries, including the United States and Europe. Avoiding the consumption of common allergens during pregnancy does not reduce the incidence of sensitization to allergens in children [12], and the restricted diet is associated with lower pregnancy weight gain [27].

DOES BREASTFEEDING PREVENT ALLERGY?

Breast milk is the gold standard for infant feeding, and has many benefits for both the mother and child. In addition to its nutritional benefits, breast milk contains many nonnutritional components, including antibodies, cytokines, and other immunomodulatory components [28]. The evidence that breastfeeding is protective against allergy, specifically food allergy, is weak; however, this may be due to methodological issues. A recent systematic review and meta-analysis including the results of 89 studies reported that breastfeeding is associated with less asthma at 5 years of age, with a greater effect in low- to middle-income countries [29]. This review reported that there is weak evidence for breastfeeding and the prevention of other atopic diseases, including eczema and rhinitis, with no effect of breastfeeding on development of food allergy [29]. The lack of protective results for breastfeeding and food allergy may be due to reverse causality, as highly atopic families are more likely to breastfeed and feed for longer, and the authors note that this should be adjusted for when reporting the results of trials investigating breastfeeding and allergy development. There are other innate issues with studies investigating the effects of breastfeeding, including the inability to randomize exposures. Additionally, the lack of evidence for breastfeeding and protection against food allergy may also be because many of the cohort studies included in the systematic review were conducted prior to the food allergy epidemic [29]. Some of the limitations of the cohort studies have been addressed by the only randomized trial of breastfeeding exposure in Belarus [30]. In the PROBIT study, women were cluster randomized via attendance at maternal health centers that followed the World Health Organization Baby Friendly Hospital Initiative (WHO BHFI) advice or other health centers [30]. Allergic sensitization was assessed using skin prick testing, and allergy symptoms were scored using standardized protocols. A total of 17,046 mother–infant pairs were randomized into the trial, and 13,889 were reviewed at 6.5 years of age. Although infants whose mothers attended the WHO BHFI health centers breastfed for longer, the results of this RCT did not support the protective effect of prolonged breastfeeding against allergic sensitization, and the development of asthma, hay fever, or eczema [30].

Food allergens consumed by the mother appear in her breast milk [31], and avoidance of allergens in the maternal diet during lactation does not reduce the incidence of sensitization in the infant. However, there are emerging data, specific to egg, that the inclusion of egg in the maternal diet during early lactation is associated with increased levels of egg-specific IgG4 in their babies, which may be important for tolerance development [32].

WHAT ABOUT INFANT FORMULA?

Not all infants are breastfed, and many breastfed infants have infant formula top-up feeds. Infant formula is usually based on modified cow's milk protein, and this is often an infant's first exposure to an allergen. For the prevention of allergy, there is insufficient evidence to recommend the use of soy-based formulas, goat-milk-based formulas, formulas containing LCPUFA, or formulas containing pre- or probiotics [17].

Partially hydrolyzed infant formula has been promoted as a means of preventing development of allergy and, until recently, infant feeding guidelines in Europe, America, and Australasia [33–35] supported the use of hydrolyzed formulas for non-breastfed infants in place of standard cow's milk formula if the infant has a family history of allergy. A recent systematic review and meta-analysis by Boyle et al. [36] investigated whether hydrolyzed cow's milk formulas can prevent allergic or autoimmune disease. This review, commissioned by the UK Food Standards Authority, included 37 eligible intervention trials of hydrolyzed formula, with over 19,000 participants. There was evidence of a conflict of interest and a high or unclear risk of bias in most studies of allergic outcomes, and evidence of publication bias for studies of eczema and wheeze. Overall, the authors reported that there was no consistent evidence that partially or extensively hydrolyzed formulas reduced the risk of allergic or autoimmune outcomes, and this is reflected in newer infant feeding guidelines for allergy prevention guidelines [37].

COMPLEMENTARY FOODS: WHEN AND WHICH FOODS?

When considering the introduction to solid foods, issues specific to the prevention of allergy relate to the timing of introduction to solid foods and the type of foods introduced [29,38–40]. In the last 10 years, there has been a reversal of recommendations regarding the introduction to solid foods for prevention of allergy. While guidelines are used to promote delayed introduction to solid foods, and staged and delayed exposure to common allergens, this was not associated with a reduction in the prevalence of food allergy and, in fact, may have contributed to the increasing prevalence. The change in guidelines has occurred as available evidence to inform these recommendations has shifted from population-based cohort studies to RCTs and systematic reviews of RCTs.

Several randomized controlled trials have investigated the timing of including common allergens in an infant's diet. Whereas most of the trials have considered single allergens, peanut [41] and egg [42–45], one trial [46] investigated the effect of adding multiple allergens into an infant's diet before one year of age. The outcomes of the single allergen trials are summarized in Table 21.1.

TABLE 21.1

Summary of Results from Single Allergen Allergy Prevention Studies

Trial	Allergen	Intervention	Result
LEAP [41]	Peanut	RCT Peanut vs. no peanut from 4–11 months Group 1: peanut SPT <1 mm: $n = 530$ Group 2: peanut SPT 1–4 mm: $n = 98$	Peanut allergy at five years: Group 1 Peanut group: 1.9% Control group: 13.7% ($p < 0.001$) Group 2 Peanut group: 10.6% Control group: 35.3% ($p = 0.004$)
STAR [43]	Egg	DBPC RCT Egg ($n = 49$) vs. no egg ($n = 37$) From 4–8 months Infants with eczema	Egg allergy at one year: Egg group: 33% Control group: 51% RR 0.65; 95% CI: 0.38, 1.11; $p = 0.11$
STEP [42]	Egg	DBPC RCT Egg ($n = 407$) vs. no egg ($n = 410$) From 4–10 months Infants with family history of atopy but without eczema	Egg allergy at one year: Egg group: 7.0% Control group: 10.3%; ARR 0.75; 95% CI: 0.48, 1.17; $p = 0.20$) Egg-specific IgG4 levels: Egg group median: 1.22 mg A/L Control group: 0.07 mg A/L; $p < 0.0001$
HEAP [44]	Egg	RCT Egg ($n = 184$) vs. no egg ($n = 199$) From 4–12 months	Sensitization to hen's egg at one year: Egg group: 5.6% Control group: 2.6% RR 2.20; 95% CI: 0.68, 7.14; $p = 0.24$ Hen's egg allergy at one year: Egg group: 2.1% Control group: 0.6% RR 3.30; 95% CI: 0.35, 31.32; $p = 0.35$
BEAT [45]	Egg	RCT Egg ($n = 165$) vs. no egg ($n = 154$) From 4–8 months Infants with family history of allergic disease and EW-SPT <2 mm	Egg sensitization at one year (EW-SPT ≥3 mm) Egg group: 11% Control group: 20% OR 0.46; 95% CI: 0.22, 0.95; $p = 0.03$ IgG4 to egg proteins and IgG4/IgE ratios higher in egg group ($p < 0.0001$)

Note: A/L, antigen per liter; ARR, adjusted relative risk; DBPC, double-blind placebo-controlled; EW-SPT, egg white skin prick test; IgE, immunoglobulin E; IgG4, immunoglobulin G4; OR, odds ratio; RCT, randomized controlled trial; RR, relative risk; SPT, skin prick test.

The Learning Early About Peanut (LEAP) allergy study [41] compared early (4–11 months) with delayed (5 years) introduction to peanut in children at high risk of peanut allergy defined as preexisting eczema and/or egg allergy. The study demonstrated an 11%–25% absolute reduction in the risk of peanut allergy in high-risk infants (and a relative risk reduction of up to 80%) if peanut was introduced between 4 and 11 months of age [41].

Four trials investigating the timing of introduction to egg have been published [42–45]. Although none of the trials reported statistically significant differences in rates of egg allergy at 1 year of age in infants fed egg early compared with later introduction to egg, there was a trend toward lower rates of egg allergy in the groups of children introduced earlier to egg. Meta-analysis of the egg trials (including 1,915 participants) showed evidence that egg introduction at 4 to 6 months was associated with a lower risk of egg allergy compared with later egg introduction (RR 0.56; 95% CI: 0.36, 0.87; $p = 0.009$) [39]. The authors concluded that there is moderate certainty that early introduction to egg, compared to delayed introduction to egg, will reduce the risk of IgE-mediated egg allergy by up to 30%. It was of concern, however, that many infants screened for these trials already had clinical egg allergy prior to the introduction of solid foods, indicating that, at least for egg allergy, sensitization and allergy development takes place very early in life, and there may need to be some level of caution for some subgroups of the population, such as babies with severe eczema, when introducing egg into the diet.

The EAT study [46] was the first randomized controlled trial to test the effect of early introduction to solid foods (from 3 months) compared with the UK guidelines on allergy development of exclusive breastfeeding until introduction to solid foods at 6 months. The study found no significant difference in food allergy rates in the primary analysis (intention to treat analysis) between the early and standard introduction groups. There was no difference in breastfeeding rates at 12 months for individuals in the early introduction group, compared with the exclusive breastfeeding group, showing that earlier introduction of allergenic foods did not have an impact on breastfeeding [47].

Meta-analysis of the LEAP and EAT studies showed evidence that peanut introduction at 4 to 11 months of age was associated with a lower risk of peanut allergy when compared to delayed introduction to peanut (RR 0.29; 95% CI: 0.11, 0.74, $p = 0.009$) [39].

UPDATED INFANT FEEDING GUIDELINES
FOR PREVENTION OF FOOD ALLERGY

Ten international allergy and immunology bodies released a joint consensus communication in 2015, highlighting new evidence from the LEAP study regarding the potential benefits of early, rather than delayed, peanut introduction during the period of complementary food introduction to prevent peanut allergy in high-risk infants [35]. This has led to the development of Australian Infant Feeding Consensus Guidelines [48] and an addendum to the USA National Institute of Health Feeding Guidelines for Reduction of Peanut Allergy [49] (Box 21.1). Whereas the Australian guidelines provide general information related to inclusion of allergens in the infant diet, the U.S. guidelines provide advice related to the inclusion of peanut into an infant's diet stratified by degree of risk, defined as the presence of eczema and/or egg allergy.

**BOX 21.1   INFANT FEEDING GUIDELINES
FOR THE PREVENTION OF FOOD ALLERGY**

Australian Infant Feeding Consensus Guidelines [37]:

- When your infant is ready, at around 6 months, but not before 4 months, start to introduce a variety of solid foods, starting with iron-rich foods, while continuing breastfeeding.
- All infants should be given allergenic solid foods including peanut butter, cooked egg, and dairy and wheat products in the first year of life. This includes infants at high risk of allergy.
- Hydrolyzed (partially and extensively) infant formula are not recommended for prevention of allergic disease.

U.S. National Institute of Allergy and Infectious Diseases Guidelines, 2010 [50]:

- Restriction of maternal diet during pregnancy or lactation is not recommended as a strategy for preventing the development of food allergy.
- It is recommended that all infants be exclusively breastfed until 4 to 6 months of age (unless contraindicated for medical reasons).
- It is recommended that introduction of solid food should not be delayed beyond 4 to 6 months. Potentially allergenic foods may be introduced at this time as well.

U.S. National Institute of Allergy and Infectious Diseases Addendum Guidelines for Introduction to Peanut, 2017 [49]:

- Addendum guideline 1—For infants with severe eczema, egg allergy or both:
  - Strongly consider evaluation by specific peanut IgE measurement and/or skin prick testing and, if necessary, an oral food challenge. Based on test results, introduce peanut-containing foods.
  - Earliest age of peanut introduction: 4 to 6 months.

CONCLUSION

The risk factors for developing childhood food allergy are complex and multifactorial, depending on a combination of genetic, nutritional, and environmental factors. Early life nutrition exposures and feeding practices play an important role, as does the maternal diet during pregnancy. The most recent public health guidelines for infant feeding to prevent food allergy promote exposure to common allergens, particularly peanut, in the first year of life. There are many other potential factors, including the influence of a variety of foods and nutrients, and the role of processing on foods, that are yet to be explored.

There are concerns regarding the potential for increasing incidence of food allergy in low- to middle-income countries as they adopt more Westernized food and lifestyle patterns, and it is possible that early intervention may prevent this increase.

REFERENCES

1. Pawankar R, Canonica GW, Holgate ST et al., editors. *World Allergy Organization white book on allergy.* Milwaukee, WI: World Allergy Organization; 2011.
2. Sicherer SH, Sampson HA. Food allergy. *J Allergy Clin Immunol* 2006; 117:S470–5.
3. Sicherer SH, Noone SA, Munoz-Furlong A. The impact of childhood food allergy on quality of life. *Ann Allergy Asthma Immunol* 2001; 87(6):461–4.
4. Osborne NJ, Koplin JJ, Martin PE et al. Prevalence of challenge-proven IgE-mediated food allergy using population-based sampling and predetermined challenge criteria in infants. *J Allergy Clin Immunol* 2011; 127(3):668–76.e2.
5. Castro-Sanchez P, Martin-Villa JM. Gut immune system and oral tolerance. *Br J Nutr* 2013; 109(Suppl 2):S3–11.
6 Prescott SL. Early-life environmental determinants of allergic diseases and the wider pandemic of inflammatory noncommunicable diseases. *J Allergy Clin Immunol* 2013; 131:23–30.
7. Boye JI. Food allergies in developing and emerging economies: Need for comprehensive data on prevalence rates. *Clin Transl Allergy* 2012; 2(1):25.
8. Prescott SL, Pawankar R, Allen KJ et al. A global survey of changing patterns of food allergy burden in children. *World Allergy Organ J* 2013; 6(1):21.
9. Koplin JJ, Peters RL, Ponsonby AL et al. Increased risk of peanut allergy in infants of Asian-born parents compared to those of Australian-born parents. *Allergy* 2014; 69:1639-47.
10. Liu AH. Revisiting the hygiene hypothesis for allergy and asthma. *J Allergy Clin Immunol* 2015; 136:860–5.
11. Hesselmar B, Hicke-Roberts A, Wennergren G. Allergy in children in hand versus machine dishwashing. *Pediatrics* 2015; 135:e590–7.
12. Netting MJ, Middleton PF, Makrides M. Does maternal diet during pregnancy and lactation affect outcomes in offspring? A systematic review of food-based approaches. *Nutrition* 2014; 30:1225–41.
13. Bager P, Wohlfahrt J, Westergaard T. Caesarean delivery and risk of atopy and allergic disease: Meta-analyses. *Clin Exp Allergy* 2008; 38:634–42.
14. Lack G. Update on risk factors for food allergy. *J Allergy Clin Immunol* 2012; 129:1187–97.
15. Fox AT, Sasieni P, du Toit G et al. Household peanut consumption as a risk factor for the development of peanut allergy. *J Allergy Clin Immunol* 2009; 123:417–23.
16. Warner JO. Early life nutrition and allergy. *Early Hum Dev* 2007; 83(12):777–83.
17. Prescott S, Nowak-Wegrzyn A. Strategies to prevent or reduce allergic disease. *Ann Nutr Metab* 2011; 59(Suppl 1):28–42.
18. Gunaratne AW, Makrides M, Collins CT. Maternal prenatal and/or postnatal n-3 long chain polyunsaturated fatty acids (LCPUFA) supplementation for preventing allergies in early childhood. *Cochrane Database Syst Rev* 2015 7:CD010085.
19. Bisgaard H, Stokholm J, Chawes BL et al. Fish oil-derived fatty acids in pregnancy and wheeze and asthma in offspring. *N Engl J Med* 2016; 375:2530–9.

20. Makrides M, Gibson RA, McPhee AJ et al. Effect of DHA supplementation during pregnancy on maternal depression and neurodevelopment of young children: A randomized controlled trial. *JAMA* 2010; 304:1675–83.
21. Almqvist C, Garden F, Xuan W et al. Omega-3 and omega-6 fatty acid exposure from early life does not affect atopy and asthma at age 5 years. *J Allergy Clin Immunol* 2007; 119:1438–44.
22. Schindler T, Sinn JK, Osborn DA. Polyunsaturated fatty acid supplementation in infancy for the prevention of allergy. *Cochrane Database Syst Rev* 2016; 10:CD010112.
23. Allan K, Kelly FJ, Devereux G. Antioxidants and allergic disease: A case of too little or too much? *Clin Exp Allergy* 2010; 40:370–80.
24. Mullins RJ, Camargo CA. Latitude, sunlight, vitamin D, and childhood food allergy/ anaphylaxis. *Curr Allergy Asthma Rep* 2012; 12:64–71.
25. Molloy J, Koplin JJ, Allen KJ et al. Vitamin D insufficiency in the first 6 months of infancy and challenge-proven IgE-mediated food allergy at 1 year of age: A case-cohort study. *Allergy* 2017; 72(8):1222–1231. doi: 10.1111/all.13122.
26. Jones AP, D'Vaz N, Meldrum S et al. 25-hydroxyvitamin D3 status is associated with developing adaptive and innate immune responses in the first 6 months of life. *Clin Exp Allergy* 2015 Jan; 45:220–31.
27. Falth-Magnusson K, Kjellman NI. Development of atopic disease in babies whose mothers were receiving exclusion diet during pregnancy: A randomized study. *J Allergy Clin Immunol* 1987; 80:868–75.
28. Victoria CG, Bahle R, Barros AJD et al. Breastfeeding in the 21st century: Epidemiology, mechanisms, and lifelong effect. *Lancet* 2016; 387:475–90.
29. Lodge CJ, Tan DJ, Lau MX et al. Breastfeeding and asthma and allergies: A systematic review and meta-analysis. *Acta Paediatrica* 2015; 104:38–53.
30. Kramer MS, Matush L, Vanilovich I et al. Effect of prolonged and exclusive breast feeding on risk of allergy and asthma: Cluster randomised trial. *BMJ* 2007; 335(7624):815.
31. Palmer DJ, Gold MS, Makrides M. Effect of maternal egg consumption on breast milk ovalbumin concentration. *Clin Exp Allergy* 2008; 38(7):1186–91.
32. Metcalfe JR, Marsh JA, D'Vaz N et al. Effects of maternal dietary egg intake during early lactation on human milk ovalbumin concentration: A randomized controlled trial. *Clin Exp Allergy* 2016; 46:1605–13.
33. Australasian Society of Clinical Immunology and Allergy. ASCIA Infant feeding advice. 2010.
34. Muraro A, Halken S, Arshad SH et al. EAACI food allergy and anaphylaxis guidelines: Primary prevention of food allergy. *Allergy* 2014; 69:590–601.
35. Fleischer DM, Sicherer S, Greenhawt M et al. Consensus communication on early peanut introduction and the prevention of peanut allergy in high-risk infants. *J Allergy Clin Immunol* 2015; 136(2):258–61.
36. Boyle RJ, Ierodiakonou D, Khan T et al. Hydrolysed formula and risk of allergic or autoimmune disease: Systematic review and meta-analysis. *BMJ* 2016; 352:i974.
37. Australasian Society of Clinical Immunology and Allergy. ASCIA guidelines for allergy prevention in infants 2016. Available at https://www.allergy.org.au/images/pcc /ASCIA_Guidelines_infant_feeding_and_allergy_prevention.pdf.
38. Grimshaw KE, Allen K, Edwards CA et al. Infant feeding and allergy prevention: A review of current knowledge and recommendations. A EuroPrevall state of the art paper. *Allergy* 2009; 64(10):1407–16.
39. Ierodiakonou D, Garcia-Larsen V, Logan A et al. Timing of allergenic food introduction to the infant diet and risk of allergic or autoimmune disease: A systematic review and meta-analysis. *JAMA* 2016; 316:1181–92.
40. Koplin JJ, Allen KJ. Optimal timing for solids introduction: Why are the guidelines always changing? *Clin Exp Allergy* 2013; 43:826–34.

41. Du Toit G, Roberts G, Sayre PH et al. Randomized trial of peanut consumption in infants at risk for peanut allergy. *N Engl J Med* 2015; 372:803–13.
42. Palmer DJ, Sullivan TR, Gold MS et al. Randomized controlled trial of early regular egg intake to prevent egg allergy. *J Allergy Clin Immunol* 2017; 139(5):1600–7; doi: 10.1016/j.jaci.2016.06.052.
43. Palmer DJ, Metcalfe J, Makrides M et al. Early regular egg exposure in infants with eczema: A randomized controlled trial. *J Allergy Clin Immunol* 2013; 132(2):387–92e1.
44. Bellach J, Schwarz V, Ahrens B et al. Randomized placebo-controlled trial of hen's egg consumption for primary prevention in infants. *J Allergy Clin Immunol* 2017; 139(5):1591–9. doi: 10.1016/j.jaci.2016.06.045.
45. Tan JW, Valerio C, Barnes EH et al. A randomized trial of egg introduction from 4 months of age in infants at risk for egg allergy. *J Allergy Clin Immunol* 2017; 139(5):1621–8. doi: 10.1016/j.jaci.2016.08.035.
46. Perkin MR, Logan K, Tseng A et al. Randomized trial of introduction of allergenic foods in breast-fed infants. *N Engl J Med* 2016; 374(18):1733–43.
47. Perkin MR, Logan K, Marrs T et al. Enquiring About Tolerance (EAT) study: Feasibility of an early allergenic food introduction regimen. *J Allergy Clin Immunol* 2016; 137(5):1477–86.
48. Netting MJ, Campbell DE, Koplin JJ et al. An Australian consensus on infant feeding guidelines to prevent food allergy: Outcomes from the Australian infant feeding summit. *J Allergy Clin Immunol Pract* 2017 May 9. pii: S2213–2198(17)30184–8. doi: 10.1016/j.jaip.2017.03.013. [Epub ahead of print]
49. Togias A, Cooper SF, Acebal ML et al. Addendum guidelines for the prevention of peanut allergy in the United States: Report of the National Institute of Allergy and Infectious Diseases-Sponsored Expert Panel. *Pediatr Dermatol* 2017; 34(1):e1–e21.
50. Boyce JA, Assa'ad A, Burks AW et al. Guidelines for the diagnosis and management of food allergy in the United States: Report of the NIAID-sponsored expert panel. *J Allergy Clin Immunol* 2010; 126:S1–58.

22 Epigenetics, Nutrition, and Infant Health

Philip T. James, Matt J. Silver,
and Andrew M. Prentice

CONTENTS

INTRODUCTION

The field of epigenetics is currently garnering a great deal of interest, exploring how our very molecular makeup in the form of modifications to the genome can be altered by factors as diverse as aging, disease, nutrition, stress, alcohol, and exposure to pollutants. Epigenetic changes have previously been implicated in the etiology of a variety of diseases [1], notably in the development of certain cancers [2], and inherited growth disorder syndromes [3], but the exploration of epigenetics' role in fetal programming is still in its infancy. This chapter focuses on how nutritional exposures during pregnancy may affect the infant epigenome, and the impact that such modifications may have on the long-term health of the child. We start by describing some keys concepts in epigenetics and discuss windows of epigenetic plasticity in the context of the *developmental origins of health and disease* (DOHaD) hypothesis. We then review some of the key mechanisms by which nutrition can affect the epigenome, with a particular focus on the role of one-carbon metabolism. We finish by outlining some of the child health outcomes that have been linked to epigenetic dysregulation, and discuss possible next steps that need to be realized if insights into the basic science of epigenetics are to be translated into tangible public health benefits.

A BRIEF PRIMER ON EPIGENETICS

Epigenetic processes describe changes to the genome that can alter gene expression without changing the underlying DNA sequence [4]. These changes are mitotically heritable, and involve the interplay of DNA methylation, histone modifications, and RNA-based mechanisms (Figure 22.1). DNA methylation most commonly occurs at loci where a cytosine is found next to a guanine on a DNA strand along its linear sequence, hence termed *cytosine-phosphate-guanine* or *CpG sites*. It involves the covalent bonding of a methyl (CH$_3$) group to the cytosine at the 5′ carbon position to form 5-methylcytosine. CpGs found in high densities are termed *CpG islands*. Roughly two-thirds of human genes contain CpG islands in their promoter regions, although repetitive elements in the genome can also contain many CpG sites [5]. CpGs are generally methylated in nonpromoter regions and unmethylated at promoter regions. Methylation at CpG sites in promoters is usually associated with transcriptional silencing, although not consistently [6]. Although methylation is the most studied chemical alteration to date, others modifications (e.g., hydroxymethylation) can occur at cytosine bases [7].

DNA methylation is catalyzed by DNA methyltransferases (DNMTs). Mammals have three types of DNMT. DNMT1 recognizes hemimethylated DNA; therefore, after DNA replication and cell division, it methylates the newly synthesized strand to

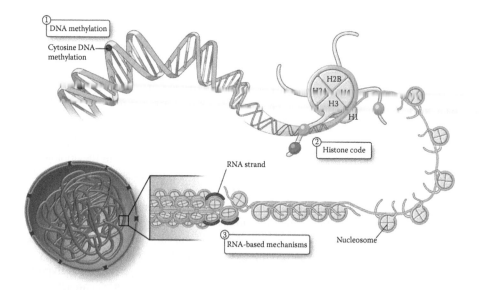

FIGURE 22.1 An overview of epigenetic mechanisms. (From Yan MS-C, Matouk CC, Marsden PA, *J Appl Physiol* 2010, 109:916–26.)

maintain methylation patterns of the original template strand (maintenance methylation). DNMT3a and 3b appear to primarily methylate fully unmethylated CpG sites (*de novo* methylation) [7]. DNA methylation is important for a host of biological processes, including transcriptional silencing, X-chromosome inactivation, genomic imprinting, and the maintenance of cellular identity by enabling tissue-specific gene expression [8].

DNA is tightly woven around histone proteins, forming compact complexes of DNA and protein called nucleosomes. Nucleosomes, in turn, are packed together to form chromatin. Within the nucleosomes, the histone proteins are arranged in an eight-part formation, comprising two copies each of histones H2A, H2B, H3, and H4. Histone modifications involve various posttranslational chemical alterations to the amino acids of the histone tails, including acetylation of lysine, methylation of lysine and arginine, phosphorylation of serine and threonine, and the ubiquitination of lysine [9]. There are several mechanisms by which chemical modifications of CpG sites and histones are thought to influence gene expression. The methyl group from 5-methylcytosine may block transcription factors either directly or through the recruitment of a methyl-binding protein. Alternatively, the DNMT enzymes acting on CpG sites may be physically linked to other enzymes, which bring about histone methylation and deacetylation [10]. Although chromatin remodeling is intricately controlled, a simplified summary is that the hyperacetylation of histones and hypomethylation of histones and CpGs is associated with a euchromatin (open) configuration, generally associated with facilitation of transcriptional activity. Conversely, hypoacetylation of histones and hypermethylation of histones and CpGs is associated with a heterochromatin (closed) structure and transcriptional repression [11].

Although noncoding RNAs do not code for proteins, many are functional and may affect gene expression [12]. Of those that influence gene expression, microRNAs (miRNAs) have been the most studied to date. These are short pieces (~22 nucleotides) of RNA that affect the epigenome through binding to target mRNAs controlling the expression of key regulators such as DMNTs and histone deacetylases [13]. In turn, CpG methylation and histone modifications can influence the transcription of certain miRNA classes [11]. MicroRNAs may also affect gene expression directly by binding to messenger RNAs, repressing their translation [14].

TIME POINTS OF PLASTICITY IN THE EPIGENOME

Times of increased cell turnover, such as during fetal development and infancy, may be particularly susceptible both to epigenetic errors and to environmental influences [15]. In this chapter, we focus on DNA methylation and the *in utero* period, to include periconception, since this falls within the first 1,000 days window, and is also a period of exceptionally rapid cell differentiation and complex epigenetic remodeling (Figure 22.2). In the first 48 hours after fertilization, there is rapid (active) demethylation of the paternal genome and a slower (passive) demethylation of the maternal genome [16]. Erasing the epigenetic marks in the zygote prior to the blastocyst stage is important to enable pluripotency of the developing cells [8]. Imprinted genes and some retrotransposons (defined later) are known to resist demethylation at this stage [17]. Remethylation then occurs in tissue-specific patterns after implantation,

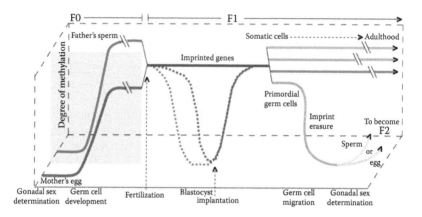

FIGURE 22.2 Epigenetic remodeling in embryogenesis. (From Perera F, Herbstman J, *Reprod Toxicol* 2011, 31:363–73.)

during the process of gastrulation and differentiation of the somatic cells through-out pregnancy. A second wave of demethylation occurs during the epigenetic repro-gramming of primordial germ cells (PGCs) in the developing embryo at the point of their migration to the genital ridge [18]. Parental imprints are erased at this stage in preparation for the laying down of sex-specific imprints in the PGCs. Remethylation of sperm cells occurs before the birth of the child, and in oocytes over the duration of their maturation [18]. The periconceptional period is therefore one of huge dyna-mism in the methylome, representing a window in which epigenetic errors could have significant consequences for the health of the child.

Which Parts of the Epigenome Are Most Susceptible to Environmental Influences?

There are some regions of the genome that demonstrate increased interindividual epigenetic variation, and may be particularly vulnerable to the impact of environ-mental influences [19]. These include imprinted genes, metastable epialleles, and transposable elements. *Imprinted genes* show monoallelic expression, whereby only the maternally or paternally inherited allele is expressed [20]. If a gene is "paternally expressed" it means the expressed allele comes from the father and the maternal allele is imprinted (silenced). In regard to growth, paternally expressed genes tend to promote *in utero* growth, whereas maternally expressed genes restrict growth, and this forms the basis of the "parental conflict theory" [21]. *Metastable epial-leles* (MEs) are genomic loci whose methylation state varies between individuals, but where variation is correlated across tissues originating from all germ layers in a single individual [22]. This indicates that the marks have been laid down in the first few days after conception before cell types start to specialize. MEs therefore provide a useful device to study the influence of the periconceptional environment, including maternal nutrition, on the offspring epigenome [23,24]. *Transposable elements* (TEs) are small pieces of DNA (usually of viral origin earlier in human history) that are mobile and can insert into new chromosomal locations throughout the genome. They are thought to make up more than half of the human genome [25]. TEs arise either through the use of RNA as an intermediate for transposition (DNA is transcribed to RNA, reverse transcribed to DNA, and then inserted by reverse transcriptase to form a retrotransposon), or through the complete DNA sequence being cut and pasted directly (forming DNA transposons, which are less common in the human genome). TEs are potentially functionally disruptive, for example, if transposed into a functional gene or when increasing copy number, and this may be one reason why most are silenced epigenetically [26]. Some TEs are thought to be vulnerable to the influence of nutrition at key time points [27], and their variable methylation patterns have been shown to affect neighboring gene expression, most notably in the Agouti mouse experiments detailed later in this chapter.

EPIGENETICS AND THE DEVELOPMENTAL ORIGINS OF HEALTH AND DISEASE

The DOHaD hypothesis describes the idea that environmental insults experienced early in life can increase the risk of adverse health outcomes throughout the life course. DOHaD grew out of David Barker's seminal work following a cohort born in Hertfordshire between 1911 and 1930. His early studies found an inverse relationship between birth weight and blood pressure at age 10, with a stronger association at age 36 [28]. His findings soon expanded to identifying associations between low birth weight and adult-onset chronic disease [29], patterns that were also seen in different cohorts such as the Nurses' Health Study I & II [30] and in Helsinki [31].

Although low birth weight provides a useful proxy for an adverse intrauterine environment, subsequent studies have attempted to pinpoint time points of vulnerability more precisely. Data from famine studies have been particularly useful in this respect. The Dutch Hunger Winter occurred toward the end of World War II, when the Western Netherlands was under German control from November 1944 to May 1945. Nazi blockades cut off food and fuel, and the siege was coupled with a harsh and early winter. Caloric intake varied from 500 to 1,000 kcal per day, depending on the area and time period. An estimated 4.5 million people were affected, 20,000 of whom died [32]. Exposure to famine during pregnancy has been associated with a wide range of offspring phenotypes, from lower birth weight [33] to increased adult blood pressure [34], obesity [35], and risk of schizophrenia [36]. Furthermore the specific timing of famine exposure *in utero* appears to have different programming effects. For example, exposure in midgestation is associated with a doubling of the prevalence of obesity for men aged 18 to 19, yet exposure in the third trimester is associated with lower obesity [35]. Many of these associations are also found from records spanning the Chinese Great Leap Forward, where famine was particularly severe from 1959 to 1961. For example, famine exposure *in utero* is associated with a doubled risk of schizophrenia in later life [37], as well as increased hyperglycemia at age 41 to 42, a trend that is exacerbated if an affluent diet was consumed later in life [38].

Although these two famine studies provide rich epidemiological evidence that nutritional exposures in early life are associated with later disease risk, they are unable to determine precise causal factors, for example, if the effects are driven by depleted maternal energy, deficient levels of certain micronutrients, or a combination of both. The Pune Maternal Nutrition Study (PMNS) in the Maharashtra State of West-Central India sheds some light on more specific nutritional exposures that may be relevant for DOHaD mechanisms. PMNS is a prospective cohort of 797 women, followed pre- and throughout pregnancy, and their offspring. Higher maternal intake of green leafy vegetables, milk, and fruit, and higher erythrocyte folate concentrations, were associated with larger size babies [39]. Higher maternal folate status at 28 weeks gestation predicted higher offspring adiposity and insulin resistance at the age of 6, low maternal vitamin B_{12} status was associated with offspring insulin resistance, and a combination of high maternal folate and low B_{12} produced the strongest associations [40].

Attention is now shifting to an investigation of the mechanisms that may mediate the observations above. In this respect, epigenetic modifications to the genome are emerging as a leading candidate that may, at least partially, explain some of the observations described in the DOHaD literature. For example, Heijmans et al. selected 60 individuals conceived during the Dutch Hunger Winter, and compared them with nonexposed siblings. Five CpG sites within *IGF2* (a maternally imprinted gene controlling fetal growth) were less methylated in individuals exposed to famine at periconception, but there was no difference in methylation for individuals exposed to famine in late gestation [41]. Indeed, the most recent study from the same group confirms that exposure to famine in the first 10 weeks of gestation shows a greater signature in the adult blood methylome (58–59 years) compared to exposure later in gestation [42]. Whilst the exact mechanisms are so far unknown, one concept being debated is the "thrifty epigenome" hypothesis. This proposes that early life insults mold an epigenetic signature to program a phenotype that is "adapted" to the intrauterine environment, which may be problematic if the environment into which the child is born then changes [43]. Under this hypothesis, malnutrition in pregnancy could program "thrifty" epigenotypes, designed to reduce metabolic rate and store energy in an attempt to adapt to a nutritionally poor environment, but may subsequently trigger symptoms of metabolic disease if the environment changes to one of relative nutritional abundance.

MATERNAL NUTRITION EXPOSURES AND THE INFANT EPIGENOME

There are a variety of factors, both nutrition-related and otherwise, which may impact the infant epigenome *in utero* through maternal exposure (Figure 22.3). Of these exposures, particular attention has been paid to the role of one-carbon metabolites in the periconceptional period and during embryonic development [44]. In the following section, we give an overview of one-carbon metabolism, and discuss evidence for the influence of one-carbon metabolites on the infant epigenome. We also briefly review other maternal nutrition exposures that have generated research in a human intergenerational setting: maternal BMI, polyunsaturated fatty acids, and vitamin C.

ONE-CARBON METABOLISM

The one-carbon pool is composed of one-carbon units and the two main carriers that activate, transport, and transfer these units: tetrahydrofolate (THF) and S-adenosyl

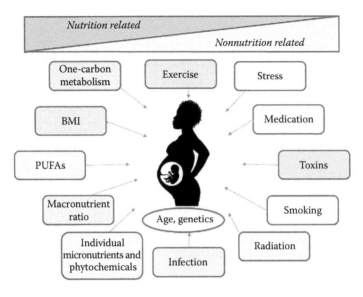

FIGURE 22.3   Potential maternal factors impacting the fetal epigenome.

methionine (SAM). One-carbon units are used as substrates for a whole range of intricate biochemical processes, including cellular biosynthesis, redox status regulation, and genome maintenance through the regulation of nucleotide pools. However, it is their role in transmethylation reactions that is central to the notion of diet-epigenome interplay, and this requires an understanding of how folate, methionine, homocysteine, transsulfuration, and transmethylation metabolic pathways interact (Figure 22.4).

S-adenosyl methionine (SAM) methylates a wide variety of acceptors in reactions catalyzed by methyl transferases. Over 200 methylation reactions are required for transcription, translation, protein localization, and signaling purposes [45], but it is the methylation of cytosine bases and amino acids on histone tails that play a role in epigenetics. The donation of SAM's methyl group forms S-adenosyl homocysteine (SAH), which is further hydrolyzed to homocysteine (Hcy), where it is maintained in an equilibrium state that thermodynamically favors SAH over Hcy [46]. A buildup of Hcy results in an increase in SAH, which in turn impedes methylation reactions since SAH competes with SAM for the active site on methyl transferase enzymes [47]. The SAM:SAH ratio is therefore often used as a proxy indicator of methylation potential [48]. In order to maintain favorable methylation conditions, Hcy has to be removed from the system. One way in which this can happen is by accepting a methyl group to form methionine, which can then in turn be condensed with ATP to form SAM and continue the cycle. Hcy can also be removed through its irreversible degradation to cystathionine and cysteine in the transsulfuration pathway requiring vitamin B_6.

The methylation of Hcy to methionine uses two distinct pathways. The major pathway is the vitamin B_{12}-dependent reaction involving folate metabolic pathways. A stepwise reduction of dietary folates and folic acid forms tetrahydrofolate (THF).

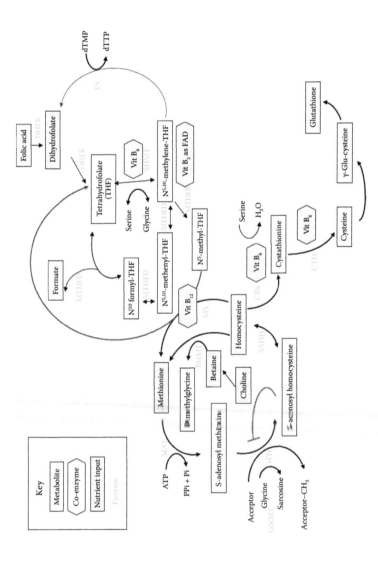

FIGURE 22.4 A simplified summary of one-carbon metabolism. BHMT, betaine homocysteine methyltransferase; CBS, cystathionine-beta-synthase; CTH, cystathionine gamma-lyase; DHFR, dihydrofolate reductase; dTMP, deoxythymidine monophosphate; dTTP, deoxythymidine triphosphate; FAD, flavin adenine dinucleotide; GNMT, glycine N methyltransferase; MAT, methionine adenosyltransferase; MS, methionine synthase; MT, methyl transferases; MTHF, methylenetetrahydrofolate; MTHFD, methylenetetrahydrofolate dehydrogenase; SAHH, S-adenosyl homocysteine hydrolase; SHMT, serine hydroxymethyltransferase; TE, thymidylate synthase.

The one-carbon units attach to the N^5 or N^{10} position, or bridge the two. THF can be converted into $N^{5,10}$-methylene-THF using a methyl group primarily from serine with vitamin B_6 as a cofactor. $N^{5,10}$-methylene-THF is either used in the thymidylate synthase pathway for DNA synthesis and repair, interconverted to $N^{5,10}$-methenyl-THF, N^{10}-formyl-THF (used in purine synthesis) and formate (an important one-carbon donor) forms, or irreversibly reduced to N^5-methyl tetrahydrofolate (methyl-THF) through the action of 5,10-methylene THF reductase (MTHFR) and vitamin B_2 as a cofactor. It is methyl-THF that then provides the methyl group for the methylation of Hcy, and in the process regenerates THF. The alternative pathway for the methylation of Hcy, predominantly used in the liver and kidneys, uses the methyl group from betaine, a product formed through the oxidation of choline. This description and Figure 22.4 give a very simplified overview of the complex and interlinking metabolic pathways involved, some of which occur with different enzymes in different cellular compartments [49]. However, this brief overview helps explain why deficiencies in the key methyl donors of choline, betaine, or folate, alongside deficiencies in vitamins B_2, B_6, and B_{12}, which play essential coenzyme roles, may disrupt the metabolic pathways that are responsible for DNA methylation. Genetic variants in proteins coding for enzymes used in key one-carbon metabolic processes can also affect activity at critical nodes and alter the flow of metabolites [50].

STUDIES INVESTIGATING ONE-CARBON METABOLITES AND EPIGENETICS

Perhaps the most famous animal experiments demonstrating how epigenetic changes driven by maternal diet in pregnancy can dramatically alter phenotype in the offspring come from the agouti mouse. In one experiment, pregnant dams were fed a diet that varied in methyl donor content (folic acid, choline, betaine, and vitamin B_{12}). Their isogenic pups showed variable methylation at an intracisternal A particle (IAP), a retrotransposon upstream of the *agouti* gene that is a metastable epiallele. The degree of methylation at this locus altered expression of the *agouti* gene, resulting in permanent phenotypic differences. The most obvious change was in fur color, but differences were also found in appetite, adiposity, and glucose tolerance, factors highly relevant to life-long chronic disease risk [19,27]. A similar experiment in a different strain of kinky-tailed mice showed that methyl donor content of the maternal diet also altered methylation at an IAP on the *Axin* gene, producing pups with varying levels of tail kink [51].

Human studies exploring associations between individual one-carbon metabolites and epigenetic effects are few and far between, especially those involving an intergenerational study design. Homocysteine and vitamin B_6 are two examples that are described in human cross-sectional studies assessing immediate effects on health at one time point within a single generation. Hyperhomocysteinemia has been associated with hypomethylation in several studies [52] supporting the hypothesis that epigenetic mechanisms may play a partial role in observed associations between hyperhomocysteinemia and decreased cardiovascular health via inflammation, free radical formation, and atherogenesis [53]. Low vitamin B_6 levels may decrease $N^{5,10}$-methenyl-THF through reduced serine hydroxymethyltransferase (SHMT) activity (Figure 22.4), stressing the thymidylate synthase pathway for DNA repair and incorporating more uracil into DNA. This reduces genome stability through increased chromosome breaks, which, in turn, have been associated with increased levels of global DNA hypomethylation alongside a putative increased risk of tumor development [54].

Of the human studies using an intergenerational design, folate, vitamin B_{12}, and choline have been most frequently investigated. Due to its role in preventing neural tube defects and involvement in some cancer therapies, folic acid has been extensively reviewed [55]. Increased periconceptional folic acid consumption has been associated with increased methylation in infants at a differentially methylated region (DMR) of the *IGF2* gene, although these associations have not been established consistently [56]. Additional studies have found that maternal folate levels in late pregnancy influence offspring methylation patterns in the imprinting control region of *IGF2* [57] and also in the imprinting regulator *ZFP57* [58]. Despite the timing of maternal folate measurements, these findings support the importance of nutritional status around conception, since many DMRs appear to have their methylation pattern established prior to gastrulation [59]. There are fewer studies assessing the effect of maternal vitamin B_{12} status on the infant epigenome. However, preliminary evidence suggests that maternal vitamin B_{12} levels at first antenatal visit are inversely correlated with infant cord blood global methylation levels [60]. For choline, one intervention study investigated the effect of supplementing mothers in the third trimester with daily choline of 480 mg versus 930 mg [61]. Higher maternal choline intake was associated with increased methylation of *CRH* and *NR3C1* promoter regions within fetal placental tissue.

The complex interlinking metabolic pathways involved in one-carbon metabolism may be more appropriately investigated by exploring the joint effects of relevant methyl donors and coenzymes. Researchers in The Gambia have been exploring human diet–epigenome interactions by exploiting a natural experimental design, in which fluctuations in energy balance and maternal nutritional exposures display a distinct bimodal seasonal pattern. The Gambia experiences a rainy ("hungry") season from July to September (Figure 22.5a), a planting season with increased energy expenditure, depleted food stores, and peaks of malarial and diarrheal diseases. The dry ("harvest") season occurs from February to April (Figure 22.5b), when

(a)   (b)

FIGURE 22.5   The Gambian (a) rainy/"hungry" and (b) dry/harvest seasons in Keneba, West Kiang. (Photo by Andrew M. Prentice.)

harvesting takes place, leading to improved food security. Despite overall house-hold food stores being more replete in the dry season, the rainy season offers the increased availability of certain micronutrients made available from rain-fed green leafy vegetables. It has been known for several decades that Gambian children born during the rainy season are up to 10 times more likely to die prematurely in young adulthood than those born during the dry season [62]. However, the processes under-lying these findings have yet to be fully delineated, and nutrition-related epigenetic regulation in early embryonic development is emerging as one plausible contributory mechanism. A recent study compared women conceiving in the peak of the dry and the peak of the rainy season [24]. Maternal periconceptional plasma concentrations of folate, vitamin B_2, methionine, betaine, and the SAM:SAH and betaine:dimethyl glycine ratios were higher in the rainy season, and concentrations of vitamin B_{12}, homocysteine, and SAH were lower. This suggested that the maternal metabolome during the rainy season contained higher concentrations of methyl donors and exhib-ited a higher methylation potential than the dry season metabolome. Indeed, off-spring of these rainy season conceptions had higher levels of CpG methylation at six MEs in peripheral blood monocytes compared to the offspring of those conceived in the dry season, with similar patterns in hair follicle DNA [24]. A subsequent study details the genome-wide search for MEs susceptible to the periconceptional environ-ment using two independent approaches [63]. Both found *VTRNA2-1* as their top hit, an imprinted gene that was also found to exhibit increased hypomethylation among Gambian offspring conceived in the dry season compared to the rainy season. Taken together, the results support the hypothesis that maternal nutrition during the peri-conceptional period can impact the infant epigenome in humans and has triggered further research into the possible phenotypic consequences.

POLYUNSATURATED FATTY ACIDS

Several studies indicate that increased consumption of ω-3 PUFAs is associated with reduced homocysteine levels [64]. In the context of early life nutrition, maternal fatty acid intake influences infant fatty acid composition via placental transfer and breast milk, with maternal PUFA intake potentially affecting infant appetite control, neuroen-docrine function, and metabolic programming [65]. Investigating whether one-carbon pathways and epigenetic mechanisms are involved is of increasing interest. In rodent models, providing a maternal diet with excess folic acid and restricted vitamin B_{12} (designed to reflect the nutritional situation in much of rural India) is associated with increased maternal oxidative stress, lower level of offspring placental and brain doco-sahexaenoic acid, and decreased placental global methylation [66]. However, maternal supplementation with PUFAs seems to partially ameliorate the disrupted one-carbon metabolism associated with this diet [67]. One potential mechanism is through PUFAs upregulating enzymes responsible for the methylation of homocysteine to methionine [68], which in theory could increase the SAM:SAH ratio and help overcome the impact of reduced MTHFR activity brought about by the restricted vitamin B_{12}. An additional recent hypothesis describes how a decreased maternal PUFA intake may increase the availability of methyl groups for DNA methylation since there would be reduced PUFAs for phosphatidylcholine synthesis and therefore less demand for methyl groups for this pathway. The disruption to the one-carbon pathways brought about by low PUFA levels is thought to alter epigenetic programming of placental genes, increasing the risk of aberrant methylation in offspring and adverse pregnancy outcome [69]. Human evidence at the intergenerational level, however, is currently limited to a study by Lee et al. [70]. Here, mothers were given either 400 mg of ω-3 PUFA (n = 131) or placebo (n = 130) daily from approximately 18 weeks gestation to birth. Among mothers supplemented with PUFAs, there was increased global methylation (measured at LINE-1 repetitive elements) in offspring of mothers who smoked (n = 26 case, 26 control) [71], and increased offspring *IGF2* promoter 3 methylation in preterm infants (n = 16 case, 20 control) [70].

BODY MASS INDEX

Maternal body mass index (BMI) shows some association with infant epigenetic patterns [72]. However, it can be difficult to define exactly what BMI represents as an exposure, since it does not always correlate well with body composition [73] and is a proxy for a wide range of potential metabolic disorders [74]. In an intergenerational setting, both maternal underweight and overweight is associated with epigenetic patterns in infant cord blood [72]. The mechanism by which BMI influences epigenetics is so far unknown. Although SAM has been shown to be independently, positively associated with fat mass in older adults, alongside a higher rate of converting methionine into SAM in obese individuals, the absence of a rise in the SAM:SAH ratio means conclusions are hard to draw [75]. Paternal BMI may also be a relevant exposure to consider, since neonates born to obese fathers in a U.S. study showed hypomethylation at some loci within imprinted genes (*MEST*, *PEG3*, *NNAT*), independent of maternal obesity [76].

VITAMIN C AND DEMETHYLATION

As outlined earlier, the periconceptional period is a time of widespread remodeling of the offspring epigenome, including rapid erasure (demethylation) of parental epigenetic marks and subsequent remethylation (Figure 22.2). In demethylation, 5-methylcytosine (5mC) can be sequentially oxidised to different states by ten-eleven translocation (TET) dioxygenases that use vitamin C (ascorbate) as a cofactor [77]. *In vitro* studies, for example, indicate that adding vitamin C to mouse or human embryonic stem cells increases activity of TET enzymes, with active demethylation seen in the germline and associated gene expression changes [78]. Vitamin C is therefore an important factor to consider in epigenetic processes during early embryogenesis, although *in vivo* studies will be required to explore whether vitamin C deficiency could play a role in aberrant demethylation.

In this section we have identified some of the key ways that maternal nutrition might influence the offspring epigenome. Other exposures, both nutrition-related and otherwise, may also be involved. These include maternal stress [79], toxin exposure [80], maternal hyperglycemia [81], the microbiome [82], dietary polyphenols [83], vitamin D [84], vitamin A [85], and infection [86]. As evidence builds, it will be necessary to explore how these multiple exposures work together to influence the infant epigenome and to more clearly define the potential consequences for human health. It is to the latter point that we now turn.

PHENOTYPES ASSOCIATED WITH THE IMPACT OF MATERNAL EXPOSURES ON THE INFANT EPIGENOME

Most population-based epigenetic studies to date have been limited to exploring either associations between maternal exposures and infant epigenetic patterns, or between infant epigenetic patterns and later phenotypes. Evidence linking maternal exposure to infant phenotypes with epigenetics as a mediating mechanism is scarce.

The first category of associations, linking maternal exposures to infant epigenetic patterns, was reviewed earlier. A number of studies linking infant epigenetic patterns with later phenotypes have focused on imprinted genes as candidate loci, due to their role in fetal growth [17] and a range of diseases [3,20]. It has been suggested that imprinted genes may be more vulnerable to the effects of epigenetic dysregulation, since epigenetic mechanisms underpin their monoallelic expression. On chromosome region 11p15.5 there are two imprinting control regions (ICRs) of interest: the H19/IGF2 domain (ICR1) and the KCNQI/CDKN1C domain (ICR2). Hypomethylation of ICR1 and hypermethylation of ICR2 are associated with Russell-Silver syndrome (RSS; an undergrowth disorder), whereas hypermethylation of ICR1 and hypomethylation of ICR2 are associated with Beckwith-Wiedemann syndrome (BWS; an overgrowth disorder) [3]. In addition to the aberrant methylation found at ICR1 and ICR2, some studies suggest patients with RSS and BWS show methylation defects at multiple gene loci [87]. Evidence for associations between growth-related phenotypes (other than RSS and BWS) and methylation patterns at several imprinted gene loci is slowly accumulating. Other phenotypes investigated to date include intrauterine growth restriction [88], small for gestational age [89], birth weight [90] and later adiposity [91].

Several studies describe associations between maternal one-carbon metabolites and infant growth-related outcomes, for example, vitamin B_{12} [92], folate [93], and homocysteine [94]. What then is the evidence that these associations are mediated through epigenetic mechanisms? Preliminary evidence from the Dutch Hunger Winter, as described earlier, suggests that exposure to famine in pregnancy, particularly around conception, is associated with differential methylation in genes linked to growth, and development [41], and that famine exposure is also related to a wide range of offspring cognitive health and cardiometabolic risk factors six decades later [95]. It is, however, difficult to establish the direction of causality, since disease states can also influence the epigenome [96]. This issue of reverse causality is particularly pertinent to studies using a retrospective cohort design. Stronger evidence comes from prospective cohorts, such as the Newborn Epigenetics Study (NEST) in the United States. Hoyo et al. describe a positive association between maternal folate levels in the first trimester and birth weight [90]. Increased maternal folate was also associated with increased methylation at *MEG3*, *PLAGL1*, and *PEG3* in infant cord blood, and decreased methylation at *IGF2*. Five differentially methylated sites were associated with birth weight, and it was speculated that the association seen between maternal folate and birth weight could be mediated by differential methylation at *MEG3*, *H19*, and *PLAGL1*. In a more recent study from the same cohort, McCullough et al. found maternal plasma concentrations of homocysteine in the first trimester were inversely associated with birth weight, particularly in males [97]. Children born to mothers with the highest quartile of plasma B_{12} showed lower weight gain between birth and 3 years. However, only maternal vitamin B_6 was positively associated with cord methylation at a DMR from *MEG3*. A further example comes from Godfrey et al., who found that higher methylation of *RXRA* and *eNOS* in umbilical cord tissue was associated with offspring adiposity at age 9, and that higher *RXRA* methylation was also associated with lower maternal carbohydrate intake in early pregnancy [91].

Given the scarcity of existing data and the many potential confounders in cohort studies, there is not yet any clear consensus on the extent to which epigenetics is involved in the etiology of suboptimal growth, particularly in the more widely investigated phenotypes of IUGR and SGA [98]. Similarly, although prepregnancy folic acid supplementation is known to prevent neural tube defects, the extent to which this protection is epigenetically mediated is not yet clear [56]. In the next few years, a number of prospective cohort studies should shed further light on the links between early life exposures and phenotypes, mediated by epigenetic mechanisms. In the meantime, we can use the literature from the range of studies described earlier to speculate that periconceptional nutritional insults may influence epigenetic mechanisms in the infant, causing disruption at multiple loci and potentially leading to a broad spectrum of phenotypic consequences in later life.

NEXT STEPS

The field of epigenetic epidemiology is in its infancy. Future challenges include the need to consider a much larger number of candidate genetic regions (a challenge that will be met as rapid improvements in technology facilitate epigenome-wide association studies with ever increasing genomic resolution); the need to focus on tissue types most appropriate to the phenotype of interest; and the need to integrate data on genetic variation, gene expression, and other epigenetic mechanisms such as histone modifications and miRNAs. Investigations will also need to assess the extent to which postnatal exposures may interact with periconceptional effects. Finally although we focus on the maternal periconceptional environment in this chapter, there is growing interest in the potential for paternal exposures to affect the infant epigenome transgenerationally via sperm methylation profiles [99]. Despite limited human evidence to date, this is an area of great potential public heath relevance.

Once disrupted epigenetic patterns can be better mapped onto phenotypes, the significance for public health impact becomes more tangible. Can we design and test nutritional interventions to "correct" a suboptimal maternal metabolome, hence reducing patterns of aberrant methylation in the infant epigenome and reducing the burden of ill health? Despite the current complexities of disentangling the web of multiple exposures influencing the infant epigenome, as technology advances, collaborations grow, and knowledge is gained, the field of nutritional epigenetics has the potential to impact the health of children not only within the first 1,000 days but also across multiple generations.

REFERENCES

1. Egger G, Liang G, Aparicio A et al. Epigenetics in human disease and prospects for epigenetic therapy. *Nature* 2004; 429:457–63.
2. Esteller M. Epigenetics in cancer. *N Engl J Med* 2008; 358:1148–59.
3. Piedrahita JA. The role of imprinted genes in fetal growth abnormalities. *Birth Defects Res A Clin Mol Teratol* 2011; 91:682–92.
4. Jaenisch R, Bird A. Epigenetic regulation of gene expression: How the genome integrates intrinsic and environmental signals. *Nat Genet* 2003; 33(Suppl):245–54.
5. Wang Y, Leung FCC. An evaluation of new criteria for CpG islands in the human genome as gene markers. *Bioinformatics* 2004; 20:1170–7.
6. Illingworth RS, Bird AP. CpG islands: "A rough guide." *FEBS Lett* 2009; 583:1713–20.
7. Hill PWS, Amouroux R, Hajkova P. DNA demethylation, Tet proteins and 5-hydroxy-methylcytosine in epigenetic reprogramming: An emerging complex story. *Genomics* 2014; 104:324–33.
8. Messerschmidt DM, Knowles BB, Solter D. DNA methylation dynamics during epigenetic reprogramming in the germline and preimplantation embryos. *Genes Dev* 2014; 28:812–28.
9. Kouzarides T. Chromatin modifications and their function. *Cell* 2007; 128:693–705.
10. Klose RJ, Bird AP. Genomic DNA methylation: The mark and its mediators. *Trends Biochem Sci* 2006; 31:89–97.
11. Richards EJ, Elgin SC. Epigenetic codes for heterochromatin formation and silencing. *Cell* 2002; 108:489–500.
12. Mattick JS, Makunin I V. Non-coding RNA. *Hum Mol Genet* 2006; 15(Spec No 1):R17–29.
13. Guil S, Esteller M. DNA methylomes, histone codes and miRNAs: Tying it all together. *Int J Biochem Cell Biol* 2009; 41:87–95.
14. Sato F, Tsuchiya S, Meltzer SJ et al. MicroRNAs and epigenetics. *FEBS J* 2011; 278:1598–609.
15. Langley-Evans SC. Nutrition in early life and the programming of adult disease: A review. *J Hum Nutr Diet* 2015; 28:1–14.
16. Seisenberger S, Peat JR, Hore TA et al. Reprogramming DNA methylation in the mammalian life cycle: Building and breaking epigenetic barriers. *Philos Trans R Soc Lond B Biol Sci* 2013; 368:20110330.
17. Reik W, Walter J. Genomic imprinting: Parental influence on the genome. *Nat Rev Genet* 2001; 2:21–32.
18. Smallwood SA, Kelsey G. De novo DNA methylation: A germ cell perspective. *Trends Genet* 2012; 28:33–42.
19. Waterland RA, Michels KB. Epigenetic epidemiology of the developmental origins hypothesis. *Annu Rev Nutr* 2007; 27:363–88.
20. Ishida M, Moore GE. The role of imprinted genes in humans. *Mol Aspects Med* 2013; 34:826–40.
21. Moore GE, Ishida M, Demetriou C et al. The role and interaction of imprinted genes in human fetal growth. *Philos Trans R Soc B Biol Sci* 2015; 370:20140074.
22. Rakyan VK, Blewitt ME, Druker R et al. Metastable epialleles in mammals. *Trends Genet* 2002; 18:348–51.
23. Waterland RA, Kellermayer R, Laritsky E et al. Season of conception in rural Gambia affects DNA methylation at putative human metastable epialleles. *PLoS Genet* 2010; 6:e1001252.
24. Dominguez-Salas P, Moore SE, Baker MS et al. Maternal nutrition at conception modulates DNA methylation of human metastable epialleles. *Nat Commun* 2014; 5:3746.

25. Lander ES, Linton LM, Birren B et al. Initial sequencing and analysis of the human genome. *Nature* 2001; 409:860–921.
26. Slotkin RK, Martienssen R. Transposable elements and the epigenetic regulation of the genome. *Nat Rev Genet* 2007; 8:272–85.
27. Waterland RA, Jirtle RL. Transposable elements: Targets for early nutritional effects on epigenetic gene regulation. *Mol Cell Biol* 2003; 23:5293–300.
28. Barker DJ, Osmond C, Golding J et al. Growth in utero, blood pressure in childhood and adult life, and mortality from cardiovascular disease. *BMJ* 1989; 298:564–7.
29. Barker DJ, Gluckman PD, Godfrey KM et al. Fetal nutrition and cardiovascular disease in adult life. *Lancet (London, England)* 1993; 341:938–41.
30. Curhan GC, Chertow GM, Willett WC et al. Birth weight and adult hypertension and obesity in women. *Circulation* 1996; 94:1310–5.
31. Eriksson JG, Forsen T, Tuomilehto J et al. Catch-up growth in childhood and death from coronary heart disease: Longitudinal study. *BMJ* 1999; 318:427–31.
32. Lumey LH, Stein AD, Kahn HS et al. Cohort profile: The Dutch Hunger Winter families study. *Int J Epidemiol* 2007; 36:1196–1204.
33. Smith CA. The effect of wartime starvation in Holland upon pregnancy and its product. *Am J Obstet Gynecol* 1947; 53:599–608.
34. Roseboom TJ, van der Meulen JH, van Montfrans GA et al. Maternal nutrition during gestation and blood pressure in later life. *J Hypertens* 2001; 19:29–34.
35. Ravelli GP, Stein ZA, Susser MW. Obesity in young men after famine exposure in utero and early infancy. *N Engl J Med* 1976; 295:349–53.
36. Susser E, Neugebauer R, Hoek HW et al. Schizophrenia after prenatal famine. Further evidence. *Arch Gen Psychiatry* 1996; 53:25–31.
37. St Clair D, Xu M, Wang P et al. Rates of adult schizophrenia following prenatal exposure to the Chinese famine of 1959–1961. *JAMA* 2005; 294:557–62.
38. Li Y, He Y, Qi L et al. Exposure to the Chinese famine in early life and the risk of hyperglycemia and type 2 diabetes in adulthood. *Diabetes* 2010; 59:2400–6.
39. Rao S, Yajnik CS, Kanade A et al. Intake of micronutrient-rich foods in rural Indian mothers is associated with the size of their babies at birth: Pune Maternal Nutrition Study. *J Nutr* 2001; 131:1217–24.
40. Yajnik CS, Deshpande SS, Jackson AA et al. Vitamin B_{12} and folate concentrations during pregnancy and insulin resistance in the offspring: The Pune Maternal Nutrition Study. *Diabetologia* 2008; 51:29–38.
41. Heijmans BT, Tobi EW, Stein AD et al. Persistent epigenetic differences associated with prenatal exposure to famine in humans. *Proc Natl Acad Sci USA* 2008; 105:17046–9.
42. Tobi EW, Slieker RC, Stein AD et al. Early gestation as the critical time-window for changes in the prenatal environment to affect the adult human blood methylome. *Int J Epidemiol* 2015; 44:1211–23.
43. Stöger R. The thrifty epigenotype: An acquired and heritable predisposition for obesity and diabetes? *BioEssays* 2008; 30:156–66.
44. Steegers-Theunissen RPM, Twigt J, Pestinger V et al. The periconceptional period, reproduction and long-term health of offspring: The importance of one-carbon metabolism. *Hum Reprod Update* 2013; 19:640–55.
45. Lu SC, Mato JM. S-adenosylmethionine in liver health, injury, and cancer. *Physiol Rev* 2012; 92:1515–42.
46. Pajares MA, Pérez-Sala D. Betaine homocysteine S-methyltransferase: Just a regulator of homocysteine metabolism? *Cell Mol Life Sci* 2006; 63:2792–803.
47. Scotti M, Stella L, Shearer EJ et al. Modeling cellular compartmentation in one-carbon metabolism. *Wiley Interdiscip Rev Syst Biol Med* 2013; 5:343–65.

48. Mason JB. Biomarkers of nutrient exposure and status in one-carbon (methyl) metabolism. *J Nutr* 2003; 133:941S–947.
49 Stover PJ, Field MS. Trafficking of intracellular folates. *Adv Nutr* 2011; 2:325–31.
50. Fredriksen A, Meyer K, Ueland PM et al. Large-scale population-based metabolic phenotyping of thirteen genetic polymorphisms related to one-carbon metabolism. *Hum Mutat* 2007; 28:856–65.
51. Waterland RA, Dolinoy DC, Lin J-R et al. Maternal methyl supplements increase offspring DNA methylation at Axin Fused. *Genesis* 2006; 44:401–6.
52. Zhou S, Zhang Z, Xu G. Notable epigenetic role of hyperhomocysteinemia in atherogenesis. *Lipids Health Dis* 2014; 13:134.
53. Beard RS, Bearden SE. Vascular complications of cystathionine β-synthase deficiency: Future directions for homocysteine-to-hydrogen sulfide research. *Am J Physiol Heart Circ Physiol* 2011; 300:H13–26.
54. Huang JY, Butler LM, Wang R et al. Dietary intake of one-carbon metabolism-related nutrients and pancreatic cancer risk: The Singapore Chinese Health Study. *Cancer Epidemiol Biomarkers Prev* 2016; 25:417–24.
55. Crider KS, Yang TP, Berry RJ et al. Folate and DNA methylation: A review of molecular mechanisms and the evidence for folate's role. *Adv Nutr* 2012; 3:21–38.
56. Gonseth S, Roy R, Houseman EA et al. Periconceptional folate consumption is associated with neonatal DNA methylation modifications in neural crest regulatory and cancer development genes. *Epigenetics* 2015; 10:1166–76.
57. Ba Y, Yu H, Liu F et al. Relationship of folate, vitamin B$_{12}$ and methylation of insulin-like growth factor-II in maternal and cord blood. *Eur J Clin Nutr* 2011; 65:480–5.
58. Amarasekera M, Martino D, Ashley S et al. Genome-wide DNA methylation profiling identifies a folate-sensitive region of differential methylation upstream of ZFP57-imprinting regulator in humans. *FASEB J* 2014; 28:4068–76.
59. Woodfine K, Huddleston JE, Murrell A. Quantitative analysis of DNA methylation at all human imprinted regions reveals preservation of epigenetic stability in adult somatic tissue. *Epigenetics Chromatin* 2011; 4:1.
60. McKay JA, Groom A, Potter C et al. Genetic and non-genetic influences during pregnancy on infant global and site specific DNA methylation: Role for folate gene variants and vitamin B$_{12}$. *PLoS One* 2012; 7:e33290.
61. Jiang X, Yan J, West AA et al. Maternal choline intake alters the epigenetic state of fetal cortisol-regulating genes in humans. *FASEB J* 2012; 26:3563–74.
62. Moore SE, Cole TJ, Poskitt EMN et al. Season of birth predicts mortality in rural Gambia. *Nature* 1997; 388:434.
63. Silver MJ, Kessler NJ, Hennig BJ et al. Independent genomewide screens identify the tumor suppressor VTRNA2-1 as a human epiallele responsive to periconceptional environment. *Genome Biol* 2015; 16:118.
64. Huang T, Zheng J, Chen Y et al. High consumption of Ω-3 polyunsaturated fatty acids decrease plasma homocysteine: A meta-analysis of randomized, placebo-controlled trials. *Nutrition* 2011; 27:863–7.
65. Kabaran S, Besler HT. Do fatty acids affect fetal programming? *J Heal Popul Nutr* 2015; 33:14.
66. Roy S, Kale A, Dangat K et al. Maternal micronutrients (folic acid and vitamin B(12)) and omega 3 fatty acids: Implications for neurodevelopmental risk in the rat offspring. *Brain Dev* 2012; 34:64–71.
67. Khot V, Kale A, Joshi A et al. Expression of genes encoding enzymes involved in the one carbon cycle in rat placenta is determined by maternal micronutrients (folic acid, vitamin B$_{12}$) and omega-3 fatty acids. *Biomed Res Int* 2014; 2014:613078.

68. Huang T, Hu X, Khan N et al. Effect of polyunsaturated fatty acids on homocysteine metabolism through regulating the gene expressions involved in methionine metabolism. *Sci World J* 2013; 2013:1–8.

69. Khot V, Chavan-Gautam P, Joshi S. Proposing interactions between maternal phospholipids and the one carbon cycle: A novel mechanism influencing the risk for cardiovascular diseases in the offspring in later life. *Life Sci* 2015; 129:16–21.

70. Lee H-S, Barraza-Villarreal A, Biessy C et al. Dietary supplementation with polyunsaturated fatty acid during pregnancy modulates DNA methylation at IGF2/H19 imprinted genes and growth of infants. *Physiol Genomics* 2014; 46:851–7.

71. Lee H-S, Barraza-Villarreal A, Hernandez-Vargas H et al. Modulation of DNA methylation states and infant immune system by dietary supplementation with ω-3 PUFA during pregnancy in an intervention study. *Am J Clin Nutr* 2013; 98:480–7.

72. Sharp GC, Lawlor DA, Richmond RC et al. Maternal pre-pregnancy BMI and gestational weight gain, offspring DNA methylation and later offspring adiposity: Findings from the Avon Longitudinal Study of Parents and Children. *Int J Epidemiol* 2015; 44:1288–304.

73. Wells JCK, Fewtrell MS. Measuring body composition. *Arch Dis Child* 2006; 91:612–7.

74. Müller MJ, Lagerpusch M, Enderle J et al. Beyond the body mass index: Tracking body composition in the pathogenesis of obesity and the metabolic syndrome. *Obes Rev* 2012; 13(Suppl 2):6–13.

75. Elshorbagy AK, Nijpels G, Valdivia-Garcia M et al. S-adenosylmethionine is associated with fat mass and truncal adiposity in older adults. *J Nutr* 2013; 143:1982–8.

76. Soubry A, Murphy SK, Wang F et al. Newborns of obese parents have altered DNA methylation patterns at imprinted genes. *Int J Obes* 2015; 39:650–7.

77. Young JI, Züchner S, Wang G. Regulation of the epigenome by vitamin C. *Annu Rev Nutr* 2015; 35:545–64.

78. Blaschke K, Ebata KT, Karimi MM et al. Vitamin C induces Tet-dependent DNA demethylation and a blastocyst-like state in ES cells. *Nature* 2013; 500:222–6.

79. Babenko O, Kovalchuk I, Metz GAS. Stress-induced perinatal and transgenerational epigenetic programming of brain development and mental health. *Neurosci Biobehav Rev* 2014; 48:70–91.

80. Anway MD, Skinner MK. Epigenetic transgenerational actions of endocrine disruptors. *Endocrinology* 2006; 147:S43–49.

81. El Hajj N, Schneider E, Lehnen H et al. Epigenetics and life-long consequences of an adverse nutritional and diabetic intrauterine environment. *Reproduction* 2014; 148:R111–20.

82. Davie JR. Inhibition of histone deacetylase activity by butyrate. *J Nutr* 2003; 133:2485S–93S.

83. Fang M, Chen D, Yang CS. Dietary polyphenols may affect DNA methylation. *J Nutr* 2007; 137:223S–8S.

84. Pereira F, Barbáchano A, Singh PK et al. Vitamin D has wide regulatory effects on histone demethylase genes. *Cell Cycle* 2012; 11:1081–9.

85. Feng Y, Zhao L-Z, Hong L et al. Alteration in methylation pattern of GATA-4 promoter region in vitamin A-deficient offspring's heart. *J Nutr Biochem* 2013; 24:1373–80.

86. Claycombe KJ, Brissette CA, Ghribi O. Epigenetics of inflammation, maternal infection, and nutrition. *J Nutr* 2015; 145:1109S–15S.

87. Azzi S, Rossignol S, Steunou V et al. Multilocus methylation analysis in a large cohort of 11p15-related foetal growth disorders (Russell Silver and Beckwith Wiedemann syndromes) reveals simultaneous loss of methylation at paternal and maternal imprinted loci. *Hum Mol Genet* 2009; 18:4724–33.

88. Einstein F, Thompson RF, Bhagat TD et al. Cytosine methylation dysregulation in neonates following intrauterine growth restriction. *PLoS One* 2010; 5:e8887.

89. Bouwland-Both MI, van Mil NH, Stolk L et al. DNA methylation of IGF2DMR and H19 is associated with fetal and infant growth: The generation R study. *PLoS One* 2013; 8:e81731.

90. Hoyo C, Daltveit AK, Iversen E et al. Erythrocyte folate concentrations, CpG methylation at genomically imprinted domains, and birth weight in a multiethnic newborn cohort. *Epigenetics* 2014; 9:1120–30.

91. Godfrey KM, Sheppard A, Gluckman PD et al. Epigenetic gene promoter methylation at birth is associated with child's later adiposity. *Diabetes* 2011; 60:1528–34.

92. Rush EC, Katre P, Yajnik CS. Vitamin B_{12}: One carbon metabolism, fetal growth and programming for chronic disease. *Eur J Clin Nutr* 2014; 68:2–7.

93. van Uitert EM, Steegers-Theunissen RPM. Influence of maternal folate status on human fetal growth parameters. *Mol Nutr Food Res* 2013; 57:582–95.

94. Hogeveen M, Blom HJ, den Heijer M. Maternal homocysteine and small-for-gestational-age offspring: Systematic review and meta-analysis. *Am J Clin Nutr* 2012; 95:130–6.

95. de Rooij SR, Painter RC, Roseboom TJ et al. Glucose tolerance at age 58 and the decline of glucose tolerance in comparison with age 50 in people prenatally exposed to the Dutch famine. *Diabetologia* 2006; 49:637–43.

96. Relton CL, Davey Smith G. Two-step epigenetic Mendelian randomization: A strategy for establishing the causal role of epigenetic processes in pathways to disease. *Int J Epidemiol* 2012; 41:161–76.

97. McCullough LE, Miller EE, Mendez MA et al. Maternal B vitamins: Effects on offspring weight and DNA methylation at genomically imprinted domains. *Clin Epigenetics* 2016; 8:8.

98. Toure DM, Baccaglini L, Opoku ST et al. Epigenetic dysregulation of Insulin-like growth factor (IGF)-related genes and adverse pregnancy outcomes: A systematic review. *J Matern Fetal Neonatal Med* 2016; 18:1–11.

99. Soubry A. Epigenetic inheritance and evolution: A paternal perspective on dietary influences. *Prog Biophys Mol Biol* 2015; 118:79–85.

23 Fetal Origins of Obesity, Cardiovascular Disease, and Type 2 Diabetes

Herculina Salome Kruger and Naomi S. Levitt

CONTENTS

INTRODUCTION

The first 1,000 days of life is a critical window of development, determining suscepti-bility to adult obesity and cardiometabolic health [1,2]. Environmental insults during this rapid development phase may result in irreversible adverse outcomes. Animal and human studies provide evidence for the fetal origins of adult noncommunicable disease hypothesis [3–5]. This hypothesis suggests that intrauterine exposures affect the fetus's development during sensitive periods, and increases the risk of noncom-municable diseases in adult life [3–5]. Systematic reviews confirm evidence of inverse epidemiological associations between birth weight and later development of hyper-tension [2] and coronary heart disease [6]. The fetal origins of disease hypothesis has been challenged as a possible statistical artifact [7], but has been confirmed by later studies with high follow-up rates and adjustment for confounders [8,9].

Early animal studies of severe undernutrition [10] and later human studies both showed a relation between fetal stressors and subsequent development of chronic diseases [3,4]. Natural experiments such as the Dutch famine cohort showed that prenatal undernutrition was associated with low birth weight, followed by adult obesity and glucose intolerance when adults were raised outside of a famine environment [11]. Birth weights and cardiovascular death rates studied in men born during 1911 to 1930 in the United Kingdom showed that low birth weight was associated with hypertension and ischemic heart disease mortality [3]. In adults born in Finland between 1924 and 1933, and who were followed up in 1971, the incidence of type 2 diabetes increased with decreasing birth size and placental weight [12]. Based on these early studies, small birth size was regarded as a marker of poor fetal nutrition and intrauterine growth restriction, independent of gestational age. Fetal programming and low fetal growth rates increased susceptibility to adult diseases [4], and promoters of pre- and postnatal growth were regarded as protective against ischemic heart disease [3].

THE CRITICAL PERIOD DETERMINING SUSCEPTIBILITY TO ADULT CHRONIC DISEASE: FETAL OR POSTNATAL LIFE?

Fetal nutrition is a more important programming stimulus affecting fetal growth than birth weight [13]. Birth weight may be an intermediate, rather than a primary indicator of the relationship between fetal growth and adult disease. Low birth weight infants who became overweight later in life were at an increased risk of several adverse events: insulin resistance at the age of 8 years [14], higher blood pressure at the age of 50 [15], and a higher incidence of coronary heart disease during late adulthood [6,16]. Thus, childhood or adult adiposity may modify the association between birth weight and later cardiovascular outcomes, indicating that catch-up growth can modify early intrauterine stressors [14,15,17].

Many early studies reported birth weight together with weight later in life. It is difficult to distinguish if later cardiovascular outcomes are triggered by either birth weight or later liver weight, or an interaction between fetal and postnatal exposures [7] Although later studies of multiple postnatal growth measures suggest that both periods are important, it is not clear whether decreased postnatal growth or catch-up growth is harmful [1]. In a Finnish cohort, both low birth weight and accelerated growth from 0 to 7 years were associated with adult hypertension [18], whereas combined low birth weight followed by suboptimal infant growth gave rise to the highest risk of adult ischemic heart disease [19]. These results indicate a need for more serial growth data to determine the contribution of growth periods to future health outcomes [1]. Three possible explanations for the association between early growth and later cardiovascular outcomes emerge from recent studies, namely, (1) fast postnatal growth itself increases risk of later obesity and cardiovascular disease [20,21]; (2) factors related to fetal growth restriction increase risk of cardiovascular disease [15]; and (3) an interaction between the two increases risk [22]. Studies of the fetal origins of adult disease should include the fetal period and repeated growth measures through the life course. A simplified diagrammatic presentation of developmental programming is shown in Figure 23.1.

	First 1,000 Days			Postnatal Programming			
	Pregnancy		Infant Outcomes	Early Infancy to Age 2 years	Childhood	Adulthood	Adult Outcomes
	In utero Programming						
	Early fetal development	Late fetal development					
Maternal health	Early fetal development	Late fetal development					
Gestational hypertension	Organ development				Genetic susceptibility	Genetic susceptibility	Hypertension
Gestational diabetes	Hyperglycemia Hyperinsulinemia		Macrosomia	Determination of appetite	Energy, protein, and micronutrient intakes	Energy, protein, and micronutrient intakes	Type 2 diabetes
Obesity	Changes in DNA methylation			Protein intake	Overweight child	Physical activity	
Smoking	Hypoxia			Rapid early weight gain			
Stress	Brain development; Structural changes in the pancreas			Brain development	Stress responsiveness, behavior; Impaired glucose tolerance	Overweight/obese adult	
Under-nutrition	Insufficient supply of nutrients; Changes in DNA methylation; Programming of obesity and cardiovascular disease; Dysregulation of glucose homeostasis		Low birth weight	Gut microbiome established		Inflammation	Cardiovascular disease
Inflammation	Oxidative stress					Oxidative stress	
Immature mother	Reduced insulin and IGF-1						Obesity
Poor placenta development	Poor placenta function						Short adult

FIGURE 23.1 The lifecycle stages of fetal programming. Factors involved in *in utero* and postnatal programming, possible mediators, and infant and adult outcomes. Factors in the same row may, but do not necessarily, indicate a direct temporal association.

A clear definition of catch-up growth is needed to differentiate between the effects of growth periods on adult disease. Catch-up growth is defined as a growth trajectory above the normal limits for age after transient growth inhibition, and refers to a beneficial realignment to genetic potential after growth faltering and not excessive infant weight gain [1]. This interpretation is important, since associations of birth weight with later outcomes span the entire birth weight spectrum, not just the low-birth-weight end [1].

CRITICISMS OF THE FETAL ORIGINS OF DISEASE HYPOTHESIS

Epidemiologists from a variety of fields have challenged the fetal origins of disease hypothesis. In the field of the fetal origins of adult disease, perinatal epidemiologists regard increasing birth weight as beneficial, while developmental origins epidemiologists regard low birth weight as less important and higher birth weight as not necessarily beneficial [1,2].

Lucas et al. [7] criticized the statistical adjustment for current body size widely applied in longitudinal studies of the fetal origins of disease, and stated that the statistical interpretation of some growth results was incorrect. They stated that failure to distinguish between the prenatal and postnatal factors affecting adult disease is problematic. Further, they argued that postnatal catch-up growth, rather than fetal programming, may primarily affect later health [7].

Other critics maintain that epidemiological studies disregard the role of genetics, because a gene mutation may be associated with low birth weight and offspring insulin resistance, resulting in type 2 diabetes and hypertension, both phenotypes of the same insulin-resistant genotype [23]. Monogenic diseases may impair the sensing of maternal hyperglycemia, decreasing insulin secretion or increasing insulin resistance, and impairing fetal growth. Polygenic influences resulting in fetal insulin resistance may result in lower birth weight. Genetic insulin resistance may cause abnormal perinatal vascular development, and explain increased adulthood risk of hypertension and vascular disease. However, it is likely that both genetic and environmental factors may predispose the infant to type 2 diabetes and hypertension [23].

EPIDEMIOLOGICAL CHALLENGES IN STUDYING THE FETAL ORIGINS OF ADULT CHRONIC DISEASE

ADJUSTMENT FOR ADIPOSITY AT THE ENDPOINT

Low birth weight is not associated with adult chronic disease risk in all studies. In some, this inverse relationship was only present after adjustment for endpoint adiposity, since adult adiposity is positively associated with adult chronic disease [15]. In studies of the relationship between birth weight and type 2 diabetes, unadjusted for endpoint body mass index (BMI), the relationship is J-shaped, with increased risk at both ends of the birth weight spectrum [9]. The underlying mechanism at the high birth weight end may be that maternal gestational diabetes is causing obesity and diabetes risk in offspring. Adjustment for endpoint BMI reveals an inverse and linear association across the birth weight spectrum, reflecting reduced risk at higher birth weights, a pattern attributed to a "thrifty phenotype" [4]. This hypothesis proposes early undernutrition, programming permanent dysregulation of glucose homeostasis, and development of type 2 diabetes. Reduced insulin secretion and insulin resistance, combined with obesity, aging, and physical inactivity, are the most important determinants of type 2 diabetes, confirmed by epidemiological evidence [14]. The relationship between poor fetal growth and insulin secretion and possible gene and environment interactions is less clear [1]. More research is necessary to better understand the contributions of maternal hyperglycemia and postnatal growth on offspring type 2 diabetes risk later in life.

SOCIAL AND ECONOMIC FACTORS AS CONFOUNDERS OR EXPLANATORY VARIABLES

Cardiovascular diseases are related to both low birth weight and poor adulthood socioeconomic circumstances. Controlling for offspring socioeconomic factors in adulthood is necessary to determine prenatal influences, particularly when social class changes across generations [2]. Poor maternal diet and strenuous physical work due to low socioeconomic circumstances may directly affect maternal energy and nutrient reserves, and have adverse effects on fetal growth and birth weight [24].

FETAL GROWTH AND SIZE AT BIRTH ARE TWO DIFFERENT INDICATORS

As birth weight is only one aspect of fetal growth, stronger associations may exist between birth length, or lean and fat masses and later outcomes, possibly due to a trimester-specific restriction of fetal growth [3]. Measures of fetal size at birth are prone to measurement error [25]; ultrasound measures of fetal growth are therefore more useful. The contribution of gestational age on adult outcomes is unclear, and determinants of preterm birth and fetal growth may differ [13,25]. Birth weight is the result of many determinants, some possibly unrelated to susceptibility to adult disease. Conversely, some prenatal determinants of adult outcomes may be unrelated to fetal growth [25].

BIOLOGICAL MECHANISMS OF THE FETAL ORIGINS OF ADULT OBESITY, CARDIOVASCULAR DISEASE, AND TYPE 2 DIABETES

MATERNAL NUTRITION DURING PREGNANCY

In animal models, maternal undernutrition during pregnancy had lasting effects on offspring metabolism, growth, and disease [10]. Pregnant rats on low-protein diets delivered offspring of lower birth weight that went on to exhibit elevated blood pressure and glucose intolerance in adult life [5]. In humans, the nutritional needs of a young, physically immature, undernourished mother compete with those of the fetus, while the placenta also competes with both for energy and protein sources [26].

Except at extremes of intake, maternal energy and macronutrient intake have relatively little impact on birth weight [25], but may be important in circumstances of prolonged negative energy balance [24]. In an animal study, insufficient glucose supply to the fetus was associated with reduced insulin and insulin-like growth factor-1

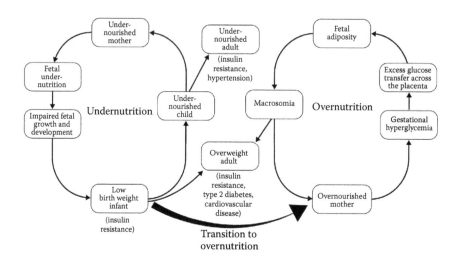

FIGURE 23.2 Fetal programming. The intergenerational insulin resistance cycle in an environment of undernutrition and overnutrition, or a transition from undernutrition to overnutrition. (Adapted from Tomar AS, Tallapragada DS, Nongmaithem SS et al., *Curr Obes Rep* 2015, 4(4):418–28.)

(IGF-1) concentrations, restricting fetal growth [27]. Infants of mothers with type 2 diabetes and gestational diabetes tend to have high birth weights, due to increased glucose availability [28]. Figure 23.2 presents the intergenerational insulin resistance cycle in an environment of undernutrition and overnutrition, or a transition from undernutrition to overnutrition and the resultant fetal programming of insulin resistance [22].

Subsistence farming populations from developing countries experience seasonal energy shortages, due to variations in food availability and energy expenditure related to food production. The impact of seasonal maternal diet and physical activity on neonatal size was examined in a prospective study in rural India [24]. Maternal energy and protein intakes were around 70% of recommended dietary allowances, and showed significant seasonal variation, with peak values during harvest time. Mean birth weight and length, adjusted for prepregnancy weight, parity, gestation, and offspring sex, was highest after longer exposure to harvest time during pregnancy and lowest after longer exposure to the lean season. Regression analysis showed that maternal energy intake at 18 weeks of gestation had a significant positive association with birth weight and length, whereas maternal physical activity at 28 weeks was negatively associated with birth weight. Higher maternal energy intakes, coupled with lower physical activity in late gestation, were associated with higher birth weight. These observations indicate that complete exposure to harvest time in late gestation could increase birth weight by 90 g, increasing further by lowering excessive maternal physical activity during harvest time [24,29].

Apart from the known roles of n-3 fatty acids in reducing the incidence of preterm birth [30], and folate deficiency in causing neural tube defects, little is known about the role of other nutrients in the programming of chronic disease risk. The relationship between maternal body size and circulating fuels, respectively, and neonatal size, was studied in the Pune Maternal Nutrition Study (1993–1996) [29]. The mothers were Indian, young, short and thin, and mostly vegetarian. At between 18 and 28 weeks gestation, fasting glucose concentrations remained stable, whereas total cholesterol and triglyceride concentrations increased and HDL-cholesterol concentrations decreased. The mean birth weight of the offspring was relatively low, at 2,666 g. Total cholesterol and triglycerides at both 18 and 28 weeks, and plasma glucose only at 28 weeks, were positively associated with birth size. The results did not change when preterm deliveries were also considered, suggesting an influence of maternal lipids on neonatal size in addition to the well-established effect of glucose during late pregnancy.

Folate is a methyl donor in the placenta for amino acid conversion and the generation of intermediates essential for cell division. Disordered one-carbon metabolism during early fetal development may increase later metabolic risk [31]. The reproducibility of the associations between maternal homocysteine concentrations and fetal growth found in the Pune birth cohort were explored in an observational study in Mysore, India. In this latter study, plasma vitamin B_{12}, folate, and homocysteine concentrations were measured at around 30 weeks gestation in the mothers, and the children's glucose and insulin concentrations, as well as neonatal anthropometry were measured at three ages: 5, 9.5, and 13.5 years [31]. Maternal homocysteine concentrations were inversely associated with all neonatal anthropometric measurements, and positively associated with glucose concentrations in the children at 5 and 9.5 years of age. Maternal serum folate concentrations, but not maternal vitamin B_{12}, were positively associated with insulin resistance in the children at 9.5 and 13.5 years of age [31].

In rural Gambia, cycles of rainy and dry seasons and a dependence on subsistence farming lead to annual variations in dietary intakes, compounded by the seasonal cycles of energy expenditure [32]. These cycles of nutrient availability and energy expenditure in mothers may affect fetal growth and development, offering a natural experiment to explore mechanisms by which early nutrient availability affects long-term pregnancy outcomes. In a prospective study of mothers' periconceptional dietary intakes and the plasma concentrations of key methyl-donor pathway substrates, 2,040 women were followed until pregnancy. Conception at the peak rainy ("hungry") season or the peak dry ("harvest") season, and predicted biomarker concentrations at conception were modeled. The offspring of rainy season conceptions, when mothers depended more on fresh green plant foods from the field, which are high in folate, had significantly higher levels of DNA methylation at the six remaining metastable epialleles (MEs) in peripheral blood lymphocytes (PBL). During the harvest season, when pregnant women had unlimited access to dry cereal foods and lower physical activity, there were increased periconceptional serum cysteine and homocysteine concentrations, predicting decreased systemic infant DNA methylation. Maternal serum riboflavin concentrations predicted increased ME methylation, while increased maternal BMI predicted decreased systemic infant DNA methylation. The consequences of these variations in methylation are unknown, but the possible implications of epigenetic variation at MEs induced by differences in maternal micronutrient status and BMI at conception may have important implications for noncommunicable disease risk later in life [32].

The effects of periconceptional multiple micronutrient supplementation on placental function in later pregnancy were assessed in a double-blind randomized placebo-controlled trial [33]. Primary outcomes were midgestational uteroplacental vascular endothelial function and placental active transport capacity. Uteroplacental vascular endothelial function was not significantly different between the two groups, but placental active transport capacity improved marginally, indicating a possible benefit of periconceptional micronutrient supplementation [33].

PLACENTA SIZE AND FUNCTION

Placenta size increases until late in gestation, supporting ongoing fetal growth. Placental vascularization during early pregnancy determines later placental nutrient transfer capacity and, ultimately, fetal growth [13]. Maternal undernutrition restricts placenta development, compromising nutrient and oxygen delivery to the fetus in both term and preterm infants. Larger placentas were associated with higher prepregnancy BMI, excessive gestational weight gain (GWG), and gestational diabetes mellitus (GDM) among the mothers. Low placental weight was associated with lower birth weight-for-gestational-age z-score in both term and preterm infants [28]. The placenta has endocrine as well as metabolic and transfer functions, and is important to understanding associations of fetal growth with adult health [13].

Women with gestational diabetes transfer excess glucose across the placenta, which contributes to fetal overgrowth. Placental size may be an important mediator between pre-pregnancy BMI, GWG, gestational diabetes, and increased fetal growth. A study showed that excessive GWG and GDM may represent a state of intrauterine overnutrition, with abundant placental nutrient supply, causing macrosomia [28]. Prepregnancy obesity and excessive GWG were positively associated with birth weight-for-gestational-age z-score at birth only among term births [28]. Other adverse risk factors may contribute to preterm birth and also impact fetal growth. Intrauterine inflammation may induce preterm birth, as well as poor fetal growth, which may in turn influence placental growth [13]. Placental weight, thickness, width, length, and cord placement position may predict neonatal outcomes. Placental size may also increase due to remodeling from a prior injury. Therefore placenta weight alone may not reflect placental function as a determinant of fetal growth [28].

Evidence suggests that the placenta plays a key role in fetal programming of cardiometabolic diseases [13,34]. The oxygen and nutrients that support fetal growth rely on the entire nutrient supply line, from maternal diet and body size to uterine perfusion, placental function, and fetal metabolism. Interruptions of this line at any point could result in programming the fetus for future risk of cardiovascular disease [34]. Progesterone and placental lactogen promote maternal glucose delivery, IGF-1 promotes fetal growth, and 11 β-hydroxy-steroid dehydrogenase type 2 inactivates glucocorticoids. These endocrine functions may play a role in fetal programming of future metabolic diseases [34].

A possible mechanism to explain the development of hypertension in later life is that pressure in the fetal circulation might be raised as a method of maintaining placental perfusion when the mother is undernourished [35]. The raised blood pressure may persist after birth. Alternatively, intrauterine growth retardation may trigger accelerated postnatal growth, accompanied by an accelerated increase in blood pressure [35].

Certain periods of fetal life may be critical for organ development. One proposed underlying causal pathway for the fetal origins of adult chronic disease is based on the hypothesis that a congenital nephron deficit underlies predisposition to hypertension in adult life [36].

MATERNAL METABOLISM DURING PREGNANCY

Maternal metabolism may influence fetal metabolic profiles directly through the placenta, or indirectly via influences of maternal hormones and/or placental metabolism. *In utero* exposures may include maternal behaviors, such as smoking and diet, or maternal metabolism, which may be associated with obesity or diabetes. The relationship of maternal midpregnancy corticotropin-releasing hormone (CRH) levels and offspring levels of adiponectin and leptin at the age of 3 years was studied in a prospective prebirth cohort study in the United States [37]. Maternal CRH blood levels were positively associated with levels of adiponectin, but not with leptin levels at age 3 years. There was no association between maternal CRH and birth weight for gestational age. The mechanism underlying the positive association between maternal levels of CRH and offspring adiponectin is unclear, but higher adiponectin may not be associated with a healthy metabolic profile in young children. The authors speculated that the increase in adiponectin was a compensatory response to increased insulin resistance in those children whose mothers had high midpregnancy CRH [37].

There is evidence from animal studies that the gut microbiome influences the diet-related metabolic profile [38]. Further research is necessary to explore metabolite profiles associated with gut microbiota in human populations during pregnancy. The gut microbiome is explored in detail in Chapter 19 of this book.

SMOKING AND POLLUTION DURING PREGNANCY

The deleterious effects of pollution and maternal smoking during pregnancy are well-known examples of prenatal exposures affecting fetal growth, but with unknown long-term health effects. A systematic review of 14 observational studies ($n = 84,563$ children) examining the association between maternal prenatal cigarette smoking and overweight offspring showed that offspring of smokers were at increased risk for being overweight at ages 3 to 33 years, compared with children of nonsmokers [38]. Differences between smokers and nonsmokers could not be explained by sociodemographic or behavioral confounders [39].

GENETIC AND EPIGENETIC FACTORS

Inherited fetal gene expression potentially underlies susceptibility to disease, but the maternal genome also affects the fetal environment and may affect fetal gene expression. The association between maternal hypertension, low birth weight, and hypertension in the offspring could be partly of genetic origin [13]. However, the subsequent pattern of development appears to be responsive to environmental influences. Developmental plasticity evolved to match an organism to its environment, but a mismatch between the resultant phenotype and the current environment increases cardiovascular risk [40]. Epigenetic processes appear to be key mechanisms in the developmental origins of chronic noncommunicable disease [40].

A study of undernutrition over 50 generations in a rat model showed low birthweight, high visceral adiposity, and insulin resistance (using hyperinsulinemic-euglycemic clamps) in undernourished rats, compared to age-/sex-matched control rats [41]. Undernourished rats also had higher serum insulin, homocysteine, endotoxin, and leptin levels, but lower adiponectin, vitamin B_{12}, and folate levels. The undernourished rats had an eightfold increased susceptibility to *Streptozotocin*-induced diabetes compared to controls. These metabolic abnormalities could not be reversed after two generations of nutrient rehabilitation. Altered epigenetic signatures in the insulin-2 gene promoter region of undernourished rats were also not reversed by nutrition, and may contribute to the persistent adverse metabolic profiles in similar multigenerational undernourished human populations [41].

In the Pune Maternal Nutrition Study and the Parthenon Cohort Study in Mysore, India (discussed earlier), evidence of causality within a Mendelian randomization framework of the association between maternal total homocysteine and offspring birth weight was studied [42]. This was assessed using a methylenetetrahydrofolate reductase (MTHFR) gene variant rs1801133 by instrumental variable and triangulation analysis, separately, and meta-analysis. Offspring birth weight was inversely related to maternal homocysteine concentration adjusted for gestational age and offspring sex in these studies and in the meta-analysis. Maternal risk genotype at rs1801133 predicted higher homocysteine concentration and lower birth weight, adjusted for gestational age, offspring sex, and rs1801133 genotype. Instrumental variable and triangulation analysis supported the causal association between maternal homocysteine concentration and offspring birth weight. These findings suggest a causal role for maternal homocysteine metabolism in fetal growth and support interventions in reducing maternal homocysteine concentrations [31,42]

A quasi-experimental study was performed to evaluate the impact of the Dutch famine of 1944–1945 during specific periods of pregnancy, or any time in gestation, on genome-wide DNA methylation levels of offspring at 59 years [43]. They compared individuals with prenatal famine exposure and time or sibling controls without prenatal famine exposure. They also studied the impact of shorter pre- and postconception exposure periods. Famine exposure during gestation weeks 1 to 10, but not during later gestation, was associated with increased DNA methylation of four specific dinucleotides. Exposure during any time in gestation resulted in increased methylation of two specific dinucleotides, while exposure around conception was associated with methylation of only one dinucleotide. This dinucleotide, cg23989336, is involved in the determination of body size in knockout mice studies [44]. All dinucleotides identified in this study were linked to genes involved in growth, development, and lipid metabolism. The authors identified early gestation, but not mid or late gestation, as a critical time period for DNA methylation changes affecting body size after prenatal famine exposure [43].

FUTURE STUDIES

Animal studies of fetal growth will continue to contribute new knowledge and are useful to study nutrient and oxygen delivery, as well as processes altering these pathways. Serial ultrasound measures to measure human fetal growth parameters throughout pregnancy will also contribute useful information due to potential issues with statistical growth trajectory models. Placental morphological pathology, as well as using specimens of placenta, maternal prenatal blood, and umbilical cord blood may be applied to measure markers of altered blood flow, endocrine and transport characteristics, or activity of specific enzymes related to later noncommunicable disease risk [34].

The fetal origins of disease theory arose from historical cohort studies, namely, the Dutch famine cohort [11]. A collaboration of five birth cohorts from low and middle-income countries (Brazil, Guatemala, India, Philippines, and South Africa) has made it possible to analyze pooled longitudinal data from Consortium for Health Orientated Research in Transitioning Societies (COHORTS) [45]. More than 22,000 mothers were enrolled before or during pregnancy and almost 20,000 children are being followed up; analyses will be adjusted for maternal variables and breastfeeding duration [45]. New cohort studies of preconceptional and pregnant women have been and are continuing to be planned to overcome the limitations of the earlier studies and to explore mechanisms of pathways gleaned from these early studies [21,39]. These studies will take decades for adult disease outcomes to occur. The associations between maternal exposures and offspring health in adolescence are being studied in the Growing Up Today Study cohort and their mothers, from the Nurses' Health Study II [39].

Metabolomics studies focus on systematic analysis of low-molecular intermediates in biological fluids. Such investigations may target specific intermediates associated with obesity or insulin resistance, or could search for novel biomarkers [46]. Metabolomics studies have potential to provide valuable information on the physiological response to nutrient intake and could be informative for fetal origins research when quantified during key developmental stages of pregnancy. Metabolite profiles associated with specific dietary patterns, or behaviors, such as smoking and physical activity can be identified [46]. Whereas birth weight and fetal growth are crude measures of the intrauterine environment, cord blood metabolomic profiling at delivery could improve assessment of adverse fetal growth outcomes, and guide future interventions to avoid such risks. The metabolomic profile could give an indication of impaired nutrient transfer to the fetus during development [46].

PUBLIC HEALTH INTERVENTIONS REGARDING FETAL PROGRAMMING

Epidemiological findings of associations between birth weight and later health outcomes provide evidence of programming of noncommunicable disease in humans. Experimental animal evidence also shows that *in utero* environmental stressors produce lifelong alterations in metabolism and pathology. These implications are important for developing countries undergoing a transition from infectious disease to noncommunicable disease burdens, as well as the nutrition transition from an active, low-calorie lifestyle to a sedentary, high-calorie lifestyle occurring globally [47]. Available data indicate that a lower birth weight combined with later higher attained BMI confers the highest risk for obesity and cardiovascular disease later in life [6,14,15,17]. Successive generations in developing countries are likely to have increasing proportions with a high cardiometabolic risk profile. Efforts to prevent the development of obesity in areas undergoing such epidemiological, economic, and nutrition transitions are paramount.

The implications for policy recommendations regarding fetal programming are not yet clear. Birth weight is only a marker for underlying etiological pathways, whereas the true etiological factors are largely unknown. More targeted interventions to modify cardiometabolic risks due to fetal programming can be designed when these factors are more clearly identified. Interventions to increase birth weight per se may not be effective and could be harmful. The focus should rather be on improving the health of women of reproductive age to improve the well-being of their offspring. Examples include the strengthening of efforts to prevent childbearing before the age of 19 years [45], and restriction of maternal weight gain to 20 kg or less for overweight and obese women to curb adolescent adiposity [39]. Strategies to reduce excessive adiposity gains during early postnatal life and in the preschool years may reduce midchildhood blood pressure, which may also affect adult blood pressure and cardiovascular disease risk [21]. These results indicate that interventions to prevent excessive weight gain during pregnancy and early postnatal life may be beneficial.

REFERENCES

1. Gillman MW. Epidemiological challenges in studying the fetal origins of adult chronic disease. *Int J Epidemiol* 2002; 31(2):294–9.
2. de Jong F, Monuteaux MC, van Elburg RM et al. Systematic review and meta-analysis of preterm birth and later systolic blood pressure. *Hypertension* 2012; 59(2):226–34.
3. Barker D, Osmond C, Golding J et al. Growth in utero, blood pressure in childhood and adult life, and mortality from cardiovascular disease. *BMJ* 1989; 298(6673):564–7.
4. Hales CN, Barker DJ. Type 2 (non-insulin-dependent) diabetes mellitus: The thrifty phenotype hypothesis. *Diabetologia* 1992; 35(7):595–601.
5. Langley SC, Jackson AA. Increased systolic blood pressure in adult rats induced by fetal exposure to maternal low-protein diets. *Clin Sci (Lond)* 1994; 86(2):217–22.
6. Frankel S, Elwood P, Sweetnam P et al. Birth weight, body-mass index in middle age, and incident coronary heart disease. *Lancet* 1996; 348(9040):1478–80.
7. Lucas A, Fewtrell MS, Cole TJ. Fetal origins of adult disease: The hypothesis revisited. *BMJ* 1999; 319(7204):245–9.
8. Rich-Edwards JW, Stampfer MJ, Manson JE et al. Birth weight and risk of cardiovascular disease in a cohort of women followed up since 1976. *BMJ* 1997; 315(7105):396–400.
9. Rich-Edwards JW, Colditz GA, Stampfer MJ et al. Birth weight and the risk for type 2 diabetes mellitus in adult women. *Ann Intern Med* 1999; 130(4 Pt 1):278–84.
10. McCance RA, Mount LE. Severe undernutrition in growing and adult animals. 5. Metabolic rate and body temperature in the pig. *Br J Nutr* 1960; 15:509–18.
11. Ravelli ACJ, van der Meulen JHP, Michels RPJ et al. Glucose tolerance in adults after prenatal exposure to famine. *Lancet* 1998; 351(9097):173–7.

12. Forsén T, Eriksson J, Tuomilehto J et al. The fetal and childhood growth of persons who develop type 2 diabetes. *Ann Intern Med* 2000; 133(3):176–82.
13. Harding JE. The nutritional basis of the fetal origins of adult disease. *Int J Epidemiol* 2001; 30(1):15–23.
14. Bavdekar A, Yajnik CS, Fall C et al. The insulin resistance syndrome in eight-year-old Indian children; small at birth, big at eight years or both? *Diabetes* 1999; 48:2422–9.
15. Leon DA, Koupilova I, Lithell HO et al. Failure to realise growth potential in utero and adult obesity in relation to blood pressure in 50 year old Swedish men. *BMJ* 1996; 312(7028):401–6.
16. Leon DA, Lithell HO, Vågerö D et al. Reduced fetal growth rate and increased risk of death from ischaemic heart disease: Cohort study of 15000 Swedish men and women born 1915–29. *BMJ* 1998; 317(7153):241–5.
17. Eriksson JG, Forsen T, Tuomilehto J et al. Catch-up growth in childhood and death from coronary heart disease: Longitudinal study. *BMJ* 1999; 318(7181):427–31.
18. Eriksson J, Forsen T, Tuomilehto J et al. Fetal and childhood growth and hypertension in adult life. *Hypertension* 2000; 36(5):790–4.
19. Eriksson JG, Forsen T, Tuomilehto J et al. Early growth and coronary heart disease in later life: Longitudinal study. *BMJ* 2001; 322(7292):949–53.
20. Salgin B, Norris SA, Prentice P et al. Even transient rapid infancy weight gain is associated with higher BMI in young adults and earlier menarche. *Int J Obes (Lond)* 2015; 39(6):939–44.
21. Perng W, Rifas-Shiman SL, Kramer MS et al. Early weight gain, linear growth, and mid-childhood blood pressure: A prospective study in Project Viva. *Hypertension* 2016; 67(2):301–8.
22. Tomar AS, Tallapragada DS, Nongmaithem SS et al. Intrauterine programming of diabetes and adiposity. *Curr Obes Rep* 2015; 4(4):418–28.
23. Hattersley AT, Tooke JE. The fetal insulin hypothesis: An alternative explanation of the association of low birth weight with diabetes and vascular disease. *Lancet* 1999; 353(9166):1789–92.
24. Rao S, Kanade AN, Yajnik CS et al. Seasonality in maternal intake and activity influence offspring's birth size among rural Indian mothers—Pune Maternal Nutrition Study. *Int J Epidemiol* 2009; 38(4):1094–103.
25. Kramer MS. Determinants of low birth-weight: Methodological assessment and meta-analysis. *Bull World Health Organ* 1987; 65(5):663–737.
26. Jackson AA. *Maternal and fetal demands for nutrients and the significance of protein restriction.* London: RCOG Press; 1999.
27. Oliver MH, Harding JE, Breier BH et al. Glucose but not a mixed amino acid infusion regulates plasma insulin-like growth factor (IGF)-I concentrations in fetal sheep. *Pediatr Res* 1993; 34:62–65.
28. Ouyang F, Parker M, Cerda S et al. Placental weight mediates the effects of prenatal factors on fetal growth: The extent differs by preterm status. *Obesity* 2013; 21(3):609–20.
29. Kulkarni SR, Kumaran K, Rao SR et al. Maternal lipids are as important as glucose for fetal growth: Findings from the Pune Maternal Nutrition Study. *Diabetes Care* 2013; 36(9):2706–13.
30. Olsen SF, Secher NJ, Tabor A et al. Randomised clinical trials of fish oil supplementation in high risk pregnancies. *BJOG* 2000; 107(3):382–95.
31. Krishnaveni GV, Veena SR, Karat SC et al. Association between maternal folate concentrations during pregnancy and insulin resistance in Indian children. *Diabetologia* 2014; 57(1):110–21.
32. Dominguez-Salas P, Moore SE, Baker MS et al. Maternal nutrition at conception modulates DNA methylation of human metastable epialleles. *Nat Commun* 2014; 5:3746.

33. Owens S, Gulati R, Fulford AJ et al. Periconceptional multiple-micronutrient supplementation and placental function in rural Gambian women: A double-blind, randomized, placebo-controlled trial. *Am J Clin Nutr* 2015; 102(6):1450–9.

34. Jansson T, Powell TL. Role of the placenta in fetal programming: Underlying mechanisms and potential interventional approaches. *Clin Sci (Lond)* 2007; 113(1):1–13.

35. Ounsted MK, Cockburn JM, Moar VA et al. Factors associated with the blood pressures of children born to women who were hypertensive during pregnancy. *Arch Dis Child* 1985; 60(7):631–5.

36. Brenner BM, Chertow GM. Congenital oligonephropathy: An inborn cause of adult hypertension and progressive renal injury? *Curr Opin Nephol Hyperten* 1993; 2(5):691–5.

37. Fasting MH, Oken E, Mantzoros CS et al. Maternal levels of corticotropin-releasing hormone during pregnancy in relation to adiponectin and leptin in early childhood. *J Clin Endincrinol Metab* 2009; 94(4):1409–15.

38. Oken E, Levitan EB, Gillman MW. Maternal smoking during pregnancy and child overweight: Systematic review and meta-analysis. *Int J Obes (Lond)* 2008; 32(2):201–10.

39. Wang Z, Klipfell E, Bennett BJ et al. Gut flora metabolism of phosphatidylcholine promotes cardiovascular disease. *Nature* 2011; 472(7341):57–63.

40. Gluckman PD, Hanson MA, Buklijas T et al. Epigenetic mechanisms that underpin metabolic and cardiovascular diseases. *Nat Rev Endocrinol* 2009; 5(7):401–8.

41. Hardikar AA, Satoor SN, Karandikar MS et al. Multigenerational undernutrition increases susceptibility to obesity and diabetes that is not reversed after dietary recuperation. *Cell Metab* 2015; 22(2):312–9.

42. Yajnik CS, Chandak GR, Joglekar C et al. Maternal homocysteine in pregnancy and offspring birth weight: Epidemiological associations and Mendelian randomization analysis. *Int J Epidemiol* 2014; 43(5):1487–97.

43. Tobi EW, Slieker RC, Stein AD et al. Early gestation as the critical time-window for changes in the prenatal environment to affect the adult human blood methylome. *Int J Epidemiol* 2015; 4(4):1211–23.

44. Skarnes WC, Rosen B, West AP et al A conditional resource for the genome-wide study of mouse gene function. *Nature* 2011; 474(7351):337–42.

45. Fall CHD, Sachdev HS, Osmond C et al. Association between maternal age at childbirth and child and adult outcomes in the off spring: A prospective study in five low-income and middle-income countries (COHORTS collaboration). *Lancet Global Health* 2015; 3(7):e366–77.

46. Hivert MF, Perng W, Watkins SM et al. Metabolomics in the developmental origins of obesity and its cardiometabolic consequences. *J Dev Orig Health Dis* 2015; 6(2). 65–78.

47. Murray CJ, Lopez AD. Alternative projections of mortality and disability by cause 1990–2020: Global Burden of Disease Study. *Lancet* 1997; 349(9064):1498–504.

Section X

*Effective Interventions
during the First 1,000 Days*

24 Effectiveness of Nutrition-Specific Interventions in Pregnancy and Early Childhood

Rebecca Heidkamp, Adrienne Clermont, and Robert E. Black

CONTENTS

INTRODUCTION

The 2013 *Lancet* series on Maternal and Child Nutrition included a comprehensive review of interventions that affect the nutritional status of women and young children in low- and middle-income countries (LMICs). The series's authors estimated that if 10 known effective interventions were scaled up to 90% coverage, we could eliminate 15% of under-5 deaths and reduce stunting by 21% globally [1]. The objective of this chapter is to review and update this work with the latest research evidence on the effectiveness of these 10 and other nutrition-specific interventions to prevent poor growth and improve child survival during the crucial first 1,000 days of life. We will focus primarily on interventions delivered at scale through the health system. Outcomes of interest include birth outcomes (preterm, small for gestational age, low birth weight), risk of subsequent growth faltering (stunting, wasting), and mortality during the first 5 years of life. We will not consider effects on anemia, developmental outcomes, or overweight/obesity risk, nor will we address the broader range of nutrition-sensitive interventions that affect more distal determinants of nutritional status (including food production systems, water and sanitation, and household socioeconomic status). Others have provided comprehensive reviews of these important strategies [2].

Interventions are presented by the continuum of care from pregnancy through early childhood (Figure 24.1). We will describe each intervention, identify the general biological mechanism through which it is understood to have an effect, and review the latest evidence on effectiveness.

MATERNAL INTERVENTIONS

Interventions to improve maternal nutritional status before and during pregnancy can have long-lasting effects on child growth and survival. Several interventions delivered to women before or during pregnancy affect neonatal outcomes (Table 24.1), including the risk of preterm and small for gestational age (SGA) births, as well as neonatal mortality [3]. Babies who are born preterm or SGA are at higher risk of stunting [4] and mortality [5] in early childhood (Table 24.2). Infants in South

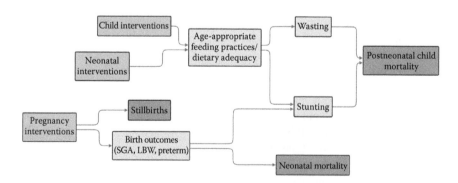

FIGURE 24.1   A conceptual map of interventions, risk factors, and outcomes.

TABLE 24.1
Maternal Intervention Effects

Intervention	Outcome of Interest	Effect Size (95% CI)
Periconceptual folic acid supplementation	Neural tube defects	RR: 0.31 (0.17, 0.58)
Iron supplementation in pregnancy	Iron-deficiency anemia	RR: 0.33 (0.16,0.69)
	Low birth weight	RR: 0.84 (0.69, 1.03)
Multiple micronutrient supplementation[a]	Low birth weight	RR: 0.88 (0.85, 0.91)
	SGA	RR: 0.90 (0.83, 0.97)
	Stillbirth	RR: 0.91 (0.85, 0.98)
Calcium supplementation in pregnancy	Preterm births	RR: 0.76 (0.60, 0.97)
Balanced energy-protein supplementation in pregnancy	SGA	RR: 0.79 (0.69, 0.90)
	Stillbirth	RR: 0.60 (0.39, 0.94)
	Birth weight	MD: +40.96 g (4.66, 77.26)

Note: RR, relative risk; MD, mean difference.

[a] Effects are compared to iron or iron-folic acid supplementation.

TABLE 24.2
Impact of Birth Outcomes on Stunting and Mortality

Birth Outcome Group	Relative Risk of Stunting at 12–60 Months (95% CI)	Relative Risk of Neonatal Mortality (95% CI)	Relative Risk of Postneonatal Mortality (95% CI)
Term, not SGA	1 (reference group)	1 (reference group)	1 (reference group)
Term, SGA	2.43 (2.22, 2.66)	3.06 (2.21, 4.23)	2.20 (1.57, 3.08)
Preterm, not SGA	1.93 (1.71, 2.18)	9.44 (5.08, 17.55)	2.85 (1.65, 4.99)
Preterm, SGA	4.51 (3.42, 5.93)	16.20 (10.00, 26.23)	3.77 (3.17, 10.49)

Asia are at a higher risk of being wasted from very soon after birth than children in other regions [6].

Since the 2013 *Lancet* series, a team at Oxford has released the INTERGROWTH–21st standards to assess fetal growth and size at birth. The standards were created using the same concept and approach as the 2006 World Health Organization (WHO) Growth Standards for children aged 0 to 59 months, following cohorts sampled from eight geographically diverse urban populations of mothers who received appropriate nutrition and health care support throughout pregnancy and delivery. The studies showed that when mothers receive adequate care and nutrition, and environmental stresses on growth are low, fetal growth is the same across geographical contexts [7]. An analysis of 16 prospective cohorts of newborns from LMICs showed that the prevalence of SGA was 23.7% (95% confidence interval [CI]: 16.5%, 31.0%) using the INTERGROWTH–21st standard, compared with 36.0% (27.0%, 45.0%) using the previous 1999–2000 U.S. birth weight reference [8].

MICRONUTRIENT INTERVENTIONS

Deficiencies in iodine, folic acid, iron, calcium, and other micronutrients are common in women of childbearing age in LMICs. Micronutrient interventions currently implemented at scale include the provision of supplements containing iron and folic acid during routine antenatal care visits, staple food fortification with folic acid, and sometimes other micronutrients.

Iron Folic Acid Supplementation

The WHO recommends daily iron and folic acid (IFA) supplementation for pregnant women to reduce maternal anemia [9]. Folate plays an important role in DNA methylation and cell replication, and is protective against neural tube defects (NTDs) when consumed preconception or during the first 12 weeks of pregnancy. A 2015 *Cochrane Review* found a 69% reduction in risk for NTDs, similar to that published in the 2013 *Lancet* series [10]. There are no available studies designed to detect an impact on neonatal mortality [10].

Another 2015 *Cochrane Review* found that consumption of iron-containing tablets (iron or IFA) during pregnancy is associated with a 67% reduction in iron-deficiency anemia at term (based on six studies with 1,088 women) [11]. Iron supplementation is associated with a 16% reduced risk of low birth weight (11 studies with 17,613 women) [11].

Multiple Micronutrient Supplementation

Multiple micronutrients (MMNs) are broadly defined as supplements that contain three or more micronutrients [12]. In 1999, the United Nations Children's Fund (UNICEF), the United Nations University, and the WHO developed a MMN tablet known as UNIMMAP, which contains one recommended daily allowance of vitamin A, vitamin B_1, vitamin B_2, niacin, vitamin B_6, vitamin B_{12}, folic acid, vitamin C, vitamin D, vitamin E, copper, selenium, and iodine, as well as 30 mg of iron and 15 mg of zinc. A new *Cochrane Review* published in 2015, looking at any MMN compared to iron or IFA, found similar effect sizes as reported in the 2013 *Lancet* series on low birth weight (12% reduction) and SGA (10% reduction), but no longer found a significant effect on preterm births. It also reported an effect of MMN on stillbirths (9% reduction) [12]. This review was based on 17 trials, including a total of 137,791 women, primarily in developing countries, and the quality of the evidence was rated as high.

Food Fortification

Salt iodization is the most common form of food fortification globally. Iodine deficiency can cause hypothyroidism and, when it occurs among pregnant women, can harm fetal neurological development. The prevention of cretinism is the primary outcome of increased iodine supply, but there is also mixed evidence that it reduces neonatal mortality. Similar to the 2013 *Lancet* series, a 2014 review on population-wide salt iodization deemed the evidence insufficient to draw any conclusions about mortality effects [13], but a 2012 review of iodized oil supplementation for pregnant women that included two randomized controlled trials with over 37,000 participants showed a significant but varied impact on infant mortality (25% to 60% reduction, depending on the severity of iodine deficiency and baseline infant mortality levels) [14].

Other staple food fortification interventions being promoted in LMICs include the fortification of cooking oil, flour, or condiments with iron, folic acid, vitamin A, vitamin D, iodine, zinc, and multiple micronutrients. Generally, studies find that consumption of fortified foods by women leads to improved micronutrient status and reduced anemia. Little evidence is available for the subsequent impact of maternal fortified food consumption on growth and mortality outcomes in young children [15].

Calcium Supplementation

Calcium supplementation in pregnancy is recommended to address hypertensive disorders of pregnancy (preeclampsia and eclampsia), which are a major cause of maternal morbidity and mortality and can lead to negative birth outcomes. A 2014 review of calcium supplementation during pregnancy found similar effect sizes as reported in the 2013 *Lancet* series [16]. Among women with low calcium intake, daily calcium supplementation of more than 1,000 mg per day significantly reduces risk of preterm births (24% reduction), as well as gestational hypertension and preeclampsia in mothers. Twice-daily consumption of 500 mg tablets delivered alongside IFA is a more feasible approach than the WHO-recommended 1,500 mg per day [17].

Balanced Protein-Energy Supplementation

In addition to the micronutrient deficiencies, pregnant women in low-resource settings may also have inadequate energy intake. Because maternal weight gain during pregnancy is important for healthy birth outcomes, food supplementation in pregnancy can be an appropriate approach for women with low prepregnancy BMI. Balanced protein-energy supplementation (in which protein provides less than 25% of the total energy content) is recommended to improve outcomes. In practice, balanced protein-energy supplementation is most commonly implemented among HIV-infected women or populations at high risk of food insecurity.

The 2013 *Lancet* series cited studies showing the impact of balanced protein-energy supplementation on SGA, stillbirths, and birth weight. A more recent 2015 *Cochrane Review* (Ota et al. [18]) showed similar impacts (21% reduction in SGA, 40% reduction in stillbirths, and an average 41 g increase in birth weight). An earlier systematic review found a larger increase in birth weight among underweight women compared to adequately nourished women [19], but Ota et al. did not find a difference between these subgroups. It is important to note that evidence supporting the Ota et al. results was only rated as being of moderate quality; many of the studies included in the review were published in the 1970s and 1980s, and are not exclusively from low-income countries. (The SGA result was based on seven trials with 4,408 women, while the stillbirth results were based on five studies with 3,408 women, and the birth weight results were based on 11 trials with 5,385 women.) The authors highlight the need for targeting balanced protein-energy supplementation specifically to undernourished women (BMI <18.5 kg/m^2), as adequately nourished women are at risk of negative birth outcomes from excessive energy and/or protein intake—an important consideration, given the growing proportion of obese and overweight women in LMICs [18].

INTRAPARTUM/NEONATAL INTERVENTIONS

A number of important nutritional interventions affecting child survival take place in the period immediately following childbirth and during the neonatal period, that is, the first month of life (Table 24.3).

EARLY INITIATION OF BREASTFEEDING

Initiation of breastfeeding is defined as "early" if it takes place within one hour of birth. Despite the health benefits of early initiation, rates are low in most countries due to a number of cultural practices including prelacteal feeding and the avoidance of colostrum. Early initiation rates are below 60% worldwide, although it is higher in LMICs than in higher-income countries [20]. The 2013 *Lancet* series concluded that "the exact scientific basis for the absolute early time window of feeding within the first hour after birth is weak," but more recent evidence has shown that early initiation appears to have an additional protective effect beyond that of exclusive or

TABLE 24.3
Neonatal Intervention Effects

Intervention	Outcome of Interest	Effect Size (95% CI)
Early initiation of breastfeeding	Neonatal mortality	Initiation within 1 hour (reference group)
		Initiation 2–23 hours—RR: 1.41 (1.24, 1.62)
		Initiation 24–96 hours—RR: 1.79 (1.39, 2.30)
Breastfeeding promotion	Early initiation of breastfeeding	OR: 1.25 (1.19, 1.32)
Neonatal vitamin A supplementation	All-cause mortality, 0–6 months	Asian populations—RR: 0.87 (0.78, 0.96)
		African populations—RR: 1.10 (1.00, 1.21)
Delayed cord clamping	Intraventricular hemorrhage	RR: 0.59 (0.41, 0.85)
	Necrotizing enterocolitis	RR: 0.62 (0.43, 0.90)

Note: OR, odds ratio; RR, relative risk.

continued breastfeeding more generally. A prospective cohort study including over 99,000 infants found that, compared to infants initiating breastfeeding within the first hour of life, those initiating within 2 and 23 hours of birth had a 41% higher risk of neonatal mortality, and those initiating within 24 and 96 hours of birth had a 79% higher risk of neonatal mortality [21]. These analyses provide the most persuasive evidence that early initiation has an effect on mortality that is independent of the effect of exclusive breastfeeding.

Recent evidence also indicates that breastfeeding promotion interventions can have a significant impact on improving early initiation behaviors. A meta-analysis by Sinha et al. (2015) found a 25% increase in breastfeeding initiation within one hour, based on all promotion interventions (49 studies) [22]. The most effective interventions were those in the community and home environments (85% increase, three studies), while those in the health facility setting were much less effective (11% increase, 29 studies).

NEONATAL VITAMIN A SUPPLEMENTATION

Although the effectiveness of periodic large-dose vitamin A supplementation among children aged 6 to 59 months has long been established (see later), neonatal vitamin A supplementation has been more recently examined, and remains controversial. The intervention consists of 50,000 IU of vitamin A administered within 72 hours of birth, which is thought to improve immune system function.

Early evidence cited in the 2013 *Lancet* series suggested a 14% reduction in the risk of mortality at 6 months of age [23] but, more recently, several large-scale randomized trials in Asia and Africa have found conflicting results. Taking the results of all studies together, it seems that Asian populations, which are more at risk of maternal vitamin A deficiency, show a benefit for neonatal supplementation (13% reduction in mortality at 6 months), whereas African populations, with much lower rates of vitamin A deficiency, do not benefit [24]. As a result, it seems that this intervention is only "justified in populations where documented levels of maternal vitamin A deficiency is moderate to high, specifically in South Asia" [25].

DELAY OF UMBILICAL CORD CLAMPING

In 2014, the WHO released recommendations to delay umbilical cord clamping by at least one minute in all neonates who do not require emergency interventions that necessitate removal from mother [26]. Delayed cord clamping leads to increased hemoglobin concentration and blood volume in the neonate. The *Cochrane Review*, cited in the 2013 *Lancet* series, was updated later in 2013, and continued to show these results [27]. In both term and preterm infants, delayed cord clamping is associated with reduced risk of iron deficiency anemia and improved iron stores up to 6 months of age. Delayed cord clamping can reduce the risk of intraventricular hemorrhage and necrotizing enterocolitis by 41% and 38%, respectively, among preterm infants, although the effect on infant mortality overall was not significant, possibly due to inadequate data [28].

CHILD INTERVENTIONS

Preventive and curative interventions to address undernutrition continue to play an important role throughout the early childhood period (Table 24.4). They have impacts on mortality both directly and through the intermediate pathways of stunting and wasting.

TABLE 24.4
Child Intervention Effects

Intervention	Outcome of Interest	Effect Size (95% CI)
Exclusive breastfeeding (0–5 months)	All-cause child mortality	Exclusive BF (reference group) Predominant BF—RR: 1.48 (1.13, 1.92) Partial BF—RR: 2.84 (1.63, 4.97) No BF—RR: 14.40 (6.13, 33.9)
Exclusive breastfeeding (0–5 months)	Infection-related child mortality	Exclusive BF (reference group) Predominant BF—RR: 1.70 (1.18, 2.45) Partial BF—RR: 4.56 (2.93, 7.11) No BF—RR: 8.66 (3.19, 23.50)
Continued breastfeeding (6–23 months)	All-cause child mortality	Continued BF (reference group) No BF (6–11 months)—RR: 1.76 (1.28, 2.41) No BF (12–23 months)—RR: 1.97 (1.45, 2.67)
Breastfeeding promotion	Exclusive breastfeeding (0–5 months)	OR: 1.44 (1.38, 1.51)
	Continued breastfeeding (6–23 months)	OR: 1.61 (1.17, 2.20)
Complementary feeding education	Height-for-age z-score (HAZ)	SMD: 0.23 (0.09, 0.36)
	Weight-for-age z-score (WAZ)	SMD: 0.16 (0.05, 0.27)
	Stunting	RR: 0.71 (0.56, 0.91)
Supplementary food provision	Weight	MD: 0.12 kg (0.05, 0.18)
	Height	MD. 0.27 cm (0.07, 0.48)
	Height-for-age z-score (HAZ)	SMD: 0.15 (0.06, 0.24)
Vitamin A supplementation	Diarrhea-specific child mortality (6–59 months)	RR: 0.70 (0.58, 0.86)
Vitamin A for treatment of measles	Measles-specific child mortality	RR: 0.62 (0.18, 0.81)
Zinc supplementation	Diarrhea incidence	RR: 0.87 (0.81, 0.94)
	Pneumonia incidence	RR: 0.81 (0.73, 0.90)
	Diarrhea mortality	RR: 0.50 (0.27, 1.00)
	Pneumonia mortality	RR: 0.51 (0.20, 1.00)
Zinc for treatment of diarrhea	Diarrhea-specific child mortality	RR: 0.72 (0.69, 0.85)

Note: BF, breastfeeding; RR, relative risk; OR, odds ratio; SMD, standard mean difference; MD, mean difference.

BREASTFEEDING PROMOTION

Only 37% of children under 6 months of age in LMICs are breastfed exclusively in accordance with WHO recommendations [29] that they receive no foods or liquids other than breast milk [20]. A 2015 review by Sankar et al. reaffirmed findings cited in the 2013 *Lancet* series that there is a dose–response relationship with mortality [30]. All-cause mortality was 14.4 times higher in nonbreastfed infants, 2.84 higher in partially breastfed, and 1.48 higher in predominantly breastfed compared to exclusively breastfed infants aged 0 to 5 months. In addition, Sankar et al. provided new relative risks (RRs) for infectious causes: infection-related mortality was 8.66 times higher in nonbreastfed infants, 4.56 times higher in partially breastfed infants, and 1.7 times higher in predominantly breastfed infants, compared to exclusively breastfed infants aged 0 to 5 months [30].

Per WHO recommendations, older children aged 6 to 23 months should continue to breastfeed while gradually introducing age-appropriate complementary foods. The updated review found smaller protective effects of breastfeeding on all-cause mortality in this older age group (RR 1.76 at 6–11 months, and RR 1.97 at 12–23 months) compared to those cited in the 2013 *Lancet* series [30].

Strategies to promote exclusive breastfeeding include facility-based counseling, home or community-based outreach, mass media interventions, and workplace policies including maternity leave protection laws. The 2015 review by Sinha et al., which included 130 intervention studies in both high-income and LMIC settings, found an overall 44% increase in exclusive breastfeeding among those who received a promotion intervention compared to those who did not [22]. Sinha et al. went further than the 2013 *Lancet* series and reported effect sizes by subcategory of intervention. Facility-based interventions had the highest impact (OR 1.46 [1.37, 1.56]) of the single intervention strategy studies, while a combination of facility and community interventions together increased the OR to 2.52 (1.39, 4.59) [22]. Impacts were higher in low-income countries than in high-income contexts. Fewer studies looked at interventions to promote continued breastfeeding from 6 to 23 months of age. The overall effect from 18 studies was a 61% increase in continued breastfeeding. Effects on continued breastfeeding were higher in urban settings than rural, higher-income than lower-income settings, and in children 12 to 23 months as compared to those 6 to 11 months [22].

PROMOTION OF AGE-APPROPRIATE COMPLEMENTARY FEEDING AND PROVISION OF SUPPLEMENTAL FOOD

Mothers are advised to begin introducing semisolid and solid foods to infants at 6 months of age, and to gradually introduce a diverse variety of age-appropriate nutrient-dense complementary foods along with breast milk until the child reaches 24 months of age. Dietary diversity is associated with better micronutrient intake and improved growth in this age group [31]. Interventions to improve complementary feeding have been evaluated, with stunting as the priority outcome [32]. There is growing interest in examining these interventions' impact on wasting, but most studies have not systematically reported wasting outcomes [33,34].

Preventive interventions for children aged 6 to 23 months fall into two general categories: education only, and provision of a supplement that contains energy and micronutrients (with or without education). Interventions within each category vary widely in terms of the content and frequency of messages, form and dose of supplements, age groups targeted, duration, and baseline nutritional status of the populations, which can make it difficult to summarize the overall effect.

Education Only

The 2013 *Lancet* series presented estimates from a 2011 review of complementary feeding interventions in both food-secure populations and food-insecure populations (measured by the percentage of the population living above or below US$1.25 per day, respectively) [32]. The pooled analysis for both groups found that educational interventions alone led to a significant improvement in height-for-age z-scores (HAZ; standard mean difference [SMD] 0.23, based on five studies), weight-for-age z-scores (WAZ; SMD 0.16, based on six studies), and stunting rates (RR 0.71, based on five studies). The HAZ and stunting rate effect sizes were of similar magnitude in both food-secure and food-insecure populations, whereas the WAZ effect was only significant among food-insecure populations.

Provision of Supplement with Energy and Micronutrients

Supplements range from providing child and/or household rations of staple foods (e.g., corn, oil, milk) to specially formulated products, including fortified flours (e.g., corn-soy blend) and lipid-based formulations (e.g., Plumpy'Doz). The 2013 *Lancet* series reported that "overall, the provision of complementary foods in food insecure populations [with or without education] was associated with significant gains in HAZ (SMD 0.39; 95% CI 0.05–0.73, seven studies) and WAZ (SMD 0.26, 95% CI 0.04–0.48, three studies), whereas the effect on stunting did not reach statistical significance (RR 0.33, 95% CI 0.11–1.00, seven studies)" [1]. A more recent review by Kristjansson et al. analyzed the evidence from nine randomized controlled trials (RCTs) that provided food supplements to children aged 6 to 59 months in LMICs, regardless of food security status [35]. The majority of studies were in children aged 6 to 23 months, and in populations with low mean growth z-scores at baseline. Six-month interventions were associated with statistically significant increases in weight (MD 0.12 kg), height (0.27 cm), and HAZ (SMD 0.15), but no significant differences in WHZ.

MICRONUTRIENT SUPPLEMENTATION

Vitamin A Supplementation

It is estimated that one-third of children worldwide are vitamin A deficient due to lack of dietary diversity, leading to decreased immune system capacity and increased mortality from a number of infectious diseases [36]. Periodic high-dose vitamin A supplementation of children aged 6 months and older is one of the few nutritional interventions that is implemented at public health scales across LMICs. More than 80 countries have programs for vitamin A supplementation [25], often delivered alongside other interventions using a campaign strategy. A large body of evidence (43 randomized trials with over 215,000 children), cited in the 2013 *Lancet* series, supports the impact of vitamin A supplementation, which reduces all-cause mortality by 24% in children 6 to 59 months, as well as diarrhea and measles incidence [37]. It is important to note that many of these studies were conducted several decades ago and, in the intervening period, causes of child mortality globally have shifted. As a result, it is more appropriate at this point to use cause-specific effect sizes (in this case, a 30% reduction in diarrhea-specific mortality due to vitamin A supplementation [38]), and apply this only to the proportion of the population that is thought to be vitamin A deficient.

Because vitamin A deficiency is a risk factor for severe measles, supplementation has also been recommended as a treatment for measles. In 2010, a meta-analysis (based on six studies, high-quality evidence) found that two or more doses (of 200,000 IU for children ≥1 year of age or 100,000 IU for infants) of vitamin A as a treatment reduced measles mortality by 62% [39].

Preventive Zinc Supplementation

Zinc supplementation consists of a daily dose of 10 mg; although many trials have tested its effect, this intervention is not currently implemented on a public health scale in any country. The 2013 *Lancet* series concluded that "preventive zinc supplementation in populations at risk of zinc deficiency reduces the risk of morbidity from childhood diarrhea and acute lower respiratory infections and might increase linear growth and weight gain in infants and young children" [1]. This evidence came from a 2011 review that showed a 13% reduction in diarrhea incidence and a 19% reduction in pneumonia incidence (based on 18 studies from LMICs), as well as an 18% reduction in all-cause mortality among children 12 to 59 months (this effect was not significant when children 6–12 months were included) [40]. Another meta-analysis showed a significant improvement in linear growth for zinc-supplemented children compared to placebo (based on 36 studies) [41]. A more recent *Cochrane Review* shows nonsignificant effects for all-cause mortality, diarrhea mortality, pneumonia mortality, and pneumonia incidence, with the only significant effect found in a 13% reduction in diarrhea incidence, but the results of this analysis cannot be used because it failed to appropriately exclude trials where iron was given with zinc [42]. Thus, results of previous analyses that included only trials that compared zinc supplements to placebo should be used and are unchanged from what was in the 2013 *Lancet* series.

As mentioned earlier regarding vitamin A, given the shift in causes of child mortality over time, it is now more appropriate to apply cause-specific mortality rates to the percentage of the population that is thought to be zinc deficient. In the case of zinc supplementation, reanalysis carried out for the *Lancet* 2013 series showed cause-specific effect sizes among zinc-deficient individuals of 50% reduction in diarrhea mortality and 49% reduction in pneumonia mortality [1].

Zinc is also recommended for treating diarrhea. Since 2004, the WHO guideline for treating diarrhea has included 20 mg zinc per day (10 mg per day for infants under 6 months) for 10 to 14 days, along with oral rehydration solution and continued age-appropriate feeding [43]. The addition of zinc to the diarrhea treatment regimen has been shown to decrease the duration and severity of diarrhea episodes, with a protective effect lasting up to 3 months. Based on diarrhea hospitalization data, zinc is estimated to decrease child mortality due to diarrhea by 23% [44].

Multiple Micronutrient Supplementation

Micronutrient powder (MNP) sachets for the home-based fortification of complementary foods are recommended by the WHO [45] and have been evaluated in several programmatic contexts. A 2013 review of 17 trials found that, although they have an impact on iron status and anemia in children 6 to 59 months, there is no effect on growth [46]. The lack of effect on growth may be explained by the relatively lower dose of zinc in MNP compared to zinc supplements [47,48], a reduction in zinc bioavailability due to high phytic acid content of foods mixed with the MNP, and/or the adverse effects of iron supplements on diarrhea and growth [49,50].

More recently, a category of supplements known as small quantity lipid-based nutrient supplements (SQ-LNS) has been introduced, which contain a complete daily requirement of micronutrients for their target age group in a lipid-based matrix that provides less than 120 kcal energy along with essential fatty acids. Trials in children 6 to 23 months are ongoing, but findings to date are mixed, with some populations demonstrating improvements in length and weight outcomes but others not [51,52]. A systematic review of these SQ-LNS studies is needed.

MANAGEMENT OF SEVERE ACUTE MALNUTRITION

Severe acute malnutrition (SAM) (defined as WHZ <–3 SD, or mid-upper arm circumference [MUAC] <115 mm, or presence of bilateral pitting edema in feet) is an acute condition, with generally low prevalence outside of emergency contexts, but a very high risk of mortality. In the past, all children with SAM were treated on an inpatient basis, but now WHO guidelines support community-based rehabilitation of SAM children age 6 to 59 months without complications, using specially formulated ready-to-use therapeutic foods (RUTFs) [53].

About 15% of all children with SAM will have complications (e.g., concurrent infections) that require the child to be stabilized in a facility with fluid management and dietary support, and treated for infections before being discharged to community-based care. A review prepared for the 2013 Lancet series used a Delphi process to systematically weigh expert opinion in the face of limited trial evidence, and reported that the overall recovery rate for these children who start as inpatient cases (most of whom transfer from facility to outpatient protocols described later) is 71% (interquartile range [IQR] 25%–95%) [54]. With improved treatment, case fatality rates have declined to 3% to 35% and are highest among HIV-infected children [1].

There have been relatively few randomized trials that test the efficacy of outpatient treatment of SAM without complications, but substantial programmatic experience has established it as the standard of care. The 2013 Delphi process review estimated the recovery rate for outpatient SAM treatment at 80% (IQR 50%–93%) [54]. A 2013 *Cochrane Review*, including four RCTs, all conducted by the same team in Malawi, found that outpatient treatment protocols are only marginally better than inpatient treatment in terms of overall wasting recovery rates, but are associated with faster recovery time and lower mortality [55]. Under ideal conditions, recovery rates of up to 90% are considered feasible, but are unlikely in contexts with relative high rates of pediatric HIV infection [56].

MANAGEMENT OF MODERATE ACUTE MALNUTRITION

Taking a "treatment" approach to rapidly rehabilitate children aged 6-59 months with moderate acute malnutrition (MAM) (defined as $-3 <$ WHZ < -2 SD, or $115 \leq$ MUAC < 125 mm, without bilateral pitting edema) remains controversial. Unlike SAM, there is currently no globally endorsed guidance for MAM treatment outside of emergency contexts. MAM treatment programs generally use food ration packages that include fortified corn-soy blend (CSB) and oil and, in some food insecure settings, a lipid-based ready-to-use supplementary food (RUSF) product with a similar macronutrient composition, but a different dose and micronutrient profile than the RUTF used in SAM treatment. The 2013 Delphi process review reports recovery rates for MAM treatment at 84% (IQR 50%–100%) [54]. In comparing supplementary feeding programs that used CSB to those using RUSF, children in the RUSF group were more likely to reach exit criteria, with WHZ that was 0.11 (0.04–0.17) greater at discharge, but there was no difference in mortality outcomes between groups [54].

CONCLUSION

The nutrition community has identified many effective interventions to address nutritional deficiencies and improve birth outcomes, growth, and survival. As described in the 2013 *Lancet* series, implementing these interventions at scale through the health system, and integrating them with nutrition-sensitive efforts in other sectors is an ongoing challenge [1,2]. Nutrition must be recognized as a core component of the maternal, neonatal, and child health agenda [25].

A second, related challenge is to improve the monitoring and evaluation of nutrition program delivery and impact. Currently, our measurement of nutrition intervention coverage at scale across LMICs is, for the most part, limited to what is captured in household surveys, such as the Demographic and Health Survey (DHS) and Multiple Indicator Cluster Survey (MICS). Little is known about the actual coverage of infant and young child feeding promotion and food supplementation interventions, acute malnutrition screening, and treatment programs, or the entire category of nutrition-sensitive interventions.

We must develop and implement new measurement approaches that enable countries to assess the impact of nutrition policies and programs and scale. Since the publication of the first *Lancet* nutrition series in 2008 and the launch of the Scaling Up Nutrition (SUN) movement in 2010, we have seen governments and donors respond to calls for increased investments in nutrition. However, if we cannot measure and demonstrate the actual progress that has been made, we will not be able to sustain these efforts.

REFERENCES

1. Bhutta ZA, Das JK, Rizvi A et al. Evidence-based interventions for improvement of maternal and child nutrition: What can be done and at what cost? *Lancet* 2013; 382:452–77.
2. Ruel MT, Alderman H, Maternal and Child Nutrition Study Group. Nutrition-sensitive interventions and programmes: How can they help to accelerate progress in improving maternal and child nutrition? *Lancet* 2013; 382:536–51.
3. Black RE, Victora CG, Walker SP et al. Maternal and child undernutrition and overweight in low-income and middle-income countries. *Lancet* 2013; 382:427–51.
4. Christian P, Lee SE, Donahue AM et al. Risk of childhood undernutrition related to small-for-gestational age and preterm birth in low- and middle-income countries. *Int J Epidemiol* 2013; 42:1340–55.
5. Katz J, Lee ACC, Kozuki N et al. Mortality risk in preterm and small-for-gestational-age infants in low-income and middle-income countries: A pooled country analysis. *Lancet Lond Engl* 2013; 382:417–25.
6. Victora CG, Onis M de, Hallal PC et al. Worldwide timing of growth faltering: Revisiting implications for interventions. *Pediatrics* 2010; 125:e473–80.
7. Villar J, Papageorghiou AT, Pang R et al. The likeness of fetal growth and newborn size across non-isolated populations in the INTERGROWTH–21st Project: The Fetal Growth Longitudinal Study and Newborn Cross-Sectional Study. *Lancet Diabetes Endocrinol* 2014; 2:781–92.
8. Kozuki N, Katz J, Christian P et al. Comparison of US birth weight references and the International Fetal and Newborn Growth Consortium for the 21st Century Standard. *JAMA Pediatr* 2015; 169:e151438.
9. World Health Organization. Guideline: Daily iron and folic acid supplementation in pregnant women. 2012. http://apps.who.int/iris/bitstream/10665/77770/1/9789241501996_eng.pdf (accessed April 4, 2016).
10. De-Regil LM, Peña-Rosas JP, Fernández-Gaxiola AC et al. Effects and safety of periconceptional oral folate supplementation for preventing birth defects. *Cochrane Database Syst Rev* 2015; 12:CD007950.
11. Peña-Rosas JP, De-Regil LM, Garcia-Casal MN et al. Daily oral iron supplementation during pregnancy. *Cochrane Database Syst Rev* 2015; 7:CD004736.
12. Haider BA, Bhutta ZA. Multiple-micronutrient supplementation for women during pregnancy. *Cochrane Database Syst Rev* 2015; 11:CD004905.
13. Aburto NJ, Abudou M, Candeias V et al. Effect and safety of salt iodization to prevent iodine deficiency disorders: A systematic review with meta-analyses. WHO eLibrary of Evidence for Nutrition Actions (eLENA). Geneva: World Health Organization; 2014. http://apps.who.int/iris/bitstream/10665/148175/1/9789241508285_eng.pdf.
14. Zimmermann MB. The effects of iodine deficiency in pregnancy and infancy. *Paediatr Perinat Epidemiol* 2012; 26:108–17.
15. Das JK, Salam RA, Kumar R et al. Micronutrient fortification of food and its impact on woman and child health: A systematic review. *Syst Rev* 2013; 2: 67.
16. Hofmeyr GJ, Lawrie TA, Atallah AN et al. Calcium supplementation during pregnancy for preventing hypertensive disorders and related problems. *Cochrane Database Syst Rev* 2014; 6:CD001059.
17. Omotayo MO, Dickin KL, O'Brien KO et al. Calcium supplementation to prevent preeclampsia: Translating guidelines into practice in low-income countries. *Adv Nutr Int Rev J* 2016; 7:275–8.
18. Ota E, Hori H, Mori R et al. Antenatal dietary education and supplementation to increase energy and protein intake. *Cochrane Database Syst Rev* 2015; 6:CD000032.

19. Imdad A, Bhutta ZA. Effect of balanced protein energy supplementation during pregnancy on birth outcomes. *BMC Public Health* 2011; 11:1–9.

20. Victora CG, Bahl R, Barros AJD et al. Breastfeeding in the 21st century: Epidemiology, mechanisms, and lifelong effect. *Lancet* 2016; 387:475–90.

21. NEOVITA Study Group. Timing of initiation, patterns of breastfeeding, and infant survival: Prospective analysis of pooled data from three randomised trials. *Lancet Glob Health* 2016; 4:e266–75.

22. Sinha B, Chowdhury R, Sankar MJ et al. Interventions to improve breastfeeding outcomes: A systematic review and meta-analysis. *Acta Paediatr* 2015; 104:114–34.

23. Haider BA, Bhutta ZA. Neonatal vitamin A supplementation for the prevention of mortality and morbidity in term neonates in developing countries. *Cochrane Database Syst Rev* 2011; 10:CD006980.

24. Haider BA, Bhutta ZA. Neonatal vitamin A supplementation: Time to move on. *Lancet* 2015; 385:1268–71.

25. Christian P, Mullany LC, Hurley KM et al. Nutrition and maternal, neonatal, and child health. *Semin Perinatol* 2015; 39:361–72.

26. World Health Organization. *Guideline: Delayed umbilical cord clamping for improved maternal and infant health and nutrition outcomes.* Geneva: World Health Organization; 2014. http://apps.who.int/iris/bitstream/10665/148793/1/9789241508209_eng.pdf (accessed April 4, 2016).

27. McDonald SJ, Middleton P, Dowswell T et al. Effect of timing of umbilical cord clamping of term infants on maternal and neonatal outcomes. *Cochrane Database Syst Rev* 2013; 7:CD004074.

28. Rabe H, Diaz-Rossello JL, Duley L et al. Effect of timing of umbilical cord clamping and other strategies to influence placental transfusion at preterm birth on maternal and infant outcomes. *Cochrane Database Syst Rev* 2012; 8:CD003248.

29. World Health Organization. 2002. The optimal duration of exclusive breastfeeding: Report of an expert consultation. http://apps.who.int/iris/bitstream/10665/67219/1/WHO_NHD_01.09.pdf (accessed April 4, 2016).

30. Sankar MJ, Sinha B, Chowdhury R et al. Optimal breastfeeding practices and infant and child mortality: A systematic review and meta-analysis. *Acta Paediatr* 2015; 104:3–13.

31. Arimond M, Ruel MT. Dietary diversity is associated with child nutritional status: Evidence from 11 demographic and health surveys. *J Nutr* 2004; 134:2579–85.

32. Lassi ZS, Das JK, Zahid G et al. Impact of education and provision of complementary feeding on growth and morbidity in children less than 2 years of age in developing countries: A systematic review. *BMC Public Health* 2013; 13:1–10.

33. Mucha N. 2014. *Preventing moderate acute malnutrition (MAM) through nutrition-sensitive interventions.* http://www.cmamforum.org/Pool/Resources/Nutrition-Sensitive-MAM-Prevention-CMAM-Forum-Dec-2014.pdf (accessed April 4, 2016).

34. Jimenez M, Stone-Jiminez M. 2014. *Preventing moderate acute malnutrition (MAM) through nutrition-specific interventions.* http://www.cmamforum.org/Pool/Resources/Nutrition-specific-MAM-prevention-CMAM-Forum-Technical-Brief-Sept-2014-.pdf (accessed April 4, 2016).

35. Kristjansson E, Francis DK, Liberato S et al. Food supplementation for improving the physical and psychosocial health of socio-economically disadvantaged children aged three months to five years. *Cochrane Database Syst Rev* 2015; 3:CD009924.

36. Stevens GA, Bennett JE, Hennocq Q et al. Trends and mortality effects of vitamin A deficiency in children in 138 low-income and middle-income countries between 1991 and 2013: A pooled analysis of population-based surveys. *Lancet Glob Health* 2015; 3:e528–36.

37. Imdad A, Herzer K, Mayo-Wilson E et al. Vitamin A supplementation for preventing morbidity and mortality in children from 6 months to 5 years of age. *Cochrane Database Syst Rev* 2010; 12:CD008524.

38. Imdad A, Yakoob MY, Sudfeld C et al. Impact of vitamin A supplementation on infant and childhood mortality. *BMC Public Health* 2011; 11 Suppl 3:S20.

39. Sudfeld CR, Navar AM, Halsey NA. Effectiveness of measles vaccination and vitamin A treatment. *Int J Epidemiol* 2010; 39 Suppl 1:i48–55.

40. Yakoob MY, Theodoratou E, Jabeen A et al. Preventive zinc supplementation in developing countries: Impact on mortality and morbidity due to diarrhea, pneumonia and malaria. *BMC Public Health* 2011; 11 Suppl 3:S23.

41. Imdad A, Bhutta ZA. Effect of preventive zinc supplementation on linear growth in children under 5 years of age in developing countries: A meta-analysis of studies for input to the lives saved tool. *BMC Public Health* 2011; 11:1–14.

42. Mayo-Wilson E, Junior JA, Imdad A et al. Zinc supplementation for preventing mortality, morbidity, and growth failure in children aged 6 months to 12 years of age. *Cochrane Database Syst Rev* 2014; 5:CD009384.

43. World Health Organization, UNICEF. 2004. WHO/UNICEF joint statement: Clinical management of acute diarrhoea. http://apps.who.int/iris/bitstream/10665/68627/1/WHO_FCH_CAH_04.7.pdf (accessed April 4, 2016).

44. Fischer-Walker CL, Black RE. Zinc for the treatment of diarrhoea: Effect on diarrhoea morbidity, mortality and incidence of future episodes. *Int J Epidemiol* 2010; 39 Suppl 1:i63–9.

45. World Health Organization. 2011. *Guideline: Use of multiple micronutrient powders for home fortification of foods consumed by infants and children 6-23 months of age.* http://apps.who.int/iris/bitstream/10665/44651/1/9789241502047_eng.pdf (accessed April 4, 2016).

46. Salam RA, MacPhail C, Das JK et al. Effectiveness of micronutrient powders (MNP) in women and children. *BMC Public Health* 2013;13 Suppl 3:S22.

47. Soofi S, Cousens S, Iqbal SP et al. Effect of provision of daily zinc and iron with several micronutrients on growth and morbidity among young children in Pakistan: A cluster-randomised trial. *Lancet* 2013; 382:29–40.

48. De-Regil LM, Suchdev PS, Vist GE et al. Home fortification of foods with multiple micronutrient powders for health and nutrition in children under two years of age. *Cochrane Database Syst Rev* 2011; 9:CD008959.

49. Gera T, Sachdev HPS. Effect of iron supplementation on incidence of infectious illness in children: Systematic review. *BMJ* 2002; 325:1142.

50. Dewey KG, Domellöf M, Cohen RJ et al. Iron supplementation affects growth and morbidity of breast-fed infants: Results of a randomized trial in Sweden and Honduras. *J Nutr* 2002; 132:3249–55.

51. Ashorn P, Alho L, Ashorn U et al. Supplementation of maternal diets during pregnancy and for 6 months postpartum and infant diets thereafter with small-quantity lipid-based nutrient supplements does not promote child growth by 18 months of age in rural Malawi: A randomized controlled trial. *J Nutr* 2015; 145:1345–53.

52. Hess SY, Abbeddou S, Jimenez EY et al. Small-quantity lipid-based nutrient supplements, regardless of their zinc content, increase growth and reduce the prevalence of stunting and wasting in young Burkinabe children: A cluster-randomized trial. *PloS One* 2015; 10:e0122242.

53. World Health Organization. 2013. *Guideline: Updates on the management of severe acute malnutrition in infants and children.* http://apps.who.int/iris/bitstream/10665/95584/1/9789241506328_eng.pdf (accessed April 4, 2016).

54. Lenters LM, Wazny K, Webb P et al. Treatment of severe and moderate acute malnutrition in low- and middle-income settings: A systematic review, meta-analysis and Delphi process. *BMC Public Health* 2013; 13 Suppl 3:S23.

55. Schoonees A, Lombard M, Musekiwa A et al. Ready-to-use therapeutic food for home-based treatment of severe acute malnutrition in children from six months to five years of age. *Cochrane Database Syst Rev* 2013; 6:CD009000.

56. Trehan I, Manary MJ. Management of severe acute malnutrition in low-income and middle-income countries. *Arch Dis Child* 2015; 100:283–7.

25 Nutrition-Sensitive Interventions for the First 1,000 Days

Jessica Fanzo and Haley Swartz

CONTENTS

INTRODUCTION

Nearly three decades of research support a strong relationship between the complex biological processes that can occur during the first 1,000 days of life and the resulting multiple burdens of malnutrition, including undernutrition (measured as stunting and wasting), overweight and obesity, and micronutrient deficiencies. Although gaps in knowledge remain, the field of nutrition has made commendable progress in identifying which policies, programs, and initiatives "work" to improve nutrition for the most vulnerable groups, particularly among women and children. To date, most evidence on programmatic successes reflect "nutrition-specific" interventions, well described in Chapter 24. Such interventions, highlighted in the 2008 and 2013 *Lancet* series on undernutrition, reflect 10 interventions in three areas, primarily in the health sector: improving micronutrient uptake through fortification or supplementation, treatment of acute malnutrition, and counseling for breastfeeding and complementary feeding.

Scaling-up coverage of these 10 nutrition-specific recommendations to 90% of the population in 34 countries, where 90% of the world's stunted children reside, would

389

result in an estimated 20% decline in stunting and a 60% reduction in wasting [1]. Such a decline in undernutrition could be much larger if the international nutrition community were to go beyond the core 10 nutrition-specific areas to include "nutrition-sensitive" approaches. Nutrition-sensitive interventions inherently involve systems engagement, incorporating multiple sectors and disciplines in the fight against undernutrition.

Nutrition-specific interventions primarily target the immediate causes of undernutrition: inadequate dietary intake and diseases, as shown in Figure 25.1 [2]. The primary objective of nutrition-sensitive interventions is not necessarily direct reductions in undernutrition, but rather indirect improvements in nutrition among program beneficiaries. Many nutrition-sensitive approaches focus on the indirect and long-term consequences of both the underlying and basic causes of childhood nutrition (in Figure 25.1, light and dark blue, respectively). The two types of interventions could be complementary and mutually reinforcing: nutrition-sensitive programs and policies may serve as the delivery mechanism for nutrition-specific interventions to mothers and children, given appropriate funding, capacity, and technical support.

This chapter will delve into the evidence on the types of nutrition-sensitive approaches and the sectors responsible for program implementation; the relationship between nutrition-sensitive interventions and the first 1,000 days of life; approaches that are effective can be scaled now; and the layered complexity of scaling-up through multiple sectors and institutional arrangements. Box 25.1 highlights the main messages of nutrition-sensitive approaches, which are applicable to both the specific needs during the first 1,000 days of life and throughout the lifespan.

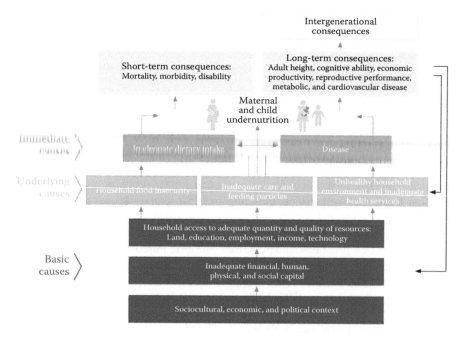

FIGURE 25.1   UNICEF framework of the determinants of maternal and child undernutrition. (From United Nations Children's Fund (UNICEF), *Improving child nutrition: The achievable imperative for global progress*, New York: UNICEF, 2013. With permission.)

> **BOX 25.1   HIGHLIGHTS OF NUTRITION-SENSITIVE
> APPROACHES IN THE FIRST 1,000 DAYS**
>
> - Nutrition-sensitive interventions are multisectoral, incorporating nutrition outcomes into health care, agricultural production, education systems, social protection transfers, and early childhood development (ECD) curricula.
> - Healthy mothers produce healthy children. Improving childhood nutrition outcomes within the first 1,000 days of life requires programming that is empowering to women, and sensitive to both gender and local contexts.
> - Interventions that directly impact health and nutrition status in the first 1,000 days of life include water, sanitation, and hygiene (WASH); women-centered agricultural projects; and access to quality antenatal care.
> - The empirical evidence evaluating the effect of nutrition-sensitive interventions on reductions in stunting, or low height-for-age, is significant, but small in scale and scope.
> - Nutrition-sensitive program evaluations must include rigorous design and measurable nutrition outcomes to further demonstrate evidence of how to reduce childhood undernutrition and to achieve the Sustainable Development Goals (SDGs).

NUTRITION OUTCOMES REQUIRE MULTIPLE SECTORS AND MULTIPLE SOLUTIONS

The first 1,000 days of life is the window of opportunity to influence child growth, nutritional status, and cognitive development [3]. A single intervention to alleviate early childhood stunting can have significant small to moderate effects [1]. In contrast, predictive models suggest that by combining micronutrient supplementation, food fortification, and disease control interventions, stunting prevalence could be reduced by one-third [1]. Further reductions in chronic undernutrition likely require a more comprehensive approach, combining nutrition-specific interventions with larger, multisector efforts [4].

In theory, multisectoral nutrition interventions can be a starting point. In practice, however, the realities of effectively working through multiple institutions make such work incredibly difficult [5]. Some argue that nutrition should incorporate agriculture, health, water and sanitation, education, social protection, gender, and environment; however, historically, multisectoral collaboration has not worked well [6]. Recent scientific discovery and operations research, have provided more evidence and tools for researchers and policy makers to assimilate sectors and use systems approaches to nutrition. Such progress is reflective of the Sustainable Development Goals (SDGs). In particular, nutrition is inextricably linked to international development in SDG2, which aims to end hunger, achieve food security and improved nutrition, and promote sustainable agriculture [7].

Working through multiple sectors simultaneously can leverage synergies and catalyze gains extending beyond achievements through sector-specific programs that work in isolation [8]. Recent research documents potential synergies between health and economic interventions, suggesting multisector approaches may generate a wider range of benefits than single-sector approaches [9,10].

As shown in Figure 25.1, many factors influence the quality of diets and disease outcomes that ultimately impact nutrition outcomes [2]. The underlying causes of malnutrition affect households and communities, including household food insecurity; inadequate care and feeding practices for children; unhealthy household environments; and inaccessible, often inadequate health and education services. Interventions to address the underlying causes thus include access to an affordable, diverse, and nutrient-rich diet; optimal maternal and child feeding and care practices; access to adequate health services; and improved sanitation and hygiene practices to ensure a healthy environment. These factors directly impact the immediate causes of malnutrition: nutrient intake and utilization and the manifestation of disease. Food, health, and care are each necessary, but not sufficient conditions for adequate childhood nutrition. Social, economic, and political factors continually impact food, health, and care, highlighting how each underlying factor differs in relative importance from country to country. Understanding the immediate and underlying causes of undernutrition in a given context is critical to delivering appropriate, effective, and sustainable solutions to meet the nutritional needs of a country's most vulnerable groups. Last, the basic causes of undernutrition focus on historical structures and country-specific processes of societies. The determinants of such causes are complex, with major factors including poverty, socioeconomic, gender and social inequities, insufficient human rights, and lack of access to essential resources. At its core, the basic causes of undernutrition encompass larger political, economic, and legal factors. Such factors affect institutional leadership and policymaking at both community and national levels, highlighting how the theoretically "best" programs to promote adequate nutrition could be defeated, unsuccessful, or deprioritized in a country's policy agenda. The factors that impact nutrition status are buried within these basic causes, deeply rooted in overarching societal norms [2].

In both programs and practice, policymakers have prioritized interventions that target immediate and underlying causes, which mainly consist of nutrition-specific approaches. Basic causes, or nutrition-sensitive interventions, are often underfunded, and overlooked as too politically sensitive or challenging to implement [11]. Such neglect has encouraged the practice of "targeting vulnerable groups" and "delivering packages" of commodities and services. Solutions to reduce the burdens of undernutrition require including basic causes and ensuring policies are sustainable to promote long-term change.

THE EVIDENCE FOR NUTRITION-SENSITIVE APPROACHES IN ADDRESSING THE FIRST 1,000 DAYS

Currently, there is no "systematic process for [the] collation of the implementation-related evidence base about how to scale up the vast array of [both] nutrition-specific and nutrition-sensitive interventions with quality and equity," indicating further research is necessary on both nutrition-sensitive and nutrition-specific interventions [12]. Further, partial empirical evidence supports a direct relationship between nutrition-sensitive interventions and health improvements during the first 1,000 days of life. In the 2013 *Lancet* series, Black et al. [13] adapted the United Nations Children's Fund (UNICEF) framework, as shown in Figure 25.1, to explicitly include nutrition-sensitive approaches to address underlying causes of undernutrition, along with factors of enabling environments that impact the basic causes of undernutrition. This new framework (Figure 25.2) outlines the necessary sectors with which the international nutrition community should engage to produce a comprehensive strategy for improving child nutrition [13]. As part of the *Lancet* series, Ruel et al. [14] further defined current evidence on nutrition-sensitive approaches and best practices. In this section, we will describe both types of intervention and the sectors responsible for implementing nutrition-sensitive approaches.

Agriculture and Food Security

Studies have shown that agricultural research, programs, and policy have put less emphasis on maximizing nutrition outputs from farming systems. Agriculture has instead had substantive impact on economic growth by enhancing farm productivity and food availability. A recent longitudinal analysis found that agricultural per capita income was more strongly associated with reductions in undernutrition than non-agricultural income [15]. After controlling for income, the analysis indicated that as economic transformation proceeded, stunting declined at a faster pace in countries that supported growth in the agricultural sector than in those that did not. However, absolute reductions in stunting were quite modest; a doubling of per capita agricultural income is associated with an approximately 15% decline in stunting [15].

Research has also demonstrated a strong association between dietary diversity, diet quality, and the nutritional status of children [16]. It is also clear that household dietary diversity is a sound predictor of the micronutrient density of the diet, particularly for young children [17]. Dietary diversity is a vital element of diet quality. The consumption of a variety of foods across and within food groups, and across different varieties of specific foods, guarantees the adequate intake of essential nutrients and important nonnutrient factors. In some communities, particularly among smallholder farmers, agriculture can make contributions to dietary diversity by increasing and/or improving diversity of landscapes and the availability of foods produced from those landscapes.

Broader nutrition-sensitive interventions focusing on the individual include the promotion of home gardens, biofortification, small-animal rearing, and women-centered agricultural practices.

Two recent reviews have examined the impact of agricultural interventions on nutrition outcomes, showing that agriculture strategies improved dietary patterns and specific micronutrient intakes (vitamin A in particular); however, there was less of an effect on growth such as stunting and wasting [18–20]. Explanations for such minimal effects include a hypothesis that agricultural interventions may best reduce short-term undernutrition rather than chronic undernutrition, or study assessments

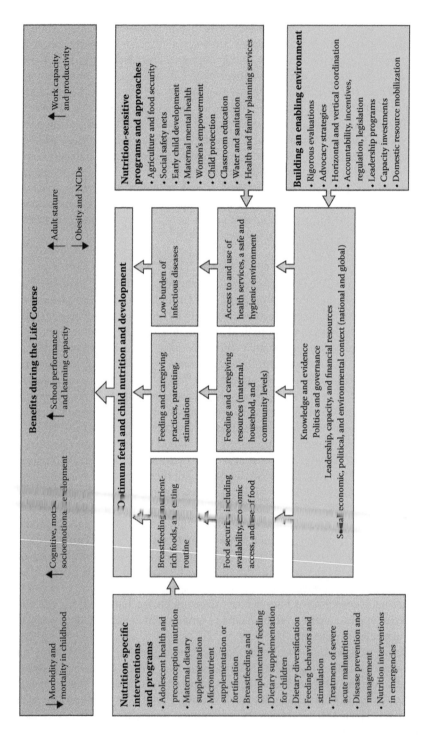

FIGURE 25.2 Framework for actions to achieve optimum fetal and child nutrition and development. (From Black RE, Victora C, Walker SP et al., *Lancet* 2013, 382(989): 427–51. With permission.)

occurred shortly after intervention, failing to capture long-term nutrition effects [20]. In a series of studies analyzed in different country contexts, overall household agricultural production had direct linkages with dietary quality and nutritional status, although impacts can widely vary [21].

Women's social status, empowerment, control over resources, time allocation, and health and nutritional status are key mediators in the pathways between agriculture inputs, intrahousehold resource allocation, and child nutrition [14]. However, there is a need for more studies to measure the effect of agriculture interventions on women's time, knowledge, practices, health, and nutritional status.

SOCIAL PROTECTION

Social protection programs target the immediate and underlying causes of maternal and child undernutrition. Conditional or unconditional cash transfers and/or an in-kind food distribution that help stabilize household income and consumption over time are examples of social protection programs [14]. They generally operate with the goal of reducing vulnerability, investing in human capital, and breaking the intergenerational cycle of poverty. Social protection has the potential to improve nutrition outcomes given the behavioral requirements of the program: targeted, nutritionally vulnerable populations, such as poor pregnant women or those who face economic, climactic, or social stresses, are required to enroll in an additional economic, social, or educational development program. For instance, school feeding programs encourage parents to enroll their children in school, and cash transfer stipends promote attendance at regular health checkups. Such components can serve as critical leverage points for nutrition-specific interventions [14].

Although programs vary in scope and scale, recent evidence indicates social protection interventions and poverty alleviation programs have direct effects on individual nutrition outcomes [14,22]. In particular, one review noted that conditional cash transfers CCT significantly improved child anthropometry but had little impact on micronutrient status [23].

WATER, SANITATION, AND HYGIENE

One of the most effective and well-known development strategies to reduce childhood stunting in low- and middle-income countries (LMICs) incorporates water, sanitation, and hygiene (WASH) interventions. Since the 1990s, even minor improvements in hygiene education, home handwashing practices, and the quality of both community drainage systems and home toilets have been associated with declining incidence of diarrhea morbidity, pneumonia, communicable diseases (CDs), and other infectious diseases.

WASH plays a fundamental role in improving nutritional outcomes. Fifty percent of malnutrition is associated with repeated diarrhea or intestinal worm infections as a result of unsafe water, inadequate sanitation, or insufficient hygiene [24]. In addition, a review found that poor WASH conditions have a significant detrimental effect on child growth and development resulting from sustained exposure to enteric pathogens [25]. A recent meta-analysis found that three individual WASH interventions—solar disinfection of water, provision of soap, and improvement of water quality—had a small, albeit significant impact on stunting reductions in children under 5 years old [26].

Diarrhea highlights the vicious cycle of undernutrition and the necessity to enhance WASH practices for children in the first 1,000 days of life. Meta-analyses, first conducted in 1990 and then updated in 2012, found that the highest incidence rate of frequent diarrhea episodes continue to be infants ages 6 to 11 months old, at 5.3 episodes per year [27]. Frequent episodes of diarrhea in children under 2 years old can significantly increase the risk of stunting and impair cognitive development [28,29].

In LMICs, young children are often afflicted by environmental enteropathy, a syndrome causing changes in the small intestine of individuals who lack basic sanitary facilities and are chronically exposed to fecal contamination [30]. This condition decreases the ability of the intestinal tract to absorb nutrients critical for optimum growth and development, leading to serious consequences in nutritional status. Rates of environmental enteropathy are high in early life, particularly among young, breastfed infants who begin to consume complementary foods. Although these children eat greater amounts of solid foods, they are exposed to the outside environment and thus face an increased risk of consuming contaminated foods. Two interventions working together—an improved food system to ensure food safety, and hygienic toilets and community-based sanitation—have been shown to reduce enteropathy rates [31,32].

EARLY CHILDHOOD DEVELOPMENT

Encompassing the first 1,000 days from birth through to primary school, early childhood development (ECD) programs aim to improve physical, cognitive, and psychosocial growth among young children. The World Bank categorizes 25 ECD interventions into "packages," representing the five stages of life before age 5 [33]. In each stage, families are provided subsidized access to food and nutrition supplements, as well as "psychosocial stimulation interventions," which include household learning activities, formal preschool education, health services, and resources for improved health care [34]. Most of the interventions essential for a child's growth and development are reflected in the six sectors discussed in this chapter, including but not limited to family planning, WASH, immunizations, deworming, and quality education [33]. Figure 25.3 provides an overview of the five standard support packages designed by the World Bank [33].

A meta-analysis comparing more than 20 small-scale ECD programs in LMICs found that integrated psychosocial stimulation interventions had a medium effect ($d = 0.42$) on the Bayley scale of childhood cognitive language and development, whereas standalone nutrition programming provided only a small effect ($d = 0.09$) [35]. In high-income countries, ECD programs—such as Head Start in the United States or the social pedagogical programs in Nordic countries—have been successful at reducing the intergenerational burdens of both undernutrition and poverty since the early 1990s [36,37]. LMICs, however, continue to face significant challenges in implementing multisectoral ECD programs. The implementation process will

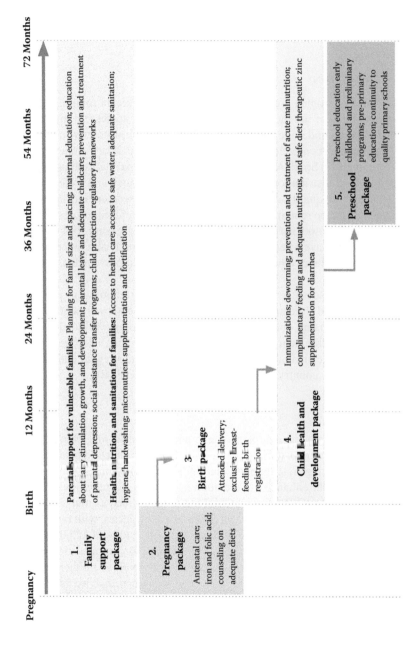

FIGURE 25.3 Five packages of essential interventions for young children and families. (From Denboba AD, Elder LL, Lombardi J et al., *Stepping up early childhood development: Investing in young children for high returns*, Washington, DC: World Bank Group, 2014. With permission.)

vary in each country, but effective ECD programs are costly. The responsibilities for program integration, targeting, and delivery frequently overlap between government ministries. Evaluations of combined nutrition and psychosocial interventions should include a country-specific analysis of the implementation process, providing sufficient data on any program features that may have benefited child growth and development [34].

WOMEN'S EMPOWERMENT

The social status of women around the world plays a significant role in determining nutritional status of the most nutritionally vulnerable: adolescent girls, young women, and children under the age of 5. Insufficient gains in women's social status can have detrimental consequences to improvements in children's nutrition outcomes. For example, in South Asia, women hold a lower social status than men [38]. With regard to food distribution, this, in turn, means that men and male offspring are given the highest quality and quantity of food, with women and female offspring allowed to eat the remaining food—resulting in higher frequencies of female undernutrition [38]. Lower social status also results in lower control of economic resources, which can have a negative impact on nutrition and health outcomes [39], as well as quality of life, psychological health, livelihood improvements, and overall earnings.

In LMICs, gender-sensitive and culturally salient programming is integral to improving maternal and child nutrition outcomes. Women's empowerment is divided into three primary domains: control of resources and autonomy, workload and time, and social support environment. All three areas influence women's decisions to have children and their associated nutrition status at birth, but the strength and direction of each effect varies by country, region, and household income [40].

FAMILY PLANNING

The ability to delay both the age of marriage and the age of first pregnancy are two significant indicators of women's health status and, in turn, child health rearing [41,42]. The child marriage rate, in particular, indicates the cultural complexities of introducing programs to enhance girl's empowerment and, thereby, improve their own and their child's nutritional status. Child marriage, defined as marriage before the age of 18, is common in six of the eight South Asian countries, with rates at 65% in Bangladesh, 47% in India, and 41% in Nepal [43]. One study of 125,000 Indian mothers and their children found that the risk of malnutrition, as measured by both underweight and stunting, was higher in young children born to mothers married as minors (<15 years), as compared to those born to women married later in life (>24 years) [44]. A meta-analysis evaluating maternal age at childbirth with child nutritional outcomes, including stunting at 2 years old, found that children born to mothers under the age of 19 are disadvantaged in both nutrition and health as compared to children born to older mothers [45].

Generally, poor pregnancy outcomes are the second leading cause of mortality in adolescent girls [46]. In mothers under 20 years of age, risk of perinatal death is 50% higher amongst newborns as compared to older mothers [47]. It is often found that children born to adolescent mothers have increased risk of low birth weight outcomes, with potentially, negative consequences into adulthood [48,49]. Improving the nutrition of adolescent girls and delaying their first pregnancy may be two key interventions, given the importance of intergenerational influences in the life cycle. Thus, improving the nutrition of adolescent girls before they become pregnant, during pregnancy and lactation, and increasing birth spacing may also help to break the vicious cycle of malnutrition and poverty.

Increasing the accessibility and quality of primary health services for women of reproductive age (WRA) has significantly improved nutrition among mothers and young infants. Programs that provide access to low-cost contraception and family planning services have consistently improved birth outcomes for both mothers and children throughout LMICs. In particular, the demand for contraception in the world's poorest countries remains high, with an estimated 222 million women in LMICs reporting unmet needs for modern methods [46]. Accessible, voluntary family planning is not only crucial to directly improve reproductive health, but is also positively associated with advancements in health, schooling, and economic indicators [50], all of which are important for adequate nutrition. Given the importance of family planning, efforts are ongoing to connect basic health services with a wider range of social support for women in LMICs.

EDUCATION

Education reflects both access to both primary and secondary schooling and nutrition education. In LMICs, maternal education tends to delay pregnancy until later in life, promote healthy hygienic practices, increase women's social status, and provide women opportunities for economic participation [51–53]. One of the strongest predictors of childhood nutrition status is maternal education [54]. Moreover, the effect of women's empowerment through education on child health is consistent in the literature. Compared with educated mothers, children of less-educated women face a higher risk of undernutrition *in utero* and a lower chance of survival during both infancy and early childhood [55].

The formal education attainment of women, particularly literacy levels, influences nutrition and health outcomes. Women who are educated and have better access to adequate and appropriate nutrition information are able to elicit appropriate health-benefitting behaviors, and have more opportunities for employment and earning power [56]. In Eastern Ethiopia, researchers found through questionnaires and health assessments that pregnant women with low and medium power of household decision-making faced a twofold higher risk of malnutrition than those with higher rates of autonomy [57]. In Bangladesh, increased maternal and paternal formal education led to decreases in child stunting [58]. In Indonesia, increased maternal and paternal education was associated with improved caregiving practices, including vitamin A capsule and immunization services, the use of iodized salt, and WASH behaviors [56].

In a cross-country analysis of 100 LMICs, Headey et al. [22] found that public investment in education was a primary source for improving women's nutrition outcomes. Specifically, women's secondary education, reductions in fertility which is, in turn, closely associated with changes in women's education, asset accumulation which is surprisingly weakly explained by economic growth, and increased access to health services were associated with stunting reductions [22]. In Nepal and Bangladesh, economic gains at the household level and rapid gains in education for both mothers and fathers both contributed to the 1.1 percentage point reduction in child underweight and stunting each year since 1997 [59]. In Nepal, asset accumulation was also a primary determinant in reductions of stunting prevalence, but only maternal education significantly predicted improvements in severe stunting and height-for-age z-scores [59].

Moreover, women's employment status, which is often dictated by education level, influences household income, which in turn impacts nutrition status. In South Africa, the chances of being undernourished during pregnancy increased two-fold for women whose household income was less than US$16 a day [56]. Similarly, in Bangladesh, poverty (directly linked to household income and employment status) was the most significant factor in predicting undernutrition among WRA [60].

HEALTH

The public health sector plays a critical role in the delivery of most of the high-impact nutrition-specific interventions, which rely heavily on functioning health facilities, services, and personnel. In countries where government health services provide wide coverage and are easily accessible, health services are a logical and sustainable delivery channel for nutrition interventions [61].

Providing universal coverage of essential interventions throughout the life cycle in an integrated primary health care system has the potential to improve nutrition during the first 1,000 days of life. The continuum of care from maternal to child health provides an opportunity to integrate nutrition services into essential primary health care packages. One aspect of primary health care that is crucial for improving the nutritional status of women and their children is antenatal and postpartum care. These critical points in time provide an opportunity for health care workers to meaningfully engage with women not only about their own health and nutrition, but also about the care of newborns and any future pregnancies. Preventative nutrition, dietary counseling, and routine physical screenings can help women in areas related to their own health needs, as well as their families. These interventions can have important implications for women's health outcomes, such as the prevention, screening, and management of infectious diseases such as malaria and diarrhea (5). Multipronged interventions often rely on behavioral change communication (BCC), a variety of approaches and tools used to design nutrition and public health programs [62]. Interpersonal communication strategies on breastfeeding and complementary feeding have been shown to increase knowledge on health behaviors that contribute to improved nutrition for infants and young children [62]. Implementing health programming with nutrition-specific interventions is likely to initiate greater reductions in child undernutrition than standalone health programming. Such projects would include the

TABLE 25.1

Integrating Nutrition-Sensitive Approaches into Sector Investments

Sector	Target Groups	Types of Interventions	Delivery Channels	Project Aims	Contextual Considerations
Agriculture and food security	Producer families Women farmers	BCC on food safety and nutrition practices Crop, breeding, and processing choices to enhance nutritional value (e.g., biofortification)	Agricultural extension rural advisory services Farmer field schools Distribution centers for technologies and inputs Microcredit and insurance	Improvements in dietary diversity and household diet quality Empower women and increase income generation	Economic and care responsibilities may be strictly gendered, highlighting that women's time and energy are limited resources.
Social protection	WRA and girls Children during the first 1,000 days	Conditional cash transfers (CCT) (e.g., take-home rations from school attendance) In-kind transfers	Health clinic services Schools Food for work, cash, or vouchers (asset program)	Improved dietary diversity of children <2 years, WRPA, pregnant and lactating women	Poverty alleviation programs must be effectively targeted toward nutritionally vulnerable groups, such as young children and WRA.
Water, sanitation, and hygiene (WASH)	Children under 2 years old Pregnant and lactating women	Safe feces disposal to reduce risk of environmental enteropathy Solar disinfection of water Providing soap	BCC community campaigns Peer counseling	Improved nutrition status of children <2 years	Programs must be designed to understand, respect, and account for social norms in each community.
Early childhood development (ECD)	Children under 5 years old Pregnant and lactating women	Psychosocial stimulation interventions (household learning activities)	Five "packages" of counseling and education: family support, pregnancy, birth, child development, preschool	Improved nutrition status of children <2 years old	The most successful ECD programs are community-specific, combining psychosocial and nutrition interventions.

(Continued)

TABLE 25.1 (CONTINUED)

Integrating Nutrition-Sensitive Approaches into Sector Investments

Sector	Target Groups	Types of Interventions	Delivery Channels	Project Aims	Contextual Considerations
Women's empowerment	Women of all ages	Control of resources and autonomy; Workload and time; Social support environment	Community centers; BCC campaigns	New balance of intrahousehold resource allocation between men and women	Programming must be both gender sensitive and culturally respectful, accounting for women's social status in the community.
Family planning	Men and WRA	Access to contraceptives; Enhance self-care practices	Health clinic services; BCC community campaigns	Delay age of both marriage first pregnancy; Spacing pregnancies	Integrating low-cost contraception into improving health care delivery is critical for improving birth outcomes by WRA.
Education	Adolescent girls	Literacy programs; Formal primary and secondary schooling	Schools; Community centers	Improve birth outcomes and childhood development	Education empowers women, providing spillover benefits onto women's health, social status, and income generation.
Health	Pregnant and lactating women; Children under 2 years old	Peer counseling to encourage exclusive breastfeeding; Emphasis on nutrition within health curricula and health professional training	Community health workers; BCC community campaigns	Improved health and nutrition status of children <2 years old	Prenatal and antenatal care are essential components of health system packages for improving maternal and child nutrition.

Source: Based on Table 6.3 from International Food Policy Research Institute (IFPRI), *Global nutrition report 2014: Actions and accountability to accelerate the world's progress on nutrition*, Washington, DC: International Food Policy Research Institute, 2014.

Note: Behavioral change communication, BCC; noncommunicable diseases, NCDs; women of reproductive age, WRA.

distribution of pneumonia, diarrhea, and deworming prevention materials along with nutrient supplement regimens such as iron folic acid supplements during pregnancy [63].

Working in tandem with nutrition-specific approaches, multisectoral programs can improve nutritional status within an infant's first 1,000 days of life. Such programs include increased access to antenatal and postnatal care, child immunizations, malaria prevention and treatment, and promotion of optimal infant and young child feeding practices (including breastfeeding promotion and timely introduction of nutritionally adequate and safe complementary foods). All improve nutritional status within an infant's first 1,000 days of life. Identifying ways to make sector investments sensitive to nutrition outcomes, the Global Nutrition Report [64] outlined target groups, types of nutrition-sensitive interventions, and delivery platforms, all of which are discussed in this chapter. A summary of this report is found in Table 25.1.

FOR NUTRITION-SENSITIVE APPROACHES, LOCAL CONTEXT MATTERS

In a 2015 study modeling the makers and markers of development on stunting rates since 1970, researchers found that the female-to-male life expectancy ratio had a significantly stronger effect on reducing the childhood stunting in South Asia than in other regions, whereas national food availability has a far stronger impact in Latin American and Caribbean countries [65]. This difference is further clarified in Figure 25.4, which displays a stark difference in the share of various nutrition-sensitive interventions between South Asian and Sub-Saharan African countries.

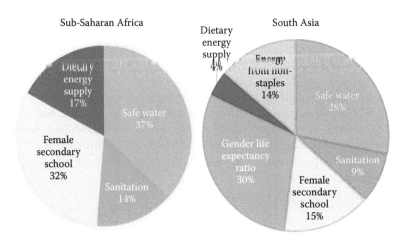

FIGURE 25.4 Comparing the contributions of underlying determinants to total reductions in childhood stunting in Sub-Saharan Africa and South Asia, 1970–2010 [65]. (From Smith L, Haddad L, *World Dev* 2015; 68:180–204. With permission.)

Improvements in water safety and the provision of secondary school for girls account for nearly two-thirds of stunting reductions in Sub-Saharan Africa from 1970 to 2010. While water safety was also a primary intervention in South Asia, an increase in the gender life expectancy ratio explained the greatest reduction in childhood stunting. Such differences between the two regions reflect the fact that South Asian countries prioritized policies to improve the underlying determinants of childhood undernutrition, namely, household access to safe water and women's care practices, while Sub-Saharan countries have realized slow progress on the basic causes of malnutrition, such as adequate sanitation, nationwide food availabilities, and dietary diversity of the food supply [65]. In LMICs, priority setting must reflect baseline measurements of childhood nutrition levels, and investments in the health, education, and agricultural sectors must correspond to local contexts.

Effective delivery platforms, coverage, and participation are major gaps for all health and nutrition services. Several delivery platforms to reach vulnerable populations exist, but there is dearth of information on what works where, and how to effectively sustain change. Some potential platforms include: the health system, especially antenatal care; community-based nutrition programs; child health days; social protection, especially conditional cash transfers the food environment; and agriculture food-based investments. Making these platforms more effective requires operations research in conjunction with capacity building and personnel training, the creation of technical guidelines, monitoring and evaluation of programs, and advocacy. Delivery mechanisms are highly dependent on the local context, which takes into account community ownership, cultural sensitivity, and the local political economy.

CONCLUSION

As countries have begun to set policy and priority agendas in pursuit of the SDGs, the international nutrition and development communities should promote and invest in targeted nutrition-sensitive, multisectoral interventions. Inextricably linked to the elimination of the multiple burdens of malnutrition is the critical window of opportunity to promote growth and development within the first 1,000 days of life. Because healthy children are born to healthy mothers, nutrition-sensitive programming must be gender-sensitive, empowering girls and women both socially and economically. It is now crucial to invest in pilot tests and interventions that integrate diverse sectors, including health, WASH, agriculture, and education in order to maximize poverty reduction and to reverse the effects of intergenerational undernutrition.

REFERENCES

1. Bhutta ZA, Das JK, Rizvi A et al. Evidence-based interventions for improvement of maternal and child nutrition: What can be done and at what cost? *Lancet* 2013; 382(9890):452–77.
2. United Nations Children's Fund (UNICEF). *Improving child nutrition: The achievable imperative for global progress.* New York: UNICEF; 2013.

3. Adair LS, Fall CH, Osmond C et al. Associations of linear growth and relative weight gain during early life with adult health and human capital in countries of low and middle income: Findings from five birth cohort studies. *Lancet* 2013; 382(9891):525–34.
4. Remans R, Pronyk PM, Fanzo J et al. Multisector intervention to accelerate reductions in child stunting: An observational study from 9 sub-Saharan African countries. *Am J Clin Nutr* 2011; 94(6):1632–42.
5. Fanzo J. Strengthening the engagement of food and health systems to improve nutrition security: Synthesis and overview of approaches to address malnutrition. *Global Food Security* 2014; 3(3–4):183–92.
6. Field J. Multisectoral nutrition planning. A postmortem. *Food Policy* 1987; 15–28.
7. United Nations, 2015. *Transforming our world: The 2030 agenda for sustainable development.* p. 13.
8. Fanzo J, Pronyk P. An evaluation of progress toward the Millennium Development Goal one hunger target: A country-level, food and nutrition security perspective. *Food Nutr Bull* 2011; 32(2):133–58.
9. Kim J, Ferrari G, Abramsky T. Assessing the incremental effects of combining economic and health interventions: The IMAGE study in South Africa. *Bull World Health Organ* 2009; 87:824–832.
10. The Millennium Villages Project. Synergies between health and economic interventions, suggesting multi-sector approaches may generate a wider range of benefits than single sector approaches. 2009. The Earth Institute, Columbia University.
11. Lamstein S, Pomeroy-Stevens A, Webb P et al. Optimizing the multisectoral nutrition policy cycle a systems perspective. *Food Nutr Bull* 2016; 37(4 Suppl):S107–114.
12. Gillespie S, Haddad L, Mannar V et al. The politics of reducing malnutrition: Building commitment and accelerating progress. *Lancet* 2013; 382:552–69.
13. Black RE, Victora C, Walker SP et al. Maternal and child undernutrition and overweight in low-income and middle-income countries. *Lancet* 2013; 382(989):427–51.
14. Ruel MT, Menon P, Habicht JP et al. Age-based preventive targeting of food assistance and behaviour change and communication for reduction of childhood undernutrition in Haiti: A cluster randomised trial. *Lancet* 2008; 371(9612):588–95.
15. Webb P, Block S. Support for agriculture during economic transformation: Impacts on poverty and undernutrition. *Proc Natl Acad Sci USA* 2012; 109(31):12309–14.
16. Arimond M, Ruel M. Dietary diversity is associated with child nutritional status: Evidence from 11 demographic and health surveys. *J Nutr* 2004; 134(10):2579–85.
17. Moursi M, Arimond M, Dewey KG et al. Dietary diversity is a good predictor of the micronutrient density of the diet of 6- to 23-month old children in Madagascar. *J Nutr* 2008; 138(12):2448–53.
18. Girard AW, Self JL, McAuliffe C et al. The effects of household food production strategies on the health and nutrition outcomes of women and young children: A systematic review. *Paediatr Perinat Epidemiol* 2012; 26(Suppl 1):205–22.
19. Reinhardt K, Fanzo J. Addressing chronic malnutrition through multi-sectoral, sustainable approaches: A review of the causes and consequences. *Front Nutr* 2014; 1:13.
20. Masset E, Haddad L, Cornelius A et al. Effectiveness of agricultural interventions that aim to improve nutritional status of children: Systematic review. *BMJ* 2012; 344:d8222.
21. Carletto G, Ruel M, Winters P et al. Farm-level pathways to improved nutritional status: Introduction to the special issue. *J Dev Stud* 2015; 51(8):945–57.
22. Headey D. Developmental drivers of nutritional change: A cross-country analysis. *World Development* 2013; 42:76–88.
23. Leroy JL, Ruel M, Verhofstadt E. The impact of conditional cash transfer programmes on child nutrition: A review of evidence using a programme theory framework. *J Dev Effect* 2009; 1(2):103–29.

24. Prüss-Üstün A, Bos R, Gore F et al. *Safer water, better health: Costs, benefits, and sustainability of interventions to protect and promote health.* Geneva: World Health Organization; 2008.
25. Cumming O, Cairncross S. Can water, sanitation and hygiene help eliminate stunting? Current evidence and policy implications. *Matern Child Nutr* 2016; 12(Suppl 1):91–105.
26. Dangour AD, Watson L, Cumming O et al. Interventions to improve water quality and supply, sanitation and hygiene practices, and their effects on the nutritional status of children. *Cochrane Database Sys Rev* 2013; 8:CD009382.
27. Fischer Walker CL, Perin J, Aryee MJ et al. Diarrhea incidence in low- and middle-income countries in 1990 and 2010: A systematic review. *BMC Public Health* 2012; 12:220.
28. Grantham-McGregor S, Cheung YB, Cueto S et al. Developmental potential in the first 5 years for children in developing countries. *Lancet* 2007; 369(9555):60–70.
29. Victora CG, Adair L, Fall C et al. Maternal and child undernutrition: Consequences for adult health and human capital. *Lancet* 2008; 371(9609):340–57.
30. Humphrey JH. Child undernutrition, tropical enteropathy, toilets, and handwashing. *Lancet* 2009; 374(9694):1032–5.
31. Guerrant RK, Oriá RB, Moore SR et al. Malnutrition as an enteric infectious disease with long-term effects on child development. *Nutr Rev* 2008; 66(9):487–505.
32. Motarjemi Y. Research priorities on safety of complementary feeding. *Pediatrics* 2000; 106(Suppl 4):1304–5.
33. Denboba AD, Elder LL, Lombardi J et al. *Stepping up early childhood development: Investing in young children for high returns.* Washington, DC: World Bank Group; 2014.
34. Yousafzai AK, Aboud F. Review of implementation processes for integrated nutrition and psychosocial stimulation interventions. *Ann NY Acad Sci* 2004; 1308:33–45.
35. Aboud FE, Yousafzai AK. Global health and development in early childhood. *Ann Rev Psychol* 2015; 66(1):433–57.
36. U.S. Department of Health and Human Services, Administration for Children and Families. *Head Start Impact Study final report.* Washington, DC: U.S. Department of Health and Human Services; January 2010.
37. Ringsmose C, Müller-Kragh G (Eds.). *Nordic social pedagogical approach to early years.* International Perspectives on Early Childhood Education and Development. New York: Springer 2017.
38. Darnton-Hill I, Webb P, Harvey P et al. Micronutrient deficiencies and gender: Social and economic costs. *Am J Clin Nutr* 2005; 81(5):1198S–205S.
39. Regasa N, Stoecker B. Contextual risk factors for maternal malnutrition in a food-insecure zone in southern Ethiopia. *J Biosoc Sci* 2012; 44(5):537–48.
40. Cunningham K, Ruel M, Ferguson E et al. Women's empowerment and child nutritional status in South Asia: A synthesis of the literature. *Matern Child Nutr* 2015; 11(1):1–19.
41. Fraser AM, Brockert JE, Ward, RH. Association of young maternal age with adverse reproductive outcomes. *N Engl J Med* 1995; 332(17):1113–8.
42. Ganchimeg T, Ota E, Morisaki N et al. Pregnancy and childbirth outcomes among adolescent mothers: A World Health Organization multicountry study. *BJOG* 2014; 121(Suppl 1):40–8.
43. Vir SC. Improving women's nutrition imperative for rapid reduction of childhood stunting in South Asia: Coupling of nutrition specific interventions with nutrition sensitive measures essential. *Matern Child Nutr* 2016; 12(Suppl 1):72–90.
44. Raj A, Saggurti N, Winter M et al. The effect of maternal child marriage on morbidity and mortality of children under 5 in India: Cross sectional study of a nationally representative sample. *BMJ* 2010; 340:b4258.

45. Fall CH, Sachdev HS, Osmond C et al. Associations of young and old maternal age at childbirth with childhood and adult outcomes in the offspring: A prospective study in five low and middle-income countries (COHORTS collaboration). *Lancet Glob Health* 2015; 3(7):e366–77.

46. World Health Association. *Unsafe abortion: Global and regional estimates of the incidence of unsafe abortion and associated mortality in 2008.* Geneva: World Health Organization; 2011.

47. Viner RM, Coffey C, Mathers C et al. 50-year mortality trends in children and young people: A study of 50 low-income, middle-income, and high-income countries. *Lancet* 2011; 377:1162–74.

48. Santhya KG. Early marriage and sexual and reproductive health vulnerabilities of young women: A synthesis of recent evidence from developing countries. *Curr Opin Obstet Gynecol* 2011; 23(5):334–9.

49. Hardy JB, Astone NM, Brooks-Gunn J et al. Like mother, like child: Intergenerational patterns of age at first birth and associations with childhood and adolescent characteristics and adult outcomes in the second generation. *Dev Psychol* 1998; 34(6):1220–32.

50. Darroch J, Signh S. Trends in contraceptive need and use in developing countries in 2003, 2008, and 2012: An analysis of national surveys. *Lancet* 2013; 381(9879):1756–62.

51. Lassi ZS, Das JK, Zahid G et al. Impact of education and provision of complementary feeding on growth and morbidity in children less than 2 years of age in developing countries: A systematic review. *BMC Public Health* 2013; 13(Suppl 3):S13.

52. Diebolt C, Perrin F. From stagnation to sustained growth: The role of female empowerment. *Am Econ Rev* 2013; 103(3):545–9.

53. Rustad C, Smith C. Nutrition knowledge and associated behavior changes in a holistic, short-term nutrition education intervention with low-income women. *J Nutr Educ Behav* 2013; 45(6):490–8.

54. Wamani H, Tylleskär T, Åstrøm AN et al. Mothers' education but not fathers' education, household assets or land ownership is the best predictor of child health inequalities in rural Uganda. *Int J Equity Health* 2004; 3(1):9.

55. Casanovas Mdel C, Lutter CK, Mangasaryan N et al. Multi-sectoral interventions for healthy growth. *Matern Child Nutr* 2013; 9(Suppl 2):46–57.

56. Semba RD, de Pee S, Sun K et al. Effect of parental formal education on risk of child stunting in Indonesia and Bangladesh: A cross-sectional study. *Lancet* 2008; 371:322–8.

57. Kedir H, Berhane Y, Worku A. Magnitude and determinants of malnutrition among pregnant women in eastern Ethiopia: Evidence from rural, community based setting. *Matern Child Nutr* 2016; 12(1):51–63.

58. Milton AH, Smith W, Rahman B et al. Prevalence and determinants of malnutrition among reproductive aged women of rural Bangladesh. *Asia Pac J Public Health* 2010; 22(1):110–7.

59. Headey D, Hoddinott J. Understanding the rapid reduction of undernutrition in Nepal, 2001–2011. *PLoS One* 2015; 10(12):e0145738.

60. Johnson FC, Rogers BL. Children's nutritional status in female-headed households in the Dominican Republic. *Soc Sci Med* 1993; 37(11):1293–301.

61. Penny ME, Creed-Kanashiro HM, Robert RC et al. Effectiveness of an educational intervention delivered through the health services to improve nutrition in young children: A cluster-randomised controlled trial. *Lancet* 2005; 365:1863–72.

62. Lamstein S, Stillman T, Koniz-Booher P et al. *Evidence of effective approaches to social and behavior change communication for preventing and reducing stunting and anemia: report from a systematic literature review.* Arlington, VA: USAID/ Strengthening Partnerships, Results, and Innovations in Nutrition Globally (SPRING) Project); 2014.

63. United Nations Children's Fund (UNICEF). *UNICEF's approach to scaling up nutrition: For mothers and their children.* New York: UNICEF; 2015.
64. International Food Policy Research Institute (IFPRI). *Global nutrition report 2014: Actions and accountability to accelerate the world's progress on nutrition.* Washington, DC: International Food Policy Research Institute; 2014, Table 6.3.
65. Smith L, Haddad L. Reducing child undernutrition: Past drivers and priorities for the post-MDG era. *World Dev* 2015; 68:180–204.

26 Global Progress in the Scaling Up Nutrition (SUN) Movement

Tom Arnold

CONTENTS

INTRODUCTION

The Scaling Up Nutrition (SUN) Movement is a movement like no other. As of 2017, the movement is led by 59 "SUN countries" and the Indian States of Maharashtra, Uttar Pradesh, and Jharkhand (Figure 26.1). Multiple stakeholders from civil society, United Nations (UN) agencies, donors, business, and academia are all united by their conviction and determination to end malnutrition in all its forms.

One in three people are affected by malnutrition in nearly every country worldwide. This is a global crisis, but the evidence needed to respond is known and well documented. Together, those in the movement are driven by the commitment and leadership of SUN countries. Progress is being made with strong political leadership, increased investments, improved alignment, and multistakeholder, multisectoral collaboration.

Many SUN countries are reporting significant progress in reducing stunting, including Benin, Cambodia, Ethiopia, Ghana, Guinea-Bissau, Kenya, Kyrgyzstan, Malawi, Tanzania, Zambia, and Zimbabwe. Several other countries in the SUN Movement are also making significant strides with preliminary data showing promising trends. United with multiple sectors and stakeholders in a truly coherent approach, defeating malnutrition is the new "normal." This belief is at the core of the SUN Movement.

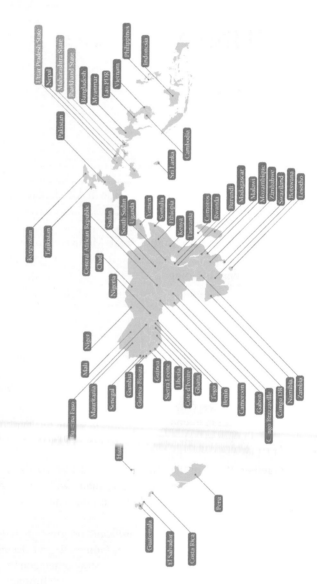

FIGURE 26.1 Countries and states that have committed to Scaling Up Nutrition.

WHY IS NUTRITION IMPORTANT?

Every man, woman, and child has the right to adequate food and nutrition. This right is enshrined in several international human rights and other treaties. Most notably, the International Covenant on Economic, Social and Cultural Rights states clearly that the "right to an adequate standard of living includes food, housing, clothing" and recognizes the "fundamental right of everyone to be free from hunger." Additionally, the Convention on the Rights of the Child obligates states parties "to combat disease and malnutrition, including within the framework of primary health care, through, inter alia, the application of readily available technology and through the provision of adequate nutritious food and clean drinking water, taking into consideration the dangers and risks of environmental pollution" [1,2].

The period of the first 1,000 days—from conception to 2 years of age—is a pivotal moment that determines a child's destiny. Strong evidence shows that eliminating malnutrition in young children has multiple benefits (Figure 26.2). For example, it can

- Boost gross national product by 11% in Africa and Asia
- Prevent child deaths by more than one-third per year
- Improve school attainment by at least one year
- Increase wages by 5% to 50%

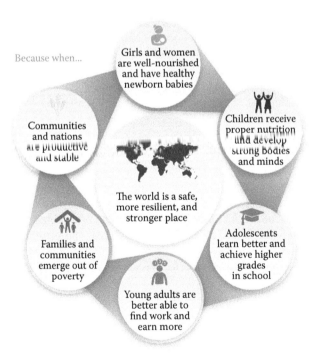

FIGURE 26.2   The virtuous circle of improved nutrition.

- Reduce poverty as well-nourished children are 33% more likely to escape poverty as adults
- Empower women to be 10% more likely to run their own business
- Break the intergenerational cycle of poverty [3]

The annual gross domestic product (GDP) losses from low weight, poor child growth, and micronutrient deficiencies average 11% in Asia and Africa—greater than the loss experienced during the 2008–2010 financial crisis [4]. Investing in nutrition has the potential to help break the poverty cycle and stimulate economic development. Every dollar invested can yield a return of US$16 [5].

Malnutrition (which includes the spectrum from undernutrition to obesity; also see Box 26.1) is linked to an estimated 45% of all under-5 mortalities [6]. It also impairs an individual's productivity, which inhibits national growth with far-reaching consequences for so many indicators of socioeconomic progress. In this sense, malnutrition represents an often invisible impediment to the successful achievement of the Sustainable Development Goals (SDGs) [7]. In other words, good nutrition is a maker and marker of development.

BOX 26.1   THE SCALE OF THE NUTRITION PROBLEM

- 2 billion people experience micronutrient malnutrition
- 1.9 billion adults are overweight or obese
- 156 million children under 5 years are too short for their age (stunted); 50 million do not weigh enough for their height (wasted), and 41 million are overweight; none of these children are growing healthily
- 794 million people are estimated to be calorie deficient
- 1 in 12 adults worldwide have type 2 diabetes
- In 14 countries, fewer than half of all children under 5 years old escape both stunting and wasting [8]

Malnutrition results not just from a lack of sufficient, adequately nutritious and safe food, but from a variety of contributing factors including health, care, education, sanitation and hygiene, access to food and resources, and women's empowerment.

Over the past decade, global and country recognition of the threat that malnutrition poses to the health and future development of children, and therefore societies, has grown exponentially. Once viewed primarily as an issue that can be tackled through the health sector, the importance of a concerted approach is now widely accepted, involving:

- *Multiple stakeholders*, supported by the UN, civil society, business, academia, and donors
- *Multiple sectors*, including health, agriculture, women's empowerment, planning, and education
- *Multiple levels*, from the highest levels of government to local community leaders

In this context, the SUN Movement is a catalyst for change. The experiences of its members are contributing to strengthened evidence on effective actions required for achieving impact, and in shaping an enabling environment for good nutrition, fit to ensure that no one is left behind and that people everywhere benefit from good nutrition.

THE SCALING UP NUTRITION (SUN) MOVEMENT: HOW IT ALL STARTED

In 2008, a global food prices crisis struck, and the World Bank estimated that at least 100 million people were likely to be pushed back into poverty [9]. This was followed by the global economic downturn in 2009, which pushed the poorest and most vulnerable to their knees, while the far-reaching implications of climate change on people's livelihoods took hold. This triple threat of unstable food prices, economic uncertainty, and volatile climate-related events led to protests and demands for action by citizens across the globe. Hunger and malnutrition were finally recognized as political issues, and given a space on the global and national agendas.

At the same time, the evidence for tackling malnutrition was made clear beyond doubt. In 2008, *Lancet* published a series of papers on maternal and child undernutrition [10], which cataloged the impact of undernutrition on development and health, highlighted proven interventions to reduce undernutrition, and called for national and international action to improve nutrition for mothers and children. This helped prompt action by drawing the world's attention to the high human cost of malnutrition, and to the inadequate response by development partners and countries alike.

It was against this backdrop that the SUN Movement was launched in 2010. Since then, the movement has catalyzed collective action to end malnutrition, bringing together governments, national and global civil society organizations, businesses, the UN system, and researchers and scientists across different sectors critical for the improvement of nutrition. It provides a space to convene, mobilize, share, learn advocate, align, and coordinate actions.

HOW THE SUN MOVEMENT WORKS

SUN countries are fostering an enabling environment for scaling up nutrition by strengthening high-level commitments, building supportive policy and legal frameworks, implementing aligned actions behind a common set of results, and not only raising but more effectively spending resources for nutrition. They are doing this by investing in a set of capabilities that will empower actors at a country level to continuously improve country planning to end malnutrition; to mobilize, advocate, and communicate for impact; build core capacities at all levels for multisectoral action; and ensure equity for all, with women and girls at the center (Figure 26.3).

SUN countries are building these capabilities through a peer-to-peer process that keeps governments in the driving seat, and national priorities and realities at the center. The SUN Movement's support system adds value to this process by offering a

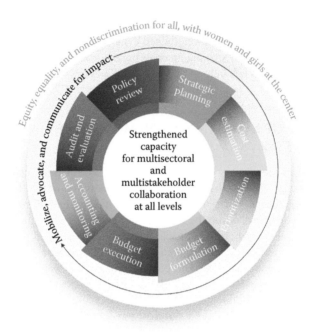

FIGURE 26.3   Strengthening core capabilities across SUN countries.

set of services that aim to unlock resources and intensify progress toward the movement's strategic objectives. This support system is comprised of the SUN Movement Secretariat and four networks—donor, UN, civil society, and business—as well as experts who offer technical support and leadership in nutrition (Figure 26.4). This support system responds and adapts to the breadth and width of SUN countries' needs, expertise, and ambitions. It leverages its members' experiences and their areas of comparative advantage.

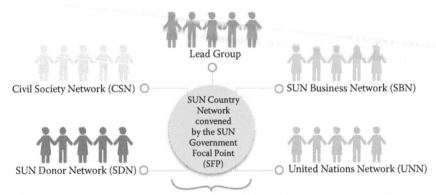

FIGURE 26.4   The SUN Movement Support System.

During the period covered by its first strategy (2012–2015), the SUN Movement grew as a space for its members to convene and advocate for strengthened multisectoral and multistakeholder collaboration to scale up nutrition. Networks were set up (comprising donors, civil society, UN, and business), stakeholders in countries were mobilized, and members aligned around a set of engagement principles. Throughout its short history, the SUN Movement has evolved and, through effective advocacy, its members have been critical catalysts of change, their experiences contributing to evidence of effective interventions, and the shaping of the nutrition landscape.

During its first few years, the movement placed an emphasis on robust, multisectoral plans for nutrition and common results frameworks. Systems and processes for learning, harnessing and sharing experiences between countries and networks were established; a means for brokering technical assistance for countries lacking providers was developed, and a process of annual joint-assessment based on progress markers tied to the movement's four strategic objectives has been established.

The SUN Movement Strategy and Roadmap 2016–2020 builds on the strengths and momentum from the movement's work to date [11]. Forging new paths ahead, the consultative process through which the second strategy was developed has positioned the movement to respond to SUN countries' demands, maximize the potential of SUN networks, and capitalize on the insight generated by the *2014 Independent Comprehensive Evaluation (ICE)* of the SUN Movement [12].

The SUN Movement has been and will remain a continuous exercise in improvement, through learning and adaptation. The 2016–2020 strategy is an expression of that. It takes stock of, and seeks to build on, the strengths and achievements of the movement's first phase, while also charting new opportunities to achieve greater impact and further optimize its contributions to improving the nutritional status of communities.

THE GLOBAL NUTRITION CONTEXT

While the focus of the SUN Movement remains primarily on the reduction of stunting, many governments are adapting their national plans to address the multiple burdens of malnutrition, especially overweight and obesity. They recognize that malnutrition, in all its forms, undermines peace and prosperity, while improved nutrition can lead to more just and sustainable futures for individuals, societies, and nations.

Momentum to improve nutrition is accelerating, and transformative change is achievable. The 2030 Agenda for Sustainable Development has committed all governments to comprehensive, integrated, and universal transformations, which include ending hunger and malnutrition by 2030. They are galvanizing action around 17 SDGs. While the ambition to "end hunger, achieve food security and improve nutrition and promote sustainable agriculture" is captured in SDG 2, at least 12 of the 17 global goals contain indicators that are highly relevant for nutrition [13]. Indeed, without adequate and sustained investments in nutrition, both directly and as part of an integrated set of interventions aimed at ending poverty and hunger, the full potential of the 2030 Agenda will not be realized.

The SDGs build on the six global nutrition targets agreed in the World Health Assembly (WHA) Resolution 65.6, which endorsed a Comprehensive Implementation

Plan on Maternal, Infant and Young Child Nutrition (MIYCN) in 2012 [14]. In 2013, the 66th session of the WHA agreed to halt the rise in diabetes and obesity as part of the adoption of the comprehensive global monitoring framework and targets for the prevention and control of noncommunicable diseases [15].

These WHA targets sit at the heart of the 2014 Rome Declaration and Framework for Action, which was the result of the Second International Conference on Nutrition (ICN2). The Rome Declaration calls for the UN system, including the Committee on World Food Security (CFS), to work more effectively together to support national and regional efforts, and enhance international cooperation and development assistance to accelerate progress in addressing malnutrition [16]. Responding to this, the CFS established an Open Ended Working Group on Nutrition (OEWG) in October 2015 to develop a clear vision for the role of the CFS on nutrition.

In April 2016, the United Nations General Assembly (UNGA) proclaimed the UN Decade of Action on Nutrition (2016–2025) to intensify action to end hunger and eradicate malnutrition worldwide, and ensure universal access to healthier and more sustainable diets—for everyone, everywhere. The Decade of Action on Nutrition coincides with the new Strategic Plan of the United Nations System Standing Committee on Nutrition (UNSCN), the dedicated platform for open, substantive, and constructive dialogue among UN agencies working on nutrition.

These commitments and agreed set of actions by the UN Member States are complemented by the "Nutrition for Growth (N4G) Compact," which mobilized US$4 billion at the first N4G Summit in 2013. A major development resulting from the N4G Compact was the establishment of the Global Nutrition Report, which records advancements in meeting global nutrition targets, documents progress on commitments made, and makes recommendations for actions to accelerate progress [4].

The SUN Movement does not seek to replace the international nutrition architecture, but it does aim to amplify the coherence of its members' actions. Its members are guided by the strengthening governance of nutrition and are acting as catalysts for change. The (now) 59 governments that have committed to scaling up nutrition, and the thousands of partners that are aligning their support behind them, are responsible and accountable for results. The lessons learned by these stakeholders are contributing to growing energy and momentum for improved nutrition, proving that more can be achieved together than is ever possible alone.

THE SUN APPROACH: SUPPORTING A MULTISECTORAL, MULTISTAKEHOLDER RESPONSE TO MALNUTRITION

Ambitious efforts to accelerate effective coverage of proven interventions are critical if SUN countries are to meet their national goals and commitments to achieve the WHA targets and the SDGs. There is now a wealth of evidence on key interventions that can significantly improve nutrition outcomes. This includes a set of 10 proven nutrition-specific interventions, including Infant Young Child Feeding, which, if sufficiently scaled up, could eliminate approximately 900,000 deaths of children under 5 years of age in the 34 high nutrition-burden countries, and reduce the number of children with stunted growth and development by 33 million [17].

But we must go further. Nutrition-sensitive approaches have enormous potential to enhance the scale and effectiveness of nutrition-specific interventions. They address the underlying determinants of fetal and child undernutrition and development, and incorporate specific nutrition goals and actions in agriculture, social safety nets, handwashing and other hygiene interventions, water and sanitation, health and family planning services, early child development, and education. In a recent budget analysis exercise conducted by 30 SUN countries, 25 were able to identify nutrition-sensitive budget allocations across more than five of the key sectors: health; agriculture; education; social protection; and water, sanitation, and hygiene (WASH) [18].

Strong policies, legislation, and frameworks are essential foundations for scaling up both nutrition-specific and nutrition-sensitive approaches. Examples include the International Code of Marketing of Breast Milk Substitutes [19], legal frameworks for maternity leave and breastfeeding protection, food fortification standards, salt iodization, and the *Codex Alimentarius* [20]. Although the SUN Movement does not set technical standards or develop normative guidance, it includes members that do. Working together, actors across the SUN Movement are working to support the development and dissemination of normative guidance, as well as to share best practices in operationalizing policies, monitoring impact, and encouraging adherence.

At a country level, national nutrition plans are the vehicles for translating policy into action and results. Countries in the SUN Movement are making progress in updating their national nutrition plans and engaging stakeholders to mobilize and align around a common set of results. They are directing negotiations among key sectors, a key process for accountability and a pathway for encouraging effective allocation of resources. Many SUN countries report significant progress in agreeing and implementing actions in line with common results frameworks and the policy management cycle. They are demonstrating that learning through implementation is key.

Current financial resources available for nutrition-specific and nutrition-sensitive interventions are woefully inadequate. In order to implement national nutrition plans at scale, SUN countries need not only to raise new money but also to better spend existing funds. A recent investment framework estimates that an additional US$7 billion per year for the next 7 years is required to reach the WHA target of reducing the prevalence of stunting by 40%. The framework proposes that US$4 billion of this financing comes from governments of countries affected by undernutrition, US$2.6 billion comes from donors, and the remaining US$400 million is raised through innovative financing facilities. The projected returns are huge: 3.7 million children's lives saved, at least 65 million fewer stunted children, and 265 million fewer women suffering from anemia [21].

The global community has committed to optimizing all financing streams toward sustainable development and coordinating it for the greatest impact. Governments will need to better align their financing frameworks, and to be mindful of the need for coherence and alignment with climate change, health, and agriculture financing. Efforts to increase the effectiveness of development cooperation need to be enhanced, based on the basic principles of country ownership, a focus on results, cost effectiveness, inclusive partnerships, greater transparency, and improved accountability. While development assistance for improved nutrition outcomes will be important, increased domestic resources are a clear means of sustainably financing nutrition. Financing streams for nutrition will need to be leveraged from domestic public and private origins, international public and private sources, and blended finance. All public funds must positively impact the poorest and most vulnerable in all societies [22].

BOX 26.2 EMPOWERING WOMEN AND GIRLS TO IMPROVE NUTRITION

How girls are treated matters for nutrition. Not surprisingly, higher levels of gender discrimination are associated with higher levels of both acute and chronic undernutrition [23]. At the same time, improvements in women's status have been shown to account for around 12% of global reductions in the proportion of children who are underweight, and improvements in women's enrollment in secondary education account for 43% of global reductions in the proportion of children who are underweight [24].

Nutrition matters for empowering girls. Improving the nutrition status of girls, adolescents, and women increases their ability to perform well at school and to become empowered in the workforce and the wider society. In short, those involved in the SUN Movement know that the virtuous circle of good nutrition and better futures starts with the girl child. Although every country is shaping its own set of solutions, some lessons are emerging from the SUN Movement.

First, women need to be at the decision-making table with adequate support and space for their voices to be heard. From state houses, parliaments, and ministries, to local authorities, communities, and households, women need equal access to opportunities, training, and information in order to contribute to the process of devising solutions to the nutrition-based challenges they face. For this reason, education and women's knowledge of nutrition is critical. This requires investments to send and keep girls in school, but also in building knowledge related to infant and young child feeding, literacy, and vocational skills.

The stories of SUN countries make it abundantly clear that a community-centered approach that ignites the power of sisterhood is essential. Women are very often best placed to decide how resources are used at home to improve nutrition. Supporting them to reach out to other women in their communities to share their experience, knowledge, and aspirations is vital. But it isn't enough to only look to the women. Men must champion and actively engage in women's empowerment. From presidents, to chiefs to husbands, fathers, and brothers, men must actively engage in ensuring that every member of their families can enjoy good nutrition [25].

WHERE THE SUN MOVEMENT IS GOING

The SUN Movement is building on existing successes. SUN countries, supported by growing networks of partners, have set themselves the ambitious goal of transforming the way in which they address undernutrition. SUN countries are beginning to put nutrition at the heart of their development policies. They have seen a dramatic increase both in high-level political commitment for nutrition and in bringing together diverse groups of people across sectors around common goals. They are setting clear targets, scaling up their programs, and putting in place the necessary legal and financial frameworks to support a nutrition revolution.

Together, stakeholders across the SUN Movement are acting to ensure that children everywhere get the best possible start in life and reach their full potential. SUN countries have proven that a large-scale social movement is possible. It is time to support the SUN Movement to transform lives and societies.

REFERENCES

1. United Nations Treaty. International Covenant on Economic, Social and Cultural Rights. Registration 3 January 1976; No. 14531.
2. United Nations Treat. Convention on the Rights of the Child. Registration 2 September 1990; No. 27531.
3. Haddad L. Child growth = Sustainable economic growth: Why we should invest in nutrition. Institute of Development Studies Sussex 2013.
4. International Food Policy Research Institute. *Global nutrition report 2016: From promise to impact: Ending malnutrition by 2030.* Washington, DC: International Food Policy Research Institute; 2016. Available from: http://dx.doi.org/10.2499/9780896295841.
5. International Food Policy Research Institute. *Global Nutrition Report 2015: Actions and accountability to advance nutrition and sustainable development.* Washington, DC: International Food Policy Research Institute; 2015. Available from: http://dx.doi .org/10.2499/9780896298835.
6. Black R, Victoria C, Walker S et al. Maternal and child undernutrition and overweight in low-income and middle-income countries. *Lancet* 2013; 82(9890):427–51.
7. United Nations Standing Committee on Nutrition. Nutrition and the post 2015 sustainable development goals: A policy brief, 2014. Available from: http://www.unscn.org/files /Publications/Nutrition__The_New_Post_2015_Sustainable_development_Goals.pdf.
8. UNICEF, World Health Organization, World Bank Group. Joint child malnutrition estimates. 2016. Available from: http://www.who.int/nutgrowthdb/estimates2015/en/.
9. World Bank. Global food crisis response program. Available from: http://www.worldbank .org/en/results/2013/04/11/global-food-crisis-response-program-results-profile (accessed June 16, 2016).
10. Black RE, Allen LH, Bhutta ZA et al. Maternal and child undernutrition: Global and regional exposures and health consequences. *Lancet* 2008; 371(9608):243–60.
11. SUN Movement Strategy and Roadmap 2016–2020. Available from: http://docs.scalingup nutrition.org/wp-content/uploads/2016/09/SR_20160901_ENG_web_pages.pdf.
12. Mokoro 2015. *Independent comprehensive evaluation of the Scaling Up Nutrition Movement: Final report—Main report and annexes.* Oxford: Mokoro Ltd; 2015. Available from: http://scalingupnutrition.org/wp-content/uploads/2015/05/SUN_ICE _FullReport-All(1-5-15).pdf.

13. United Nations Office of Economic and Social Affairs. Sustainable development knowledge platform. Available from: https://sustainabledevelopment.un.org/?menu=1300.
14. Sixty-Fifth World Health Assembly. Resolution 65.6: Maternal, infant and young child nutrition. Available from: http://apps.who.int/gb/ebwha/pdf_files/WHA65/A65_R6-en.pdf (accessed February 7, 2017).
15. Sixty-Sixth World Health Assembly. Resolutions and decisions. Available from: http://apps.who.int/gb/ebwha/pdf_files/WHA66-REC1/A66_REC1-en.pdf#page=25.
16. Food and Agriculture Organization. Second International Conference on Nutrition. Available from: http://www.fao.org/about/meetings/icn2/en/.
17. Black R, Alderman H, Bhutta Z et al. Maternal and child nutrition: Building momentum for impact. *Lancet* 2013; 382(9890):372–5.
18. Scaling Up Nutrition. *Investigating nutrition in national budgets.* Available from: http://docs.scalingupnutrition.org/wp-content/uploads/2015/02/SUN-Budget-Analysis-Short-Synthesis-Report-SUNGG-version-EN.pdf.
19. World Health Organization. *International code of marketing of breast-milk substitutes.* Geneva: World Health Organization; 1981. Available from: http://www.who.int/nutrition/publications/code_english.pdf.
20. Food and Agriculture Organization. Codex Alimentarius Food Standards: Food grade salt (CODEX STAN 150-1985). Amended 2006.
21. Shekar M, Kakietek J, D'Alimonte M et al. *Investing in nutrition: The foundation of development—An investment framework to reach the global nutrition targets.* Available from: http://thousanddays.org/tdays-content/uploads/Investing-in-Nutrition-The-Foundation-for-Development.pdf.
22. United Nations General Assembly. Addis Ababa Action Agenda of the Third International Conference on Financing for Development. Resolution 69/313, adopted by the General Assembly. July 2015. http://www.un.org/ga/search/view_doc.asp?symbol=A/RES/69/313/
23. Mucha N. Enabling and equipping women to improve nutrition. Bread for the World Briefing Paper 2012; Number 16. Available from: http://www.bread.org/sites/default/files/downloads/briefing-paper-16.pdf.
24. Smith L, Haddad H. *Explaining child malnutrition in developing countries: A cross-country analysis.* Discussion Paper No. 60. Washington, DC: International Food Policy Research Institute; 1999. Available from: http://ageconsearch.umn.edu/bitstream/94515/2/explaining%20child%20malnutrition%20in%20developing%20countries.pdf.
25. Scaling Up Nutrition. Empowering girls and women to improve nutrition: Building a sisterhood of success, SUN Movement In Practice Brief 2016 (May). Available from: http://scalingupnutrition.org/wp-content/uploads/2016/05/IN-PRACTICE-BRIEF-6-EMPOWERING-WOMEN-AND-GIRLS-TO-IMPROVE-NUTITION-BUILDING-A-SISTERHOOD-OF-SUCCESS.pdf.

Section XI

Before and Beyond
the 1,000 Days

27 A Role for Preconception Nutrition

Taylor Marie Snyder, Homero Martinez,
Sara Wuehler, and Luz Maria De-Regil

CONTENTS

INTRODUCTION

Although the 1,000-day period between conception and a child's second birthday is paramount in building a strong foundation for the child's life, the significance of the preconception period must not be ignored. A woman's weight, dietary habits, and nutrient stores practiced prior to and during early conception can influence pregnancy outcomes by altering both maternal and/or fetal metabolism during preconception, conception, implantation, placentation, and embryo- or organogenesis [1].

Women's preconception nutrition influences pregnancy outcomes by affecting the supply of nutrients early in pregnancy, which in turn influences fetal development. However, some less intuitive risk factors that result from the inadequate, and possibly transgenerational, nutritional status of the women in childhood and during adolescence (e.g., maternal short stature) also increase maternal and infant perinatal mortality and morbidity. Therefore, it is critical to ensure adequate nutritional status for women starting in childhood, before the reproductive years [2]. Nutrition interventions need to be integrated with other health and social interventions to maximize women's human potential. The following sections elaborate on all these factors.

THE ROLE OF NUTRITION IN GROWTH AND DEVELOPMENT

PRECONCEPTION WEIGHT AND PREGNANCY OUTCOMES

Women of reproductive age are vulnerable to malnutrition in any of its forms: undernutrition, vitamin and mineral deficiencies, or overweight and obesity. Preconception underweight or overweight, short stature, and micronutrient deficiencies can all contribute independently or simultaneously to increase maternal and fetal complications during pregnancy [3].

Prepregnancy overweight (body mass index [BMI] ≥ 25 kg/m^2) approximately doubles the risk for the two highest causes of maternal mortality: gestational diabetes mellitus and hypertensive disorders of pregnancy, including preeclampsia [4]. Additionally, it is linked to a spectrum of adverse pregnancy outcomes, including obstetric anesthesia-related complications, maternal morbidity due to infection, poor lactation practices, prolonged gestation, preterm births, and decreased success with vaginal childbirth [3,5,6]. Maternal overweight is positively associated with large for gestational age (LGA) infants, macrosomia, childhood overweight, and childhood obesity [4,7,8]. It is also inversely associated with infants born low birth weight (LBW) and small for gestational age (SGA). Maternal obesity also confers an increased risk of fetal and neonatal death as well as childhood disease [4,7,8]. The risks associated with a high BMI can be reduced significantly during the postpartum and interpregnancy periods by practicing exclusive breastfeeding during an infant's first 6 months and increasing physical activity [2].

At the other end of the spectrum, prepregnancy underweight, generally considered BMI <18.5 kg/m^2, increases the risk of several negative outcomes: intrauterine growth restriction, stillbirth, preterm birth, SGA, and LBW [3,6]. Several observational studies from countries including China and Vietnam have evaluated the relationships between prepregnancy BMI and infant birth size, in addition to other health outcomes. These studies found that infants born to underweight women were at increased risk of being SGA [1]. It is also important to note that maternal underweight, which generally reflects preconceptional underweight, is one form of maternal malnutrition, which in turn is independently associated with LBW [9].

MICRONUTRIENT DEFICIENCIES

Micronutrient deficiencies (MNDs) can result from an inadequate dietary intake, low bioavailability of dietary nutrients, and increased nutrient requirements. The bioavailability of nutrients can be positively or negatively affected by dietary factors, such as vitamin C or phytate, or lifestyle habits such as smoking and consuming alcohol.

In this line, physiological or pathological conditions like repid growth, menstrual bleeding, parasitic or other infactions (malaria, HIV) increase the daily requirements [2] (Table 27.1). Globally, an estimated 2 billion people (over 30% of the population)

TABLE 27.1

Common Micronutrient Deficiencies and Their Role, Potential Causes, and Risk Factors in Women of Reproductive Age [10–11]

Nutrient	Role and Function	Causes and Risk Factors of Deficiency
Iron	Serves as a carrier of oxygen to the tissues from the lungs by red blood cell hemoglobin, as a transport medium for electrons within cells, and as an integrated part of important enzyme systems in various tissues. Deficiency can: • Cause microcytic hypochromic anemia • Inhibit intellectual capacity • Affect motor development, physical performance and productivity • Increase the risk of LBW and maternal and infant mortality	Nutritional iron deficiency develops when the diet cannot supply enough iron to meet the body's requirements. Risk factors: • Menstrual blood loss • Pregnancy demands • Adolescent growth spurt • Vegetarian diet • Malnutrition • Malaria • Hookworm or other parasites that lead to blood loss (more frequent in tropical countries)
Folate	Coenzyme in the metabolism of nucleic and amino acids. Deficiency can: • Interrupt physiological neural tube formation, neurological function, and brain development • Cause NTDs and megaloblastic anemia • Increase risk of colorectal cancer and cardiovascular disease and stroke	Inadequate dietary intake of folate and folic acid, which is highly common globally. Risk factors: • Diets low in folate and/or folic acid • High prevalence of the MTHFR C677T polymorphism • Malabsorption conditions (e.g., coeliac disease and/or tropical sprue)
Iodine	Responsible for the synthesis of thyroid hormones by the thyroid gland. Thyroid hormones impact bodily metabolic processes, growth, and development. Deficiency occurring during fetal and neonatal growth and development can lead to: • Irreversible damage of the brain and central nervous system. This damage may lead to irreversible mental retardation, most seriously to cretinism. It can also lead to retarded physical growth. • Goiter, juvenile hypothyroidism, and iodine-induced hyperthyroidism	The iodine content of food depends on the iodine content of the environment in which it was grown or raised. Risk factors: • Diets low in iodine • Lack of consumption of adequately iodized salt

(Continued)

TABLE 27.1 (CONTINUED)

Common Micronutrient Deficiencies and Their Role, Potential Causes, and Risk Factors in Women of Reproductive Age [10–11]

Nutrient	Role and Function	Causes and Risk Factors of Deficiency
Calcium	Plays an essential role in blood clotting, muscle contraction, nerve transmission, and the formation of bones and teeth. Deficiency can lead to: • Development of corticosteroid osteoporosis	Inadequate consumption of calcium, often associated with low consumption of dairy products. Risk factors: • Diets low in milk or milk-based products • Vegan diet • Adolescent growth spurt (when an increase in the rate of skeletal calcium accretion occurs)
Vitamin B_{12}	Acts as a coenzyme in nucleic acid metabolism. Deficiency can lead to: • Megaloblastic anemia	Inadequate consumption and absorption of vitamin B_{12}. Risk factors: • Diets low in B_{12} • Vegan diet (which offers no intake, as plants do not synthesize vitamin B_{12}) • Vegetarian diet (which offers a low intake, unless eggs, milk, and other dairy foods are consumed) • Malabsorption disorders (rarely due to autoimmune disease) • Hypochlorhydria associated with atrophic gastritis
Vitamin D	Required to maintain normal blood levels of calcium and phosphate needed for the normal mineralization of bone, muscle contraction, nerve conduction, and general cellular function in all cells of the body. Deficiency can lead to: • Osteoporosis and blood loss • Certain types of cancer • Heart disease • Dementia	Inadequacy is common in those with darkly pigmented skin or with limited sun exposure. Risk factors: • Diets low in vitamin D • Being obese • Lack of sun exposure • Adolescent growth spurt • Dress codes that do not permit the skin to receive direct sunlight

Source: Data from World Health Organization (WHO), *Vitamin and mineral requirements in human nutrition*, 2nd ed., Bangkok, Thailand: WHO, 2004; and Save the Children, *Adolescent nutrition: Policy and programming in SUN+ countries*, London: Save the Children, 2015.

Note: NTDs, neural tube defects.

are experiencing at least one micronutrient deficiency [2], with vitamin A, iodine, zinc, iron, and folate being the most widespread [12].

A mother's micronutrient status is a strong determinant of her baby's nutritional status, as micronutrients are transferred from mother to fetus across the placental barrier [2,4] (Table 27.2). The extent of transfer depends on the mechanism used by the micronutrient to cross the placental barrier, and is, therefore, micronutrient-specific [2]. In the preconception period, micronutrients play an important role in fertility and reproductive function [4]. Through several biological pathways, micronutrients are involved in the early stages of gestation during which organogenesis occurs [4]. Iron, iodine, zinc, and long-chain n-3 polyunsaturated fatty acids are essential to the development of the brain and nervous system, while insufficiencies of vitamins A, B_6, B_{12}, and folate during embryogenesis may lead to fetal malformations and pregnancy loss [1]. Although the role of vitamin D and calcium have been studied during pregnancy, the function of their status during the preconceptional period on later pregnancy outcomes remains to be studied.

Given its public health significance and effects on health, anemia (particularly due to iron deficiency) and folate insufficiency have attracted the attention of academia and policy makers. Anemia is defined as a low level of hemoglobin in the blood, which impairs oxygen delivery to the tissues by red blood cells [13]. Blood hemoglobin concentration is the most common indicator of anemia at the population level [12]. Worldwide, approximately 528 million women of reproductive age had anemia in 2011, of which 32.4 million were pregnant [14]. This means that the global prevalence of anemia was approximately 29% for all women of reproductive age and 38% for pregnant women [14].

Causes of anemia vary across countries and populations. Although there are wide ranges in the proportion of anemia that is related to iron deficiency, on average about half of all anemia is attributed to iron deficiency [2,14]. Additional causes of anemia include deficiencies in other micronutrients (e.g., folate, and vitamins A and B_{12}), acute and chronic infection and/or inflammation, and inherited or acquired blood disorders [14]. Iron deficiency anemia adversely affects cognitive and motor development, and can affect physical performance and productivity [14]. Anemia during pregnancy is associated with LBW and an increased risk of both maternal and infant mortality [14].

There are currently no accurate global estimates of the prevalence of folate insufficiency among women of reproductive age [15]. However, there is strong evidence from a number of randomized controlled trials that taking folic acid around the time of conception and to about the first 28 days reduces the risk of neural tube defects (NTDs) by 69% [16]. Globally, it is estimated that approximately 303,000 newborns die within 4 weeks of birth every year due to congenital anomalies, including NTDs [17]. Between days 21 and 27 postconception, the neural plate closes to form what will be the spinal cord and cranium [10]. NTDs, such as spina bifida or anencephaly, result from the improper closure of the spinal cord and cranium, leading to death or varying degrees of paralysis.

Just as over and underweight during preconception can have consequences for pregnancy outcomes, micronutrient excesses can be just as detrimental as their deficiencies. Safe upper limits have been established for all life stages for many

TABLE 27.2

Recommended Amounts of Key Micronutrients Influencing Fetal Growth and Pregnancy Outcomes for Women of Reproductive Age [18–20]

Nutrient	WHO Recommendation						IOM Recommendation		
Iron	Age	Mean Body Weight (kg)	RNI for Bioavailability of Dietary Iron (mg/day)				Age	RDAª and AIᵇ (mg/day)	ULᶜ (mg/day)
			15%	12%	10%	5%			
	11–14 years	45	21.8	27.7	32.7	65.4	9–13 years	8	40
	15–17 years	55	20.7	25.8	31	62	14–18 years	15	45
	18+ years	62	19.6	24.5	29.4	58.8	19–50 years	18	45
							19–50 years	18	45
							51–70 years	8	45
Folate (Folic Acid)	Age	EAR (µg/day)	RNI (µg/day)				Age	RDA/AI (µg/day)	UL (µg/day)
	10–18 years	330	400				9–13 years	300	600
	19–65 years	320	400				14–18 years	400	800
							14–18 years	400	800
							19–70 years	400	1,000
Iodine	Age	RNI (µg/day)	Dosage (µg/kg/day)	UL (µg/kg/day)			Age	RDA (µg/day)	UL (µg/day)
	13+ years	150	2	30			9–13 years	120	600
							14–18 years	150	900
							19–70 years	150	1,100

(Continued)

TABLE 27.2 (CONTINUED)
Recommended Amounts of Key Micronutrients Influencing Fetal Growth and Pregnancy Outcomes for Women of Reproductive Age [18–20]

Calcium

	WHO Recommendation	IOM Recommendation		
	Age	North American and Western European RNI (mg/day)		
	19–65 years	1,000		
			Age	RDA/AI (mg/day)	UL (mg/day)

Age	RDA/AI (mg/day)	UL (mg/day)
9–18 years	1,300	3,000
19–50 years	1,000	2,500
51–70 years	1,200	2,500

Vitamin B$_{12}$

WHO Recommendation:

Age	EAR (µg/day)	RNI (µg/day)
10–18 years	2	2.4
19–65 years	2	2.4

IOM Recommendation:

Age	RDA/AI (µg/day)	UL (µg/day)
9–13 years	1.8	Not Determined
14–70 years	2.4	Not Determined

Vitamin D

WHO Recommendation:

Age	RNI (µg/day)
10–50 years	5
51–65 years	10

IOM Recommendation:

Age	RDA (µg/day)	UL (µg/day)
9–70 years	15	100

Source: Data from World Health Organization (WHO): *Vitamin and mineral requirements in human nutrition,* 2nd ed., Bangkok, Thailand: WHO, 2004; Institute of Medicine, *Dietary Reference Intakes: The essential guide to nutrient requirements,* Otten JJ HJ, Meyers LD, editors, Washington, DC: The National Academies Press, 2006; and Institute of Medicine Standing Committee on the Scientific Evaluation of Dietary References, *Dietary reference intakes for thiamin, riboflavin, niacin, vitamin B$_6$, folate, vitamin B$_{12}$, pantothenic acid, biotin, and choline,* Washington, DC: National Academy of Sciences, 1998.

Note: AI, adequate intake; EAR, estimated average requirement; IOM, Institute of Medicine; RDA, recommended dietary allowance; RNI, recommended nutrient intake; UL, tolerable upper intake level.

[a] The Recommended Dietary Allowance (RDA) is the average daily level of intake sufficient to meet the nutrient requirements of nearly all (97%–98%) healthy people (IOM).

[b] Adequate Intake (AI) is established when evidence is insufficient to develop an RDA and is set at a level assumed to ensure nutritional adequacy (IOM).

[c] Tolerable Upper Intake Level (UL) is the maximum daily intake unlikely to cause adverse health effects (IOM).

micronutrients [16]. In the periconceptional period, for example, an excess of vitamin A can affect embryonic development and result in teratogenesis, and excess iodine can be harmful to the thyroid [10].

THE ROLE OF EPIGENETICS

Nutritional factors can affect both male and female cells during embryonic and fetal development [21,22]. Although a woman's oocytes (eggs) are formed exclusively during her own fetal development, their quality can be affected by events or exposures experienced any time between her own conception and that of her offspring, including nutritional deficiencies [2]. Epigenetic processes allow one genotype to display multiple phenotypes depending on environmental cues that can induce epigenetic and other responses that modify the growth and metabolic trajectory in the oocyte and embryo [2]. These cues can include maternal hyperglycemia or dietary imbalances, such as under- or overconsumption of micronutrients, or an imbalance of nutrients involved in the methylation cycle [2].

PRECONCEPTION CARE AND NUTRITION

PRECONCEPTION CARE

Clinicians and researchers emphasize that many effective prenatal care interventions would achieve better outcomes if started during the preconceptional period [2]. The preconceptional period has been divided into proximal and distal periods. The proximal period is the time immediately preceding pregnancy (up to 2 years prior to conception). The distal preconception period focuses on the mother's early years, her adolescence, and the 2-plus years prior to a pregnancy (not necessarily the first pregnancy) [4]. Thus, effective preconception interventions are tailored to either the proximal or distal periods.

Preconception care is defined as any preventive, promotive, or curative health care intervention provided to women of reproductive age in the preconceptional period [4,5]. Care is aimed at improving health-related outcomes in all women, and her newborns and children holistically [4,5]. This encompasses initiatives that go beyond nutrition such as women's education and empowerment. Preconception care includes targeted health interventions, such as receiving vaccinations and consuming vitamin and mineral supplements.

A SPECIAL EMPHASIS ON THE TEENAGE GIRL

As mentioned earlier, maternal risk factors are rooted in childhood and adolescence. Adolescence represents the second major growth phase in an individual's life. While no longer experiencing the changes associated with early childhood, adolescents undergo a period of rapid growth and maturation as the ability to reproduce is developed [23]. Cultural practices in which girls are fed last, after male family members, might, among other poor eating habits, increase the risk of developing malnutrition in this life stage [23]. Insufficient nutrition during childhood and adolescence can also result in vitamin and mineral deficiencies which are associated with impaired physical and cognitive performance, and reduced immunological response [2,24]. The International Federation of Gynecology and Obstetrics recommends the promotion of a varied and healthy diet as the first step toward meeting the nutrient needs of adolescents [2]. If a healthy diet cannot be achieved, adolescent girls can consume additional nutrients through supplementation or fortification of food staples and condiments, as described later.

An important aspect of adolescent health, one that complements nutrition interventions, is the prevention of teenage pregnancy and early attention to pregnancy care, if needed [4]. Prevention of adolescent pregnancy is crucial to avoiding competition for nutrients between the growing adolescent and fetus, competition that may impair adolescent growth in benefit of the fetus [25,26]. As mentioned earlier, maternal short stature increases the risk of obstetrical intervention, maternal mortality, and childhood mortality [2].

External support of adolescents to ensure marriage and pregnancy do not occur too early may be necessary, because adolescent girls often lack the necessary education, support, and access to health care to make decisions about their nutrition and reproductive health care [5]. High rates of child marriage are linked to poor family planning, higher fertility, unintended pregnancies, and an elevated risk for complications during childbirth [27]. Pregnant adolescents are at increased risk of complications at delivery, and of having preterm babies, or giving birth to babies of LBW or SGA [2,11,24]. Furthermore, very young mothers (under the age of 15) are often not physically mature enough to deliver a healthy, full-term baby and they have an increased risk of death or disability from maternal anemia, obstructed labor, fistula, preterm birth, and LBW [4,28].

Despite increased risks, married and unmarried adolescents are less likely than adults to use any form of contraception during sexual activity [5]. To prevent unintended pregnancy, preconception care must include access to contraceptives, and education about their appropriate use, along with information about reproductive rights, including the right to abstinence [29,30].

MALE PRECONCEPTION HEALTH

The preconception health of the male partner is an incredibly important, yet often overlooked, element of preconception health. In addition to a woman's eggs, male sperm can also be impacted by environmental factors experienced from the father's conception through to the production of mature sperm [2,21,22]. Paternal preconception epigenetic exposures can influence the developmental path of the embryo, as spermatozoon (motile sperm cell) affect fetal growth in utero [18]. Paternal age, smoking habits, and exposure to environmental toxins are associated with the development of malformations, mental health disorders, and cancer in offspring [21,22].

Similarly, paternal micronutrient deficiencies, obesity, and poorly controlled diabetes may negatively impact sperm and egg quality, fertility, and the general health of offspring [2,21,22]. Both human and rodent studies have confirmed that paternal obesity impairs basic spermatozoon function, sex hormones, and molecular composition

[21,22]. These results disrupt embryonic development and health, and thus increase the disease burden of their offspring, with the greatest impact on female offspring [21,22]. Furthermore, it has been hypothesized that paternal obesity at conception may increase the likelihood of offspring experiencing obesity and the corresponding negative consequences for multiple generations [21,22].

KEY INTERVENTIONS FOR IMPROVING MICRONUTRIENT STATUS

NUTRITION-SPECIFIC INTERVENTIONS

A variety of interventions are used to address micronutrient deficiencies in women of reproductive age, which can broadly be categorized as dietary modification, supplementation, biofortification, and fortification of staple foods and condiments at a central level or at the point of use. Factors such as the economy and health infrastructure of a society can impact the feasibility and effectiveness of interventions, and therefore must be considered when tailoring solutions for the target population. Table 27.3 lists preconception health interventions that have the ability to improve the nutritional status of women of reproductive age, leading to indirect and direct pathways to optimize health and nutrition.

Many interventions geared toward improving nutritional status are closely tied to adapting human behaviors to embrace recommended preventative health measures. Thus, it is important to acknowledge the likelihood that following recommended preventive health measures is dependent on an array of factors influencing human behavior. Elements impacting the efficacy of solutions include an individual's perception of how susceptible they are to a given condition, and their perception of the seriousness of a given condition [31]. These perceptions are factored into their weighing of perceived benefits of preventive action against the perceived barriers, which ultimately determines their likelihood of taking the recommended preventative health actions [31].

DIETARY MODIFICATION

Dietary changes often focus on an increase in dietary variety and encouraging behaviors that increase the intake, absorption, and utilization of micronutrients to satisfy requirements. For example, to prevent or alleviate iron deficiency, interventions may include increasing intake of foods rich in bioavailable iron, increasing intake of foods in combination that enhance iron absorption, or decreasing foods that inhibit iron absorption [2,10].

Launching public health campaigns aimed to inform and influence behavioral change in all individuals is a strategy used to encourage dietary modifications, as well as to optimize other health and nutrition practices, thus also addressing preconception. One benefit of public health campaigns is the broad reach of messages beyond individuals who are specifically seeking care. Large-scale strategies should address beneficial and detrimental local cultural beliefs, reach a broad age range of women, target those who may be considering pregnancy, develop preemptory awareness and knowledge, and take into account various other demographics and personal characteristics [27,31].

TABLE 27.3
Preconception Health Interventions That Can Improve Nutritional Status

Interventions	Components of Interventions
General preconception care and routine health promotion	• Physical activity • Adequate nutritional status, ranging from healthy weight to dietary diversity • Appropriate weight control • Specific nutrient intake • Folic acid and other micronutrient consumption
Provide medically accurate and comprehensive information for reproductive health and family planning	• Pregnancy prevention • Birth spacing (interpregnancy intervals) • Considerations during advanced maternal age
Prevention of negative outside influences	• Female genital mutilation • Intimate partner violence
Prevention and treatment of sexually transmitted infections (STIs)	• Prevention components include medically accurate, age appropriate, sexual health education and access to effective methods of contraception • Treatment components include access to medications, like antiretroviral for HIV/AIDS
Prevention and management of other infections (immunizations, vaccinations)	• Immunizations against vaccine-preventable diseases (e.g., tetanus, rubella, polio) • Seasonal influenza vaccines • Tuberculosis • Deworming
Screen for, diagnose, and manage chronic diseases and cancers	• Diabetes • Asthma • Hypertension • Breast cancer • Cervical cancer
Screen for, diagnose, and manage mental health disorders	• A range of mental health disorders, including depression and anxiety
Lifestyle modifications (cessation of harmful behaviors)	• Secondhand smoke exposure • Substance abuse, including alcoholism and smoking cigarettes • Eating disorders
Environmental hazard prevention and exposure reduction	• Reducing exposure to lead, pesticides, and chemicals, including toxic medicines • Provision of sanitation • Access and use of clean water • Elimination of disease vectors • Prevention of overcrowding and indoor air pollution
Targeting needs of special populations	• Support for immigrants • Emergency services for refugees

(Continued)

TABLE 27.3 (CONTINUED)

Preconception Health Interventions That Can Improve Nutritional Status

Interventions	Components of Interventions
Preventing the spread of global public health threats	• Zika • Ebola

Source: Data from Dean S, Imam A, Lassi Z et al., A systematic review of preconception risks and interventions, The Aga Khan University, 2011; World Health Organization (WHO), *Global strategy for women's, children's and adolescent's health 2016–2030*, Italy, 2015; and Moos MK, Dunlop AL, Jack BW et al., *Am J Obst Gynecol* 2008; 199(Suppl. 2):S280–9.

Note: AIDS, acquired immunodeficiency syndrome; HIV, human immunodeficiency virus; STIs, sexually transmitted infections.

In spite of being the ultimate goal, achieving an optimal diet is not feasible in many settings where access, availability, and affordability of foods are limited. Thus, other solutions are needed.

SUPPLEMENTATION

Oral supplementation refers to the direct provision of micronutrients in the form of a pill, tablet, liquid, or other formulation. Supplements are given daily or intermittently (e.g., weekly or every 6 months) depending on the formulation and intervention [32]. Supplementation is probably the most widespread public health intervention to address micronutrient deficiencies. It has proven to be effective, but challenges to implementation still remain.

Iron and folic acid are the nutrients most commonly supplemented among women of reproductive age prior to pregnancy or during the postpartum period. However, weekly supplementation is preferred over a daily dose as a preventive public health measure to prevent anemia due to iron deficiency [32]. In addition to iron, iodine supplementation may be necessary in settings where salt iodization coverage is lower than 90% [33]. Women of reproductive age in low- to middle-income countries often have multiple coexisting micronutrient deficiencies [24]. However, there is no recommended supplementation scheme or composition of a supplement to address several deficiencies concomitantly in nonpregnant women.

FOOD FORTIFICATION

Food fortification is the addition of one or more essential nutrients to a food for the purpose of preventing or correcting a demonstrated deficiency in the general population or in specific population groups [2]. Fortification is considered to be one of the most cost-effective nutrition interventions. In addition, it requires little to no active participation or behavior change by the end users [34].

Staple foods such wheat and maize flours, and rice are fortified in many countries [12,31]. The addition of iron and folic acid to wheat and maize flours has proven to be an effective and safe strategy to both increase iron stores and prevent NTDs [31]. For example, fortification of flour and ready-to-eat cereals in countries such as the United States, Canada, Chile, and Costa Rica has been linked to significant reductions in the incidence of NTDs in these countries [1]. Some sectors have expressed concern about the amount of folic acid that men are receiving through fortification. However, to date, there is no evidence that such an approach is not safe [35].

The most successful case of fortification is the addition of iodine to salt as a means of improving iodine status [33,36]. Over 120 countries have implemented this intervention, resulting in 71% of households worldwide having access to adequately iodized salt [37].

Point-of-use fortification, also known as "home fortification," involves the addition of micronutrients in powder form to energy-containing foods [38,39]. These micronutrient powders can be added to foods during or after cooking. Point-of-use fortification is most commonly used for children under 2 years of age [38,39] and to our knowledge it has not been researched in the preconceptional period so as to inform a clinical or public health intervention.

Biofortification has emerged as another important intervention to correct selected micronutrient deficiencies. Biofortification can improve the nutrient density of staple food crops through conventional plant breeding, agronomic management, or genetic engineering [40]. Currently, the most common targeted micronutrients are iron, zinc, and vitamin A, due to the high prevalence of deficiencies of these micronutrients among children under the age of 5 and women of reproductive age in developing countries [41]. As staple foods predominate in the diets of the poor, and because it implicitly targets low-income households, biofortification can reach relatively remote areas and vulnerable populations [42]. Once established, a biofortified crop system can be sustained because nutritionally improved varieties continue to grow and be consumed without substantial additional inputs, even if donor attention and funding fades [42]. Furthermore, farmers often favor these mineral-packed seeds, because the trace minerals help their crops resist disease and other environmental stressors [42].

CONCLUSION AND RECOMMENDATIONS

A woman's nutritional status influences important pregnancy outcomes that have long-term, intergenerational effects on women and their offspring. Critically, research indicates that maternal risk factors are rooted in childhood and adolescence. Therefore, preconception care interventions must begin long before adolescence and continue throughout women's reproductive years, including during the intervals between pregnancies. The ultimate aim is to break the intergenerational cycle of undernutrition, where undernourished and stunted mothers give birth to premature or SGA infants, which in turn repeat the cycle.

Additional research is needed to strengthen and improve the evidence on the most effective interventions and delivery platforms to reach women. As preconception care is preventive rather than curative, interventions should address population-wide improvements on micronutrient and nutritional status that are evidence-based and scalable. Also, efforts should be made to systematically explore the effects of nutrition in the preconceptional period on the incidence of noncommunicable conditions in the adult life of the offspring.

REFERENCES

1. Ramakrishnan U, Grant F, Goldenberg T et al. Effect of women's nutrition before and during early pregnancy on maternal and infant outcomes: A systematic review. *Paediatr Perinat Epidemiol* 2012; 26(Suppl. 1):285–301.

2. Hanson M, Bardsley A, De-Regil L et al. The International Federation of Gynecology and Obstetrics (FIGO) recommendations on adolescent, preconception, and maternal nutrition: "Think Nutrition First." *Int J Gynaecol Obstet* 2015; 131(Suppl. 4):S213–53.

3. Dean SV, Lassi ZS, Imam AM et al. Preconception care: Nutritional risks and interventions. *Reprod Health* 2014; 11(Suppl. 3):S3.

4. Dean S, Imam A, Lassi Z et al. A systematic review of preconception risks and interventions. The Aga Khan University; 2011.

5. Dean S, Mason EM, Howson C et al. Born too soon: Care before and between pregnancy to prevent preterm births: From evidence to action. *Reprod Health* 2013; 10(Suppl. 1):1–16.

6. Johnson K, Posner SF, Biermann J et al. Recommendations to improve preconception health and health care—United States. A report of the CDC/ATSDR Preconception Care Work Group and the Select Panel on Preconception Care. Centers for Disease Control; 2006.

7. Li N, Liu E, Guo J et al. Maternal prepregnancy body mass index and gestational weight gain on pregnancy outcomes. *PLoS One* 2013; 8(12):e82310.

8. Li N, Liu E, Guo J et al. Maternal prepregnancy body mass index and gestational weight gain on offspring overweight in early infancy. *PLoS One* 2013; 8(10):e77809.

9. Pan Y, Zhang S, Wang Q et al. Investigating the association between prepregnancy body mass index and adverse pregnancy outcomes: A large cohort study of 536098 Chinese pregnant women in rural China. *BMJ Open* 2016; 6(7).

10. World Health Organization (WHO). *Vitamin and mineral requirements in human nutrition.* 2nd ed. Bangkok, Thailand: WHO; 2004.

11. Save the Children. *Adolescent nutrition: Policy and programming in SUN+ countries.* London: Save the Children; 2015.

12. Dailey RL, West Jr KP, Black RE. The epidemiology of global micronutrient deficiencies. *Ann Nutr Metab* 2015; 66(Suppl. 2):22–33.

13. Kozuki N, Lee AC, Katz J. Moderate to severe, but not mild, maternal anemia is associated with increased risk of small-for-gestational-age outcomes. *J Nutr* 2012; 142(2):358–62.

14. World Health Organization (WHO). *The global prevalence of anaemia in 2011.* Geneva: World Health Organization; 2015.

15. World Health Organization (WHO). *Optimal serum and red blood cell folate concentrations in women of reproductive age for prevention of neural tube defects.* Geneva: World Health Organization; 2015.

16. De-Regil LM, Peña-Rosas JP, Fernández-Gaxiola AC et al. Effects and safety of periconceptional oral folate supplementation for preventing birth defects. *Cochrane Database Syst Rev* 2015; 14(12):CD007950.

17. World Health Organization (WHO). WHO Congenital anomalies fact sheet [Internet]. Available from: http://www.who.int/mediacentre/factsheets/fs370/en/.

18. Institute of Medicine. *Dietary Reference Intakes: The essential guide to nutrient requirements.* Otten JJ HJ, Meyers LD, editors. Washington, DC: The National Academies Press; 2006.

19. Institute of Medicine Committee to Review Dietary Reference Intakes for Vitamin D, Calcium. *Dietary reference intakes for calcium and vitamin D*. Washington, DC: National Academy of Sciences; 2011.

20. Institute of Medicine Standing Committee on the Scientific Evaluation of Dietary References. *Dietary reference intakes for thiamin, riboflavin, niacin, vitamin B_6, folate, vitamin B_{12}, pantothenic acid, biotin, and choline*. Washington, DC: National Academy of Sciences; 1998.

21. McPherson NO, Owens JA, Fullston T et al. Preconception diet or exercise intervention in obese fathers normalizes sperm microRNA profile and metabolic syndrome in female offspring. *Am J Physiol Endocrinol Metab* 2015; 308(9):E805–21.

22. McPherson NO, Fullston T, Aitken RJ et al. Paternal obesity, interventions, and mechanistic pathways to impaired health in offspring. *Ann Nutr Metab* 2014; 64(3–4):231–8.

23. Lenders CM, McElrath TF, Scholl TO. Nutrition in adolescent pregnancy. *Curr Opinion Ped* 2000; 12(3):291–6.

24. Haider BA, Bhutta ZA. Multiple-micronutrient supplementation for women during pregnancy. *Cochrane Database Syst Rev* 2015; 11:CD004905.

25. Casanueva E, Roselló-Soberón ME, De-Regil LM et al. Adolescents with adequate birth weight newborns diminish energy expenditure and cease growth. *J Nutr* 2006; 136(10):2498–2501.

26. Scholl TO, Hediger ML, Schall JI et al. Maternal growth during pregnancy and the competition for nutrients. *Am J Clin Nutr* 1994; 60:183–8.

27. World Health Organization (WHO). *Global strategy for women's, children's and adolescent's health 2016–2030*. Italy; 2015.

28. Gibbs CM, Wendt A, Peters S et al. The impact of early age at first childbirth on maternal and infant health. *Paediatr Perinat Epidemiol* 2012; 26(Suppl 1):259–84.

29. Moos MK, Dunlop AL, Jack BW et al. Healthier women, healthier reproductive outcomes: Recommendations for the routine care of all women of reproductive age. *Am J Obst Gynecol* 2008; 199(Suppl. 2):S280–9.

30. Ravindran TK, Balasubramanian P. "Yes" to abortion but "no" to sexual rights: The paradoxical reality of married women in rural Tamil Nadu, India. *Reprod Health Matters* 2004; 12(23):88–99.

31. Rofail D, Colligs A, Abetz L et al. Factors contributing to the success of folic acid public health campaigns. *J Public Health* 2012; 34(1):90–9.

32. World Health Organization (WHO). *Guidelines for combined iron and folic acid supplementation in menstruating women*. Geneva. World Health Organization; 2011.

33. World Health Organization (WHO), United Nations Children's Fund (UNICEF). *Reaching optimal iodine nutrition in pregnant and lactating women and young children*. Geneva: World Health Organization; 2007.

34. Allen L, de Benoist B, Dary O, Hurrel R (editors). *Guidelines on food fortification with micronutrients*. Geneva: World Health Organization; 2006.

35. Vollset SE, Clarke R, Lewington S et al. Effects of folic acid supplementation on overall and site-specific cancer incidence during the randomised trials: Meta-analyses of data on 50,000 individuals. *Lancet* 2013; 381(9871):1029–36.

36. Mason JB SR, Saldanha LS, Ramakrishnan U et al. The first 500 days of life: Policies to support maternal nutrition. *Glob Health Action* 2014; 7.

37. Iodine Global Network (IGN). Global Scorecard 2014: Number of iodine deficient countries more than halved in past decade. *IDD Newsletter* 2015; 43(1):5–7.

38. De-Regil LM, Suchdev PS, Vist GE et al. Home fortification of foods with multiple micronutrient powders for health and nutrition in children under two years of age (Review). *Evid Based Child Health* 2013; 8(1):112–201.

39. Suchdev PS, Peña-Rosas JP, De-Regil LM. Multiple micronutrient powders for home (point-of-use) fortification of foods in pregnant women. *Cochrane Database Syst Rev* 2015; 19(6):CD011158.
40. Garcia-Casal MN, Peña-Rosas JP, Pachón H et al. Staple crops biofortified with increased micronutrient content: Effects on vitamin and mineral status, as well as health and cognitive function in the general population (Protocol). *Cochrane Database Syst Rev* 2016; 8:CD012311.
41. La Frano MR, de Moura FF, Boy E et al. Bioavailability of iron, zinc, and provitamin A carotenoids in biofortified staple crops. *Nutr Rev* 2014; 72(5):289–307.
42. Nestel P, Bouis HE, Meenakshi JV et al. Biofortification of staple food crops. *J Nutr* 2006; 136(4):1064–7.

28 Interpregnancy Intervals and Birth Spacing

Amanda Wendt and Usha Ramakrishnan

CONTENTS

INTRODUCTION

The World Health Organization (WHO) has recognized the first 1,000 days of life as a critical window for the improvement of child growth and development. More recently, the preconception period has also been recognized as crucial, especially in populations where there are high levels of undernutrition. In addition to nutrition-specific interventions, counseling that encourages both recommended birth intervals and a delayed age of first conception, have been emphasized as important factors indirectly affecting maternal and fetal nutritional status. Since the early 1900s, researchers have reported associations of the time between two consecutive pregnancies and adverse birth outcomes [1]. Observational studies then and since have shown adverse outcomes to be associated with both short and long time intervals.

Short interpregnancy intervals (IPIs) are common in both high- and low-income countries around the globe (see Table 28.1). Two analyses, using Demographic and Health Survey (DHS) data from 72 and 61 countries, respectively, found that almost a quarter of all births followed short IPIs (<24 months) [2,3], with one reporting 12% following long IPIs (>60 months) [3]. Even in the United States, only half of all pregnancies were conceived in the optimal IPI, defined as 18 to 59 months; a third of pregnancies began within 18 months of the previous birth and 15% after long IPIs (≥60 months) [4].

Recent systematic reviews have found that women with short IPIs are at increased risk for several adverse birth outcomes, namely, small for gestational age (SGA), preterm birth (PTB), low birth weight (LBW), stillbirth, early neonatal death [5,6], and uterine rupture in women attempting vaginal delivery after a cesarean section (VBAC) [7]. Long IPIs were also reported to be associated with PTB, LBW, SGA, labor dystocia, and preeclampsia [8,9].

However, for many outcomes, the relationship is less clear. Typically, studies assessing birth spacing have been observational; thus, unmeasured or residual confounding is an issue that must be considered [5]. Women with short IPIs are more likely to have a lower socioeconomic status (SES), use antenatal care services less, and are more likely to have an unplanned pregnancy; all of these factors are also associated with adverse pregnancy outcomes. Longer IPIs are correlated with advancing maternal age and possible infecundity, both shown to be independently associated with poor birth outcomes. Finally, studies often differ in their exposure definition, in terms of both the interval measured and interval length. IPI is most commonly defined as the time from one birth outcome (e.g., live birth, stillbirth, abortion) until

TABLE 28.1
Estimated Prevalence of Short and Long Interpregnancy Intervals by Region

	<24 months (%)	≥60 months (%)	Number of Intervals[a]
Central Asia	33	12.8	3,699
Latin America and Caribbean	30.4	12.4	61,772
North Africa, West Asia, and Europe	30.1	15.7	32,400
Sub-Saharan Africa	19.9	9.6	198,747
South and Southeast Asia	24.7	14.8	75,150
Total	24.6	11.8	371,768

Source: Data from Rutstein SO, Trends in birth spacing, DHS Comparative Reports No. 28, Calverton, MD: ICF Macro, 2011.

[a] Weighted by number of intervals.

the subsequent conception. Birth interval is also used, measuring the time from one live birth to the next. Interdelivery interval (IDI) similarly measures this interval, but includes other pregnancy outcomes. Birth interval or IDI measurement are therefore possible when the date of conception and gestational age are unknown. However, PTBs may artificially shorten the interval in these instances, overestimating the LBW associated with these intervals. None of these definitions take into account breastfeeding practices, which could further deplete maternal reserves in those who breastfeed and also quickly become pregnant. To account for this, some researchers measure the duration of the "recuperative interval" defined as the nonpregnant, non-lactating interval between pregnancies [10].

Another issue is the variation of cutoff values used to define IPI duration as "short," "long," and even "optimal," which makes comparisons challenging. The WHO currently recommends IPIs of longer than 24 months, with the strongest evidence supporting avoidance of IPIs less than 18 months and longer than 59 months. Limited evidence of possible adverse effects associated with 18 to 27 month intervals led the consultation participants to compromise on a 24-month recommendation. This was also done to simplify the public health message (2 years), and to complement UNICEF's recommendation of breastfeeding for the first 2 years of life. For pregnancies occurring after an abortion (spontaneous or induced), the recommended IPI was ≥6 months [11]. Many studies define short IPI using these categorizations or subsets, often with the shortest IPI defined as <6 months. However, other short or long IPI definitions include the WHO "optimal" interval recommendations or in some cases, the reference group also contains intervals the WHO has recommended to avoid [12,13].

Despite the inherent challenges of varying definitions of exposure and comparison groups between studies, evidence has demonstrated, with varying degrees of strength, the association of birth spacing with various maternal and infant outcomes. In this chapter, we first discuss hypothesized biological mechanisms to explain how short or long IPIs may result in adverse maternal and child outcomes, followed by a description of the evidence linking IPIs to these outcomes.

BIOLOGICAL MECHANISM HYPOTHESES

In 2012, Conde-Agudelo and colleagues reviewed existing hypothesized causal mechanisms relating short interpregnancy intervals to adverse maternal and child outcomes [14]. In the following, we outline mechanisms reviewed by Conde-Agudelo et al., as well as others relating to both short and long intervals (see Figure 28.1).

Maternal Nutrient Depletion

Pregnancy and lactation are critical periods characterized by increased growth and development which, in turn, greatly increase requirements for energy, protein, calcium, essential fatty acids, and several micronutrients, such as iron, folic acid, and vitamin A. The maternal nutrition depletion hypothesis posits that during pregnancy and lactation, maternal energy and nutrient stores are depleted, and a recuperative interval (nonpregnant, nonlactating) of adequate length is needed to

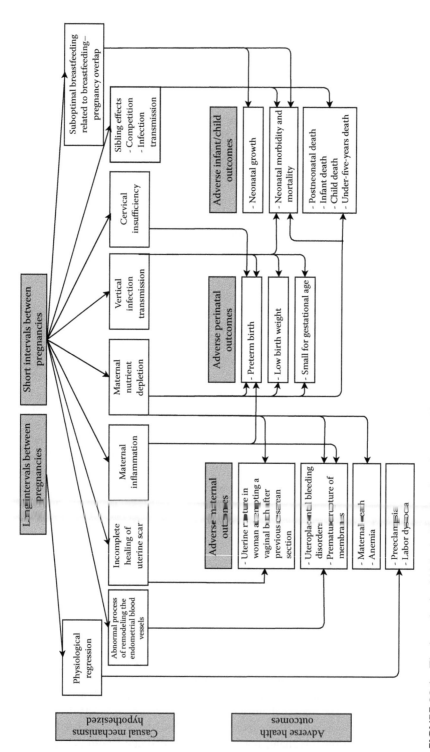

FIGURE 28.1 The hypothesized biological mechanisms between birth spacing and maternal, perinatal, infant, and child outcomes. (Adapted with permission from Conde-Agudelo A, Rosas-Bermúdez A, Castano F et al., *Stud Fam Plann* 2012, 43:93–114.)

recover these stores before the next pregnancy [15,16]. This hypothesis was first introduced over 50 years ago as "maternal depletion syndrome," described as the cumulative effect of multiple pregnancies over a lifetime on nutritional status. In 1992, the theory was redefined to focus on spacing from one pregnancy to the next and authors highlighted that both pregnancy and lactation were "depletion periods." In addition, Winkvist described four types of women, defined by weight change over the reproductive cycle, and posited that only the "marginally malnourished" would have maternal depletion syndrome, defined as a state in which a "nutrient replete" status could be achieved with an adequate recuperative interval. Those who gained or did not change weight during the reproductive cycle were considered already replete; and severely malnourished women, though also depleted, would need improved nutrition to become replete as only extending the recuperative interval would not be enough [16].

In these situations, where women are not consuming adequate nutrients during gestation, it was previously assumed that the fetus took priority and received nutrients at the mother's expense. However, this concept was challenged by animal studies, which found that upon food restriction during gestation, if there was maternal weight loss, even greater losses were observed in fetal birth weight [17]. This trend was further confirmed through a reanalysis of data collected during the Dutch famine (1944–1945), resulting in diet restriction lasting 7 months. Women affected during their second to third or third trimesters who lost weight gave birth to infants with even greater birth weight losses. A pregnancy weight loss of 3% from preconception weight resulted in a 10% loss in infant birth weight [18]. This further supported the pattern seen in animal studies that, under maternal dietary restriction, infant growth is actually inhibited to a greater extent than maternal weight gain.

This differential partitioning of nutrients, which depends upon initial maternal nutritional status and consumption, was further explored by Winkvist et al. in 1994 among women in Pakistan. Authors found malnourished women gained weight over the reproductive cycle, but infant weight decreased (from one pregnancy to the next). Marginally malnourished women lost weight over the reproductive cycle although infant weight increased. Well-nourished women did not show a difference on average of weight gain or infant birth weight changes over the reproductive cycle. Further analysis found that periods of overlap (when breastfeeding continued during the subsequent pregnancy) led to weight gain in both malnourished groups. Authors hypothesized that high reproductive stress may lead to partitioning of nutrients favoring maternal rather than fetal nutrition [19].

A recent systematic review examining this hypothesis found "unclear evidence," as results of studies on this topic have been mixed [14]. Differences in the populations studied, time of measurement (e.g., birth interval versus full reproductive cycle), initial nutritional status of the mother, and dietary intake during gestation all play an important role and different measures across studies make true comparisons challenging. Furthermore, to date, the majority of studies have only focused on anthropometric measures or hemoglobin. However, this could soon be examined further through several recent and ongoing randomized controlled trials, which are using periconceptional whole food or micronutrient supplementation to examine undernutrition effects on maternal and child outcomes [20]. These studies, if previous IPI is documented, could provide further insight into the impact of IPI on micronutrient status, as well as how improved nutrition in the periconceptional phase influences this relationship.

There is some evidence suggesting that the effect of birth interval on poor pregnancy outcomes such as PTB and SGA may be explained by folate depletion. Findings from a large Dutch cohort study suggest that folic acid appeared to mediate the effect of birth spacing on birth weight and SGA [21]. "Early users" were defined as those reporting preconception folic acid consumption, "late users" reported consuming postconception, and "nonusers" did not report folic acid supplementation. For nonusers, a 1-month increase in IPI increased mean birth weight by 165 g (±40 g) and reduced the likelihood of SGA (OR: 0.38; 95% CI: 0.24–0.60) [22]. For both early and late folic acid users, these relationships were not statistically significant, suggesting that perinatal folic acid consumption may improve birth weight following a short IPI. Although the authors' analysis was adjusted for several confounders, they noted that residual confounding was possible due to the different maternal characteristics of women who consume versus do not consume folic acid supplements during pregnancy that were not captured in the analysis. Furthermore, it was possible that this effect was due to different micronutrients in the supplements consumed; however, 65% of the women reported consuming only folic acid [21].

BREASTFEEDING–PREGNANCY OVERLAP

As previously mentioned, several studies have shown that overlapping periods of breastfeeding and pregnancy are common across many countries, and are logically associated with shorter IPIs. This overlap has been hypothesized to adversely affect child growth in the womb, as well as to affect the lactation volume and milk quality of the previous and subsequent child. In 1990, Merchant et al. classified women who participated in the original Institute of Nutrition of Central America and Panama (INCAP) longitudinal study conducted in Guatemala from 1969 to 1977, according to birth interval and measures of overlap [23]. In this report, "long recuperative interval" was defined as at least a 6 month nonlactating, nonpregnant period between pregnancies, while a "short recuperative interval" indicated less than 6 months. Those with an overlap of breastfeeding during pregnancy were classified as having a "short overlap" if the child was weaned in the first trimester, and a "long overlap" if weaning took place in the second or third trimester. This study showed a borderline significant trend ($p = 0.10$) of increasing birth weight across the four groups as the duration of overlap decreased and recuperative interval increased [24]. Studies examining breastfeeding in association with overlap have also found evidence of reduced nutrient concentrations in breast milk [22].

PHYSIOLOGICAL RECOVERY FROM PREVIOUS PREGNANCY

Various adverse effects associated with short IPIs may also be due to detrimental effects associated with an incomplete physiological recover from the prior pregnancy. Brief descriptions of some proposed mechanisms are provided next.

Infection Transmission

Women who acquire a "persistent" infection during their previous pregnancy followed by a short birth interval may be at a higher risk of still being infected. Cheng et al. found that, compared to women with a birth interval of >36 months, those with a short birth interval (<12 months) were 60% more likely to test positive for group B streptococci infection in their subsequent pregnancy, which is associated with neonatal infection [23].

Maternal Inflammation

Another proposed mechanism posits that inflammatory processes present during a pregnancy may be carried over to the subsequent pregnancy, especially if they are in close succession. A U.S. study found recurrent PTB to be associated with placental inflammatory lesions, which were more common among women who experienced lesions in the previous pregnancy [25]. Elevated inflammation has also been posited as an explanation for the increased risk of preterm premature rupture of membranes (PPROM) associated with short IPI [26] leading to PTB [27].

Cervical Insufficiency

Cervical insufficiency, defined as early dilation or effacement of the uterine cervix before contractions or labor, could occur after short IPIs due to insufficient time for reproductive tissue muscle tone recovery and could lead to PTB [28]. One study found that normalization of the cervical tissue collagen content following spontaneous delivery occurred only after 12 months postpartum [29].

Abnormal Remodeling of Endometrial Blood Vessels

Short IPIs may not allow sufficient time for remodeling of endometrial blood vessels, which may lead to increased risks of placenta previa and abruption placenta [30]. Wax et al. also found an increased risk of placenta accreta, increta, and percreta in women with short cesarean to birth intervals [31], although no studies have examined this hypothesis further.

Uterine Scar Healing Following Cesarean Section

Short IPIs may increase the risk of uterine rupture among women attempting a vaginal birth after caesarean section (VBAC) [8]. Studies examining the thickness of the lower uterine segment (indicating scar strength from the previous cesarean incision) found that IPIs under 6 months were associated with insufficient scar strength [32], and that complete uterine restoration did not occur until 6 to 9 months [33].

EFFECTS OF CLOSELY SPACED SIBLINGS

Short birth intervals additionally mean that siblings of similar ages will be present in the household, and thereby could increase the risk of infectious disease transmission, especially in resource-poor settings. Shorter birth intervals have been associated with increased diarrhea [34] and *Helicobacter pylori* infection [35].

PHYSIOLOGICAL REGRESSION: LONG INTERPREGNANCY INTERVALS

Longer IPIs have also been associated with adverse outcomes, such as PTB, LBW, SGA, preeclampsia, and labor dystocia [8,9]. Conde-Agudelo and Belizan found a lower risk of preeclampsia in subsequent pregnancies; however, this effect was not seen when IPI was 5 years or more, when the risk was equivalent to nulliparous women [30]. Overall, it is probable that the association between duration of IPI and adverse maternal and infant outcomes are due to multiple factors that likely vary depending on the woman's nutritional status, breastfeeding practices, and previous birth outcome(s). All of the above factors typically vary at an individual and population level in different settings, and may explain the conflicting evidence described in the following sections (see Table 28.2).

INTERPREGNANCY INTERVALS AND MATERNAL OUTCOMES

MATERNAL MORBIDITY AND MORTALITY

Several studies have examined the association between IPI and maternal morbidities, such as hemorrhage, PPROM, maternal mortality, preeclampsia/eclampsia, and hypertensive disorders. However, there are often few studies to compare for a single outcome. In a 2012 systematic review, authors found only five moderate-quality studies examining short IPI and which measured in total 10 maternal morbidity outcomes. No single outcome was reported adequately across three different studies, the criteria required to perform a meta-analysis [5].

Conde-Agudelo and Belizan analyzed data from over 450,000 pregnancies in Latin America and found a significant increased risk of third trimester bleeding and PPROM among women with IPIs of 0 to 5 months, compared to 18 to 23 months after adjustment for several potential confounders, including maternal age, parity, pregnancy outcome, and health care use [30]. Other studies have not found a significant association between IPI and hemorrhage, bleeding, or PPROM outcomes [5]. For maternal death, Conde-Agudelo found an increased risk (OR: 2.5; 95%CI: 2.1, 5.4) among women with IPIs of 0 to 5 months [30]. In Bangladesh, Rahman et al. did not find a significantly increased risk; however, the shortest IPI group examined was <12 months [36]. Overall, it appears that short IPI could be associated with adverse maternal outcomes; however, to allow comparison, further high-quality research and multiple studies assessing similar outcomes are needed.

For women attempting a VBAC, the evidence is more consistent. Two U.S. studies found an increased risk of uterine rupture following an interdelivery interval of <19 months [37] and IPI of <6 months [38]. In Canada, authors found a dose-response effect, as those with interdelivery intervals of <12 months had the greatest incidence of uterine rupture followed by 12 to 23 months with 24 to 35 months as the reference [39].

The evidence of adverse outcomes associated with longer IPIs also showed a clearer picture. A 2016 meta-analysis found increased odds of preeclampsia in IPIs of >4 years compared to 2 to 4 years (OR: 1.10; 95% CI: 1.102, 1.19) [9]. Conde-Agudelo et al. (2007) also reported that longer IPIs were associated with an increased

TABLE 28.2
Estimates and Strength of Evidence of Links between Interpregnancy Interval and Adverse Outcomes

Adverse Outcomes	Short Interpregnancy Interval OR	Short Interpregnancy Interval (95% CI)	Long Interpregnancy Interval	Number of Studies[a] [Reference]	Strength of Evidence
Maternal Outcomes					
Maternal morbidities/ mortality[b]	*		*	4 [9], 5 [5], 22 [8]	Low
Anemia	*			3 [5], 5 [8]	Low
Uterine rupture following cesarean delivery	+			4 [8]	Low
Preeclampsia			+	4 [9], 6 [8]	Low
Labor dystocia			+	1 [8]	Low
Child Outcomes					
Stillbirth	1.35 [5]	(1.07, 1.71)		3 [5]	Moderate
Early neonatal death	1.29 [5]	(1.02, 1.64)		4 [6]	Moderate
Preterm birth[c,e]				12 [5], 16 [6]	
	1.58 [5]	(1.40, 1.78)			Moderate
	1.41 [5]	(1.20, 1.65)			Moderate
Low birth weight[c]	1.44 [5]	(1.30, 1.61)		6 [5]	
				10 [6]	Moderate
Small for gestational age	1.26 [6]	(1.18, 1.33)		13 [6]	Moderate
Birth defects	+				Low
Child Growth and Development					
Stunting	+			8 [10]	Low
Malnutrition[d]	*			35 [10]	Low
Autism Spectrum Disorder	+		+	7 [40]	Low
Schizophrenia[f]	+			2 [41,42]	Low

Note: +, trend of increased risk; *, unclear/mixed results.

[a] Studies included in reviews that assessed strength of the evidence of the respective association unless otherwise indicated. Numbers of studies where estimates are provided include only the number of studies used in meta-analyses.

[b] Maternal morbidities/mortality: Postpartum hemorrhage, premature rupture of membranes, third trimester bleeding, preeclampsia (for short IPI), eclampsia, hypertensive disorders, maternal infection, uteroplacental bleeding disorders (placenta previa and placental abruption), gestational diabetes, placenta accreta, puerperal endometritis, composite morbidity measure, maternal mortality, proteinuria, high blood pressure, maternal death.

[c] Preterm birth and low birth weight: Odds ratios reported considering IPIs <6 months as compared to study reference IPIs of <12 months or a subset.

[d] Child malnutrition: Stunting, underweight, wasting, weight change, MUAC.

[e] Number of studies on preterm birth (extreme and moderate combined).

[f] No review has assessed the relationship between IPI and schizophrenia.

risk of preeclampsia and labor dystocia in a systematic review [8]. All six included studies examining preeclampsia in the 2007 systematic review showed an increased risk with longer IPIs. Two studies also reported a 10% to 12% increased risk for each additional interpregnancy or birth interval year [8]. Labor dystocia was evaluated in one high-quality study including 650,000 pregnancies. The authors found a dose-response relationship, with intervals beyond two years showing increased odds of labor dystocia [43].

MATERNAL NUTRITIONAL STATUS

The association between IPI and maternal nutritional status has also been explored with mixed results. In a cross-sectional study in Latin America of over 450,000 pregnancies, Conde-Agudelo and Belizan found that women with an IPI of <6 months had a 30% increased risk of anemia as compared to 18- to 23-month intervals [30]. In Nigeria, a study found women with IPIs of <24 months had an increased anemia risk [44]. Two other studies found no significant association, though one only assessed clinical anemia [45,46].

INTERPREGNANCY INTERVALS AND CHILD OUTCOMES

STILLBIRTH AND EARLY NEONATAL DEATH

There are many studies examining the relationship of birth spacing to stillbirth and early neonatal death. However, only three studies on stillbirth and three for early neonatal death were included in a recent meta-analysis, as many poor-quality studies were excluded, such as those that did not consider any confounding variables or define IPI as <12 months or a subset of this group. The meta-analysis found an overall random-effects OR of 1.35 (95% CI: 1.07, 1.71) for stillbirth among pregnancies following an IPI of <7 months (or a subset) with "unexposed" groups' IPI between 12 and 50 months (or a subset) [5]. Stillbirth was defined as fetal death ≥20 weeks of gestation in one included study [47] and as ≥28 weeks in the other two [48,49]. The meta-analysis on early neonatal death (death in the first week of life) resulted in an OR of 1.29 (95% CI: 1.02, 1.64), with similar IPI categorizations. Evidence for the association of both outcomes and short IPIs was assessed to be of "moderate" quality [5]. A 2006 review did not create pooled estimates on these outcomes; however, they created meta-regression curves, which appeared to show higher risks for both outcomes in IPIs <6 months or >50 months [6].

PRETERM BIRTH

Preterm birth has been associated with short IPIs across many studies of higher quality. Conde-Agudelo et al. (2006) reported a J-shaped meta-regression curve and pooled estimates that showed highest risk of PTB among those with a low IPI (<6 months) (OR: 1.77; 95% CI: 1.54, 2.04) as well as an elevated risk for IPIs of 6 to 11 months (OR: 1.23; 95% CI: 1.16, 1.31) and longer IPIs (>60 months) (OR: 1.27; 95% CI: 1.17, 1.39) [6]. Similar results were obtained by Wendt et al. (2012), in which short IPIs (<6 months; 6–11 months) were significantly associated both with extreme preterm birth (<33 weeks) and moderate preterm birth (<37 weeks). The authors concluded that the evidence for short IPI and PTB was of moderate quality [5].

Low Birth Weight

Low birth weight, which can be due to PTB or restricted growth *in utero*, has also long been associated with shorter IPIs. A 2012 systematic review reported higher risks of LBW for IPIs less than 6 months, and a weaker but still significant elevated risk for 6 to 11 month IPIs compared to >12 months (<6 months OR: 1.44; 95% CI: 1.30, 1.61; 6–12 months OR: 1.12; 95% CI: 1.08, 1.17) [5]. Conde-Agudelo et al. (2006) also reported similar pooled estimates for these IPI groups, and longer IPIs (≥60 months) were found to be associated with LBW (OR: 1.43; 95% CI: 1.27, 1.62) as compared with an 18- to 23-month interval [6].

Small for Gestational Age

Conde-Agudelo and colleagues's meta-analyses for SGA found an increased risk in shorter and longer IPIs, with the lowest risk group having an 18- to 23-month IPI [6]. For short IPIs (<6 months), the pooled odds ratio was 1.26 (95% CI: 1.18, 1.33). The odds ratio for long IPIs (≥60 months) was 1.29 (95% CI: 1.20, 1.39). In the meta-regression curves constructed for PTB, LBW, and SGA, the risk of the outcome increased with IPIs of >6 months or <20 months [6].

Birth Defects

Fewer studies have examined the association between IPI and birth defects, and most reports come from high-income countries. Existing evidence suggests a possible increase in birth defects, especially among women with IPIs <6 months. Grisaru-Granovsky et al. reported an increased risk of congenital malformations (OR: 1.14; 95%CI: 1.04, 1.24) in IPIs of <6 months, compared to IPIs of 12 to 23 months in a prospective study of over 400,000 pregnancies in Israel [50]. An increased risk of birth defects with both shorter IPIs (0–5 months OR: 1.15; 95% CI: 1.03, 1.28) and longer IPIs (≥60 months OR: 1.15; 95% CI: 1.04, 1.26) when compared to a reference IPI of 18 to 23 months was also found in a case control study in Washington state [51]. Another U.S. case control study identified gastroschisis cases through the National Birth Defects Study in 10 U.S. states. Women with IPIs of <12 months had an increased risk (OR: 1.7; 95% CI: 1.1, 2.5) compared to an 18- to 23-month interval. Notably, a stronger association was found among women living in northern states (above 37°N latitude), especially with a winter/fall conception, suggesting a possible role for vitamin D depletion, which is also associated with inflammation [52].

Child Growth and Development

Dewey et al. (2007) reviewed data from 52 studies, and found overall that longer birth intervals were associated with a reduced risk of malnutrition, but not in all populations [10]. Outcomes reviewed included stunting, underweight, wasting, weight change, and mid-upper arm circumference (MUAC), and the evidence between short IPIs and child malnutrition was mixed [10]. However, a recent study, which used DHS data from 61 low- and middle-income countries, found that, compared to birth intervals of 24 to 35 months, women with short intervals of <12 months or 12 to 23 months had a small, increased risk of their child being stunted (<12 months: RR: 1.09; 95% CI: 1.06, 1.12; 12–23 months: RR: 1.06; 95% CI: 1.05, 1.06) [2]. A major limitation, however, is the cross-sectional nature of the studies and absence of data on the overlap between pregnancy and breastfeeding.

The relationship between birth spacing and child development has often received less attention in contrast to maternal and pregnancy outcomes. However, this is an important area [11], especially for developmental disorders such as autism spectrum disorder (ASD), which is a major concern [40]. In a recent systematic review, Conde-Agudelo et al. (2016) estimated that an IPI of less than 12 months was associated with a twofold increased risk of any ASD (pooled OR: 1.90; 95% CI: 1.16, 3.09) and a stronger association when only autistic disorder was considered (OR: 2.62; 95% CI: 1.53, 4.50) [50]. Three of the included studies also found longer IPIs to be associated with increased risk of ASD. The authors also reported emerging evidence that short IPIs were associated with developmental delay and cerebral palsy. The association between birth spacing and ASD was not attenuated by the typical confounding factors such as SES and parental characteristics or mediated by PTB and LBW [40]. The maternal folate depletion hypothesis may be a plausible explanation. Periconceptional folic acid consumption has been associated with a reduced risk of autistic disorder, although not Asperger disorder or pervasive developmental disorder not otherwise specified [53], and folic acid supplementation may attenuate the relationship between IPI and ASD [54]. Additional hypotheses included the pathways of maternal stress, maternal inflammation, and residual confounding.

A few studies have also examined the association between birth spacing and schizophrenia, which has been hypothesized to occur through maternal folate depletion [41]. Smits et al. (2004) reported an increased risk of schizophrenia, particularly among shorter birth intervals (15–20 months or, assuming term births, IPIs of 6–11 months) in a large cohort of 1.43 million people born from 1950 to 1983 in Denmark [55]. Women with a birth interval of ≤14 months (an estimated IPI of 0–5 months) did not show an increased risk compared to women with birth intervals of ≥45 months (IPI of approximately ≥36 months). The authors surmised that this was due to nonlactating women, who may recover both folate stores and fertility faster (although no information on breastfeeding was collected) [41]. A 2011 study of a Swedish population (born from 1973 to 1980) also found that, compared to IPIs of 13 to 24 months, IPIs of <6 months and 7 to 12 months were also associated with increased risk of schizophrenia [42]. Overall, the evidence base for the association between IPI and schizophrenia is quite small, and further research is required to understand this connection.

INTERPREGNANCY INTERVALS FOLLOWING OTHER PREGNANCY OUTCOMES

Most of the birth-spacing research to date has focused on pregnancies following live births, although that is unfortunately not the only pregnancy outcome possibility. Studies that have examined pregnancy outcomes following miscarriages, stillbirths, or abortions have overall recommended shorter IPIs than the 18 to 23 month interval, which the data supports for IPIs following live births. The WHO recommendation

TABLE 28.3

Interpregnancy Interval Recommendations Following Different Pregnancy Outcomes and Strength of Evidence Estimates

Pregnancy Outcome	IPI Recommendation	Number of Studies Included	Strength of Recommendation[a]
Live birth	18–23 months	20	Moderate
Stillbirth	15–24 months	3	Low
Miscarriage			
Spontaneous	<6 months	3	Moderate
Ectopic/MTX	3–6 months	2	Low
Abortion			
Medical	>6 months	1	Low
Surgical	3–6 months	2	Low
Special Populations			
Cesarean delivery	6–12 months	5	Moderate
Gestational trophoblastic disease (GTD)	12 months	1	Low

Source: Bigelow CA, Bryant AS, *Obstet Gynecol Surv* 2015, 70:458–64.

[a] Bigelow and Bryant assigned values of B and C were changed to moderate and low, respectively.

is to wait at least 6 months before the subsequent pregnancy to decrease the risk of adverse maternal and perinatal outcomes following a spontaneous or induced abortion [11], based largely on a large cross-sectional analysis conducted in Latin America [7]. A 2015 review for clinicians examined literature addressing birth spacing following other pregnancy outcomes and gave a moderate rating of B to the strength of the evidence regarding IPI recommendations following a live birth, spontaneous miscarriage, and cesarean delivery. Other pregnancy outcomes evaluated received a low C rating due largely to the lack of available evidence and study heterogeneity (see Table 28.3) [56].

CONFOUNDING

When examining the relationship between IPI and maternal and child outcomes, it is important to note that this association can be confounded by many factors. Several characteristics, such as lower SES, less access to or utilization of health care services, unintended pregnancy, postpartum stress, and unstable lifestyles, have been independently associated with shorter IPIs and adverse pregnancy outcomes. Longer IPIs may be associated with infecundity, a change of partner or other social/familial disruptions, advanced age, maternal illness, or unplanned pregnancies, which again are associated with adverse outcomes [5,8,10]. Many of these variables are quite challenging to measure accurately and, even when adjusted for, residual confounding remains a possibility. Studies that were included in the major systematic reviews on this topic did exclude research that did not adjust for any confounders. However, even including only those with adjusted analyses, studies of different populations and adjusting for different sets of confounders continue to add to the data heterogeneity of this topic.

POLICY IMPLICATIONS

Birth spacing is a modifiable risk factor and has the potential to greatly affect maternal and perinatal outcomes at the population level. Due in part to the WHO's recommended birth-spacing intervals, some countries have integrated birth spacing into their maternal and reproductive health policy. For example, in India, the Ministry of Health and Family Welfare established a campaign called "Ek teen do" (one, three, two), including mass media messaging and health-worker incentives to encourage a spacing of 3 years between the first and second births [57]. In Vietnam, national guidelines dictate only two children, spaced 3 to 5 years apart, although this is not always followed at the population level [58]. However, overall, few countries have implemented birth-spacing programs or successfully addressed the unmet need for family planning services, which would include effective birth-spacing strategies [55]. Birth spacing has been recommended to be included as one of the key evidence-based interventions to address preconception care in order to decrease preventable maternal and infant deaths, and improve maternal and child nutrition [59]. Awareness of the importance of birth spacing by both health providers, women, and their families will be crucial to improving counseling and adoption of optimal birth-spacing practices.

REFERENCES

1. Woodbury RM. Causal factors in infant mortality: A statistical study based on investigations in eight cities. Children's Bureau Publication No. 142. Washington, DC: Government Printing Office; 1925, p. 245.
2. Fink G, Sudfeld CR, Danaei G et al. Scaling-up access to family planning may improve linear growth and child development in low and middle income countries. *PLoS One* 2014; 9:e102391.
3. Rutstein SO. Trends in birth spacing. DHS Comparative Reports No, 28. Calverton, MD: ICF Macro; 2011.
4. Gemmill A, Lindberg LD. Short interpregnancy intervals in the United States. *Obstet Gynecol* 2013; 122:64–71.
5. Wendt A, Gibbs CM, Peters S et al. Impact of increasing inter-pregnancy interval on maternal and infant health. *Paediatr Perinat Epidemiol* 2012; 26(Suppl 1):239–58.
6. Conde-Agudelo A, Rosas-Bermudez A, Kafury-Goeta AC. Birth spacing and risk of adverse perinatal outcomes: A meta-analysis. *JAMA* 2006; 295:1809–23.
7. Conde-Agudelo A, Belizan JM, Breman R et al. Effect of the interpregnancy interval after an abortion on maternal and perinatal health in Latin America. *Int J Gynaecol Obstet* 2005; 89(Suppl 1):S34–40.
8. Conde-Agudelo A, Rosas-Bermudez A, Kafury-Goeta AC. Effects of birth spacing on maternal health: A systematic review. *Am J Obstet Gynecol* 2007; 196:297–308.
9. Cormick G, Betran AP, Ciapponi A et al. Inter-pregnancy interval and risk of recurrent pre-eclampsia: Systematic review and meta-analysis. *Reprod Health* 2016; 13:83.

10. Dewey KG, Cohen RJ. Does birth spacing affect maternal or child nutritional status? A systematic literature review. *Matern Child Nutr* 2007; 3:151–73.
11. World Health Organization (WHO). *Report of a WHO technical consultation on birth spacing*. Geneva: WHO; 2007.
12. Kleijer ME, Dekker GA, Heard AR. Risk factors for intrauterine growth restriction in a socio-economically disadvantaged region. *J Matern Fetal Neonatal Med* 2005; 18:23–30.
13. Dechering WH, Perera RS. A secondary analysis of determinants of low birth weight. *Ceylon Med J* 1991; 36:52–62.
14. Conde-Agudelo A, Rosas-Bermudez A, Castano F et al. Effects of birth spacing on maternal, perinatal, infant, and child health: A systematic review of causal mechanisms. *Stud Fam Plann* 2012; 43:93–114.
15. King JC. The risk of maternal nutritional depletion and poor outcomes increases in early or closely spaced pregnancies. *J Nutr* 2003; 133:1732S–6S.
16. Winkvist A, Rasmussen KM, Habicht JP. A new definition of maternal depletion syndrome. *Am J Public Health* 1992; 82:691–4.
17. Berg BN. Dietary restriction and reproduction in the rat. *J Nutr* 1965; 87:344–8.
18. Rosso P. Nutrition and maternal-fetal exchange. *Am J Clin Nutr* 1981; 34:744–55.
19. Winkvist A, Jalil F, Habicht JP et al. Maternal energy depletion is buffered among malnourished women in Punjab, Pakistan. *J Nutr* 1994; 124:2376–85.
20. King JC. A summary of pathways or mechanisms linking preconception maternal nutrition with birth outcomes. *J Nutr* 2016; 146:1437S–44S.
21. van Eijsden M, Smits LJ, van der Wal MF et al. Association between short interpregnancy intervals and term birth weight: The role of folate depletion. *Am J Clin Nutr* 2008; 88:147–53.
22. Merchant K, Martorell R, Haas J. Maternal and fetal responses to the stresses of lactation concurrent with pregnancy and of short recuperative intervals. *Am J Clin Nutr* 1990; 52:280–8.
23. Cheng PJ, Chueh HY, Liu CM et al. Risk factors for recurrence of group B streptococcus colonization in a subsequent pregnancy. *Obstet Gynecol* 2008; 111:704–9.
24. Merchant K, Martorell R, Haas JD. Consequences for maternal nutrition of reproductive stress across consecutive pregnancies. *Am J Clin Nutr* 1990; 52:616–20.
25. Himes KP, Simhan HN. Risk of recurrent preterm birth and placental pathology. *Obstet Gynecol* 2008; 112:121–6.
26. Getahun D, Strickland D, Ananth CV et al. Recurrence of preterm premature rupture of membranes in relation to interval between pregnancies. *Am J Obstet Gynecol* 2010; 202:570.e571–576.
27. Shachar BZ, Lyell DJ. Interpregnancy interval and obstetrical complications. *Obstet Gynecol Surv* 2012; 67:584–96.
28. Haaga JG. How is birthspacing related to infant health? *Malays J Reprod Health* 1988; 6:108–20.
29. Sundtoft I, Sommer S, Uldbjerg N. Cervical collagen concentration within 15 months after delivery. *Am J Obstet Gynecol* 2011; 205:59.e51–53.
30. Conde-Agudelo A, Belizan JM. Maternal morbidity and mortality associated with interpregnancy interval: Cross sectional study. *BMJ* 2000; 321:1255–9.
31. Wax JR, Seiler A, Horowitz S et al. Interpregnancy interval as a risk factor for placenta accreta. *Conn Med* 2000; 64:659–61.
32. Ait-Allah A, Abdelmonem A, Rasheed S. Pregnancy spacing after primary Cesarean section: Its impact on uterine scar strength and mode of delivery. *Int J Gynaecol Obstet* 2009; 107:S435.
33. Dicle O, Kucukler C, Pirnar T et al. Magnetic resonance imaging evaluation of incision healing after cesarean sections. *Eur Radiol* 1997; 7:31–34.

34. Manun'ebo MN, Haggerty PA, Kalengaie M et al. Influence of demographic, socio-economic and environmental variables on childhood diarrhoea in a rural area of Zaire. *J Trop Med Hyg* 1994; 97:31–38.

35. Goodman KJ, Correa P. Transmission of *Helicobacter pylori* among siblings. *Lancet* 2000; 355:358–62.

36. Rahman M, Da Vanzo J, Razzaque A et al. *Demographic, programmatic, and socio-economic correlates of maternal mortality in Matlab.* Watertown, MA: Pathfinder International; 2009.

37. Shipp TD, Zelop CM, Repke JT et al. Interdelivery interval and risk of symptomatic uterine rupture. *Obstet Gynecol* 2001; 97:175–7.

38. Esposito MA, Menihan CA, Malee MP. Association of interpregnancy interval with uterine scar failure in labor: A case-control study. *Am J Obstet Gynecol* 2000; 183:1180–3.

39. Bujold E, Mehta SH, Bujold C et al. Interdelivery interval and uterine rupture. *Am J Obstet Gynecol* 2002; 187:1199–1202.

40. Conde-Agudelo A, Rosas-Bermudez A, Norton MH. Birth spacing and risk of autism and other neurodevelopmental disabilities: A systematic review. *Pediatrics* (Available from: https://www.ncbi.nlm.nih.gov/pubmed/27244802). doi: 10.1542/peds2015-3482.

41. Smits L, Pedersen C, Mortensen P et al. Association between short birth intervals and schizophrenia in the offspring. *Schizophr Res* 2004; 70:49–56.

42. Gunawardana L, Smith GD, Zammit S et al. Pre-conception inter-pregnancy interval and risk of schizophrenia. *Br J Psychiatry* 2011; 199:338–9.

43. Zhu BP, Grigorescu V, Le T et al. Labor dystocia and its association with interpreg-nancy interval. *Am J Obstet Gynecol* 2006; 195:121–8.

44. Orji EO, Shittu AS, Makinde ON et al. Effect of prolonged birth spacing on maternal and perinatal outcome. *East Afr Med J* 2004; 81:388–91.

45. Razzaque A, Da Vanzo J, Rahman M et al. Pregnancy spacing and maternal morbidity in Matlab, Bangladesh. *Int J Gynaecol Obstet* 2005; 89(Suppl 1):S41–49.

46. Singh K, Fong YF, Arulkumaran S. Anaemia in pregnancy—A cross-sectional study in Singapore. *Eur J Clin Nutr* 1998; 52:65–70.

47. Conde-Agudelo A, Belizan JM, Norton MH et al. Effect of the interpregnancy interval on perinatal outcomes in Latin America. *Obstet Gynecol* 2005; 106:359–66.

48. Stephansson O, Dickman PW, Cnattingius S. The influence of interpregnancy interval on the subsequent risk of stillbirth and early neonatal death. *Obstet Gynecol* 2003; 102:101–8.

49. DaVanzo J, Hale L, Razzaque A et al. Effects of interpregnancy interval and outcome of the preceding pregnancy on pregnancy outcomes in Matlab, Bangladesh. *BJOG* 2007; 114:1079–87.

50. Grisaru-Granovsky S, Gordon ES, Haklai Z et al. Effect of interpregnancy interval on adverse perinatal outcomes—A national study. *Contraception* 2009; 80:512–8.

51. Kwon S, Lazo-Escalante M, Villaran MV et al. Relationship between interpregnancy interval and birth defects in Washington State. *J Perinatol* 2012; 32:45–50.

52. Getz KD, Anderka MT, Werler MM et al. Short interpregnancy interval and gastroschi-sis risk in the National Birth Defects Prevention Study. *Birth Defects Res A Clin Mol Teratol* 2012; 94:714–20.

53. Suren P, Roth C, Bresnahan M et al. Association between maternal use of folic acid supplements and risk of autism spectrum disorders in children. *JAMA* 2013; 309:570–7.

54. Gunnes N, Suren P, Bresnahan M et al. Interpregnancy interval and risk of autistic disorder. *Epidemiology* 2013; 24:906–12.

55. Rizvi A, Khan A. Birth spacing as a health intervention. *Ann Pak Inst Med Sci* 2011; 7:113–4.

56. Bigelow CA, Bryant AS. Short interpregnancy intervals: An evidence-based guide for clinicians. *Obstet Gynecol Surv* 2015; 70:458–64.
57. National Health Mission, Ministry of Health & Family Welfare, Government of India. Maintain 3 year gap. http://nrhm.gov.in/mediamenu/fp-mass-media-campaign/birth -spacing-campaign/maintain-three-year-gap.html.
58. Hoa HT, Toan NV, Johansson A et al. Child spacing and two child policy in practice in rural Vietnam: Cross sectional survey. *BMJ* 1996; 313:1113–6.
59. Bhutta ZA, Das JK, Bahl R et al. Can available interventions end preventable deaths in mothers, newborn babies, and stillbirths, and at what cost? *Lancet* 2014; 384:347–70.

29 Adolescent Nutrition
A Critical Opportunity for Intervention

Emily Mates and Anne Bush

CONTENTS

INTRODUCTION

Adolescents are defined by the World Health Organization (WHO) as people aged between 10 and 19 years. The Lancet Commission on Adolescents 2016 analysis used a range of 10 to 24 years of age to cover the groups defined by the United Nations Department of Economic and Social Affairs (UNDESA) as adolescents (ages 10–19 years), youths (ages 15–24 years), and young people (ages 10–24 years) [1].

As of 2012, there were 1.2 billion adolescents in the world, of which 90% live in low- and middle-income countries (LMICs). In industrialized countries, adolescents make up 12% of the population, as compared to 19% in LMICs [2]. As more children are surviving beyond their fifth birthday, the population of adolescents is increasing [3].

There is growing awareness of the importance of adolescent health and nutrition, and increasing recognition that investing in adolescents' health and well-being is essential to achievement of the post-2015 development agenda. This is captured by the inclusion of adolescents within the United Nations Secretary General's Global Strategy for Women's and Children's Health. The recent Lancet Commission on Adolescents [1] has also galvanized attention.

Understanding and evidence of the global situation on adolescent nutrition is limited. Most health and nutrition surveys start from age 15 years, resulting in a lack of information about those 10 to 14 years, for example, National Demographic and Health Surveys (DHS) typically focus on women of reproductive age (15–49 years). Anthropometric nutrition surveys almost exclusively focus on children under 5 years of age.

Nutrition intervention during adolescence is critical to influence current, future, and intergenerational nutrition and health, for the following reasons:

1. Adolescents are a nutritionally vulnerable group, with high nutritional requirements due to rapid growth and sexual maturation.
2. There is the potential to address nutritional deficits from the first decade of life and for catch-up growth.
3. Adolescence provides an opportunity to break the intergenerational cycle of malnutrition.
4. Improving nutrition/dietary practices at this age may have a positive impact on adult health.
5. Adolescence is a period of openness to new ideas [1,4,5] and therefore presents an opportunity to target interventions behaviors and practices that will influence health outcomes in adulthood.

Each of these is discussed in more detail next.

ADOLESCENCE IS A TIME OF INCREASED NUTRITIONAL REQUIREMENTS

Increased nutritional requirements during adolescence reflects the fact that growth during adolescence is faster than at any other time in an individual's life, with the exception of the first year. During this period adolescents gain up to 50% of their adult weight and skeletal mass and 15%–25% of their adult height [4], with changes in body shape and composition. Increased nutritional requirements (e.g., for iron and vitamin A) also result from adolescence being the period of sexual maturation [6,7].

MACRONUTRIENTS: ENERGY AND PROTEIN

The rapid rate of growth during and following puberty results in adolescents having some of the highest energy and protein requirements of any age group. Energy needs for growth have two components: (1) energy used to synthesize growing tissues, and (2) the energy deposited in those tissues, mainly as fat and protein [8].

These high requirements mean that adolescents are vulnerable to undernutrition. The term *undernutrition* is used to encompass low height-for-age (stunting), low weight-for-height (wasting), and low weight-for-age (underweight). It also includes micronutrient deficiencies. If protein and energy intakes are inadequate, the biological changes of adolescence may be affected. Undernutrition can result in a number of negative consequences for adolescents, including: delayed physical growth and sexual maturation, reduced work capacity, impaired cognitive ability, and heightened risk of chronic diseases in adulthood [9]. Adolescent girls who are stunted and become pregnant have an increased risk of complications such as obstructed labor and vesicovaginal fistula (a tract that develops between the vagina and the bladder as a result of prolonged, obstructed labor, causing uncontrollable leaking of urine through the vagina). Their infants are at an increased risk of low birth weight and preterm birth and, in turn, are more at risk of becoming stunted and giving birth to small infants themselves [1].

In some countries, up to half of all adolescent girls 15 to 19 years are stunted [2]. Based on information from 64 countries with available data, in 10 countries more than one-quarter of adolescent girls 15 to 19 years are underweight (BMI <18.5 kg/m^2), while in India 47% are underweight [10].

Evidence points to the effectiveness of balanced protein-energy supplementation in addressing undernutrition in children and preventing adverse perinatal outcomes in undernourished pregnant women [11]. Other interventions to improve household food security, such as cash transfers, which provide a predictable direct transfer to increase household disposable income of vulnerable families, also show promise [11]. However, few interventions have specifically targeted adolescents or have been evaluated for their effectiveness during adolescence. As a result, little is known about the benefits or potential negative side effects (e.g., an increased risk of obesity) of interventions to address protein-energy malnutrition in this age group [1]. Although school-based feeding programs are increasingly being broadened to include adolescents, there are few studies of its effectiveness in this age group. A systematic review in 2007 found only one study that included children up to 13 years, and results indicated no effect on height or weight [12]. In only one subsequent study of its kind, school feeding has been shown to reduce anemia in girls aged 10 to 13 years, but did not have an impact on anthropometric indicators such as BMI [13].

At the other end of the scale, prevalence rates of obesity and overweight in adolescents are increasing in LMICs, as well as in high income countries (HICs). This likely, in part, reflects the greater consumption of foods that are high in added sugars and fats, combined with decreases in physical activity and possibly because they are stunted, lessening their energy requirements. Based on a reanalysis of DHS data from 2005 to 2010 from 58 countries with available data in 11 countries, the United Nations Children's Fund (UNICEF) reported that more than one-fifth of adolescent girls 15 to 19 years were overweight (BMI >25 kg/m^2) [10]. Obesity in adolescence is a strong predictor of adult obesity and associated morbidity [1]. Obesity is a risk factor for cardiovascular diseases and type 2 diabetes during adolescence and adulthood, and is associated with complications during pregnancy, such as gestational diabetes [2].

Thus, interventions targeted at adolescents to prevent overweight and obesity are critical, and should be focused on the adoption of healthy eating habits and physical exercise. A recent systematic review found evidence that interventions to promote good nutrition and prevent obesity can have a marginal effect on reducing BMI in adolescence [7]. However, with a lack of data from LMICs, the findings in this review are limited to 10 studies from HICs only, so little is understood about what works in adolescence across other settings. The Lancet Commission on Adolescents concludes that multicomponent approaches, which include policy measures, education strategies, and environmental changes to promote increased physical activity and healthy eating habits, are more likely to be effective than single-component approaches [1]. Some studies highlight specific barriers to increasing physical activity among adolescent girls, for example, suitable place and social support, and suggest school-based approaches may also be effective [7]. Overall there is a lack of studies in this age group and more research to generate high-quality evidence of what works is required.

MICRONUTRIENTS

As well as increasing requirements for energy and protein, the high velocity of growth during adolescence increases the requirements for many micronutrients, particularly those with a role in energy metabolism and tissue synthesis, including skeletal tissue, red blood cells, and muscle mass. Of particular importance are iron, folate, calcium, vitamin D, and zinc.

Iron

Iron requirements increase dramatically during adolescence in both girls and boys, and can exceed those for adult menstruating women [8]. This is a result of an expansion in total blood volume, including a marked increase in hemoglobin mass and concentration, the increase in lean body mass, and the onset of menses in young females.

Low intakes combined with low bioavailability of iron during the period of increased requirements means that both adolescent males and females are vulnerable to iron deficiency anemia (IDA), defined as low hemoglobin and low iron status. IDA reduces physical work capacity, affects cognitive function, and may depress physical growth [4], yet little is known about the effect of IDA on learning and educational attainment in adolescents [1]. Globally, IDA is the third highest cause of years lost to death and disability (DALYs) among adolescents [4]. The problem is particularly acute in Africa, where DALY rates due to IDA are the highest.

Interventions to address IDA during adolescence are therefore crucial to address these negative consequences. Adolescence also provides an opportunity to intervene to prevent future negative consequences of anemia in pregnancy. Even if an adolescent girl has adequate iron status, low dietary iron intakes of foods with low iron bioavailability may result in the inability to maintain adequate stores of iron. Low iron stores in women of reproductive age may indicate an increased risk of IDA in future pregnancies, due to the additional requirements during pregnancy. IDA has also been associated with adverse maternal and fetal outcomes [15]. The risk of an infant being born with low birth weight (LBW) is significantly greater in the event of moderate maternal preconception anemia [11]. Poor iron and folic acid status has been linked with preterm births and fetal growth restriction [16].

The significant negative consequences of IDA mentioned above provide a strong case for intervening in adolescence to prevent current problems and subsequent problems in pregnancy, although direct evidence of effective interventions in this age group remains limited. A recent systematic review [7] found that iron/iron folate supplementation alone or in combination with other micronutrients reduced the prevalence of anemia among adolescents, and that school-based delivery had the greatest effect on anemia reduction. However, 22 of the 31 studies included in the review were focused on females and 23 were conducted in LMICs, so findings were not generalizable across sexes or contexts. Beyond this, there is again a dearth of evidence for interventions specifically targeting adolescents, but population-wide interventions, including iron fortification of staple foods (such as wheat flour, maize), improving food security and nutrient density, regular deworming where intestinal helminths are common, and the prevention and control of malaria, will also benefit adolescents. The WHO recommends daily iron supplementation as a public health intervention for all menstruating adolescent girls and adult women living in settings where anemia is highly prevalent (>40% prevalence) [17]. During pregnancy, supplementation with iron and folic acid [18] or multiple micronutrient supplements [19] are associated with improved outcomes, including a reduced risk of LBW and small for gestational age (SGA). However, it should be noted that the latest guidelines from the WHO only recommend iron/iron-folate and not multiple micronutrient supplementation during pregnancy [20].

Folate

Folate has an integral role in DNA, RNA, and protein synthesis, and can protect against neural tube defects as folate is required for neural tube closure during the first trimester of pregnancy. Recommended intakes of folate in adolescents (400 µg/day) are the same as for adults, and are based on the intake required during both preconception and pregnancy to prevent neural tube defects. In view of the high requirements and low bioavailability, folic acid supplementation is recommended in at-risk groups. A 2015 *Cochrane Review* found a 69% reduction in the risk of neural tube defects with oral folic acid supplementation in doses ranging from 300 µg to 4,000 µg a day, with and without other micronutrients, before conception and up to the first 12 weeks of pregnancy [21]. Thus, in contexts of early marriage and pregnancy, supplementation with folic acid during adolescence is an important nutrition intervention. The WHO recommends daily folic acid supplementation with 400 µg/day (in combination with daily iron supplementation) commenced as early as possible, ideally before conception, to prevent neural tube defects [20].

Calcium and Vitamin D

Adolescence is a period with rapid accrual of bone mass. Approximately 40% to 60% of adult bone mass is accumulated during adolescence, with over 25% developed during the 2-year period of peak skeletal growth [22]. Calcium and vitamin D are important nutritional factors influencing this process. As bone mineral density is a primary determinant of later-life osteoporosis and its complications, interventions to ensure adequate intakes of calcium and vitamin D during adolescence is a critical opportunity to reduce the risk of osteoporosis later in life. Furthermore, randomized controlled trials have shown that exercise has the potential to improve bone health under conditions of adequate calcium intake [23]. Data from different countries suggests that many adolescents are not achieving their recommended calcium intakes of 1300 mg/day. For example, in the United States, Average Daily Intakes among adolescent girls are 876 mg/day (less than 67% of the recommended dietary allowance [RDA]), and less than 15% of this group meets the RDA [22]. In developing countries, calcium intakes of children and adolescents are typically a third to a half of the RDA [24]. However, the benefits of supplementation have been found to be minimal, resulting in only small increases in bone mass that are short-lived once supplementation is discontinued. The reasons for such small effects of supplementation are unknown. A better approach is to support the achievement of adequate amounts of calcium from dietary sources (which have better bioavailability than supplements) as part of an overall approach to promote lifelong healthier eating habits in this age group [22].

The rapid growth of the skeleton during puberty increases requirements for the active form of vitamin D. Insufficient stores during the period of increased growth during adolescence may lead to vitamin D deficiency [8], and severe vitamin D deficiency in adolescence is associated with reduced bone mass [22].

Vitamin D is synthesized in the skin through exposure to sunlight containing sufficient ultraviolet B (UVB) radiation, and this is the main source for most people. However, the optimum sunlight exposure time to make adequate vitamin D is difficult to quantify due to the wide variety of factors affecting synthesis. These include latitude, season, time of day, amount of skin exposed, skin pigmentation, and use of sunscreen. In populations where children and adolescents are spending more time indoors and/or increasingly wearing sun protection when they do go out, this reduced exposure to UVB radiation means dietary sources of vitamin D become more important. Some countries (e.g., United Kingdom) have recently revised recommendations on vitamin D intakes to reflect this [25].

Calcium is also critical for pregnancy during adolescence, whereby maternal–fetal competition for calcium means that adolescents are more at risk of gestational hypertension and preeclampsia than older mothers; calcium supplementation is therefore beneficial, especially in populations with low calcium intakes [26].

Zinc

Zinc is an essential component of a large number of enzymes and has a central role in cell division, protein synthesis, and growth. Adequate zinc intake is therefore crucial in adolescence, as pubertal growth spurts considerably increase requirements. In contexts where diets are nutrient-poor, lack diversity, and have poor bioavailability due to high phytate and fiber, zinc status may be compromised. Furthermore, during the third trimester of pregnancy, zinc requirements are approximately twice that of nonpregnant females [8].

There is little or no direct evidence for many micronutrients, on which to base requirements during adolescence, and therefore Reference Nutrient Intakes (RNIs) are extrapolated from adult values. RNIs are principally based upon populations of healthy individuals. Where relevant and possible, RNIs allow for variations in micronutrient bioavailability and utilization [8], but due to gaps in data and understanding, they may not always be applicable to adolescents living in LMICs where diets often lack diversity and are nutrient-poor, and where there is typically a high burden of disease [9].

Micronutrient interventions during adolescence are critical to ensure that higher needs are being met to prevent negative consequences both during adolescence and later in life and to preempt additional demands of entering parenthood.

POTENTIAL FOR CATCH-UP GROWTH

There is increasing interest in adolescence as an additional opportunity for catching up on growth deficits from childhood [27,28]. This is mediated through a delay to maturation and/or an extended pubertal growth phase [27,29]. However, questions remain as to the extent to which catch-up growth is possible and its mechanisms: whether it occurs before or during the pubertal growth spurt; how complete it is (i.e., whether it will affect the final adult height attained or just an accelerated growth spurt over a shorter time frame); and whether it is spontaneous or requires additional interventions [29]. Longitudinal data from the Young Lives Study shows that 36% of stunted children aged 8 years managed to catch up with their peers by age 15 [30]. Further empirical evidence on pubertal catch-up growth is provided by longitudinal data from rural Tanzania [31]. It may be that the effects of growth-promoting interventions initiated years before puberty are realized during adolescence. Another question is whether catch-up growth interventions in adolescence are likely to improve other consequences of early growth retardation, such as cognitive impairment [29]. Again, results from the Young Lives Study show that children who caught up in height by age 15 had smaller deficits in cognitive function than children who remained stunted [30].

In summary, catch-up growth during adolescence offers potential in addition to the established window of opportunity for intervention during the first 1,000 days. However, Patton et al. [1] make the point that "any opportunity for catch-up will be restricted by early pregnancy and for that reason, delaying first pregnancy is essential in stunted adolescent girls."

AN OPPORTUNITY TO INFLUENCE THE PRECONCEPTION PERIOD AND PREVENT THE INTERGENERATIONAL CYCLE OF MALNUTRITION

Adolescence is a critical preconception period when it is possible to influence the negative impacts of becoming pregnant early and/or in a less-than-optimal nutritional state. It provides a timely opportunity to prepare for the additional demands of pregnancy and lactation, and to prevent early pregnancy with all its associated risks for the mother and the infant. It also provides an opportunity to enhance nutrition skills and behaviors that can have lasting benefits on household dietary practices.

Nutrition interventions in adolescent girls that will benefit adult pregnancy outcomes include increasing prepregnancy weight and body nutrient stores; improving iron status to reduce the risk of IDA and associated complications; and improving folate status to reduce the risk of neural tube defects in newborns and megaloblastic anemia in mothers [4].

However, of all the interventions, delaying teenage marriage and pregnancy will have the most significant impact on nutritional status and intergenerational cycle (see the section "Early Marriage and Adolescent Pregnancy") [11].

NUTRITION BEHAVIORS AND PRACTICES DURING ADOLESCENCE HAVE CONSEQUENCES FOR PRESENT AND FUTURE ADULT HEALTH AND NUTRITION

Nutrient intakes during adolescence can affect health later in adulthood. For example, calcium status and bone deposition during adolescence are important factors of bone mineral mass in adult. Optimizing calcium intake during adolescence reduces risk of osteoporosis later in life. Adequate iron and folic acid intake during adolescence is important to prevent anemia, ensure adequate iron stores prepregnancy, and protect the health of future offspring (e.g., through the prevention of neural tube defects). Obesity in early adolescence not only compromises adolescent development, but also predicts obesity later in life [32]. Furthermore, negative practices acquired during adolescence may continue through to adulthood.

ADOLESCENCE IS A PERIOD OF SIGNIFICANT DEVELOPMENTAL TRANSITION

Adolescence is a period of significant developmental transition, a time of openness to new ideas, and a time of making personal choices rather than following family patterns. It therefore presents a key opportunity for health and nutrition education interventions, with the potential to impact lifetime eating habits and health. Moreover, adolescents are often the first to adopt new technologies, particularly through the Internet [5]. This creates opportunities for innovative delivery mechanisms; for example, through social media rather than traditional health education services [1].

As described earlier, there are compelling reasons why adolescent nutrition is a critical time for intervention. However, there is limited evidence of the benefit of specifically targeting adolescents or of the effectiveness of interventions that have been proven effective in other populations or age groups [1]. The use of social media and information technologies, and interventions such as cash transfers, social protection, and microfinance initiatives are all potential approaches. However, there is a current lack of rigorous evaluations in all these areas and therefore a need for further high-quality research [3].

In terms of delivery platforms, there are known difficulties in reaching adolescents; for example, adolescent girls do not typically access standard maternal health services. Adolescents, particularly those in LMICs, are often disempowered and have low social status. School-based delivery strategies have been the most highly evaluated and show promise as platforms for nutrition interventions, as well as sexual health and substance abuse prevention [3], but poor enrollment is a major limitation in terms of overall coverage. What seems to be crucial is for specific nutrition interventions to be integrated into a broader package of actions, which focus on delaying pregnancy as a priority, empowering young women through access to education and health care, and increasing their control over household resources [1].

EARLY MARRIAGE AND ADOLESCENT PREGNANCY

Too often, adolescent girls are married before the onset of puberty and before they reach adulthood. Early marriage occurs in many LMICs, but is particularly common in parts of Asia and Sub-Saharan Africa. Evidence shows that approximately one in three girls in developing countries is married before the age of 18 and, importantly, one in nine is married before the age of 15 [33]. This translates into an estimated 15 million adolescent marriages each year [34].

A complicating factor for interventions targeted at young women is that when adolescent girls are better nourished, it is common for menarche to start at younger ages. Earlier onsets of menarche may also lead to premature marriage and first childbirth, particularly in settings where age at menarche is positively associated with age at marriage and first childbirth [35]. While it is important, therefore, to improve the nutritional status of adolescent girls, interventions need to be mindful of potential risks, and it will be helpful to ensure that robust policies and community sensitization activities to delay marriage and first childbirth among adolescent girls are implemented alongside any interventions. Legislation and laws can assist in protecting adolescents from the harms of child marriage; however, there is wide variation at national levels in their implementation, and customary or religious laws can take precedence. Guaranteeing 18 years as the minimum age for marriage requires the education and engagement of community leaders and professionals within the justice system, and governments have a duty to ensure the enforcement of legislation prohibiting child marriage, as it can be considered a priority health action for adolescents [1].

The sexual health risks that result in teenage pregnancy have considerable effects on the health and well-being of young women across the life course. Pregnancy (and early marriage) might signal the end of formal education, denying a young woman not only the opportunities that education provides for her own life (improved mental health, greater cognitive abilities, better opportunities for employment, etc.), but also the enhanced knowledge and confidence that schooling can give her to mother her own children. Recent data from South Africa (which has high rates of adolescent pregnancy and very high rates of coerced sex and violence against women and girls) showed that 75% of school-aged girls leave school because they are pregnant, and less than 50% return to complete their education [36]. This can also have an economic cost, with a country losing the annual income a young woman would have earned over her lifetime if she had not had an early pregnancy.

Sex education is lacking in many countries, and some girls do not know how to avoid getting pregnant. They may feel too inhibited or ashamed to seek contraception services, contraceptives may be too expensive, or they may not be widely or legally available. Even when contraceptives are widely available, sexually active adolescent girls are less likely to use them, compared to adults. Girls married young are more vulnerable to intimate partner violence and sexual abuse than those who marry later, and they may be unable to refuse unwanted sex or resist coerced sex, which tends to be unprotected [14].

Early independence and self-sufficiency for a young married girl away from her family is likely to increase health risks, the consequences of which may negatively influence the health of the next generation: her own children [1]. Teenage pregnancy and early marriage also raise risks of poverty, and might limit undernourished girls' growth. The United Nations Population Fund (UNFPA) estimated that pregnancy- and birth-related complications are the second leading cause of death of adolescent girls, and yet an overwhelming number of these deaths (74%) are preventable [37]. Adolescent girls are two to five times more likely to die from pregnancy-related causes than women aged 20 to 29 years [7].

Adolescents are particularly vulnerable in humanitarian and conflict situations to exploitation and gender-based violence, which may affect their access to food, nutrition, and health services, including reproductive health. In addition, the incidence of early marriage and pregnancy can increase during these challenging situations [5].

The risks of adolescent births extend beyond those of the mother. Although there has been a marked, although uneven, decrease in the birth rates among adolescent girls since 1990, approximately 16 million adolescent girls still give birth each year, which equates to roughly 11% of all births worldwide; almost 95% of these births occur in LMICs [14]. It is estimated that there is a 50% increased risk of stillbirth and neonatal death, and an increased risk of preterm birth and LBW compared to older mothers [11]. Infants born SGA are three times more likely to die during the neonatal period [2], and infants with LBW are estimated to be twice as likely to die as those weighing 2.5 kg or more [38]. Those infants who do survive are at a greater risk of suffering continued nutritional deficits during childhood, and are more likely to grow up to be stunted mothers or fathers themselves. For women, this means perpetuating the "intergenerational cycle of undernutrition" [39], as short maternal stature has been estimated to be the most important risk factor for stunting in infancy and early childhood [40]. The risks for an infant born with LBW continue into adult life, with a higher chance of developing noncommunicable diseases, such as cardiovascular disease and type 2 diabetes [41]. Additionally, Patton et al. described the possibility of transgenerational epigenetic inheritance, whereby preconception influences alter patterns of gene expression that might pass to the next generation [1].

Where breast milk quantity and quality is concerned, although maternal undernutrition (unless it is severe) has a limited effect on the volume of breast milk, research indicates that maternal intakes/stores of essential fatty acids and key micronutrients do affect the concentrations of these nutrients in breast milk [42]. Recent research suggests significant effects of maternal deficiency on the concentration of B vitamins in breast milk, in particular for vitamin B_{12} [43]. The risk of infant depletion is therefore higher as a result. This is also relevant for vitamin A, where adequate content in breast milk is vital for infant status as infant stores are low at birth [38]. Although this research has not been conducted on adolescents, it can be hypothesized that these issues may be particularly relevant for adolescent mothers who already have higher requirements of key nutrients themselves and who are likely to become further depleted during pregnancy. Adolescent girls are likely to have poorer access to and uptake of antenatal care, including micronutrient supplementation. Those girls who do get pregnant are more likely to give birth without a skilled attendant, raising the health and nutritional risks both for themselves and for their infants [44].

In summary, undernourished girls are more likely to grow up to become stunted mothers and give birth to smaller, sicker babies, particularly if they become pregnant at a young age. Improving adolescent girls' nutrition and delaying pregnancy is a critical opportunity to break this negative intergenerational cycle of malnutrition. Ensuring the nutritional well-being of adolescent girls, enhancing women's and girls' equality and empowerment, and reducing rates of early marriage and adolescent pregnancy, will therefore help countries reach the global nutrition targets proposed by the World Health Assembly for 2025 and the Sustainable Development Goals by 2030 [44].

ACKNOWLEDGMENT

The authors were supported by Irish Aid.

REFERENCES

1. Patton GC, Sawyer SM, Santelli JS et al. Our future: A Lancet commission on adolescent health and well being. *Lancet* 2016; 387(10036):2423–78.
2. Black RE, Victora CG, Walker SP et al. Maternal and child undernutrition and overweight in low-income and middle-income countries. *Lancet* 2013; 382(9890):427–51.
3. Lassi ZS, Salam RA, Das JK et al. An unfinished agenda on adolescent health: Opportunities for interventions. *Semin Perinatol* 2015; 39(5):353–60.
4. World Health Organization (WHO). *Nutrition in adolescence: Issues and challenges for the health sector: Issues in adolescent health and development.* Geneva: World Health Organization. 2005.
5. United Nations Children's Fund (UNICEF). The state of the world's children 2011: Adolescence an age of opportunity. 2011. http://www.unicef.org/sowc2011/ (accessed January 25, 2017).
6. Brabin L, Brabin BJ. The cost of successful adolescent growth and development in girls in relation to iron and vitamin A status. *Am J Clin Nutr* 1992; 55(5):955–8.
7. Salam R, Hooda M, Das JK et al. Interventions to improve adolescent nutrition: A systematic review and meta-analysis. *J Adolesc Health* 2016; 59(4S):S29–S39.
8. FAO/WHO. 1998. *Human vitamin and mineral requirement: Report of a joint FAO/WHO expert consultation Bangkok, Thailand 21–30 September 1998.* 2nd ed. Rome: World Health Organization.
9. Thurnham DI. Nutrition of adolescent girls in low and middle income countries. *Sight Life* 2013; 27(3):26–37.
10. UNICEF. 2012. *Progress for children: A report card on adolescents.* www.unicef.org/publications/index_62280.html (accessed January 23, 2017).
11. Bhutta ZA, Das JK, Rizvi A et al. Evidence-based interventions for improvement of maternal and child nutrition: What can be done and at what cost? *Lancet* 2013; 382:452–77.
12. Kristjansson B, Petticrew M, MacDonald B et al. School feeding for improving the physical and psychosocial health of disadvantaged students. *Cochrane Database Syst Rev* 2007; 1:CD004676.
13. Adelman S, Gilligan D, Konde-Lule J et al. *School feeding reduces anaemia prevalence in adolescent girls and other vulnerable household members in a cluster randomized controlled trial in Uganda.* Washington, DC: International Food Policy Research Institute, 2012.

14. World Health Organization. *Health for the world's adolescents: A second chance in the second decade.* 2014. http://apps.who.int/iris/bitstream/10665/112750/1/WHO_FWC _MCA_14.05_eng.pdf?ua=1 (accessed December 2016).
15. Beard J. Iron requirements in adolescent females. *J Nutr* 2000; 130:440S–2S.
16. King JC. The risk of maternal nutritional depletion and poor outcomes increases in early or closely spaced pregnancies. *J Nutr* 2003; 133:1732S–6S.
17. World Health Organization. *Guideline: Daily iron supplementation in adult women and adolescent girls.* 2016. http://apps.who.int/iris/bitstream/10665/204761/1/9789241510196 _eng.pdf?ua=1 (accessed on January 25, 2017).
18. Peña-Rosas JP, De-Regil LM, Garcia-Casal MN et al. Daily oral iron supplementation during pregnancy. *Cochrane Database Syst Rev* 2015; 7:CD004736.
19. Haider BA, Bhutta ZA. Multiple-micronutrient supplementation for women during pregnancy. *Cochrane Database Syst Rev* 2015; 11:CD004905.
20. World Health Organization (WHO). *WHO recommendation on antenatal care for a positive pregnancy experience.* 2016. http://apps.who.int/iris/bitstream/10665/250796 /1/9789241549912-eng.pdf?ua=1 (accessed on January 25, 2017).
21. De-Regil LM, Peña-Rosas JP, Fernández-Gaxiola AC et al. Effects and safety of periconceptional oral folate supplementation for preventing birth defects. *Cochrane Database Syst Rev* 2015; 12:CD007950.
22. Golden N, Abrams S. Optimizing bone health in children and adolescents. *Pediatrics* 2014; 134:e1229–43.
23. Julian-Almarcegui C, Gomez-Cabello A, Huybrechts I et al. Combined effects of inter-action between physical activity and nutrition on bone health in children and adoles-cents: A systematic review. *Nutr Rev* 2015; 73:127–39.
24. Pettifor JM. Calcium and vitamin D metabolism in children in developing countries. *Ann Nutr Metab* 2014; 64(Suppl 2):15–22.
25. Scientific Advisory Committee on Nutrition. *Vitamin D and health.* 2016. https://www .gov.uk/government/groups/scientific-advisory-committee-on-nutrition (accessed on January 25, 2017).
26. World Health Organization (WHO). *Guideline: Calcium supplementation in pregnant women.* Geneva: World Health Organization; 2013.
27. Golden, MH. Is complete catch-up possible for stunted malnourished children? *Eur J Clin Nutr* 1994; 48(Suppl 1):S58–S70.
28. Prentice AM, Ward KA, Goldberg GR et al. Critical windows for nutritional interven-tions against stunting. *Am J Clin Nutr* 2013; 97(5):911–8.
29. Martorell R, Khan LK, Schroeder DG. Reversibility of stunting. Epidemiological findings in children from developing countries. *Eur J Clin Nutr* 1994; 48(Suppl 1): S45–S57.
30. Fink G, Rockers PC. Childhood growth, schooling and cognitive development: Further evidence from the Young Lives Study. *Am J Clin Nutr* 2014; 100:182–8.
31. Hirvonen K. Measuring Catch Up growth in malnourished populations. *Ann Hum Biol* 2014; 41(1):67–75.
32. Singh AS, Mulder C, Twisk JW et al. Tracking of childhood overweight into adulthood: A systematic review of the literature. *Obes Rev* 2008; 9(5):474–88.
33. United Nations Population Fund. *Marrying too young: End child marriage.* 2012. http://www.girlsnotbrides.org/reports-and-publications/marrying-too-young-end-child -marriage/ (accessed October 19, 2016).
34. World Health Organization (WHO). *The partnership for maternal, newborn and child health: Reaching child brides.* 2012. http://www.who.int/pmnch/topics/part_publications /knowledge_summary_22_reaching_child_brides/en/ (accessed March 25, 2015).
35. Rah JH, Shamim AA, Arju UT et al. Age of onset, nutritional determinants and seasonal variations in menarche in rural Bangladesh. *J Health Popul Nutr* 2009; 27(6):802–7.

36. Lawn J. Progress for family planning but 16 million adolescent pregnancies left behind. *The Huffington Post* [blog], July 11, 2013. www.huffingtonpost.co.uk/professor-joy -lawn/family-planning-16million-adolescent-pregnancies_b_3573967.html (accessed November 12, 2016).

37. United Nations Population Fund. *Adolescent pregnancy: A review of the evidence.* 2013. https://www.unfpa.org/sites/default/files/pub-pdf/ADOLESCENT%20PREGNANCY _UNFPA.pdf (accessed December 10, 2016).

38. Black RE, Allen LH, Bhutta ZA et al. Maternal and child undernutrition: Global and regional exposures and health consequences. *Lancet* 2008; 371(9608):243–60.

39. Victora CG, Adair L, Fall C et al. Maternal and child undernutrition: Consequences for adult health and human capital. *Lancet* 2008; 371(9609):340–57.

40. Özaltin E, Hill K, Subramanian SV. Association of maternal stature with offspring mortality, underweight, and stunting in low-to-middle-income countries. *JAMA* 2010; 303(15):1507–16.

41. van Abeelen AF, Elias SG, Bossuyt PM et al. Cardiovascular consequences of famine in the young. *Eur Heart J* 2012; 33(4):538–45.

42. Food and Agriculture Organization. *Fats and fatty acids in human nutrition: Report on an expert consultation.* FAO Food and Nutrition Paper. Rome: FAO; 2011.

43. Allen L. B vitamins in breast milk: Relative importance of maternal status and intake, and effects on infant status and function. *Adv Nutr* 2012; 3:362–9.

44. International Food Policy Research Institute. *Global Nutrition Report 2014: Actions and accountability to accelerate the world's progress on nutrition.* Washington, DC; 2014.

Section XII

Discovery Research

30 Metabolomics and Proteomics

Methodological Advances to Increase Our Knowledge of Biology during the First 1,000 Days

Richard D. Semba and Marta Gonzalez-Freire

CONTENTS

INTRODUCTION

Proteomics (the study of the structure and function of proteins expressed by an organism) and metabolomics (the study of small, low molecular-weight metabolites and their cellular processes) are rapidly evolving disciplines that have great potential to provide fundamental insights into the biology of the first 1,000 days of child growth and development. The study of individual proteins and metabolites has a long tradition, but a collective approach to their study came recently. The terms *proteome* and *metabolome* were first mentioned in the published scientific literature in 1996 [1] and 1998 [2], respectively. German botanist Hans Winkler (1877–1945) originated the related term *genome* more than seven decades earlier [3].

PROTEOMICS AND METABOLOMICS IN THE POSTGENOMIC ERA

In 2003, the Human Genome Project, which had the goal of mapping all the genes of the human genome, was declared complete [4]. In the postgenomic era, two major challenges in the life sciences include the elucidation of all the proteins and metabolites in the human body. The proteome and metabolome have a level of complexity that far exceeds the genome. In humans, ~20,000 protein-coding genes give rise to ~100,000 proteins and an estimated 1 million different protein-modified forms [5,6]. The many forms of proteins arise from mutations, RNA editing, RNA splicing, post-translational modifications, and protein degradation; the proteome does not strictly reflect the genome. Proteins function as enzymes, hormones, receptors, immune mediators, structure, transporters, and modulators of cell communication and signaling. The metabolome consists of amino acids, amines, peptides, sugars, oligonucleotides, ketones, aldehydes, lipids, steroids, vitamins, and other molecules. These metabolites reflect intrinsic chemical processes in cells, as well as environmental exposures such as diet and gut microbial flora. The current Human Metabolome Database contains more than 40,000 entries [7]—a number that is expected to grow quickly in the future.

The goals of proteomics include the detection of the diversity of proteins, their quantity, their isoforms, and the localization and interactions of proteins. The goals of metabolomics include mapping the function of metabolic pathways, many of which remain partially or completely uncharacterized [8], as well as detecting and measuring the diversity and dynamic changes of metabolites. This fundamental work should help lead to the discovery of new biological mechanisms, biomarkers, drug targets, and pathways of disease. Proteomics and metabolomics are vital steps in the progress of science toward translational research, clinical trials, and personalized medicine and nutrition. The research fields of cancer, neurology, endocrinology, and cardiovascular disease have been in the vanguard in using proteomic and metabolomic approaches in scientific investigation. In contrast, the field of nutrition has been slow in applying these powerful techniques.

The technology to investigate the immense complexity of the proteome and metabolome has advanced rapidly over the last several years. Newer mass spectrometers have greater sensitivity, higher reproducibility, better comprehensiveness, and more rapid throughput, allowing the identification and quantification of thousands of proteins and metabolites in tissues and samples. Mass spectrometers are the heart of the laboratory and the drivers of innovation. The Orbitrap mass analyzer was commercially available in 2005 [9] and gave rise to subsequent generations of Orbitrap mass spectrometers. The comprehensive analysis of proteins and lipids has been increased dramatically by sequential windowed data independent acquisition of the total high-resolution mass spectra (SWATH-MS) on triple time-of-flight mass spectrometers [10,11]. Selected reaction monitoring (SRM; also known as multiple reaction monitoring), a targeted mass spectrometry technique, can use triple quadrupole, or QTrap, mass spectrometers to measure sets of proteins [12]. SRM allows the precise, antibody-free quantitation of proteins in a multiplexed fashion. Targeted metabolomic approaches can be conducted using QTraps. Some examples of these applications to the first 1,000 days are presented later. Nuclear magnetic resonance

spectroscopy can also be used to measure metabolites, with the advantage of minimal sample preparation but the major disadvantage of low sensitivity.

TECHNOLOGICAL BREAKTHROUGHS ACCELERATE RESEARCH

Currently, it is possible to measure >10,000 proteins in a single biological sample,
something that was not feasible several years ago. These advances are due not only
to new innovations in mass spectrometry instrumentation, but also to improvements in sample preparation and standardized guidelines for sample collection.
The depletion of highly abundant proteins, such as albumin and immunoglobulins,
facilitates the detection of low abundant proteins (the "deep proteome"). More
effective electrophoresis and chromatography protocols and tools, such as stable
isotope labeling with amino acids in cell culture (SILAC), isobaric tags for relative
and absolute quantitation (iTRAQ), and tandem mass tags (TMTs), have facilitated
the quantification of proteins [13–15]. Methods are also improving for the detection
of posttranslation modifications (PTMs), such as phosphorylation, glycosylation,
acetylation, ubiquitination, sumoylation, and citrullination. PTMs are important
to study since they reflect the diversity of protein function. Many PTMs are difficult to study because they are labile to sample processing and mass spectrometry.
For example, O-GlcNAcylation, an important PTM that rivals phosphorylation in
abundance and distribution, has been especially challenging to detect and measure. Many proteins have functions that are unknown or not well understood. By
studying the proteins with which a particular protein interacts, it is possible to
deduce biological functions and pathways. Protocols have recently been developed
for proteomic analysis of dried blood spots and formalin-fixed, paraffin-embedded
tissues [16].

Many recent metabolomic studies have used so-called targeted approaches, in
which a panel of well-characterized and validated metabolites is measured using
liquid chromatography-tandem mass spectrometry (LC-MS/MS). Given the complexity of various metabolites, there is no single analytical platform to measure the
complete metabolome. At present, about 200 to 1,000 metabolites can be measured
using LC-MS/MS in commercial labs or some academic labs, depending upon the
type of sample.

Bioinformatics has played a vital role in the acceleration of proteomics and
metabolomics. Raw MS data from proteomic analyses can be analyzed using open
source search engines such as X!Tandem and OMSSA, or proprietary databases
such as Mascot and Sequest. The software assigns sequence information for peptides based upon the spectra, and then protein identifications based upon the specific
peptides. Authoritative and comprehensive protein databases include neXtProt for
human proteins [5]. Annotated databases such as Gene Ontology (GO) [17] and pathway databases such as *Kyoto Encyclopedia of Genes and Genomes* (KEGG) [18] and
Database for Annotation, Visualization and Integrated Discovery (DAVID) [19] are
particularly useful for the identification of biological pathways in the resulting data
from proteomic and metabolomics investigations. Online resources and databases of
metabolites include Metabolomics Workbench, METLIN, and BiGG [20].

WHAT CAN WE LEARN FROM PROTEOMICS
ABOUT THE FIRST 1,000 DAYS OF LIFE?

In using a proteomic approach to understanding health, the investigator can pose a question about a certain disease phenotype. The first step can be a hypothesis-free discovery phase study. As an example, let us say we would like to characterize the differences in circulating proteins among stunted and nonstunted children at 12 months of age. A cross-sectional study at 12 months of age would be subject to all the caveats and limitations about linear growth retardation, which begins *in utero* and is the sum of many biological factors prior to the phenotype of stunting at 12 months of age. However, we need an initial point of entry into a complex problem. An important reality is that mass spectrometry-based proteomics is expensive and difficult to fund in the discovery phase. Sample processing can involve the depletion of high abundance proteins using expensive affinity purification columns. The sample depleted of high abundance proteins is often split into several different fractions. Each sample fraction may require 2 to 4 hours of instrument time on a mass spectrometer; many university core facilities charge around US$100 per hour for instrument time. The proteomic analysis of a single blood sample obtained from a child in a rural field site could easily cost several hundred dollars, taking into consideration reagents, columns, and instrument time alone. One advantage of mass spectrometry-based proteomic approaches is that they generally require a small amount of plasma (~10 μL), in contrast to conventional antibody-based assays that require larger sample volumes.

There are three general approaches that are used in mass spectrometry-based proteomics: (1) data-dependent acquisition (DDA) usually done using Orbitrap mass analyzers; (2) targeted proteomics using SRM; and (3) data independent acquisition (DIA) using SWATH [21] (Figure 30.1). Samples are typically prepared using extraction and digestion with an enzyme such as trypsin. Large epidemiological studies can utilize a 96-well plate format and robotics for high throughput and lower variability in processing [22]. In DDA, a full spectrum of peptides is required of all the ion species that coelute at a certain point during the gradient and then fragmented at the MS1 level. This is followed by the collection of fragmentation spectra at the MS2 level that are used to identify the peptides. The instrument alternates between MS1 and MS2, but between cycles the ion species that are not fragmented in MS1 are lost and not detected. The Orbitraps are currently limited to a capacity of ~1 million ions with a restricted dynamic range. Targeted proteomics using SRM is based upon the fragmentation of peptides based upon their known mass-to-charge ratio (m/z) in the first quadrupole, followed by the collection of fragmentation spectra in the third quadrupole. Recently, as reported by Kusebauch and colleagues, it has theoretically been possible to detect and quantify any of the 20,277 proteins of the protein-coding genes in the human proteome using SRM [23]. The Human SRMAtlas currently consists of 166,174 proteotypic peptides representing the human proteome [23]. SRM assays can be multiplexed to simultaneously detect and quantify large sets of proteins in plasma [24]. Conventional antibody- or aptamer-based approaches that rely upon the recognition of epitopes often cannot differentiate highly homologous proteins or protein variants based upon a single amino acid substitution, and these

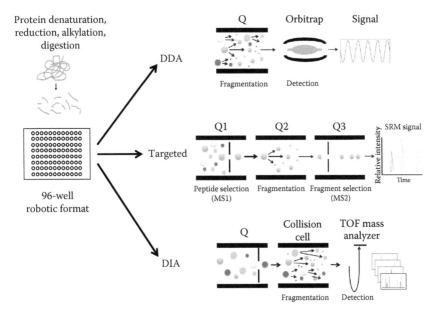

FIGURE 30.1 Examples of proteomic workflows. Proteins are extracted and digested with an enzyme (usually trypsin). Robotic sample handling in a 96-well format can greatly reduce variability. Peptides are separated and then ionized using electrospray. In data-dependent acquisition (DDA), a full spectrum of peptides is acquired in the quadrupole and fragmented. An Orbitrap mass analyzer is used to detect fragmentation spectra. In targeted selected reaction monitoring, peptides with known mass-to-charge ratio (*m/z*) are selected in the first quadrupole and then fragmented, and then specific fragmentation spectra are used to characterize each peptide. In data-independent acquisition (DIA), which can utilize sequential window acquisition of all theoretical fragment-ion spectra (SWATH)-MS, ranges of *m/z* values, using fixed or variable ranges, are selected and peptides are fragmented. Fragmentation spectra are collected continuously in a time-of-flight mass spectrometer. (Adapted from Aebersold R, Mann M, *Nature* 2016, 537:347–55.)

limitations can be overcome by use of SRM [25]. In DIA, all peptides are fragmented continuously across the entire mass range, giving rise to large multiplexed spectra. The dynamic range of DIA is about 4 to 5 orders of magnitude. There are now SWATH-MS libraries that can facilitate the identification of >10,000 plasma proteins [26]. One advantage of using SWATH is that once SWATH-MS data have been collected in an experiment, the data can be interrogated again in the future, as SWATH-MS libraries continue to evolve in their coverage.

Returning to the question we posed earlier, in the discovery phase, either DDA or DIA approaches could be used to identify differences in the expression of plasma proteins between children with or without a particular phenotype. In the past, proteomic studies used small sample sizes and technology such as two-dimensional gel electrophoresis (2-DIGE), followed by mass spectrometry, to distinguish phenotypes in infants and children. These methods are relatively insensitive and have generally provided limited insights into disease pathogenesis. Recently, West and colleagues applied the more advanced proteomic approaches discussed earlier in order to identify circulating protein markers for micronutrient status, such as vitamin E status [27].

An alternative to mass spectrometry-based proteomic approaches is the use of a platform based upon SOMAmers. SOMAmers are short, single-stranded deoxyoligonucleotides that have been selected from large screening libraries for their ability to bind to specific epitopes in peptides or proteins [28]. A commercially available platform, SomaScan (SomaLogic, Inc., Boulder, Colorado) can measure >1,000 proteins in a single, multiplexed assay. The advantage of the SOMAmer platform is rapid throughput and the lack of need for fractionation or other more laborious sample preparation. The potential disadvantages are the high cost, which is similar to that of mass spectrometry-based proteomic assays, the larger plasma volume required, the more limited ability of SOMAmers to distinguish between highly homologous proteins and sequence variants [29], and the semiquantitative nature of the results, which are expressed in relative fluorescent units [28]. With recent new advances, it should be possible to measure a few thousand different proteins using the SomaScan platform in the near future and achieve more accurate quantification with use of protein standards. The SomaScan platform has great potential for providing high-throughput, rapid results for proteomic studies.

WHAT CAN WE LEARN FROM METABOLOMICS ABOUT THE FIRST 1,000 DAYS OF LIFE?

Metabolomics has been more widely applied to infant and child health than proteomics. Plasma and urine metabolomics studies have been used to study prematurity, intrauterine growth retardation, inborn errors of metabolism, infectious diseases, and congenital malformations [30]. The first 1,000 days of life are associated with rapid bone and skeletal muscle growth, increases in organ size, and large developmental changes in the brain and nervous system. Alterations in specific metabolic pathways during this critical period may be revealed through metabolomics.

We recently used a targeted metabolomics approach to gain insight into the serum metabolome in over 300 stunted and nonstunted children aged 12 to 59 months in rural Malawi. This study revealed that stunted children had significantly lower serum concentrations of all nine essential amino acids and three conditionally essential amino acids (Figure 30.2) [31]. The mechanistic target of rapamycin complex 1 (mTORC1) pathway is the master regulator of growth and is extremely sensitive to the availability of amino acids. If essential amino acids are lacking, mTORC1 will inhibit growth. These findings suggest that the relative scarcity of essential amino acids in the diet may be adversely affecting child growth through mTORC1 [32]. Foods with the highest quality protein, that is, those containing sufficient amounts of all nine essential amino acids, are animal source foods, which are relatively expensive and rarely consumed by poor families in rural Africa.

Another finding from the metabolomics study of children in Malawi was the association of stunting with lower serum phosphatidylcholines and sphingomyelins [31]. These findings were suggestive that choline may be limited in the diet of stunted children. Serum choline was not specifically measured in the targeted metabolomics

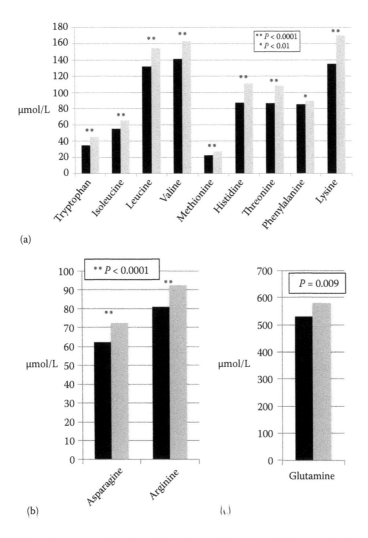

FIGURE 30.2 Serum essential and conditional essential amino acids in stunted and nonstunted children in Malawi. Mean serum (a) essential amino acids, (b) conditionally essential amino acids asparagine and arginine, and (c) conditionally essential amino acid glutamine in 194 stunted (black bars) and 119 nonstunted (gray bars) children, aged 12 to 59 months in rural Malawi. (Data from Semba RD, Shardell M, Sakr Ashour FA et al., *EBioMedicine* 2016, 6:246–52.)

platform that we used. In order to expand these investigations, we developed a new LC-MS/MS assay in our laboratory to measure serum choline. Stunting was also associated with lower serum choline concentrations (Figure 30.3) [33]. These findings suggest that stunted children are not receiving adequate amounts of choline in their diet. The richest dietary sources of choline are animal source foods, especially eggs, which are rarely consumed by children in poor families from rural Africa.

Our targeted metabolomics studies also implicate another conditionally essential nutrient, carnitine, in the pathogenesis of stunting [31]. Carnitine is essential for the

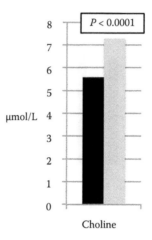

FIGURE 30.3   Serum choline in stunted and nonstunted children in Malawi. The median serum choline in 201 stunted (black bars) and 124 nonstunted (gray bars) children, aged 12 to 59 months, in rural Malawi. (Data from Semba RD, Zhang P, Gonzalez-Freire M et al., *Am J Clin Nutr* 2016, 104:191–7.)

β-oxidation of fatty acids in the mitochondria and in peroxisomes for maintaining the intracellular balance of coenzyme A, and for normal catabolism of branched chain amino acids. Overall, in the study population, serum carnitine concentrations were much lower than the normal range for healthy children of 40 to 60 µmol/L, and consistent with carnitine deficiency [31]. Serum carnitine concentrations were significantly lower in stunted compared with nonstunted children [31] (Figure 30.4). Carnitine is found in animal source foods and is only present in plant foods in negligible amounts.

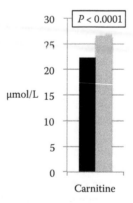

FIGURE 30.4   Serum carnitine in stunted and nonstunted children in Malawi. The mean of serum carnitine in 194 stunted (black bars) and 119 nonstunted (gray bars) children, aged 12 to 59 months, in rural Malawi. (Data from Semba RD, Shardell M, Sakr Ashour FA et al., *EBioMedicine* 2016, 6:246–52.)

We also applied a targeted metabolomics approach to examine the relationship between serum metabolites and environmental enteric dysfunction (EED) in the same children from Malawi. EED is an asymptomatic condition characterized by small intestinal inflammation, villous atrophy, and increased gut permeability. EED is common among young children in developing countries and is considered to be a major factor contributing to child stunting. These studies showed that increased gut permeability was associated with lower serum tryptophan, phosphatidylcholines, sphingomyelins, ornithine, and citrulline, and elevated serum glutamate, taurine, and serotonin [34]. This study provided evidence that EED is associated with abnormalities in the kynurenine pathway, urea cycle, and transsulfuration pathway of sulfur amino acids. In the same children, we also characterized serum bile acids using LC-MS/MS. These results show that children with EED have altered bile acid metabolism [35].

A targeted metabolomics approach was used to gain insight into the pathophysiology of kwashiorkor and marasmus in children in Malawi [36]. Metabolic profiling could distinguish between children with kwashiorkor or marasmus. Kwashiorkor was associated with lower serum concentrations of amino acids; acylcarnitines; and phosphatidylcholine diacyls C36:4, C38:4, and C40:6 compared to children with marasmus. After nutritional rehabilitation, amino acids and biogenic amines increased, but circulating phosphatidylcholines and sphingomyelins generally did not recover, suggesting persistent metabolic alterations still exist in children after treatment for severe acute malnutrition [36].

The gut microbiota play an important role in amino acid synthesis, bile acid metabolism, vitamin and short-chain fatty acid production, and the metabolism of dietary polyphenols [37]. Alterations in the gut microbiome can be reflected by changes in the serum metabolome in polyphenol metabolites [38], carnitine and choline [39], amino acids [40], and lipids [37]. EED is common among young children in developing countries and is associated with alterations in the gut microbiota [41]. An abnormal gut microbiome associated with EED could potentially alter the metabolism of nutrients and affect growth and development of children in the first 1,000 days

CONCLUSION

The main findings from metabolomic investigations from the first 1,000 days are that stunted children have low serum concentrations of essential amino acids, conditionally essential amino acids, choline, and carnitine. The richest dietary sources of these molecules are animal source foods. As noted, the consumption of animal source foods is limited among children in developing countries [42]. Animal source foods provide <5% of total energy intake in many countries in Sub-Saharan Africa, and 5% to 10% of total energy intake in most other African and southern Asian countries. In Europe, the United States, Canada, and Australasia, animal source foods provide >20% of total energy intake [42]. The prospect of increasing the availability of animal source foods in low-income countries raises many issues about social justice and environmental sustainability. Food production systems face the challenge of reducing the environmental impact of food production, while at the same time ensuring that all people have physical and economic access to nutritious foods [43,44], including animal source foods that are good sources of essential amino acids, choline, and carnitine.

REFERENCES

1. Wilkins MR, Pasquali C, Appel RD et al. From proteins to proteomes: Large scale protein identification by two-dimensional electrophoresis and amino acid analysis. *Biotechnology (NY)* 1996; 14:61–65.
2. Tweeddale H, Notley-McRobb L, Ferenci T. Effect of slow growth on metabolism of *Escherichia coli*, as revealed by global metabolite pool ("metabolome") analysis. *J Bacteriol* 1998; 180:5109–116.
3. Winkler H. *Verbreitung und Ursache der Parthenogenesis im Pflanzen - und Tierreiche.* Jena, Germany: Verlag Fischer; 1920.
4. Collins FS, Green ED, Guttmacher AE et al. A vision for the future of genomics research: A blueprint for the genomic era. *Nature* 2003; 422:835–47.
5. Gaudet P, Michel PA, Zahn-Zabal M et al. The neXtProt knowledgebase on human proteins: 2017 update. *Nucleic Acids Res* 2017; 45(D1):D177–82.
6. Fu W, O'Connor TD, Jun G et al. Analysis of 6,515 exomes reveals the recent origin of most human protein-coding variants. *Nature* 2013; 493:216–20.
7. Wishart DS, Jewison T, Guo AC et al. HMDB 3.0—The human metabolome database in 2013. *Nucleic Acids Res* 2013; 41:D801–7.
8. Mulvihill MM, Nomura DK. Metabolomic strategies to map functions of metabolic pathways. *Am J Physiol Endocrinol Metab* 2014; 307:E237–44.
9. Hu Q, Noll RJ, Makarov A et al. The Orbitrap: A new mass spectrometer. *J Mass Spectrom* 2005; 40:430–43.
10. Gillet LC, Navarro P, Tate S et al. Targeted data extraction of the MS/MS spectra generated by data-independent acquisition: A new concept for consistent and accurate proteome analysis. *Mol Cell Proteomics* 2012; 11(6):O111.016717.
11. Simons B, Kauhanen D, Sylvänne T et al. Shotgun lipidomics by sequential precursor ion fragmentation on a hybrid quadrupole time-of-flight mass spectrometer. *Metabolites* 2012; 20:195–213.
12. Picotti P, Aebersold R. Selected reaction monitoring-based proteomics: Workflows, potential pitfalls, and future directions. *Nat Methods* 2012; 9:555–66.
13. Evans C, Noirel J, Ow SY et al. An insight into ITRAQ. Where do we stand now? *Anal Bioanal Chem* 2012; 404:1011–27.
14. Hoedt E, Zhang G, Neubert TA. Stable isotope labeling by amino acids in cell culture (SILAC) for quantitative proteomics. *Adv Exp Med Biol* 2014; 806:93–106.
15. Rauniyar N, Yates JR 3rd. Isobaric labeling-based relative quantification in shotgun proteomics. *J Proteome Res* 2014; 13:5293–309.
16. Wiśniewski JR, Duś K, Mann M. Proteomic workflow for analysis of archival formalin-fixed and paraffin-embedded clinical samples to a depth of 10000 proteins. *Proteomics Clin Appl* 2013; 7:225–33.
17. Ashburner M, Ball CA, Blake JA et al. Gene ontology: Tool for the unification of biology. *Nat Genet* 2000; 25:25–29.
18. Kanehisa M, Araki M, Goto S et al. KEGG for linking genomes to life and the environment. *Nucleic Acids Res* 2008; 36:D480–4.
19. Huang da W, Sherman BT, Lempicki RA. Systematic and integrative analysis of large gene lists using DAVID bioinformatics resources. *Nat Protoc* 2009; 4:44–57.

20. Tzoulaki I, Ebbels TMD, Valdes A et al. Design and analysis of metabolomics studies in epidemiologic research: A primer on -omic technologies. *Am J Epidemiol* 2014; 180:129–39.
21. Aebersold R, Mann M. Mass-spectrometric exploration of proteome structure and function. *Nature* 2016; 537:347–55.
22. Zhu M, Zhang P, Geng-Spyropoulos M et al. A robotic protocol for high-throughput processing of samples for selected reaction monitoring assays. *Proteomics* 2017 Mar; 17(5). doi: 10.1002/pmic.201600339. Epub 2016 Dec 23.
23. Kusebauch U, Campbell DS, Deutsch EW et al. Human SRMAtlas: A resource of targeted assays to quantify the complete human proteome. *Cell* 2016; 166:766–78.
24. Domanski D, Percy AJ, Yang J et al. MRM-based multiplexed quantitation of 67 putative cardiovascular disease biomarkers in human plasma. *Proteomics* 2012; 12:1222–43.
25. Zhang P, Zhu M, Geng-Spyropoulos M et al. A novel, multiplexed targeted mass spectrometry assay for quantification of complement factor H (CFH) variants and CFH-related proteins 1-5 in human plasma. *Proteomics* 2017 May 16; doi: 10.1002/pmic .201600237. [Epub ahead of print.]
26. Rosenberger G, Koh CC, Guo T et al. A repository of assays to quantify 10,000 human proteins by SWATH-MS. *Sci Data* 2014; 1:140031.
27. West KP Jr, Cole RN, Shrestha S et al. A plasma α-tocopherome can be identified from proteins associated with vitamin E status in school-aged children of Nepal. *J Nutr* 2015; 145:2646–56.
28. Kraemer S, Vaught JD, Bock C et al. From SOMAmer-based biomarker discovery to diagnostic and clinical applications: A SOMAmer-based, streamlined multiplex proteomic assay. *PLoS One* 2011; 6:e26332.
29. Egerman MA, Cadena SM, Gilbert JA et al. GDF11 increases with age and inhibits skeletal muscle regeneration. *Cell Metab* 2015; 22:164–74.
30. Noto A, Fanos V, Dessì A. Metabolomics in newborns. *Adv Clin Chem* 2016; 74:35–61.
31. Semba RD, Shardell M, Sakr Ashour FA et al. Child stunting is associated with low circulating essential amino acids. *EBioMedicine* 2016; 6:246–52.
32. Semba RD, Trehan I, Gonzalez-Freire M et al. Perspective: The potential role of essential amino acids and the mechanistic target of rapamycin complex 1 (mTORC1) pathway in the pathogenesis of child stunting. *Adv Nutr* 2016; 7:853–65.
33. Semba RD, Zhang P, Gonzalez-Freire M et al. The association of serum choline with linear growth failure in young children from rural Malawi. *Am J Clin Nutr* 2016; 104:191–7.
34. Semba RD, Shardell M, Trehan I et al. Metabolic alterations in children with environmental enteric dysfunction. *Sci Rep* 2016; 6:28009.
35. Semba RD, Gonzalez-Freire M, Moaddel R et al. Environmental enteric dysfunction is associated with altered bile acid metabolism. *J Pediatr Gastroenterol Nutr* 2017 Apr;64(4):536–540. doi: 10.1097/MPG.0000000000001313.
36. Di Giovanni V, Bourdon C, Wang DX et al. Metabolomic changes in serum of children with different clinical diagnoses of malnutrition. *J Nutr* 2016 Dec;146(12):2436–2444. doi: 10.3945/jn.116.239145. Epub 2016 Nov 2.
37. Vernocchi P, Del Chierico F, Putignani L. Gut microbiota profiling: Metabolomics based approach to unravel compounds affecting human health. *Front Microbiol* 2016; 7:1144.
38. Lees HJ, Swann JR, Wilson ID. Hippurate: The natural history of a mammalian-microbial cometabolite. *J Proteome Res* 2013; 12:1527–46.
39. Koeth RA, Wang Z, Levison BS. Intestinal microbiota metabolism of L-carnitine, a nutrient in red meat, promotes atherosclerosis. *Nat Med* 2013; 19:576–85.

40. Poesen R, Claes K, Evenepoel P et al. Microbiota-derived phenylacetylglutamine associates with overall mortality and cardiovascular disease in patients with CKD. *J Am Soc Nephrol* 2016; 27:3479–87.

41. Ordiz MI, Stephenson K, Agapova S et al. Environmental enteric dysfunction and the fecal microbiota in Malawian children. *Am J Trop Med Hyg* 2017 Feb 8;96(2):473–476. doi: 10.4269/ajtmh.16-0617.

42. Dror DK, Allen LH. The importance of milk and other animal-source foods for children in low-income countries. *Food Nutr Bull* 2011; 32:227–43.

43. Capper JL, Bauman DE. The role of productivity in improving the environmental sustainability of ruminant production systems. *Annu Rev Anim Biosci* 2013; 1:469–89.

44. Leinonen I, Kyriazakis I. How can we improve the environmental sustainability of poultry production? *Proc Nutr Soc* 2016; 75:265–73.

Index